The Restless Clock

The Restless CLOCK

A History of the Centuries-Long Argument over What Makes Living Things Tick

Jessica Riskin

The University of Chicago Press
Chicago and London

The University of Chicago Press, Chicago 60637
The University of Chicago Press, Ltd., London

Published 2016
Paperback edition 2017
Printed in the United States of America

23 22 21 20 19 18 17   4 5 6 7 8

ISBN-13: 978-0-226-30292-8 (cloth)
ISBN-13: 978-0-226-52826-7 (paper)
ISBN-13: 978-0-226-30308-6 (e-book)
DOI: https://doi.org/10.7208/chicago/9780226303086.001.0001

Library of Congress Cataloging-in-Publication Data

Riskin, Jessica, author.
The restless clock : a history of the centuries-long argument over what makes living things tick / Jessica Riskin.
pages ; cm
Includes bibliographical references and index.
ISBN 978-0-226-30292-8 (cloth : alk. paper) — ISBN 978-0-226-30308-6 (e-book)
1. Vitalism. 2. Mechanism (Philosophy) 3. Life (Biology) 4. Science—Philosophy. I. Title.
Q175.32.V65R57 2016
147—dc23

2015019941

∞ This paper meets the requirements of ANSI/NISO Z39.48-1992 (Permanence of Paper).

For Madeleine and for Oliver,

my *vis viva*, my *pouvoir de la vie*

In German, the name for the balance of a clock is *Unruhe*—that is to say *disquiet*. One could say that it is the same thing in our body, which can never be perfectly at ease: because if it were, a new impression of objects, a little change in the organs, in the vessels and viscera, would change the balance and make these parts exert some small effort to get back to the best state possible; which produces a perpetual conflict that is, so to speak, the disquiet of our Clock, so that this appellation is rather to my liking.
—G. W. Leibniz, *Nouveaux essais* (1704)

Now, to make the comparison of a watch better suited to a living body and less imperfect, one must compare the exciting cause of organic movements with the spring of the watch; and consider the supple containing parts as well as the essential fluids contained by them as the works of the movement of the instrument in question. Then one can see, first of all, that the spring (the exciting cause) is the essential motor, without which, in fact, everything would remain inactive, and that its variations in tension must cause variations in the energy and rapidity of the movements.
—J.-B. Lamarck, *Philosophie zoologique* (1809)

Let us analyse the motion of a real clock accurately. It is not at all a purely mechanical phenomenon. A purely mechanical clock would need no spring, no winding. Once set in motion, it would go on forever. A real clock without a spring stops after a few beats of the pendulum, its mechanical energy is turned into heat. This is an infinitely complicated atomistic process. The general picture the physicist forms of it compels him to admit that the inverse process is not entirely impossible: a springless clock might suddenly begin to move, at the expense of the heat energy of its own cog wheels and of the environment. The physicist would have to say: The clock experiences an exceptionally intense fit of Brownian movement.
—E. Schrödinger, *What Is Life?* (1944)

~ CONTENTS ~

ILLUSTRATIONS

Figures

Plates

following page 210

INTRODUCTION

Huxley's Joke, or the Problem of Agency in Nature and Science

On a Sunday evening in November 1868, the English naturalist Thomas Henry Huxley, professor of natural history at the Royal School of Mines and of anatomy and physiology at the Royal College of Surgeons in London, friend and defender of Charles Darwin, made a joke about which people continue to chuckle almost a century and a half later, and whose humor perfectly captures what this book is about.

Huxley had been invited to Edinburgh by a renegade clergyman, the Reverend James Cranbrook, to inaugurate a new series of "lectures on non-theological topics." Huxley chose as his non-theological topic, *protoplasm* or, as he defined it for the uninitiated, "the physical basis of life." His main point was simple: we ought, he said, to be able to understand the properties of protoplasm, including its quite extraordinary property of being alive, simply in terms of its component parts, without invoking any special *something*, any force or power called "vitality."[1]

After all, Huxley pointed out—here's the joke—water has extraordinary properties too, but we know that it is made of hydrogen and oxygen combined in certain proportions within a range of temperatures, and we do not "assume that something called 'aquosity' entered into and took possession of the oxide of hydrogen . . . then guided the aqueous particles to their places." To be sure, Huxley continued, we do not presently understand just how water's properties follow from its composition any more than we understand how protoplasm can be alive, yet "we live in the hope and in the faith that . . . we shall by-and-by be able to see our way as clearly from the constituents of water to the properties of water, as we are now able to deduce the operations of a watch from the form of its parts and the manner in which they are put together."[2]

Huxley's lecture was a huge hit. When it appeared in print as the lead article in the *Fortnightly Review* the following February, several editions of the issue sold out immediately and John Morley, the review's editor, reckoned no article for a generation had "excited so profound a sensation."[3] The quip about aquosity continues almost a century and a half later to reappear regularly in biology textbooks and works of popular science.[4] A successful joke condenses layers of implicit argument and assumption into a very few words. In violation of the principle that one should never explain a joke (and in confirmation of the general feeling that the simpler the joke, the longer the explanation), this book offers an extended explanation of Huxley's joke. In particular, *The Restless Clock* addresses three of its aspects.

First, the joke assumes a founding principle of modern science, namely, that a scientific explanation must not attribute will or agency to natural phenomena: no active powers such as "aquosity" that "take possession" of things and "guide" them along their way. This rule also disallows, for example, explaining the falling weight driving a clock by saying that the weight wants to move closer to the center of the earth, or explaining the expansion of steam in a steam engine by saying that the steam intends to move upward toward the sky.

Second, Huxley's joke plays upon the uncertainties and hesitations involved in extending this principle banning agency to the explanation of living phenomena: in affirming that "vitality" is no more useful or scientific a concept than "aquosity."

Finally, in place of explanations invoking mysterious powers such as "aquosity," Huxley recommended mechanist scientific explanations that took as their model of nature the workings of an artificial machine such as a watch.

The Restless Clock examines the origins and history of the principle banning agency from science and this principle's accompanying clockwork model of nature, in particular as these apply to the science of living things. The *Restless Clock* also tells the story of a tradition of dissenters who would have rejected Huxley's punchline since they embraced the opposite principle: that agency is an essential and ineradicable part of nature.

You have probably already noticed that "agency" is a key word in this book.

Therefore let me begin by saying what I mean by it. I mean something like consciousness but more basic, more rudimentary, a primitive, prerequisite quality. A thing cannot be conscious without having agency, but it can have agency without being conscious. For example, one might consider a plant's phototropic capacity to seek sunlight to be a kind of agency, without meaning to ascribe consciousness to the plant. One might see certain electrical phenomena as exhibiting agency, such as the movement of electrons to maintain a conservation of charge.

By "agency," then, I mean simply an intrinsic capacity to act in the world, to do things in a way that is neither predetermined nor random. Its opposite is passivity. The reader will encounter in this book many scientific ascriptions to natural things—and many denials—of various forms of agency: living forces, sensitive capacities, vital fluids, and self-organizing tendencies. A common feature unites these ascriptions and denials: in each case, the ostensible force or tendency or capacity would originate *within* the natural form in question. A thing with agency is a thing whose activity originates inside itself rather than outside. A billiard ball that starts to roll when another billiard ball smacks into it looks passive: its movement appears to originate outside itself. What about a compass needle swinging around to point north? An asparagus fern sending a shoot across the room overnight? One might consider that many things in nature, if not most, exhibit agency: an activity that appears to originate within themselves.

However, the scientific principle banning ascriptions of agency to natural things supposes a material world that is essentially passive. This principle came into dominion around the middle of the seventeenth century, during the period that historians generally identify as the origin moment of modern science, or the New Science as its inventors called it. It is the informing axiom of a mechanistic approach to science. Mechanism, the core paradigm of modern science from the mid-seventeenth century onward, describes the world as a machine—a great clock, in seventeenth- and eighteenth-century imagery —whose parts are made of inert matter, moving only when set in motion by some external force, such as a clockmaker winding the spring. According to this originally seventeenth-century model, a mechanism is something lacking agency, produced and moved by outside forces; and nature, as a great mechanism, is similarly passive. Assuming that living beings are part of nature, according to this model, they too must be rationally explicable without appeal to intentions or desires, agency or will.

This ideal of explanation is standard in the natural sciences, and even the human and social sciences frequently strive for natural-scientific explanations in which agency is absent. The ban on agency seems as close to the heart of what science *is* as any scientific rule or principle. To violate it seems tantamount to lapsing out of science into mysticism.

Yet historical scrutiny reveals that this model of science itself had a theological origin. A material world lacking agency assumed, indeed required, a supernatural god. The seventeenth-century banishment of agency, perception, consciousness, and will from nature and from natural science gave a monopoly on all of these attributes to an external god. The classical mechanist approach to science, with its attendant mechanical model of nature and of living creatures, relied crucially as it was developing from around the mid-seventeenth century upon an accompanying theology, namely the argument from design. The authors of the argument from design sought proof of the existence of God in the evidence of mechanical design in nature, God's artifact. For example, physiologists who scrutinized the structure of the eye described a close resemblance to a lens instrument such as a microscope or a telescope. One cannot have a lens instrument without an instrument maker, they argued—a microscope does not put itself together from parts—so likewise, one cannot have an eye without a divine Optician.

A purely passive artifact world devoid of agency would not have been a plausible account of living nature on its own and it won no converts on its own. This mode of science, call it theological mechanism, relied upon a divine Designer to whom it outsourced perception, will, and purposeful action. In other words, the principle banning agency from nature and natural science was not only an informing principle of modern science, but was simultaneously an informing principle of modern theology.

The Protestant Reformation, which starkly distinguished God from His works, was the theological sea change that preceded the modern model of nature as passive machinery. The Reformation transformed the world not just for Protestants but for everyone: this story features a mix of Catholics, Protestants, Deists, and others: Jews, Unitarians, Muslims, Latitudinarians. Despite their cultural and theological differences, from the seventeenth century onward these actors oriented their work in relation to a prevailing model of nature. This model represented a nature composed of intrinsically inert mechanisms whose passivity indicated a supernatural source of action.

In short, a contradiction sits at the origin of modern science. The central principle responsible for defining scientific explanations as distinct from religious and mystical ones was the prohibition on appeals to agency and will. This principle itself relied for its establishment upon a theological notion, the divine Engineer, and a theological program, the argument from design. To put it another way, when the inventors of modern science banished mysterious agencies from nature to the province of a transcendent God, they predicated their rigorously naturalist approach on a supernatural power. They bequeathed to their heirs a dilemma that remains active over three centuries later.

Current scientific accounts of living phenomena are permeated by officially disallowed appeals to agency. I talked about this with a friend who is a biologist, and she agreed that it is absolutely against the rules in her field to attribute agency to a natural entity such as, say, a cell or a molecule, but she also agreed that biologists do it constantly, just as a manner of speaking: they speak and write *as if* natural entities expressed all sorts of purposes and intentions, but they don't mean it literally. "Sure, we do it all the time, when we're teaching, in lectures, even in published articles. But it's just a sort of placeholder for things we don't know yet. The more we get to know, the less the phenomena will seem purposeful. In the meantime we talk as if natural entities had intentions and desires just to make it easier to talk about them." (This sounds to me like Huxley's projection of a future complete understanding of water in terms of its component parts.)

Certain verbs, my friend further specified, are worse than others: those that seem "anthropomorphizing," such as "want," are only permissible in casual settings. Biologists can say, and allow their doctoral students to say, that "cells want to move toward the wound" in conversation but never in print. In contrast, other active verbs do not seem anthropomorphizing. The example my friend chose was "regulate": proteins "regulate" cell divisions. She said she does not see this sort of verb as ascribing agency in any bad, anthropomorphic way—it does not attribute human desires to a cell, for example —but rather as shorthand for a complex process that would be cumbersome to spell out on each occasion and that anyway often contains elements beyond the current reach of biologists' understanding.[5] This sort of active verb is permissible and even widespread in journal articles and textbooks. Proteins "control" chemical reactions; muscle cells "harvest" energy; genes "dictate" the production of enzymes.[6]

Still, while "regulate," "control," "harvest," and "dictate" do not ascribe human emotions to genes or proteins, they do imply purposeful action. Furthermore, I asked my friend, isn't it really an article of faith, this conviction that if you knew everything about the systems you study, the things that look purposeful would turn out to be entirely rote? There was a brief silence while she generously pondered the question. Then she laughed and said, "Yes, OK, you're right: it's a matter of faith. And, as with any matter of faith, I am absolutely unwilling to consider the possibility that it could be wrong. I *know* that if I knew everything about the processes I study, I would have no reason to appeal to agencies of any kind, even as a manner of speaking, let alone as a means of explanation."

I think that biologists' figures of speech reflect a deeply hidden yet abiding quandary created by the seventeenth-century banishment of agency from nature: do the order and action in the natural world originate inside or outside? Either answer raises big problems. Saying "inside" violates the ban on ascriptions of agency to natural phenomena such as cells or molecules, and so risks sounding mystical and magical. Saying "outside" assumes a supernatural source of nature's order, and so violates another scientific principle, the principle of naturalism.

Many before me have identified this quandary. Beginning in the seventeenth century, some sought to avoid it by rejecting the argument from design and the passive-mechanist model of natural science that went with it. The title of this book comes from a work that epitomizes the competing, active-mechanist view of natural machinery, and even of artificial machinery such as clocks. The German philosopher, mathematician, and inventor Gottfried Wilhelm Leibniz wrote the clockwork passage that provided this book's title as he was struggling to find a different model for nature and science from the passive machinery of his contemporaries. He described clockwork and, by analogy, human beings in this way: "In German," he wrote (he was writing in French), "the word for the balance of a clock is *Unruhe*—which also means disquiet; and one can take that for a model of how it is in our bodies, which can never be perfectly at their ease."[7] As Leibniz saw it, the balance of a clock was in a constant state of agitated motion, and so too were human bodies.

To be clocklike, to Leibniz, was to be responsive, agitated, and restless. How different this is from what people generally understand by the clockwork metaphor! The clockwork universe with its clockwork creatures has familiarly signified regularity and constraint, not agitation and responsiveness. In Leibniz's alternative notion of machinery and mechanist science, however, machinelike meant forceful, restless, purposeful, sentient, perceptive. Mechanical meant lifelike, and vice versa: living beings were the most mechanical things in the universe.

Since the classical mechanists were by and large the victors who wrote the histories, their opponents have had a bad reputation in historical and philosophical writing as mystics and even superstitious reactionaries. I should say that by "classical mechanists," I mean Cartesians, Newtonians, Robert Boyle and his followers: the groups that played the dominant role during the seventeenth century to define modern scientific principles and practices. Although they disagreed with one another on many matters, including the source of action in nature's machinery, they agreed that the material world needed to be set in motion by an external power. Their critics argued that the machinery was self-moving.

Despite their reputation, critics of classical mechanism and the argument from design included a distinct group who objected not out of a commitment to traditional, religious accounts of nature, but rather out of a rigorous naturalism: a determination to establish science as fully autonomous. As Leibniz pointed out, if one wanted to disallow appeals to a supernatural god, then passive clockwork would not work as a model of living nature. One needed a different model: active, restless clockwork. Such a model would naturalize the very phenomena that the argument from design outsourced to a divine creator: perception, will, purpose, agency. All of these had to be integral to the natural world and its creatures.

From this impulse to naturalize rather than to outsource agency, there emerged a different mechanist science: not classical mechanism—brute, passive—but active mechanism. This alternative science was still mechanist, in that it offered rational, systematic accounts of natural phenomena in terms of component parts and their functions. It invoked no magical or miraculous properties, only natural ones. However, active mechanists such as Leibniz described the machinery of nature as containing its own sources of action inside itself: as self-constituting and self-transforming machinery.

Modern scientific accounts of life have been shaped by a struggle between these two competing mechanisms, two scientific principles. One, passive mechanism, the overtly victorious and therefore more visible, evacuates agency from nature (initially to the province of a supernatural god). It informs, for example, the physiology of the eye viewed as a lens instrument such as a microscope or telescope. The opposite principle, active mechanism, eclipsed but still working from the shadows, avoids the supernaturalism of the first approach by viewing agency as a primitive feature of the natural world like force or matter, an aspect of the very stuff of nature's machinery, and especially its living machinery. This competing principle informs, for example, the physiology of the eye as practiced by the nineteenth-century German physiologist and physicist Hermann von Helmholtz, who refuted the telescope analogy by arguing that the eye was a perceiving mechanism, the functioning of which rested upon its capacity for perception.[8]

The Restless Clock follows this struggle in modern science from its inception. The story begins (chapter 1) with the lifelike machines or "automata" that spread across the landscape of late medieval and Renaissance Europe from churches to palace gardens to town squares. These machines inspired the mechanistic sciences of life that emerged in the seventeenth century through the work of intellectual radicals such as René Descartes and G. W. Leibniz. From the start, these sciences were torn between active and passive models of nature's machinery (chapters 2 and 3). The new mechanistic sciences of life in turn gave rise to a new breed of lifelike machines (chapter 4): philosophical, experimental, simulative machines that actually performed animal and human processes such as playing a flute, writing a message, breathing, bleeding, speaking, sketching. Accompanying these experimental models of living beings was a hypothetical figure, the Enlightenment man-machine or "android" (chapter 5), whose authors invoked him to propose that human beings might be material entities through and through. The authors of this thought experiment drew conclusions that were at once physiological, social, moral, economic and political.

A major development of the experiments and thought experiments of this period, the mid- to late eighteenth century, came to fruition in the work of the French naturalist Jean-Baptiste Lamarck. This development was the momentous idea that living beings might be not just active but also *self-making* and *self-transforming* machines whose structures changed over time (chapter 6). Charles Darwin, when he adopted this Lamarckian idea, inherited the active-mechanist model of life that it assumed. However, Darwin also inherited a passive-mechanist model of living beings, because this model was implicit in another idea essential to his theory, an idea that had developed within the passive-mechanist tradition of arguments from design: the notion that living creatures were perfectly "fitted" and "adapted" to their environments. Hence Darwin's theory of evolution was torn between active- and passive-mechanist models of living beings (chapter 7).

Around the turn of the nineteenth to twentieth centuries, Darwinists in the German-speaking world, negotiating the intellectual, religious, and institutional politics of the new research universities, offered a reinterpretation of Darwin's theory in which they aimed to eradicate all traces of active mechanism: a passive-mechanist neo-Darwinism (chapter 8). During the first half of the twentieth century, culminating in the decades after World War II, this neo-

Darwinist approach, officially passive-mechanist with buried strains of active mechanism, informed the philosophical, scientific and engineering movement called "cybernetics" (chapter 9) and through cybernetics, the founding of new scientific approaches and disciplines including artificial intelligence, cognitive science, and mathematical biology.

And so the old contradiction, buried in history, maintains a subterranean activity in current science. It lies at the root, for instance, of ongoing skirmishes among biologists and their critics over the appearance and implications of apparent design in nature and the role of teleology in scientific explanation. These battles have generated influential scientific approaches and principles such as Richard Dawkins's notion of the "selfish gene"[9] and Daniel Dennett's campaign to eliminate "skyhooks" (a "skyhook" being a purposeful "force or power or process") from evolutionary biology.[10] In both cases, the seventeenth-century, contradictory approach to agency in nature continues to exert a powerful seismic pressure from below the surface. The same centuries-old contradiction has been at work in the roboticist Rodney Brooks's and others' "embodied," "evolutionary," and "behavioral" approaches to artificial intelligence.[11] This book's final chapter (chapter 10) examines some instances of these recent and current scientific debates and programs in the light of their hidden history.

The philosopher and historian of science Thomas Kuhn, in his 1962 book *The Structure of Scientific Revolutions,* described science as shaped, at each stage, by a dominant "paradigm" (a model or approach). This paradigm would inform all scientific research until its limitations began to undermine it, and then a new paradigm would emerge to overthrow it, for example, the way the heliocentric (sun-centered) model of the universe overthrew the geocentric (earth-centered) model in the sixteenth to seventeenth centuries.

The story told here, in contrast, is not about a single paradigm shaping scientific research, but rather about an engagement between competing principles and approaches. The people involved in this competition have been ambivalent in their commitments, and the losing principle has not disappeared from science. Instead it has remained, obscured from view by the winning principle, but still active. Thus a conflict between two competing principles has shaped the development of modern scientific accounts of life. This book traces the development of the eclipsed scientific principle, the naturalization of agency, and its confrontations with the principle that eclipsed it, the banishment of agency from nature. To identify this struggle is to recognize intellectual possibilities that have been hidden by the course of history.

Intellectual possibilities are not the sole fruits of this investigation, though, nor could they have been, since ideas are inseparable from the world in which they arise. Social and political engagements, as well as intellectual and cultural ones, have all along been inextricable from the competition between scientific models of living and human beings. The classical brute-mechanist approach to the science of life and the active-mechanist approach have developed, as we shall see, in close conjunction with mechanical and industrial arrangements such as the automatic loom and the transformed world of production that accompanied it; with economic policies including the division of various kinds of labor; with taxonomies and rankings of human beings by sex, race, class, geographical origin, and temperament; and with projects of imperial conquest and governance. In what follows, investigating this centuries-old dialectic in science will mean uncovering the hidden action of forces that are at once intellectual and political, scientific and social.

One major purpose of *The Restless Clock* has been to demonstrate the importance of historical understanding to current thinking about the sciences of life and mind. Historical analysis, by revealing the now-hidden forces that shaped current scientific problems and principles, can reopen foreclosed ways of thinking. Investigating the origins and development of current scientific principles means rediscovering alternative possibilities for what it has meant, and what it can mean, to offer a scientific model of a living being.

Along the Route de l'Horlogerie (The Clockmaker's Way) through the Jura Mountains in Switzerland, mechanical creatures two and three centuries old remain in the alpine villages where they were first created, attended by curators and watchmakers who are often the direct descendants of the original builders. I traveled there in the course of writing this book. Among the clockwork beings I encountered is a peasant teaching his pig to hunt truffles. Holding a truffle in one hand, and his pig on his opposite knee, the peasant is apparently in the midst of explaining that you find a truffle by its smell. Raising the truffle to his nose, he inhales (his chest rises), he shakes his head from side to side, and he simultaneously closes his eyes, giving an irresistible display of sentience. The machine is strikingly persuasive. It seems to suggest that sentience and living agency might just consist of movements of passive mechanical parts. Or else it suggests that mechanical parts are anything but passive. In fact, I think it suggests both things at once. The story lies in the journey to and fro between these possibilities. If aquosity were not a compelling possibility, the joke would not have been funny.

~ 1 ~

Machines in the Garden

Once upon a time, Sir Lancelot's castle, Joyous Guard, was a cursed and miserable place known as Dolorous Guard. Lancelot changed its name after he captured it from the evil lord Brandin of the Isles, defeating three knights of copper. The first knight stood above the castle gate, "big and sturdy, in full armour and holding a great axe in his hands." This knight, however, was easily dispatched: according to the enchantment that had placed him above the gate, he would crash to the ground when the one destined to conquer the castle first caught a glimpse of him. Lancelot gazed upon the copper knight, "big and strange," and down he obligingly crashed.[1]

Further on, however, in the castle cemetery, Lancelot came upon two more copper knights who put up more of a fight. They guarded a door through which Lancelot had to pass to find the key to the castle's enchantments. Holding heavy steel swords, they waited to clobber anyone attempting to pass through. Unafraid, Lancelot raised his shield and leapt between the knights. One smote him on the right shoulder, breaking his shield and piercing his hauberk "so cruelly that the red blood ran down his body," but Lancelot persevered. Next, he encountered "a copper damsel, very finely cast" holding the key he sought. With it, Lancelot opened a copper chest from which thirty copper tubes emanated, releasing a whirlwind of devils. The copper damsel and knights collapsed to the ground: the enchantments were broken.[2]

The automaton knights and damsels of Arthurian legend were accompanied by gold, silver and copper children, satyrs, archers, musicians, oracles and giants.[3] These fictional artificial beings had plenty of real counterparts. Actual mechanical people and animals thronged the landscape of late medieval and early modern Europe, and like the fictional beings, the real automata were responsive, engaging, and frequently given to attacking human trespassers, mostly in good fun.

Automata were familiar features of daily life, originating in churches and cathedrals, and spreading from there. Jesuit missionaries carried them to China as offerings to dramatize the power of Christian Europe. Wealthy estate owners installed automata in their palaces and gardens, where they became major tourist attractions for travelers from across Europe.

Our story begins with these lifelike machines. They provided the material context for the new scientific and philosophical model of living beings as machines that would emerge around the middle of the seventeenth century. If "mechanical" subsequently came to signify passive and rote, in the age of these earlier machines, it meant no such thing. On the contrary, the automata we are about to consider exhibited a vital and even a divine agency.

Deus qua Machina

A mechanical Christ on a crucifix, known as the Rood of Grace, drew flocks of pilgrims to Boxley Abbey in Kent during the fifteenth century (see figure 1.1). This Jesus, which operated at Easter and the Ascension, "was made to move the eyes and lipps by stringes of haire."[4] Moreover, the Rood was able

> to bow down and lifte up it selfe, to shake and stirre the handes and feete, to nod the head, to rolle the eies, to wag the chaps, to bende the browes, and finally to represent to the eie, both the proper motion of each member of the body, and also a lively, expresse, and significant shew of a well contented or displeased minde: byting the lippe, and gathering a frowning, forward, and disdainful face, when it would pretend offence: and shewing a most milde, amiable, and smyling cheere and countenaunce, when it woulde seeme to be well pleased.[5]

Before approaching the Rood for benediction, one had to undergo a test of purity administered by a remote-controlled saint:

Figure 1.1 Pilgrim souvenir of the Rood of Grace, fourteenth century, © Museum of London.

> Sainct Rumwald was the picture of a pretie Boy sainct of stone . . . of it selfe short, and not seeming to be heavie: but for as much as it was wrought out of a great and weightie stone . . . it was hardly to be lifted by the handes of the strongest man. Neverthelesse (such was the conveighance) by the helpe of an engine fixed to the backe thereof, it was easily prised up with the foote of him that was the keeper, and therefore, of no moment at all in the handes of such as had offered frankly: and contrariwise, by the meane of a pinne, running into a post . . . it was, to such as offered faintly, so fast and unmoveable, that no force of hande might once stirre it.[6]

Having proven your "cleane life and innocencie" at the hands of the rigged Saint Rumwald, you could proceed to the mechanized Jesus. Automaton Christs—muttering, blinking, grimacing on the Cross—were especially popular.[7] A sixteenth-century Breton Jesus rolled his eyes and moved his lips while blood flowed from a wound in his side. At his feet, the Virgin and three attendant women gesticulated, while at the top of the Cross, a head symbolizing the Trinity glanced shiftily from side to side.[8]

Mechanical devils were rife. Poised in sacristies, they made dreadful faces, howled, and stuck out their tongues to instill fear in the hearts of sinners. The Satan machines rolled their eyes and flailed their arms and wings; some even had moveable horns and crowns (see plate 1). One sixteenth-century life-size wooden devil burst from its cage, "horrible, twisted, horned, rolling furious eyes, sticking out a blood-red tongue, seeming to throw itself upon the spectator, spitting in his face and letting out great howls," while making an obscene gesture with its left hand. This devil was moved by a weight that also powered a set of bellows, forcing air and water through a copper tube in the neck and mouth, allowing the creature to howl and spit.[9] A muscular, crank-operated devil with sharply pointed ears and wild eyes remains in residence at the Castello Sforzesco in Milan (see plate 2).[10]

There were also automaton angels. In one Florentine festival, a host of these carried the soul of Saint Cecilia up to Heaven.[11] For the feast of the Annunciation at San Felice, the fifteenth-century Florentine architect Filippo Brunelleschi sent the archangel Gabriel in the reverse direction in a mechanical "mandorla," an almond-shaped symbol in which two merging circles represent Heaven and earth, matter and spirit. Brunelleschi, a master of holy mechanics, mechanized Heaven too. His mechanical Paradise was "truly marvellous . . . for on high a Heaven full of living and moving figures could be seen as well as countless lights, flashing on and off like lightning."[12]

Brunelleschi was outdone in the second half of the century by Cecca (Francesco D'Angelo), who engineered Christ's Ascension at the Church of Santa Maria del Carmine. Here, where Christ was borne aloft on "a Mount very well made of wood" the "said Heaven was somewhat larger than that of S. Felice in Piazza." The festival planners added a second Heaven over the chief tribune, with "certain great wheels" that "moved in most beautiful order ten circles standing for the ten Heavens." These were filled with stars: little copper lamps suspended from pivots so that they would remain upright as the heavens turned. Two angels stood on a platform suspended from pulleys. They were arranged to come down and announce to Christ that he was to ascend into Heaven.[13]

The heavenly machinery was balanced below by elaborately engineered hells. The Passion play at Valenciennes in 1547 featured a hell with a monstrous mouth that gaped open and shut, revealing devils and tormented sinners.[14] The mechanical infernos with moving gates were accompanied by rumbling thunder and flashes of lightning, and writhing automaton demons and dragons.[15]

A menagerie of mechanical beasts played roles in religious theater. A mechanical bear menaced David's sheep.[16] Daniel's lions gnashed their teeth[17] and more lions knelt before Saint Denis.[18] Balaam's ass balked and swerved before the angel of the Lord.[19] The serpent twined itself round the trunk of the Tree of Knowledge to proffer its apple to Eve.[20] A wild boar tracked by hunters, a leopard that sniffed Saint André, a dromedary that wagged its head, moved its lips, and stuck out its tongue, a host of dog- and wolf-shaped devils surging up from the underworld, and serpents and dragons spewing flames from their mouths, noses, eyes, and ears rewarded the devoted spectators at the forty-day performance of the *Mystère des actes des apôtres* in Bourges in 1537.[21] The machines were commissioned from local artisans, usually clockmakers.[22]

Mechanical enactments of biblical events spread across the European landscape, during the late fifteenth and early sixteenth centuries.[23] The holy machinery was not only to be found in cities. In May 1501, an engineer in the village of Rabastens, near Toulouse, was engaged to build an endless screw that could propel the Assumption of the Virgin. The following August, the Virgin rose heavenward, attended by rotating angels, and disappeared into Paradise, its entrance hidden in clouds. Meanwhile a golden, flaming sun also rotated, carrying more angels on its rays.[24] Another mechanical Assumption of the Virgin took place annually in Toulouse, moving in alternate years between the Eglise Notre-Dame de la Daurade and the Eglise Saint-Etienne.[25] At home, in the region around Toulouse, children built small replicas of the Virgin elevator for the Assumption in the same way that they arranged crèches at Christmas.[26]

The Eternal Father appeared in mechanical reenactments. In Dieppe, he loomed at the top of the Eglise Saint-Jacques, a "venerable old man" astride a cloud in an azure, star-sprinkled canopy of Heaven. Mechanical angels flew about him, flapping their wings and swinging their censers. Some played the *Ave Maria* in time to the organ on handbells and horns at the end of each office. After the service, the angels blew out the altar candles.[27] At the feast of Whitsuntide, the Holy Ghost, in the form of a white dove, flew down from the main vault of Saint Paul's Cathedral in London, breathing a "most pleasant Perfume" over the congregation.[28]

The earliest modern mechanical figures were found mostly in churches and cathedrals and exhibited religious themes. Many figures were connected to clocks, outgrowths of the Church's drive to improve timekeeping for the sake of a reformed calendar and better prediction of feast days,[29] or with organs. A mechanical man gripping a mallet to ring the hour became a familiar sight on clock towers across Europe in the mid-fourteenth century. He went by the name "Jack" in England; "Jean" in Flanders; "Jacquemart" in France; and "Hans" in Germany.[30] Over the next century, the bell-ringer acquired company. On the clock in the Piazza San Marco in Venice, beginning in 1499, two giant shepherds struck the hour while an angel playing a horn emerged, followed by the three Magi (see figure 1.2). The Magi bowed before the Virgin and Child and removed the crowns from their heads with one hand while using the other to extend their gifts. They then stood, replaced their crowns, and exited through an automatic door.[31] The scene of the Magi was a common motif on church clocks, which also often included calendars indicating feast days; the positions, oppositions, and conjunctions of the stars; the signs of the zodiac; the phases of the moon; and, as in the San Marco clock, astronomical models of a Ptolemaic cosmos.[32]

There were also roosters. Mechanical cocks crowed and flapped their wings on clocks across Europe from about the mid-fourteenth century.[33] Perhaps the earliest, built around 1340, flapped and crowed on the hour at Cluny Abbey, near Macon. Meanwhile, an angel opened a door to bow before the Virgin; a white dove representing the Holy Spirit flew down and was blessed by the

Figure 1.2 Automaton Magi on the Piazza San Marco clock, courtesy Fausto Maroder.

Eternal Father; and fantastic creatures emerged to stick out their tongues and roll their eyes before retreating inside the clock.[34] Another rooster did its flapping and crowing on the town clock in Niort from about 1570. This bird presided over three separate scenes involving some forty figures. Care appeared in a window to exhort Servitude to come out and strike the hour. An automaton Gabriel enacted the Annunciation with a mechanical Mary, Holy Ghost, and Eternal Father. A mechanical choir of angels sang while their Kapellmeister, holding the music and beating time, inclined successively toward each group in the choir as its members rang their appointed carillons. Saint Peter appeared from behind a door, looked about, opened another door and, at the admonition of two children, disappeared back into his chamber to make way for the twelve Apostles. These arrived holding hammers with which they rang the hour while the children nodded their heads in time. The clock depicted a door with two automaton Hercules on either side, ready to drop their clubs on anyone who tried to enter; above them, Vulcan with his hammer also stood guard.[35]

The Cluny, Niort, and other roosters all were outdone by the renowned Rooster of Strasbourg Cathedral. For nearly five centuries, the Strasbourg Rooster cocked its head, flapped its wings, and crowed on the hour atop the Clock of the Three Kings, originally built between 1352 and 1354, and refurbished by the clockmaker brothers Isaac and Josias Habrecht between 1540 and 1574 (see figure 1.3). Beneath the Rooster, the astrolabe turned and the Magi scene played out its familiar sequence. In the Habrecht version, the Rooster, Magi, Virgin, and Child were joined by a host of other automata: a rotation of Roman gods who indicated the day of the week; an angel who raised her wand as the hour was rung, and another who turned her hourglass on the quarter hour; a baby, a youth, a soldier, and an old man representing the four stages of life, who rang the quarter hours; and above them, a mechanical Christ came forth after the old man finished ringing the final quarter hour, but then retreated to make way for Death to strike the hour with a bone.[36]

Apart from church clocks, the other prime spot for mechanical figures was the church organs.[37] Organ-driven mechanical angels came in whole choirs of bustling figures, sometimes accompanied by flocks of singing birds. Automaton angels lifted horns to their mouths and played drums and carillons.[38] At the cathedral in Beauvais, Saint Peter towered atop an organ of the late fourteenth or early fifteenth century and blessed the congregation on his feast day by nodding his head and moving his eyes.[39] Strasbourg Cathedral was hectic with mechanical activity, having automata connected to its organ as well as its clock. Three

Figure 1.3 Strasbourg astronomical clock, engraving by Isaac Brunn, "Horloge astronomique de la Cathédrale," Cabinet des Estampes de Strasbourg. Photo © Musées de Strasbourg, Mathieu Bertola.

moving figures, known as Rohraffen, were attached to the strings of the organ in the late fifteenth century (where they remain): Samson opening and closing the jaws of a lion; the Herald of the village, lifting his trumpet to his lips; and the Bretzelmann (pretzel seller) in a red and black cape.

The Bretzelmann, still at Strasbourg Cathedral, has long hair and a shaggy beard, an aquiline nose, and an evil look. Set in motion, he seems to speak with great emphasis, opening and shutting his mouth while shaking his head and gesticulating with his right arm.[40] At Pentecost, throughout the service, the Bretzelmann mocked the priest, laughing, hurling insults and coarse jokes, and singing nasty songs:

> Bellowing forth profane and bawdy songs in a raucous voice accompanied by lewd gestures, [he] drowns the hymns of the people entering and mocks them in derisive pantomime, with the result that he not only turns people's devotion into discord and their lamentation into guffaws, but also hinders even the clerics singing the divine services; nay more, he causes disturbance in the divine solemnities of the masses . . . [a disturbance] long abominable and detestable to the zealot for ecclesiastical, nay more, divine reverence.[41]

Other organs sported disembodied heads that frowned, contorted their faces, rolled their eyes, stuck out their tongues and opened and closed their mouths as the music played.[42] A colossal automaton head animated the church organ in Neustadt-an-der-Harth in Bavaria, and others were to be found across Germany and the Low Countries from the fifteenth century.[43] From the organ gallery of the cathedral in Barcelona, the head of a moor hung by its turban. It made mild facial expressions when the music played softly; when the strains grew louder, it rolled its eyes and grimaced as though in pain.[44] And in the Cloître des Augustins in Montoire, in the Loire valley, a mechanical head on the organ gallery gnashed its teeth with a noisy clatter.[45]

In sum, Europe during the later Middle Ages and the Renaissance was alive with mechanical beings and the Catholic Church was their main patron. The Church was also a primary sponsor, between the late fifteenth and late sixteenth centuries, of the translation and printing of a flood of ancient texts on mechanical and hydraulic automata, which then informed the construction of new devices. The first printed edition of Vitruvius's *De Architectura*, for example—containing descriptions of the third-century BCE engineer Ctesibius's water organ and other automata—appeared in 1486 as a key part of the Renaissance popes' project to build a Christian Rome.[46]

Automata also appeared in secular settings: on town halls, municipal clock towers,[47] and the grounds of noble estates. Early modern engineers mechanized political icons as well as religious ones. From the late Middle Ages, automata were part of a lively civic and urban culture.[48] An example is the clock that Charles IV commissioned for the Frauenkirche in Nuremberg to commemorate his Golden Bull, which established the constitutional structure of the Holy Roman Empire and set the number of electors at seven. On the clock, which was inaugurated in 1361, seven figures known collectively as the Männleinlaufen (parade of little men) emerge at noon to bow before the emperor (see plate 3).[49] Another legendary automaton was the lion built by Leonardo da Vinci in 1515 for a banquet hosted by Florentine merchants in Lyon in honor of Francis I: "wherefore Leonardo being asked to devise some bizarre thing, made a lion which walked several steps and then opened its breast, showing it full of lilies."[50] The lion represented Lyon and the lilies the French throne.

Clockwork automata were the playthings of princes, especially the Holy Roman emperors, from the late fifteenth century. Hans Bullmann of Nuremberg built android musicians, for which Ferdinand I summoned him to Vienna.[51] Henry VIII, according to a 1542 inventory, had an automaton clock at Westminster.[52] Hans Schlottheim, a clockmaker in Augsburg, designed automaton-embellished utensil holders to sit on banquet tables. These were wrought in gold, silver, or brass, typically in the form of a ship. One, which Schlottheim made for Rudolph II around 1580 and is now at the British Museum, has figures moving around a sundial and passing before a throne. Schlottheim also devised two automaton crayfish—one crept forward, the other backward—bought by the Prince Elector of Saxony in 1587.[53]

Noble houses hummed and whirred with clock automata that were miniaturizations of the ones in churches and designed by the same people. The Habrecht brothers, who renovated the Strasbourg Cathedral clock in the mid-sixteenth century, also did a brisk business in household automaton clocks.[54] Automata figured too in lay theater.[55] In 1547, John Dee, the future magus and court philosopher to Queen Elizabeth I, but then a nineteen-year-old reader in Greek at Trinity College, Cambridge, built what seems to have been a mechanical flying dung beetle for an undergraduate production of Aristophanes's *Pax*. At the point in the play when Trygaeus, a peacemaking Athenian, determined to reach Jupiter's Olympian palace, leaps onto his unlovely Pegasus and exhorts it to fly, Dee's artificial insect took to the air, inspiring "a great wondring, and many vaine reportes spread abroad of the meanes how that was effected."[56]

But automata were first and most extensively to be found in churches and cathedrals. Indeed, even before the age of clock and organ automata, as early as the mid-thirteenth century, the sketchbook of Villard de Honnecourt included rope-and-pulley controlled mechanisms, one for a mechanical angel that turned to point its finger at the sun and another for an eagle, the caption to which reads: "How to make the eagle face the Deacon while the Gospel is being read."[57] Later, automaton Christs, angels, devils, and Virgins prepared the ground for mechanical animals of every variety and clockwork models of the cosmos itself.

A Franciscan monk of iron and linden wood built around 1560 and attributed to a man named Juanelo Turriano offers a final example (see plate 4).[58] Turriano's life is a tale in itself. Clockmaker, architect, and engineer to the Holy Roman Emperor Charles V, and then to his son and heir, King Philip II of Spain, Turriano went into retreat with the former, after his abdication in 1556, at the monastery of Yuste, near Plasencia.[59] There, the clockmaker built automata to comfort a gouty ex-emperor: an automaton lady who danced and played a tambourine; a flight of wooden sparrows that fluttered and "flew about the room as if alive"; a miniature army of prancing horses and soldiers playing diminutive trumpets.[60] According to legend, Philip's son, Don Carlos, made a miraculous recovery following a head injury, cured when the relics of a fifteenth-century Franciscan monk, Diego de Alcalà, were brought to the prince's bed at the moment of crisis. To express his eternal gratitude, the king asked Turriano to build the mechanical monk.

The monk, wearing a tunic, cowl, and sandals, with his mechanism hidden beneath his habit, is a fully self-contained device. Turning his head, moving his eyes, he steps forward, raising the crucifix and rosary clutched in his left hand as he performs a series of devotions, striking his breast with his right hand, then kissing the cross. Sixteen inches and five pounds, the monk is somehow formidable. Perhaps even more than his contemporaries—the muttering Christs, the horn-playing angels, the eye-rolling devils, the teeth-chattering heads—he embodies the power of an image, the peculiar power of a moving image, and the extraordinary presence of a moving, devotional image.[61] The Catholic Church was the cradle of the clockwork universe and its mechanical inhabitants.

Mechanization has come to be so closely associated with modernization that it is difficult to think one's way back into what automatic machinery might have signified in the late Middle Ages. The major looming obstacle that obstructs the view of the late medieval period, as seen from the present, is the Protestant Reformation. The Reformation marked a sea change in the way people understood the relations between matter and spirit, nature and divinity.[62] Among the many implications of these changed relations—among the less-remarked implications, but important nevertheless—was that machines, and mechanism in general, came to signify something new: an artificial mechanism, composed of material parts, became definitively empty of spirit.

To understand why this was so, one must consider that in each of their main doctrines, the Reformers distinguished God from nature and spirit from matter. The doctrinal heart of their movement was their denial of the miracle of the Eucharist, transubstantiation: although they disagreed among themselves about just what happened during the Sacrament, they all maintained that the communion bread and wine remained bread and wine. That is, the sacramental bread and wine *represented*, but did not *become*, the body and blood of Christ, as Catholic doctrine said they did. In fact, the Reformers denied the occurrence of miracles in general. The sacraments and rituals of Christian practice, such as the Eucharist, became merely symbolic representations of spiritual grace rather than miraculous events. The Reformers also rejected icons and other representations of the deity as idolatry. In each of these ways, they asserted a new set of distinctions between the material world of terrestrial life and the divine realm of spirit.[63]

Before these developments, the automaton icons—the Christs, Virgins, Holy Fathers, angels, and devils that animated churches and cathedrals—confronted their viewers in a medieval Catholic world that held no sharp distinction between the material and the spiritual, earthly and divine. Rolling their eyes, moving their lips, gesturing and grimacing, these automata dramatized the intimate, corporeal relation between representation and divinity, icon and saint. The holy machines existed in a tradition of imagery that extended the tangible, visible, earthly manifestations of Christian lore and doctrine ever further.[64] The icons were mechanical but neither passive nor rote. On the contrary, they were representations in motion, inspirited statues: they were mechanical *and* divine.

With the Reformation, the same machines began to look very different. Reformism and clockmaking developed side by side from Augsburg to Strasbourg to Geneva. The flood of mechanized religious images coincided both in time and place with an opposing impulse: the spread of the Reformers' view that representational religious images were sacrilegious because they blurred what was properly a sharp boundary between symbol and deity, matter and spirit. The mechanical icons went from being divine, inspirited statues to being deceitful, fraudulent: material contraptions masquerading as their antithesis, spiritual beings.

Over the humming, groaning, chirping, whistling, chattering ecclesiastical machinery, the Reformation cast a partial hush. The uncouth Bretzelmann of Strasbourg Cathedral was silenced along with many of his fellow organ automata and, indeed, with many of the church organs themselves, which became emblematic of Catholic ritual.[65] Henry VIII, in establishing the Anglican Church, banned mechanical statues from English churches.[66] The grimacing Rood of Boxley Abbey gave its last performance in 1538, after being snatched from Boxley by Geoffrey Chamber as part of his commissioned defacement of the abbey. Chamber wrote to Thomas Cromwell that he had found in the Rood "certain engines and old wire, with old rotten sticks in the back, which caused the eyes to move and stir in the head thereof, 'like unto a lively thing,' and also, 'the nether lip likewise to move as though it should speak,' which was not a little strange to him and others present."[67] But can it have been any surprise that the Rood was made of wood and wire? It and its many cousins had been built and maintained by local artisans—clockmakers, carpenters—and treated by local beholders with great familiarity, inspiring, by the accounts of contemporary chroniclers, at least as much laughter as awe. The Bretzelmann of Strasbourg Cathedral was . . . well, funny. Similarly, in the case of the lever-and-pulley-operated Saint Rumwald, "many times it mooved more laughter than devotion, to beholde a great lubber to lifte at that in vaine, which a young boy (or wench) had easily taken up before him."[68]

That mechanical icons were mechanical cannot have been big news. Still Chamber and his fellow iconoclasts introduced the idea that, by virtue of being mechanical, such icons were deceptions. Machinery, that is, could not represent divinity other than deceitfully. One could no longer know a thing to be mechanical and simultaneously believe it to be divine, because the relations of the divine to the material had changed. Whereas these realms had before existed on a continuum, they had become definitively separate and distinct in the theology of the Reformers. The destruction of mechanized icons represented only small swells inside the larger surges of iconoclasm that spread across Europe during the middle decades of the sixteenth century.[69] But the demolition of the Rood and its kind reveals that one core logic of iconoclasm, the rigorous distinction between the realm of divine spirit and the world of brute matter, brought with it a fundamentally transformed view of machines. These went, in the stroke of an iconoclast's ax, from being manifestations of spirit and liveliness to being fraudulent heaps of inert parts.

The abbot and monks, when Chamber questioned them, denied any knowledge of the mechanical Rood.[70] Yet it had inspired great devotion in the people of Kent, as well as pilgrimages from across the realm,[71] so Chamber deemed it an immediate danger and promptly removed it to Maidstone. There he displayed it in the public market and instilled in the townspeople a "wondrous detestation and hatred [of the Rood] so that if the monastery had to be defaced again they would pluck it down or burn it."[72] Although the material workings of the Rood could not have surprised anyone, Chamber taught his audiences to see these workings as evidence of a deception. The chronicler Charles Wriothesley described the events as follows:

> Allso the sayde roode was sett in the market place first at Maydstone, and there shewed openlye to the people the craft of movinge the eyes and lipps, that all the people there might see the illusion that had bene used in the sayde image by the monckes of the saide plaace of manye yeares tyme out of mynde, whereby they had gotten great riches in deceiving the people thinckinge that the sayde image had so moved by the power of God, which now playnlye appeared to the contrarye.[73]

The Rood was then transported to London where John Hilsey, bishop of Rochester, exhibited it during a sermon at Saint Paul's Cross, after which it was torn apart and burned before a crowd of duly admonished onlookers.[74] Again, Wriothesley recorded the occasion:

> This yeare, the 24th daie of Februarie, beinge the Soundaie of Sexagesima and the Sainct Mathias daie, the image of the roode that was at the Abbey of Bexley, in Kent, called the Roode of Grace, was brought to Poules Crosse, and their, at the sermon made by the Bishopp of Rochester, the abuses of the . . . engines, used in old tyme in the said image, was declared, which image was made of paper and cloutes from the legges upward; ech legges and armes were of timber; and so the people had bene eluded and caused to doe great adolatrie by the said image.[75]

Three decades later, the lawyer and historian William Lambarde gave a caustic account of the Rood and "the Monkes, which were in love with the Picture." Of the Rood, Lambarde wrote sarcastically, "it needed not Prometheus fire to make it a lively man, but onely the helpe of the covetous Priestes of Bell, or the aide of some craftie College of Monkes." As for the Rood's colleague, Saint Rumwald, Lambarde revealed it to have been operated by "a religious impostor standing out of sight." He recalled Cromwell's triumph over the monks and their machines: "But what? I shall not neede to reporte, howe lewdly these Monkes, to their own enriching and the spoile of Gods people, abused this wooden God . . . because a good sort be yet on live that sawe the fraude openly detected at Paules Cross."[76]

As with other Reformist initiatives, both sides of the confessional divide participated in this rejection of mechanized religious images. By the mid-seventeenth century, certain Catholic monarchs had developed a distaste for automaton angels and mechanical Ascensions. In 1647, Louis XIV and the Queen Mother came to view the automaton angels of Dieppe and found them not to their liking; that was the end of the angels.[77] An interdiction in 1666 put an end to the Virgin's annual mechanical Ascension in Toulouse on the grounds that it distracted the congregation and caused "irreverent reflections."[78]

Yet mechanized devotional objects did not disappear; on the contrary, they survived and flourished. So did the great theological and philosophical problem these machines had come to dramatize: the problem of the relations between body and soul, matter and spirit, mechanism and agency. During the late sixteenth century and throughout the seventeenth, the proliferating and elaborating machines coexisted with proliferating and elaborating theological and philosophical suspicions of them. In other words, an older, originally Catholic tradition of mechanical representation persisted in intense conflict with a newer, originally Protestant assertion of utter distinctness between the divine and material realms. This conflict itself shaped the thinking of Catholics and Protestants alike.

The problem of sacred images was a core question for the Council of Trent, an ecumenical council of the Catholic Church that met in Trento, Italy, between 1545 and 1563 to mount a theological response to the Reformers, in part by issuing condemnations on what it decided were Protestant heresies. In 1563 the Council issued a decree on the use of sacred images that placed a ban on "unusual" images except those approved by a bishop.[79] This ban helped to motivate a thematic shift in mechanical icons. For example, in the wake of the Council's decree, the three-dimensional nativity scene (*presepio*) rose to prominence in Catholic settings as an acceptable representation of the divinity and an answer to the Lutheran Christmas tree.

The Jesuit Order, assigned by the Council of Trent the task of defeating Protestant theology, embraced the *presepio* and made it its own; the Order also heightened its dramatic power by mechanizing it. Within a few decades, in aristocratic and wealthy bourgeois homes as well as in churches, a fad was in full swing for mechanical and talking *presepi*. The sixteenth-century architect Bernardo Buontalenti built a clockwork *presepio* for his pupil, Francesco, son of Cosimo I de' Medici, a machine featuring opening and closing heavens, flying angels, and figures walking toward the manger. And Schlottheim built an elaborate mechanical crèche around 1589 for the Court of Saxony. The crèche, now displayed in the Museum für Sächsische Volkskunst in Dresden, includes shepherds and kings proceeding past the manger while angels fly down from Heaven, Joseph rocks the cradle, and an ox and an ass rise up to stand before the holy Infant.[80]

A prominent representative of the Jesuitical love of mechanical devotional images was the polymath Athanasius Kircher, who served as a nexus of philosophical conversation and activity during the middle decades of the seventeenth century. Among many other devices, Kircher designed a hydraulic machine to represent the Resurrection of the Saviour and another "to exhibit Christ walking on water, and bringing help to Peter who is gradually sinking, by a magnetic trick." In this contraption, the operative features were a strong magnet placed in Peter's chest and the iron of Christ's outstretched hands "or any part of his toga turned toward Peter." The two figures, propped on corks in a basin of water, would then be inexorably drawn together: "The iron hands of Christ soon feel the magnetic power diffused from the breast of Peter. . . . The artifice will be greater if the statue of Christ is flexible in its middle, for in this way it will bend itself, to the great admiration and piety of the spectators."[81]

More generally, the Jesuits made clockwork automata a principal tool in their promulgation of Christianity. They arrived before a succession of Chinese emperors bearing gifts of automata. One such offering, dispatched in 1618 by Nicolas Trigault, the Jesuit ambassador of the Chinese Mission, was an elaborate mechanized nativity scene. The works were fully internal and spring-driven. As Trigault described it, the scene included the three Magi giving homage with bows, the Holy Virgin responding with gracious gestures, Joseph rocking the cradle where the Holy Child lay, an ass and an ox thrusting their heads toward the cradle, the Holy Father making a benediction, two angels continuously ascending and descending, and even moving shepherds.[82]

The Jesuits' automatic offerings included worldly themes as well as religious ones. The Jesuit priest Gabriel de Magalhaens, who arrived in China in 1640, presented to the Emperor Kangxi a spring-driven android knight that marched about with a drawn sword for a quarter of an hour.[83] The Jesuits spread explicitly Christian automata as well as secular ones around the world. Many of the same clockmakers and engineers who designed religious automata for churches also built secular ones for private patrons or public settings. In the clockmaking region of southern Germany during the late sixteenth and early seventeenth centuries, mechanical animals such as Schlottheim's mechanical crayfish became popular: automaton spiders; Neptune astride a creeping bronze tortoise; a life-size bear, wearing real fur and beating on a drum.[84] In the 1680s and '90s, clockmakers began to fabricate animated paintings (*tableaux mécaniques*) depicting hunting parties and other rustic scenes.[85]

Among the most elaborate nonreligious automata were waterworks on the grounds of palaces and estates. The wealthy and powerful found in lifelike machinery an endless source of comedy, and of the most bawdily uproarious, knee-slapping variety. Here again, the lifelike machines appeared neither passive nor rote but full of agency and spirit, though in this case not really divine, but of a decidedly earthy sort. From the sublime, onward to the ridiculous.

Frolicsome Engines

The palace and garden automata of the late Middle Ages and early Renaissance were as lively, responsive, and engaging as the Church automata, and even more amusing. Over a period of several centuries, automatically spraying their unsuspecting guests with water and perpetrating other mechanized acts of hospitable abuse were favorite pastimes of Italian, French, and German aristocrats.[86] "Frolicsome engines" (*engiens d'esbattement*)[87] were to be found as early as the late thirteenth century at the chateau of Hesdin (in present-day Pas-de-Calais), seat of the comtes d'Artois. The machines are mentioned, beginning in 1299, in the account books of Robert II (Robert the Noble), comte d'Artois. The following year, the family appointed a castle "Master of Engines" (Maistre des engiens du chastel). After that, the *engiens* make regular appearances in the accounts, continuing through the reign of Robert II's successor, Mathilde (known as Mahaut), comtesse d'Artois. From these entries, we learn that the engines included mechanical monkeys with real skins (regularly replaced)[88] and, after 1312, sporting horns;[89] "an elephant and a he-goat";[90] and a machine referred to as "the boar's head."[91] The comtesse Mahaut's descendant, Philippe le Bon, Duke of Burgundy from 1419 until his death in 1467, conducted a thorough refurbishment of the stock he had been left by his forebears and expanded it considerably. His own account books contain a meticulous catalogue of the many mechanized tricks he inflicted on visitors. These included:

> painting of 3 personages that spout water and wet people at will . . . a machine for wetting ladies when they step on it . . . an "engien" which, when its knobs are touched, strikes in the face those who are underneath and covers them with black or white . . . another machine by which all who pass through will be struck and beaten by sound cuffs on their head and shoulders . . . a wooden hermit who speaks to people who come to that room . . . 6 personages more than there were before, which wet people in various ways . . . eight pipes for wetting ladies from below and three pipes by which, when people stop in front of them, they are all whitened and covered with flour . . . a window where, when people wish to open it, a personage in front of it wets people and closes the window again in spite of them . . . a lectern on which there is a book of ballades, and, when they try to read it, people are all covered with black, and, as soon as they look inside, they are all wet with water . . . [a] mirror where people are sent to look at themselves when they are besmirched, and, when they look into it, they are once more all covered with flour, and all whitened . . . a personage of wood that appears above a bench in the middle of the gallery and fools [people] and speaks by a trick and cries out on behalf of Monsieur le Duc that everyone should go out of the gallery, and those who go because of that summons will be beaten by tall personages dressed like "sots" and "sottes," who will apply the rods aforesaid, or they will have to fall into the water at the entrance to the bridge, and those who do not want to leave will be so wetted that they will not know where to go to escape from the water . . . a window in which there is a box suspended in the air, and on that box there is an owl which makes various faces in looking at people and gives an answer to everything that one wishes to ask it, and its voice can be heard in that box.[92]

The Hesdin *engiens d'esbattement*, in all their malicious glory, achieved great notoriety and inspired many imitations in the following century.[93]

By 1580 and 1581, when the French essayist and statesman Michel de Montaigne was traveling through Europe, hydraulic automata had grown so commonplace in noble palaces and on the grounds of bourgeois estates that he grew bored with them. Outside Augsburg, at the summer palace of the rich banking family Fugger, Montaigne saw sprays of water from hidden brass jets activated by springs. "While the ladies are amused seeing the fish play, one simply releases some spring: suddenly all these jets spurt thin, hard streams of water to the height of a man's head, and fill the petticoats and thighs of the ladies with this coolness." Elsewhere, hidden jets could be triggered to gush directly into the face of a visitor who stopped to admire a particular fountain.[94] According to one source, the Fuggers palace also had an automaton lion that sprang forward when a door was opened.[95]

At Pratolino, a palace of Francesco I de' Medici, Grand Duke of Tuscany, Montaigne marveled at Buontalenti's elaborate installations. In one "miraculous" grotto he witnessed statues dancing to harmonious music, mechanical animals dipping their necks to drink, all caused by flowing water. Lulled into complacency by the charming scene, he became the victim of a surprise attack.

> At a single movement, all the grotto is full of water, all the seats gush water on your buttocks; and, fleeing the grotto, climbing back up the castle stairs, there comes out of them . . . a thousand jets of water that will give you a bath right up to the top of the house.[96]

The grotto at Pratolino also had singing birds and an automaton girl who emerged from behind a door to fill a bucket with water. She placed it on her head and walked away, glancing flirtatiously at a nearby shepherd.[97] Another of the Grand Duke's residences boasted a grotto bustling with hydraulically driven "water mills and windmills, little church bells, soldiers of the guard, animals, hunts, and a thousand such things."[98]

Montaigne was unimpressed by the already famous Villa d'Este in Tivoli. The Tivoli palace and gardens had been built during the 1550's and '60's by Cardinal Ippolito II d'Este, then Governor of Tivoli, as consolation after an unsuccessful campaign to win the papacy. Completed in 1572, the grottoes were already old news, and Montaigne, arriving 1580, declined to write a lengthy description of them, since there were already "published books and pictures on the subject." Moreover, the "gushing of an infinity of spouts of water checked and launched by a single spring that one can work from far away, I had seen elsewhere on my trip."[99]

He did provide a meticulous, if jaded, account of the water organ, detailing how the water fell into a chamber, thereby forcing the air out through the organ pipes. A second current of water turned a toothed wheel that caused the keyboard to "get struck in a certain order." The mechanism triggered the sound of trumpets while elsewhere one could hear birdsong from "little bronze flutes." The whole included a little scene enacted by an automaton owl and group of birds: the owl, appearing at the top of a rock, seemed to frighten the birds into silence, but they resumed their song as the owl receded from view. "All these inventions, or similar ones, produced by the same rules of nature," Montaigne observed with an implicit yawn, "I have seen elsewhere."[100]

In 1598, almost twenty years after Montaigne's travels, when Henri IV decided his palaces needed embellishment, he lured away Tommaso and Alessandro Francini, engineers to Ferdinando I de' Medici, then Grand Duke of Tuscany, to supply the requisite waterworks. The Francinis began at Saint-Germain-en-Laye, where they mechanized a small throng of classical gods and heroes and other moving figures all in bronze.[101]

There were grottoes devoted to Neptune, Mercury, Orpheus, Hercules, Bacchus, Perseus, and Andromeda. John Evelyn, the gardener and diarist, visited the palace at Saint-Germain-en-Laye in 1644 and recorded in his diary what he had seen there.[102] He and other visitors described an automaton Neptune with a streaming blue beard, brandishing his trident, naked astride a chariot pulled by seahorses, accompanied by three round-bellied, horn-playing tritons. Farriers, "their faces black with filth and sweat," hammered iron on an anvil and—"that which is most pleasant and seems made to provoke laughter"—drenched their eager audiences with surprise sprays of water. Mercury posed by a window with one foot carelessly propped, "loudly intoning a trumpet." Elsewhere, Orpheus played his lyre for an audience of animals and trees who, including the trees, stretched and craned toward him.[103] A towering Perseus descended upon a mighty dragon arising from beneath the waves. Perseus swung his sword to behead the fearsome beast, sending its corpse back down into the watery depths; whereupon farther back in the grotto, Andromeda promptly lost her chains. Meanwhile, busy figures of artisans—blacksmiths, weavers, millers, carpenters, knife-grinders, fishermen—went about their sundry tasks.[104]

Another dragon appeared in the Dragon Grotto, shaking its terrible head and wings while belching steam. Despite its ferocity, this dragon was surrounded by "various little birds, which really one would say were not painted & counterfeit, but living and fluttering their wings, which make the air resound with a thousand sorts of song; and above all the Nightingales there vie to make music in several choirs." There were cuckoos too and in yet another grotto, a nymph played at an organ.[105] The Grotto of Torches—a subterranean chamber lit only by flames—displayed a heady sequence of scenes "by force of water": first, an idyllic, island-dotted sea in which fishes and sea monsters sported happily beneath a rising sun; then, a violent storm, thunder and lightning, wrecked ships heaved up on shore. Next came a calm and fertile vista, a flowerbed in bloom and trees filled with fruit. In the distance, the king and his family strolled, all except the Dauphin who arrived from on high in a chariot carried by two angels. The angels crowned the prince with a glittering coronet. Finally, there was a desolate landscape, a desert littered with ruins where reptiles, insects, and other wild creatures crawled about. At the last, a fairy emerged playing a flute and the animals gathered round to listen.[106]

What was it like to live amidst such machines, to be familiar with them, to have them shape one's earliest intuitions about machinery: how it works, what it does, how it compares to living creatures? We can form a reasonable impression thanks to a meticulous daily record of the life of a child who grew up with the hydraulic grottoes of Saint-Germain-en-Laye in his garden. The record includes every passing fancy, every lisping pronouncement, the menu at each meal down to the numbers of prunes or grapes consumed, and careful descriptions of all bowel movements. To be sure, it was no ordinary kind of experience: the child was the future Louis XIII, son of Henri IV and Maria de' Medici, born just when the Francinis were working on his father's fountains. The Dauphin's birth was recorded by his doctor and caretaker, Jean Hérouard, on September 27, 1601, as having taken place at "ten-thirty and a half quarter according to my watch made in Abbeville by M. Plantard."[107] The prince would spend his childhood mostly at Saint-Germain-en-Laye where he developed a passion for mechanical things.

As a toddler, the Dauphin watched the workers from his windows[108] and, from the age of three, in the spring of 1605, he began visiting the grottoes several times each week.[109] Hérouard's diary describes him in bed one morning instructing a chambermaid, "Pretend that I am Ofus [Orpheus] and you are the fountainee [fountaineer], you sing da canaries."[110] Soon afterward, he was working the grotto faucets, spraying himself and everyone else with water.[111] The prince plagued Tommaso Francini with visits to his workshop, demanding the name of each instrument and explanations of how they worked.[112] At home, he talked continuously about Francini and pretended to be Francini, making wax models, working the fountains, collecting his pay. He played fountains in bed, in his gilt washbasin and under the dining table—"*fssss*" and "*dss*"—making believe he was spraying people with water. On one occasion, he was rebuked by a nurse for climbing under the table to play fountains to the neglect of a visiting dignitary.[113] Francini built a small wooden fountain for the Dauphin, which was installed near his rooms on his fourth birthday.[114] While work on the fountain was under way, the prince went continually to the workshop to see it, begging "let's go see my fountain at Francino's place."[115]

At first, the Dauphin could not be persuaded to enter the Orpheus grotto. Finally his governess, Madame de Montglat, enticed him in with a handful of sugared peas, having first covered the figure of Orpheus with a drape. Thereafter, the prince boasted that he had been to the very back of the grotto and was not afraid even to touch Orpheus himself.[116] In addition to occasional notes of fear, the passion also contained more than a hint of childish eroticism. Hérouard dutifully recorded on one occasion: "Says he has a faucet in his ass and another in his willie: '*fs fs*.'" The future absolutist—who was given to exposing himself to the servants and whose "willie" was the focus of much teasing attention from all members of the household including the king and queen—was especially fond of the willie-fountain joke, which he frequently repeated.[117]

The day the Dauphin's governor, Monsieur de Souvé (Gilles, marquis de Courtenevaux), arrived at Saint-Germain-en-Laye, shortly before the prince's seventh birthday, Louis insisted on taking the tired traveler on an immediate tour of the grottoes, where he worked the faucets himself.[118] As a child king, having ascended to the throne at age nine after his father's assassination, Louis XIII continued to visit Francini, going straight to his workshop upon arriving at the palace, and amusing himself for hours at a time by forging, soldering, and filing fountain pipes.[119]

Louis XIII liked clockwork as well as hydraulic automata. Hérouard's journal describes the Dauphin, age four, beating his spoon against his plate and announcing to his governess: "Maman ga [Mme. de Montglat] I am ringing da hour *dan, dan,* it rings like da jackamart who beats on da anvil."[120] Here he is at six, shopping in Paris along the rue Saint Honoré, choosing a spring-driven toy carriage on offer for 15 écus.[121] Later in the same year, the Dauphin was given a cabinet fabricated in Nuremberg with "a great number of personages doing diverse actions by the movement of sand." The personages enacted Christ's Passion and the taking of Jerusalem. The prince played fervently with the instrument, quickly grasping how to make it stop and go, demonstrating it to everyone in the palace and discoursing about the works with mispronunciations that charmed his guardian: "*contrepès*, pour countrepoids."[122]

This predilection for mechanical games persisted through generations of French princes. Louis XIV was born at Saint-Germain-en-Laye and received mechanical toys— automaton clocks, a carriage and company of soldiers, a mechanical theater that enacted an opera in five acts—well into his dotage.[123] His son, Louis XIII's grandson, had an arsenal of automaton toys including another mechanical army of a hundred soldiers.[124]

You didn't need to be a king or a prince: the popes too competed in the game of hydraulic trickery. When Ippolito Aldobrandini became Pope Clement VIII in 1592, he assigned his nephew, Cardinal Pietro Aldobrandini, the task of building a villa of unprecedented magnificence. Aldobrandini engaged the hydraulic engineers Orazio Olivieri and Giovanni Guglielmi to design what Edith Wharton, on her tour of Italian villas, would describe as "the inevitable *théatre d'eau*."[125] At the Villa Aldobrandini, the waterworks included a room of hydraulic and pneumatic marvels, the Stanza dei Venti (Room of Winds), which would draw visitors throughout the seventeenth and eighteenth centuries (see figure 1.4). Water from hidden, spring-triggered spouts, it should go without saying, leapt out to spray hapless visitors. Other spouts of water and water-powered jets of air played organ and fife music and produced eerie sounds—thunder, wind, rain, whistles, shrieks—while wooden globes danced magico-mechanically across the floor.[126]

The popes, their nephews, and their grand-nephews, all the little cardinals and archbishops wanted their own hydro-mechanical toys. Markus Sittikus von Hohenems, sovereign and archbishop of Salzburg from 1612 until his death in 1619, installed waterworks at his Schloss Hellbrunn that remain in operation almost four centuries later.[127] When he was elected archbishop, Sittikus was already a connoisseur of automata. He had lived briefly at the Villa Aldobrandini; moreover, his uncle, Cardinal Marco Sittico Altemps, nephew of Pope Pius IV, had built the Villa Mondragone, which had a renowned Water Theater designed by the engineer Giovanni Fontana.[128] In Sittikus's garden, visitors are still invited to seat themselves around a stone table, on stone benches with hidden spouts that release jets of water on command, drenching the obedient from below.

In the Neptune Grotto to which they proceed, dripping and uproarious, guests gape at the Germaul, a stone gargoyle that rolls its eyes menacingly and sticks out its tongue. Fleeing the Germaul, the visitors are again watered down

STANZA DE VENTI NEL TEATRO DI BELVEDERE DI FRASCATI CON LA FAMOSA FONTANA DEL MONTE PARNASO CON APOLLINE ET LE MVSE CHE SVONANO CON INSTRVMENTI HIDRAVLICI À FORZA D'ACQVA ARCHITETTVRA DI GIACOMO DELLA PORTA.
Gio Batta Falda del et sculp
Gio Iac Rossi le stampa in Roma alla Pace con Priu del S P
7

Figure 1.4 Etching of the Stanza dei Venti, Villa Aldobrandini, by Giovanni Battista Falda, with a surprise "wind" in the foreground. Courtesy Metropolitan Museum of Art, www.metmuseum.org.

from spring-triggered spouts concealed in the walls. Arriving remoistened in the Birdcall Grotto, they are surrounded by the hydraulically produced sound of chirping and twittering birds. Afterward, they are led along the Royal Way past five small grottoes, each housing a scene enacted by automata: a miller grinding his wheat; a potter working at his wheel; a scissors grinder and his wife sharpening blades on a wheel while their child plays at their feet; Perseus freeing Andromeda from the dragon; Apollo flaying Marsyas. Next, present-day visitors arrive at an elaborate water-driven Mechanical Theater displaying a town square populated by more than a hundred moving figures: carpenters, innkeepers, musicians and other street performers, a barber giving a client a shave, a butcher slaughtering an ox, a farmer pushing an old woman in a wheelbarrow, a marching military guard, a dancing bear. The Mechanical Theater, completed in 1752, was the contribution of Archbishop Andreas Jakob Graf von Dietrichstein; it replaced an earlier hydraulically powered mechanical scene representing a forge.

At the time Sittikus's waterworks were being installed, princes across the land were importing hydraulic engineers to install automata on their palace grounds; it was one of their first acts as sovereign. The adolescent Palatine Elector, Frederick, brought his hydraulic engineer along with his seventeen-year-old bride, Elizabeth, daughter of King James I. Elizabeth traveled to Heidelberg for her wedding in 1613 accompanied by Salomon De Caus, an engineer from northern France and Huguenot refugee at her father's court.[129] De Caus would remain at Heidelberg as Frederick's engineer until 1620 when the Elector, then also king of Bohemia, would lose his crown to the Holy Roman Emperor Ferdinand II and have to flee with his family to The Hague. The brevity of Frederick's Bohemian reign, which lasted a single winter, earned him the nickname "Winter King." But De Caus had time to transform the palace gardens into yet another hydraulic wonderland.

The waterworks' creator described grottoes in which fabulous creatures performed magico-mechanical feats.[130] In one, water poured from the breasts of a woman in the middle of the cavern, and from the mouth of a fish held by a man seated beside her. The couple was serenaded by a Satyr playing a flute, and opposite him, by the nymph Echo, softly repeating each phrase. In the Grotto of Orpheus, the minstrel played his cello, charming the beasts around him—leopard, ram, lion, boar, stag, sheep, rabbit and snake—who danced in time to the music. The Grotto of Neptune contained the God of the Sea himself and some attendant creatures—a pair of swimming horses whose reigns he gripped,

Figure 1.5 Isaac De Caus, Grotto of Neptune from *Les raisons des forces mouvantes* (1615), reproduced in *Nouvelle invention de lever l'eau* (1644), courtesy Department of Special Collections, Stanford Libraries.

a couple of wading nymphs playing horns, and a cherub astride two dolphins—all turning in stately circles around a great, Gothic rock upon which a siren held a jug spouting water (see figure 1.5).[131]

A burgeoning literature on automatic machinery informed and accompanied such installations as the Palatine gardens waterworks. This literature began with a series of ancient texts on mechanical and hydraulic automata, as we have seen. These included principally, in addition to Vitruvius's *Ten Books*, the treatises of the ancient Greek engineer Hero of Alexandria (first century CE), which were repeatedly translated and printed over the course of the sixteenth century.[132] In turn, these inspired modern works that borrowed extensively from the classical ones. An influential example is Agostino Ramelli's *Le diverse e artificiose machine* (1588). Ramelli was an Italian engineer who moved to France to serve in the war against the Huguenots under the duc d'Anjou, later King Henri III, to whom Ramelli dedicated his work on machines. The treatise contains a plan for an "ingenious and delightful" fountain of twittering birds, based closely on designs from Hero's *Pneumatica*. Within the fountain, a nest of compartments is joined by a network of siphons. The siphons are connected above, through pipes, to little figures of birds with flutes in them. As water descends through the fountain, the siphons begin to siphon, emptying certain compartments and filling others, forcing air up through the various pipes in turn. The air, as it comes out the tops of the pipes into the birds with their flutes, makes them flutter and trill.[133]

De Caus was the author of another such work, *Les raisons des forces mouvantes avec diverses machines tant utiles que plaisantes* (1615), which has trees full of automaton birds, including one, in direct imitation of a design by Hero, just like the one Montaigne had noted at the Villa d'Este: the birds flutter and chirp while an owl turns slowly toward them. When the intimidating owl faces the birds, they fall silent, but as he turns away, they resume their ruckus (see figure 1.6).[134] De Caus's treatise also contains meticulous accounts of the mechanisms of hydraulic grottoes like those of the Palatine gardens. In one, Galatea rides astride a big seashell drawn by two dolphins (see figure 1.7). Behind her, a Cyclops has put his club aside to play on a flageolet, while sheep gambol about. The mechanism is made entirely of wood, driven by two waterwheels. These are put in motion by jets of water from two pipes that emerge from a common reservoir. The pipes have valves that open and close alternately by means of a system of counterpoises, so that the wheelwork turns one way and then the other as Galatea and her dolphins move back and forth across the scene. A third

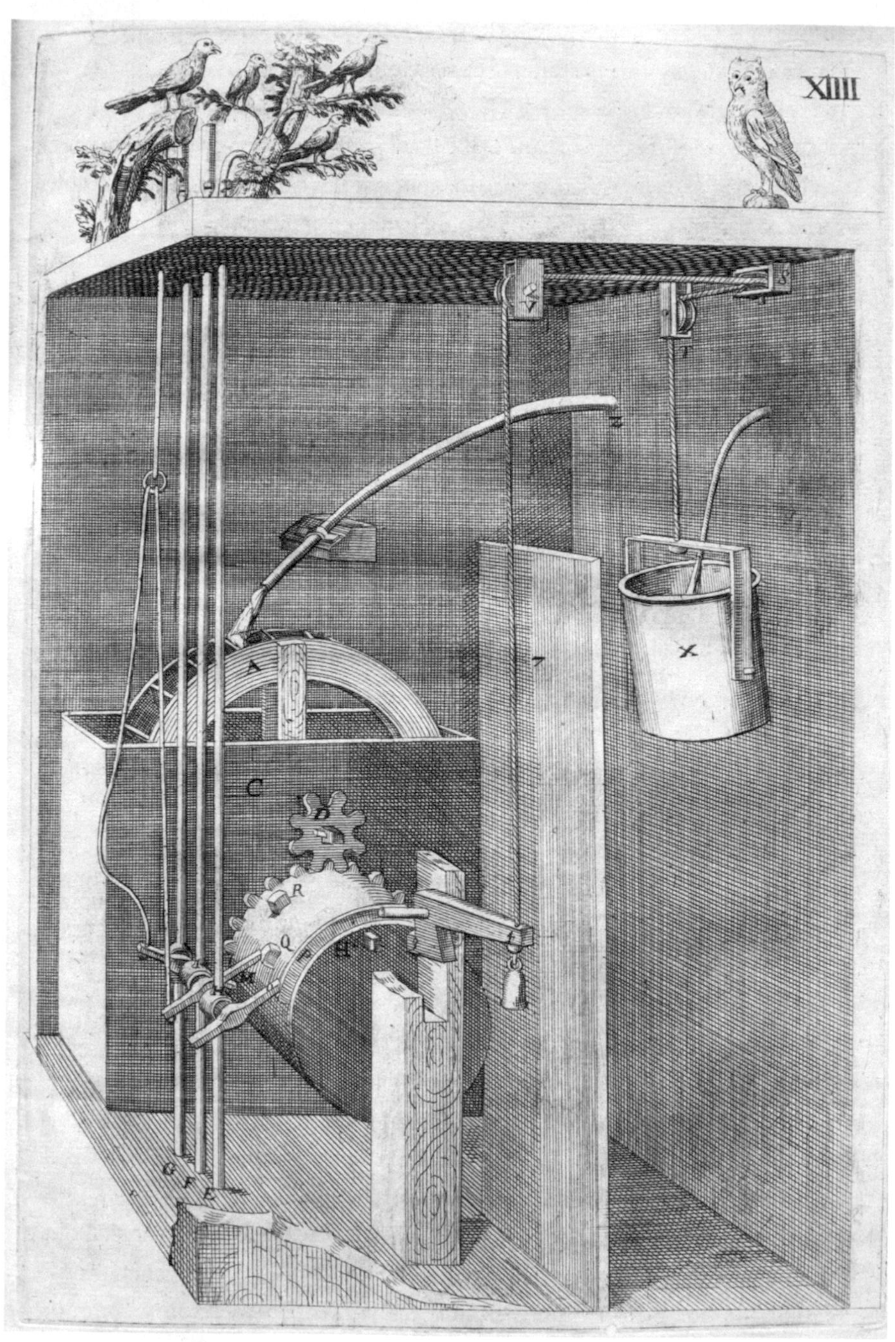

Figure 1.6 Isaac De Caus, menacing owl and frightened birds, from *Les raisons des forces mouvantes* (1615), reproduced in *Nouvelle invention de lever l'eau* (1644), courtesy Department of Special Collections, Stanford Libraries. Photo by Andrew Schupanitz.

Figure 1.7 Isaac De Caus, Grotto of Galatea, from *Les raisons des forces mouvantes* (1615), reproduced in *Nouvelle invention de lever l'eau* (1644), courtesy Department of Special Collections, Stanford Libraries. Photo by Andrew Schupanitz.

waterwheel, through a train of gear wheels, drives a pinned barrel that is in turn connected with the keys of the flageolet.[135]

By the 1660s, when John Evelyn was at work on his gardening manuals, he took it as a matter of course that an essential part of the business would be to instruct "our docill Gardiner, how he may himselfe make & contrive these wonderfull Automats, . . . which at present so celebrate the Gardens of the greatest Princes; . . . & many other famous Gardens of the most illustrious persons of the World." It was not just an added flourish but actually "necessary," Evelyn counseled, ". . . in these Inventions, to give some motion to the living creatures . . . that they may the {better} imitate nature." The possibilities were legion:

> We may . . . people our Rocks with *Fowle, Conies, Capricornes, Goates* {& rapitary beasts, with} *Hermites, Satyres,* {Masceras} *Shepheards,* {rustic workes river gods Antiqs etc} and with divers *Machines* or *Mills* made to move by the ingenious placing of wheels, painted & turned by some seacret pipes of waters; The *Figures* above named may be formed of Potters earth, well

> moulded and baked; but if the statues must be larger, of stone or Mettal: By these motions, histories, {*Andromedas*} and *sceanes* may be represented.[136]

In addition to the elaborate networks of levers, wheels, gears, flowing fluids, and falling weights, the advent of organ-barrel programming helped in the building of complex systems of lifelike motions. Kircher—who designed and described many automata including an "automatic organ machine which utters the voices of animals and birds"—published a systematic account of the camshaft in 1650 (see figure 1.8).[137] But by then, camshafts or pinned cylinders had already been in use for centuries.[138] The Mesopotamian engineer, mathematician, and astronomer al-Jazari used them in his automata in the early thirteenth century, including a band of automaton musicians in a boat.[139] In 1599, Queen Elizabeth presented an organ clock driven by a camshaft to the sultan of Turkey.[140]

During the first decades of the seventeenth century, the use of camshafts spread everywhere. De Caus adopted them to organize the motions in his reproductions of Hero's singing and fluttering birds.[141] The Augsburg clockmaker Achilles Langenbucher put the new technology to work in mechanical musical

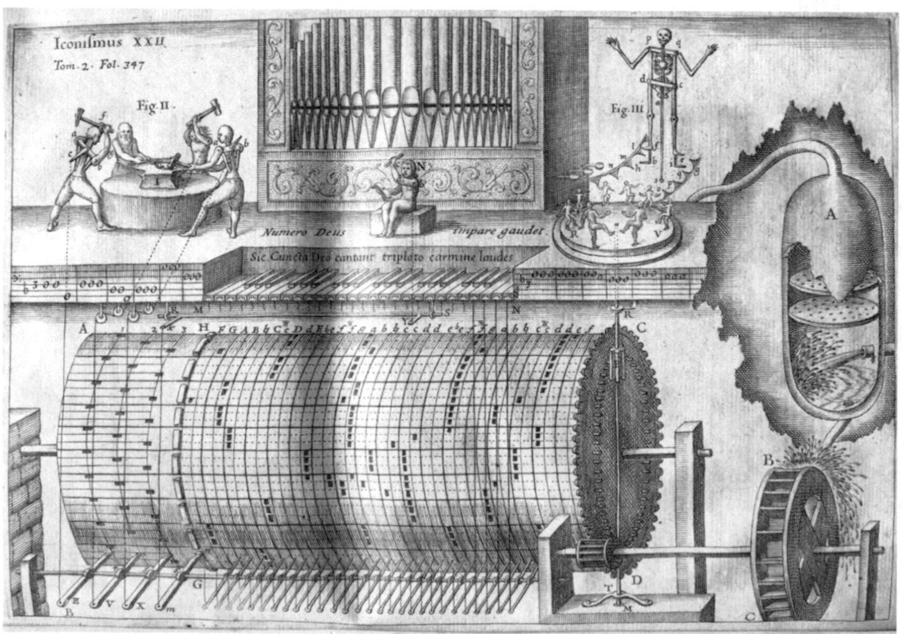

Figure 1.8 Athanasius Kircher's rendition of a camshaft from *Musurgia Universalis* (1650), courtesy Department of Special Collections, Stanford Libraries. Photo by Andrew Schupanitz.

ensembles composed of many player-less instruments.[142] Evelyn included an extensive description of the camshaft (the "Phonotactic Cylinder") in *Elysium Britannicum*.[143] His discussion included explicit instructions for making such a device, which, like the construction of automata more generally, he considered to be essential to the art of gardening:

> A *Cylinder* may be fitted so as to move, take out, & change the Teeth at pleasure, to place other in their stead: and so a new *Composition* may at any tyme be applied; . . . For example of this: Divide a *Cylinder* into 24 *Measures*, each of these {full} divide againe into 8 equal spaces, as we noted for *Quavers*; you shall bore holes, at every point of these divisions; so as being{furnished} with a greate number {of} *Teeth* (as the *Printers* box is with Letters) for all sorts of *Notes*, which you may keepe in a divided Drawer contrived some where about the Organ, you may insert a new *Composition* or Tunes at pleasure in your *Cylinder* which, the more large & ample it is, will be so much the better for our purpose.[144]

By means of a camshaft, a single flow of fluid could work myriad effects. Evelyn singled out, as most "expeditious" and "ingenious," those waterworks that "onely with the {precipitation} of water alone produce wind sufficient for all our motions." A single "Artificiall Ventiduct" created by filling a chamber with water, thereby forcing out the air, could be "sufficient either to refrigerate a roome in Summer, or to animate any . . . Bird, blow the Fire, [or] turne any Image or wheele." Similarly, through the "rarifaction" of air by heating, one could create a stream of wind; this wind could then turn a cogwheel that could pluck wires to play a tune or make another patterned sound, as in the case of the "celebrated statue of Memnon, which is reported to have spoaken & uttered a voice like a man, so soone as the Sun arose & darted his rayes upon it." The same wind, Evelyn noted, might "also serve to make artificiall Eyes & hands move; And Birds furnished with proper calls & whistles, will be heard to sing, to move their tailes, heads & clap their wings."[145]

Hydraulic and mechanical figures had become routine. Treatises such as De Caus's and Evelyn's helped to spread familiarity with hydraulic antics below the sphere of popes and princes. Martin Löhner, a hydraulic engineer and the Master of Wells (*Brunnenmeister*) for Nüremberg, established a much-visited host of automata at his own comparatively humble house: Vulcan laboring at his forge; Hercules bludgeoning his dragon; Actaeon surprising Diana and her nymphs in their bath, whereupon Diana threw water at Actaeon, who turned away, grew antlers on his head, and was attacked by his own dogs; Cerberus spitting fire at Hercules; a lion emerging from his cave to drink from a basin, then retiring; the nine Muses, each engaged at her appointed art.[146] Waterworks were de rigueur not only for popes, cardinals, archbishops, and kings but for ministers too. Richelieu had his own at his residence at Reuil. Evelyn, visiting in 1644, pronounced that garden "so magnificent, that I doubt whether Italy has any exceeding it." He recorded having been shot by streams of water, on his way out of one of Richelieu's grottoes, from muskets held by "two extravagant [automaton] musketeers."[147]

One might think the joke would wear thin. One would be wrong. The sport proceeded right through the seventeenth century. Evelyn described with malicious satisfaction, circa 1660, the "wayes of contriving seacret pipes to lie so as may wett the {gazing} Spectators, underneath, behind, in front and at every side according as the Fontaneere is pleased to turne & governe these clandestine & prepostrous showers." Evelyn included, for example, a design for making "*a chaire which shall wett those that sit upon it, though no water appeare.*" The functional features are a water-filled cushion attached to a pipe that rises through the back of the chair and has an opening, concealed in "the carvd head of a Lyon or some other beast," at the top. Thus when the victim sits down on the cushion, he unknowingly squeezes water up into the pipe to "spurt into his neck immediately." This "waggish invention," Evelyn said, he had found in the garden of the pope's cross-bearer.[148]

The gulled continued to take their licks with unflagging surprise and delight. Anne-Louise d'Orléans, duchesse de Montpensier, the memoirist and wayward cousin of Louis XIV, cheerfully recorded her experience at the Essonnes estate of the master of finances for the royal household, where she visited with her friend, Madame de Lixein, in the summer of 1656:

> As I passed through a grotto, they released the fountains, which came out of the pavement. Everyone fled; Madame de Lixein fell and a thousand people fell on her. . . . We saw her being led out by two people, her mask muddy, and her face the same; her handkerchief torn, her clothes, her oversleeves, in short, disconcerted in the funniest way in the world, and I cannot remember it without laughing. I laughed in her face and she started laughing too, finding that she was in a state to inspire it. She took this accident as a person of humor. She took no meal and went right to bed. . . . Upon returning, I visited her: we laughed a lot again, she and I.[149]

The historian Robert Darnton has recommended taking note of the mystifying jokes of the past, as these indicate "where to grasp a foreign system of meaning in order to unravel it."[150] To what exotic tapestry do these mischievous machines in their endless funniness connect? The philosopher Henri Bergson described the quintessential comic situation as "some mechanism stuck onto the living": the appearance of a human being as an automaton. We laugh, Bergson claimed, as a "corrective": to reassert the distance between machinery and life.[151] But, as Darnton's recommendation assumes, humor has a history,[152] and the need to establish that human beings are not machines cannot have had the same urgency in 1500 or 1600 as it had in 1900. Rabelais's, not Chaplin's, was the sense of humor at play.

The frolicsome engines catalogued in this chapter represented something like the opposite of Bergson's scenario: not people as rote automata but machines as responsively alive. The machines' human targets, laughing at the machines' whimsical vitality, do not seem to me to have been reasserting their own transcendence of machinery. I think they were doing something more like delighting in a base corporeality that they took to anchor even the very highest of human lives in an actively material world.

Arriving, then, at the mid-seventeenth century, when the idea of animal- and human-machinery began to flourish in philosophical and scientific discussion, we find that mechanical images of living creatures were already everywhere. They were familiar, not only to the nobility and the wealthy bourgeoisie, but also to their servants and to the engineers and artisans who built the machines, as well as to the audiences who flocked to witness them and the literate who read about them. The culture surrounding these lifelike machines had initially assumed no antithesis between machinery and agency, either vital or divine. The idea that a fully material entity could have no agency, but must be purely passive, arose with the Reformation, when the Reformers asserted a categorical distinction between matter and spirit, setting their clockmaker God rigorously apart from his clockwork Creation and assigning him a monopoly on agency. The medieval Catholic God had enjoyed no such monopoly; he had presided over a cosmos permeated by spirit.

Proponents of this new idea, that the material world was intrinsically inert and passive, inaugurated an intense and world-transforming conflict with a persisting older tradition in which matter and mechanism remained active and vital, and in which automata represented spirit in every corporeal guise available and life at its very liveliest.

~ 2 ~

Descartes among the Machines

A bathing automaton Diana flees to hide in the reeds, and when spectators try to pursue her, a mechanical Neptune advances to threaten them with his trident. Retreating, the spectators are next pursued by a sea monster emerging to spew water in their faces. This account of a visit to a hydraulic grotto is by René Descartes, philosophical revolutionary and one of the principal founders of the New Science (see figure 2.1). Descartes probably lived in Saint-Germain-en-Laye for a time and almost certainly visited the same waterworks that had so amused the young Louis XIII.[1] Amid the spirited, lively machines of the churches and the playgrounds of the wealthy, Descartes introduced the modern scientific notion of the "animal-machine": the idea that animals, and also human bodies, were essentially machinery.[2]

Descartes's proposal that an animal is a machine has sounded to most people, from the seventeenth century until the present, like saying that an animal is essentially inanimate. We will see many examples of people who have thought this.[3] But it is a misreading of Descartes: the key point about his animal-machine was that it was *alive*; it was a *living* machine. The responsive, engaging, lifelike machines that were all around him, if one takes them into account, make it easier to imagine how he arrived at this idea. To be alive was the whole purpose of Descartes's animal-machine. Not as if alive, not apparently alive, but *actually alive*. By describing animals as automata, Descartes did not mean to

Figure 2.1 Portrait of René Descartes, engraving by W. Holl after painting by Franz Hals, Library of Congress Prints and Photographs Division.

reduce them to lifelessness. On the contrary, he meant to declare that one could explain every aspect of life in terms of machinery, and so could understand the workings of living beings as fully as a clockmaker understands a clock. Rather than to reduce life to mechanism, he meant to elevate mechanism to life: to explain life, never to explain it away.

I began with the actual machines, the great bustling population of lifelike devices that enlivened the world in which Descartes lived, before arriving at his radical philosophical and scientific proposal, because having these machines in mind changes how one reads his philosophical writing about animal-machinery, as indeed the machines themselves shaped how Descartes conceived the idea during the 1630s and '40s. He had the machines in mind as he wrote, and what the machines told him, as they should tell us, was that machinery seemed anything but passive and rote. Machines appeared, on the contrary, full of agency: they acted, engaged, and responded.

The Animal-Machine Was Once Alive

To justify his radical suggestion that the body was nothing but "a statue, or a machine made of earth," Descartes invoked the machines that were everywhere, that had become familiar features of the landscape. Just look, he exhorted his readers, at all the clocks, fountains, and mills. If people could build such machinery, God must surely be able to do it even better.[4] Considering the royal hydraulic grottoes, Descartes noticed the same thing that John Evelyn had emphasized: a single force could work many operations. A flow of water could move several machines, play various instruments, even cause words to be pronounced.[5] Descartes drew an analogy between the water that created these diverse effects and the "animal spirits" that he supposed drove living bodies. The nerves through which the animal spirits ran were like fountain pipes, the muscles and tendons like "engines and springs." The constant and involuntary motions of the animal-machine, such as respiration or the beating of the heart, would then be like the steady ticking of a clock or the turning of a mill that underlay the other intermittent functions.[6]

The heart's movement required no soul whatsoever, Descartes argued, but only the same sort of "fire without light" familiar from rotting hay and fermenting wine, since he saw the heart as the source of the body's heat. Descartes recommended that the reader have a large animal dissected before him in order to grasp why the heart's movement required no soul. It would be obvious from the shape and arrangement of the vessels and chambers that the heart moved in its accustomed way purely because of how its parts were constituted. When the heart's own cavities were not filled with blood, they would necessarily be replenished "from the vena cava into the right cavity and from the venous artery into the left," since these vessels were always filled with blood and open to the heart. The motion of the heart accordingly followed from the arrangement of the organs "as that of a clock from the force, position and shape of its counterweights and wheels."[7]

Regarding the particular mechanism of the heart, Descartes partially disagreed with his contemporary, the English doctor William Harvey, who ascribed a pumping function to the heart, implying that it resembled a water pump.[8] Descartes endorsed Harvey's claim that the blood circulated through the body, but he rejected the notion that the heart was a pump and retained instead the traditional, widely established view of the heart as a kind of furnace to heat the blood.[9] He explained the motion of the blood in traditional Aristotelian terms as due to heating rather than pumping (and rather ungenerously insisted that Aristotle had hit on the right answer "just by chance").[10] Still, the analogy to artificial hydraulic systems suggested itself to both Descartes and Harvey despite their disagreement about the heart's function.

But to Descartes the crucial implication of artificial waterworks, even more than the possibility of working multiple effects by a single flow of fluid, was their demonstration that machinery could be responsive. These machines did not just carry out a predetermined sequence of movements like a clock. They moved in response to human beings arriving in their midst. Responsive machinery implied that sensation, the way in which living creatures respond to the world, could be understood in mechanical terms. The responses of sense organs to external objects, Descartes argued, were like what happened in the hydraulic grottoes when visitors arrived. Diana hid; Neptune threatened; the sea monster pursued. The hydraulic automata fled, menaced, interacted, attacked. By engaging with their audiences, they implied, as Descartes saw it, that animal sensory responsiveness might also be a matter of machinery.[11]

For example, one might feel a pain in one's foot when the nerves "stretched like cords" from the foot to the brain exerted a tug there (see figure 2.2).[12] According to this marionette-like account of sensation, nerves were very fine threads stretching from the brain to all the other parts of the body and were arranged such that touching any part of the body would set a nerve ending in motion, and the movement would pass along the nerve to the brain. The resulting sensations depended upon the particular nerves set in motion, the sorts of motion they received, and the shapes of the particles moving them. Objects touching the skin could cause several different sorts of motion in its nerves corresponding to their hardness, weight, temperature, and humidity. In the nerves of the tongue, differently shaped particles of food caused different sensations of taste.

Smells resulted from particles in the air sufficiently fine and active to enter the nerves of the nose. Sounds arose from the impact of the air itself against the nerves of the ear. Vision, the subtlest of the senses, took place when particles of light set the nerves of the eye in motion. Nerves to the stomach and throat transmitted "appetites," internal sensations such as hunger and thirst. Mean-

Figure 2.2 Illustration of the mechanism of sensation from Descartes's *Traité de l'homme*: heat from the fire sets a nerve in motion, and this motion, transmitted from foot to brain, allows the figure to feel heat. Courtesy Department of Special Collections, Stanford Libraries. Photo by Andrew Schupanitz.

while, the "little nerves" of the heart and diaphragm carried other internal sensations, affections such as joy, sadness, love, and anger.[13]

Later, Descartes extended the list to include all the passions: admiration, esteem, contempt, generosity, liberality, malice, timidity, vanity, humility, veneration, disdain, love, hate, desire, hope, fear, jealousy, confidence, despair, timidity, courage, sadness, boldness, malice, remorse, joy, sorrow, gratitude, envy, pity, indignation, ruthlessness, anger, pride, shame, disgust, regret, among others. Each began with the impingement of external objects upon the sense organs, which set the animal spirits in motion through the nerves, which in turn caused an "agitation" in the "little gland at the center of the brain" (the pineal gland, which Descartes named "gland H").[14] The movements associated with the different passions depended upon the size, shape, and activity of the particles of animal spirits coursing through the tubes and pores of the body-machine.[15]

Descartes's main purpose in representing all of these features of life as mechanical was to change how people understood the natural world, and to change what it meant to offer an explanation of a natural phenomenon. Henceforth, Descartes argued, an explanation should be a rational account of the motions of material parts. In particular, he wanted to do away with the "forms" and "faculties" of Aristotelian philosophy, terms designating essences and powers that Descartes found uselessly mysterious and sought to eliminate from philosophical explanations.

If life, movement, sensation, and all the passions were explicable just in terms of material parts in motion, then two of Aristotle's three souls were unnecessary: the vegetative soul present in plants, animals, and humans, responsible for life and growth; and the sensitive soul, present in animals and humans, responsible for sensation and motion. Descartes dismissed both of these Aristotelian souls as superfluous. Aristotle had described a third soul, the rational soul, present in humans alone, endowing humans with the capacity for rational thought. This uniquely human, rational soul, Descartes retained. But he eliminated the vegetative and sensitive souls altogether, even explaining reproduction without reference to souls, forms, or faculties but purely by the heating, expanding, pressing, and jostling of the "two liquids" mixed together in copulation.[16]

In general, Descartes attributed all the functions of life and sentience, apart from rational thought, to animal-machinery alone: digestion; circulation; nutrition; growth; respiration; waking and sleeping; the sense organs' reception of light, sound, smell, taste, heat, and other qualities; the impression of the ideas of these qualities in the brain; the retention of the ideas in memory; the "internal movements of the appetites and passions"; and the external movements that come in response both to these internal appetites and passions and to external objects.[17]

In human beings, of the old three souls, only the rational soul remained, residing in the brain like the "fountaineer" stationed at the tanks where the fountain pipes met, able to redirect the waterworks at will.[18] Every other function, apart from rational thought, originated in the working of the body-machinery, just as the "movements of a clock, or other automaton, follow from that of the counterweights and wheels."[19]

Animals, having no souls, were fully machines and in that sense comparable to clocks.[20] But, careful, this did not mean that they *were* clocks.[21] To say that the body, animal or human, was *like* a watch, in the sense of being a fully material composition of parts, was not to say that it *was* a watch. Clocks and watches did not breathe, eat, or walk, all functions that Descartes attributed to animal-machinery. Clocks and watches had no vital warmth, another quality for which Descartes thought the animal-machine, rather than a soul, was responsible.

The warm, mobile, responsive machine became still and cold at death, not because the soul left it, but just the reverse: in humans, the rational soul departed because the machine stopped working, growing still and cold.[22] Descartes never suggested that actual clocks or watches lived or sensed, whereas he assumed that the animal-machine was alive and that in terms of sentience, it had everything but the ability to reason. "I am now dissecting the heads of various animals," Descartes wrote to his main confidante and intellectual interlocutor, the mechanist, mathematician, and Minim friar Marin Mersenne, "to explain what imagination, memory, etc., consist of."[23]

To explain how one could account for such things as imagination and memory mechanically, Descartes drew upon a further comparison with artificial mechanisms: church organs. These produced different sounds according to the pipes into which the organist channeled the air by pressing keys on the keyboard. By analogy, a brain produced different feelings according to which pores filled with animal spirits. The heart and arteries were like the bellows of the organ, pushing animal spirits into the cavities of the brain. External objects, acting upon the sense organs, moved certain nerves such that they would receive a flow of animal spirits from the brain: these objects were like the fingers of the organist as they selectively pressed the keys.[24] The tones of an organ, with their unarguably physical nature and origin, provided an apt metaphor for the feelings of Descartes's animal-machine: resonant despite being mechanically produced.

Organs, especially hydraulic organs, offered a more suitable metaphor than clocks and watches for Descartes's animal-machinery, which was essentially hydraulic.[25] In this sense, his animal-machinery was in keeping with the humoral theories of ancient medical tradition, which located life in the interacting fluids of living creatures.[26] Descartes had in mind an old, established model of bodily machinery, as well as many examples of lifelike machines, in which the machinery was both active and responsive. Along with airs and heat, Descartes's hydraulic body-machine ran on fluids including a vital, nervous fluid, the "animal spirits." Fluids, even ordinary ones, behave differently from solids. They pursue equilibria, climb siphons, exhibit a kind of purposefulness. Claude Perrault, architect of the eastern wing of the Louvre, brother of the fabulist

Charles Perrault, doctor and Cartesian-leaning natural philosopher, remarked that a flowing river "seems to seek the valley."[27] Fluids appeared to offer a material basis for agency, purposeful action, in Descartes's body-machine.

There was no need for an immaterial soul to explain nutrition, sensation, or movement: the sheer force of particles in fluid motion, Descartes thought, could account for these. In nutrition, the blood, rarefied by the heart's heat, passed so forcefully out into the body that it left some particles behind in the members while expelling others, depending upon the shape of the body's pores. The most agitated and penetrating parts of the blood, meanwhile, formed the "animal spirits" responsible for sensation and motion. These were "like a subtle wind, or rather like a very pure and lively flame" rising continually from the heart to the brain, and passing thence through the nerves to the muscles. A hydraulic system for the distribution of animal spirits could account for wakefulness, sleep, dreams, hunger, thirst, and the other "internal passions"; for the imprinting of ideas of external objects upon the brain by means of the senses; for the memory that preserves these ideas; for the imagination that composes new ones; and for responsive movement. None of this, Descartes argued, could surprise those aquainted with automata and their intricate mechanisms and motions.[28]

In all of this, it is once again important to keep in mind that Descartes was working from an ancient, active-mechanical model of living beings, just as he was working from actual lifelike machines that seemed active and responsive. To be sure, he dismissed two of the three Aristotelian souls, the vegetative and the sensitive, but apart from that, Descartes's view of animal and human life did not depart much from the ancient physiological tradition established by Aristotle and Galen, the predominant medical philosopher of classical antiquity (second century CE).[29] This ancient tradition was filled with analogies between living bodies and artificial machines.

Aristotle habitually drew parallels between animal physiology and artificial devices. For example, he compared animals to automatic puppets that transmitted motion through their parts by means of pegs striking one another. Aristotle also drew an analogy between animal movements and a toy wagon that goes in a circle when the child pushes it straight, thanks to wheels of unequal diameter. Likewise, he thought, animals moved according to the disposition of their parts: "The bones are like the pegs and the iron; the tendons are like the strings."[30] He found that the organs of respiration must be "like the bellows in a smithy, for both heart and lungs conform pretty well to this shape." When an animal breathed, its chest raised up "in the manner of a forge-bellows," expanded by the heat of the heart. As the cool air entered, it caused the lungs to contract and expel the breath "like the bellows once more."[31] This comparison of the lungs to a pair of bellows recurred throughout ancient medical and physiological writing.

The very creation of life seemed to Aristotle analogous to the connected operations of automatic machinery. He saw the development of the embryo from its initial state of potentiality as corresponding to the motions transmitted through the parts of an automaton. It is possible, he wrote, that in the development of the embryo "A should move B, and B move C" in the same way as in automatic puppets, whose parts have "a sort of potentiality of motion in them," ready to move and to set one another in motion. A single force could therefore move parts with which it was never in direct contact. In the same way, "that which made the semen" could set all the parts of the embryo into developmental motion by just a brief and limited contact.[32] The comparison worked both ways: while Aristotle compared the embryo in development to an automaton set in motion, he also saw the machine as like an embryo, its parts pregnant with dynamic potentiality.

Analogies between the body and various sorts of machines appear in the collection of ancient medical writings by followers of Hippocrates (fifth to fourth centuries BCE), known as the Hippocratic corpus, and in the writings of Galen. The heart, for example, observed a Hippocratic author, had "bellows," namely the lungs, just like furnaces. About the membranes of the heart, the author wrote, "In my opinion these serve as the guy-ropes and stays of the heart and its vessels."[33]

To Galen, the larynx resembled a flute. He developed the analogy in detail, arguing that just as the pipe in question owed its sound to the combination of its shape and its reedlike "tongue," so the larynx required a narrowing of the part containing the sounding "cartilages" in order to produce the voice.[34] Galen also applied the common analogy to bellows to describe the functioning of the lungs. He compared the blood vessels to a network of water conduits in a garden[35] and the arteries to aqueducts.[36] In ancient physiology, irrigation structures were a standard source of metaphors for the circulatory system.[37] Plato made the comparison in the *Timaeus*. In devising the human body, he said, the "superior powers" had cut channels through the body as one would through a garden, in order to water it from a stream. The primary two "channels or veins" extended down along the backbone "in order that the stream coming down from above might flow freely to the other parts, and equalise the irrigation."[38]

Aristotle used the same analogy in *De partibus animalium* (On the Parts of Animals). The "water-courses in gardens," he wrote, were constructed so as to distribute water from a single source into many channels, which divided and subdivided throughout the garden. "Now just after the same fashion has nature laid down channels for the conveyance of the blood throughout the whole body." Aristotle considered that the blood produced the body's flesh in the same way that the water in irrigation channels laid down mud. In irrigation channels the largest remained clear while the smallest "soon fill up with mud and disappear"; likewise, the largest blood vessels remained open while the smallest were "converted actually into flesh."[39]

Descartes did not have just the ancient tradition of body-machine analogies to draw upon, but also the medieval one that grew out of it.[40] As in classical philosophy, living machinery in the medieval philosophical tradition was imbued with vitality and agency. In the new universities founded between the late eleventh and thirteenth centuries, beginning with Bologna, Oxford, and Paris, the Scholastic philosophers (so called because of their association with the new schools) worked to reconcile the teachings of the ancient philosophers, especially Aristotle, with Christian theology. Thomas Aquinas, the pivotal figure in the Scholastic tradition, studied and taught at the University of Paris in the middle of the thirteenth century. He considered that animals might be regarded as machines some four centuries before Descartes and in strikingly similar terms. Aquinas argued that animals were moved by reason although they themselves lacked reason, just as an arrow is moved by the motion of the archer, or a clock or other engine by the person who built it. Likewise, the appearance of reason, prudence, or sagacity in animals' movements originated in the divine art that had produced them. However, animals did move in accordance with their own faculties of sense, appetite, imagination, and desire.[41] They were devices lacking reason, but not life or sentience.

This sort of argument was deeply familiar to Descartes. Scholastic philosophy was the basis of his education and, indeed, of virtually everyone else's, being the official philosophy of the Catholic Church. At the age of ten, Descartes entered a prestigious Jesuit school, the College Royal Henri-le-Grand in the town of La Flèche. There he studied the Coimbran commentaries, a series of eleven books on Aristotle gathered from the teachings of Jesuits at the University of Coimbra and published during the 1590s as the definitive doctrine of the Jesuit order.[42] Jean Fernel, a sixteenth-century French doctor, coiner of the term "physiology" and the primary authority on whom the Coimbran commentators based their physiological teaching, recalled Galen in comparing the trachea to a flute and the epiglottis to "the tongue or reed of panpipes." Fernel wrote that the human skeletal structure was like the "base and pillar that support a house" and that "the ancients compared [the spine] to the keel, the first part of a ship to be laid down." The recurrent nerves, meanwhile, Fernel observed, drawing once again on Galen's corpus of comparisons, "mimic[ked] the movement that is carried out in building machinery by the use of pulleys."[43]

A final, especially pertinent example of pre-Cartesian representations of animal physiology as machinery appears in the work of Eustachius a Sancto Paulo, a member of the Cistercian Feuillant order and a professor of philosophy at the Sorbonne in Paris. Descartes judged Eustachius's *Summa philosophiae quadripartita* (Compendium of Philosophy in Four Parts) (1609) to be the finest that Scholastic philosophy had to offer. His admiration was such that he considered designating Eustachius his representative Scholastic interlocutor by printing the *Summa philosophiae* alongside his own philosophical system in the same volume.[44] Eustachius, as Descartes later would, compared the animal faculty of common sense to a "fountain" pouring forth animal spirits.[45]

Descartes's animal-machinery closely resembled that of the ancient and Scholastic writers described above: bellows, furnaces, irrigation systems, fountains.[46] Recognizing that he was working with an already established model, just as he was working from a plethora of physical examples in the form of actual machines, is crucial to understanding what he meant by his idea of the animal- or body-machine.[47] He assumed it had vitality and various kinds of agency, just as it had in ancient and medieval tradition. Descartes insisted to Mersenne, "I have supposed nothing new in anatomy, nor anything that could be the least bit controversial."[48]

This insistence that he was proposing nothing new in anatomy might seem odd: in that case, why bother writing an anatomical treatise? The explanation is that Descartes's main revolutionary purpose concerned method, and not necessarily substance. It was fine, even advantageous, for his accounts of anatomical systems to resemble those in ancient and medieval anatomy. The point was that he was applying a different method to understanding them: one in which he referred to nothing but material parts in motion, the sorts of things a clockmaker needed to know to make a clock. Descartes's revolution in method, as I have been suggesting, meant changing how people went about understanding and explaining natural phenomena.

This methodological or *epistemological* revolution, by which I mean a revolution in how people thought they should go about understanding the world, brought with it a profound *ontological* revolution, by which I mean a revolution in what people thought the world essentially *was*: in the eyes of Descartes's followers, all the world (except human, rational thought) became just moving bits of uniform matter. But establishing that the world was made of moving bits of uniform matter was not Descartes's primary purpose. Rather, he was committed to *understanding* the world in mechanist terms. Machinery, in Descartes's usage, meant intelligibility.[49]

Descartes frequently characterized his method by saying he considered all things in the physical world as machinery: a philosopher's view of the physical world was precisely an artisan's view of a machine.[50] For the living body to be a machine meant principally that it was fully comprehensible in materialist terms, without recourse to the souls or forms or faculties of the Scholastics. For instance, as we have seen, Descartes replaced the vegetative soul with the movements of blood and nutritive particles to account for growth; and he dismissed the vegetative and sensitive souls to explain reproduction instead by a mixing of seminal fluids. Thus, in the frequent references to clockwork throughout Descartes's physiological writing, clockwork represented a mode of intelligibility rather than the particular machinery of life.

This use of clockwork to represent full comprehensibility became commonplace around the middle of the seventeenth century. The English philosopher and medical doctor John Locke, for example, compared a full, divine understanding of nature to that of someone "who knows all the springs and wheels, and other contrivances within, of the famous clock at Strasburgh"; whereas the human view of nature was that of "a gazing countryman . . . who barely sees the motion of the hand, and hears the clock strike, and observes only some of the outward appearances."[51]

For Locke too, as for Descartes, the analogy to clockwork did not imply that the phenomenon in question resembled a clock. It meant rather that the comprehension to be achieved was comparable to a clockmaker's understanding of a clock. Just as Descartes used clockwork to represent the intelligibility of a materialist physiology whose machinery was not clocklike, Locke invoked clockwork to epitomize the comprehensibility of chemical processes. Imagine, he wrote, if we understood "the mechanical affections" of the particles of rhubarb, hemlock, opium, and human bodies in the way that a watchmaker understands the workings of a watch, and how he can alter its operations by filing the wheels. In that case, we would know in advance, even without experiencing it, "that rhubarb will purge, hemlock kill, and opium make a man sleep," just as a watchmaker knows that a little piece of paper laid on the balance will stop the watch until it is removed, or that filing a part will incapacitate the mechanism altogether.[52]

In short, clockwork meant intelligibility in terms of material parts, not literal clockwork. Descartes's animal-machinery resembled ancient and medieval animal-machinery in many respects: it was warm, fluid, responsive, mobile, sentient, and full of agency. Its salient difference was that it was fully material and so completely intelligible in Descartes's new science.

Given its cultural and theological familiarity, it makes sense that the animal-machine seemed to Descartes a safe vehicle for promoting his new science.[53] Safety became crucially important in 1633, when he learned of the trial and condemnation of Galileo Galilei for espousing Copernican (heliocentric) astronomy.[54] Descartes had just completed his *Traité de l'homme* (Treatise on Man), in which he laid out the above model of living bodies as machines, as well as a treatise on light, *Traité de la lumière*, which presented a mechanist, Copernican cosmology.[55] The two treatises were intended as parts of a single work to be entitled *Le Monde* (The World) which would also include a third part, an account of the rational soul. The news of Galileo's condemnation alarmed Descartes and caused him to reconsider this philosophical program. In late November 1633, he wrote in panic to Mersenne:

> I was planning to send you my *World* for the New Year . . . but I will tell you that, having made inquiries in that time at Leyden and at Amsterdam whether the *System of the World* of Galileo was not out, since I thought I had learned it was going to be printed in Italy last year, I learned that it was true it had been printed, but that all the copies had been burned at once in Rome and he had been condemned to some fine: which so astonished me, that I more or less resolved to burn all my papers, or at least not to let anyone see them. Because I could not imagine that he, an Italian, and in the Pope's good esteem, as I understand it, could have been criminalized for anything other than that he doubtless wanted to establish the movement of the earth . . . and I confess that if it is false, all the foundations of my philosophy are too, since it follows clearly from them. And it is so linked with all the parts of my treatise, that I

> would not know how to detach it without rendering the rest defective. But since I would not for anything in the world want to put forth a discourse, in which the least word could be disapproved by the Church, I prefer to suppress it, then to let it appear in crippled form.

Descartes concluded with a plea to Mersenne to "send me what you know of the Galileo affair."[56] Over the subsequent months, he filled his letters to Mersenne with worried references to Galileo's condemnation and repetitions of his decision to be cautious, announcing that he had taken as his motto "*bene vixit, bene qui latuit* [he has lived well who has been well hidden]."[57]

Descartes's fear may have been disproportionate. In the Netherlands, where he had been living for several years, it was not dangerous to publish Copernican writings.[58] Indeed, in 1636 the Dutch publisher Louis Elsevier visited Galileo under house arrest at his villa in Arcetri and arranged to have his *Discourses on Two New Sciences* smuggled out of Italy and published in Leiden.[59] At least, there was no grave danger to one's person. But Descartes had good reason to be concerned for the survival and success of his doctrine. For anyone who wanted to see his philosophy widely adopted, the Jesuit Order held the keys to the philosophical kingdom in seventeenth-century Europe. The Jesuits were the scholars, the teachers; they ran the schools. They had trained Descartes himself and virtually all of his educated contemporaries. The best way to accomplish a philosophical coup of the sort Descartes was after was to persuade the Jesuits to adopt his system.[60]

Descartes was clearly keen to do this. When he returned to cosmology and physics, primarily in his *Principes de philosophie* (Principles of Philosophy) (1644), it was not to push Copernicanism but, on the contrary, to justify his world system in a way that would guarantee its compatibility with orthodox doctrine.[61] In contrast, Descartes's writing continued to feature the conceit of the body- or animal-machine. To the end of his life, Descartes returned to the notion of bodily machinery and continued to try to extend it as far as it would go. He seems to have seen the idea of the body-machine as a way to demonstrate his revolutionary philosophical method without risk of offending the Church.

The importance of overcoming theological objections to his philosophical system clearly preoccupied him. When, with Mersenne acting on his behalf, he solicited objections to his philosophical *Méditations*, Descartes lavished attention on the theological ones—especially those that came from Antoine Arnauld, the Jansenist Sorbonne theologian and logician, and from the Jesuit

Pierre Bourdin—while paying significantly less attention to the two sets of philosophical objections offered by the English materialist Thomas Hobbes and the mathematician Pierre Gassendi.[62] (Gassendi himself, in fact, explained organic matter in terms of special atoms that he pictured as tiny machines built by the divine Engineer.)[63] Moreover, the only set of objections that Descartes himself solicited were from a Catholic theologian, a Dutch priest named Johannes Caterus (Johan de Kater).[64] Mersenne also compiled two sets from "theologians and philosophers."[65]

If Descartes judged that the body-machine was a safe context in which to demonstrate his philosophical method, he was right. The five sets of primarily theological objections, along with the responses he got to the *Discours de la méthode* (Discourse on Method) (1637) following its publication, show that Descartes's body-machine did not initially pose a theological problem.[66] Fromondus and Plempius, Jesuits who sent some of the first objections to the *Discours,* quarreled with aspects of the animal-machinery Descartes described, but not with the idea of animal-machinery itself. Regarding Fromondus, Descartes reported to the mathematician, astronomer, and physicist Christiaan Huygens that "the dispute between us was like a game of chess; we have remained good friends." Plempius, who disagreed with Descartes's account of the movement of the heart, had likewise expressed his disagreements "but as a friend, in order to better discover the truth." Finally, Antoine Vatier, a Jesuit teacher at La Flèche, where Descartes had been a student, sent him "as much approbation as I could wish from anyone."[67]

Caterus did not mention the idea of the animal-machine at all, nor did Mersenne. Arnauld referred to it only in passing, in mild and not especially theological terms. "As far as the souls of the brutes are concerned," he wrote, Descartes was already well known for denying that they had any. Arnauld expected people would have trouble believing it. It seemed hard to fathom that a sheep would flee a wolf simply because the light reflected from the wolf's body into the sheep's eye, moving the little filaments of its optic nerves, would send animal spirits from its brain outward through its nerves.[68] Arnauld found the idea of animal automatism unlikely but not shocking: he did not include it in the section of his objections entitled "Things That Might Trouble the Theologians." There, he singled out (as it turned out rightly) the miracle of the Eucharist as the probable catch: "That which I predict the Theologians will find the most offensive is that, according to his principles, it does not seem that the things the Church teaches regarding the sacred mystery of the Eucharist could subsist and remain in their entirety.[69]

The orthodox Thomist (associated with Thomas Aquinas) view of transubstantiation, to which the Jesuits were committed as the appointed protectors and promulgators of Church philosophy, had been established by the Council of Trent (1545–63). This account held that the communion bread continued after its transformation to seem like bread because its accidental qualities—shape, color, taste, smell—remained through consecration, while its substance transubstantiated into the body of Christ. The trouble was that according to Descartes's theory of matter, the accidental qualities all reduced to the extension (size and shape) of the basic particles, the only essential properties of matter. Therefore, a persistence of accidental qualities, despite a transformation in basic substance, would be impossible. It was this conflict between Cartesian matter theory and the orthodox view of the miracle of the Eucharist that was primarily responsible for incurring the official condemnations issued at Louvain in 1662, which in turn led to Descartes's writings being put on the Index of Prohibited Books in 1663.[70]

The Louvain condemnations identified five problematic aspects of Cartesian philosophy, all having to do with Descartes's account of the relations among matter, form, and extension and his rejection of the possibility of a plurality of worlds, an apparent infringement on God's omnipotence (if an omnipotent God wanted to create several worlds, the authorities at Louvain felt, he ought to be allowed to have done so). None of the condemnations concerned animal automatism or the mechanical model of human life.[71]

Other Catholics too continued to offer animal-machine analogies in which the machinery retained its traditional vitality. For example, the English Catholic Cartesian Kenelm Digby drew carefully observed parallels between living things and two engines that he had seen on a trip to Spain. Plants, he said, were like the pumping apparatus that raised water from the Tagus River—seventeen thousand liters daily to a height of ninety meters—to supply the Alcazár palace in Toledo.[72] King Philip II had commissioned the clockmaker, architect, and engineer Juanelo Turriano (probable builder of the automaton monk mentioned in chapter 1) to design the system in 1565.[73] Digby meticulously described the apparatus, which was driven by a big waterwheel in the Tagus River.

The system consisted of two columns of swinging, "ladle"-shaped buckets attached to timber frames that were alternately raised and lowered by the waterwheel "like two legges that by turnes trode the water; as in the vintage, men presse grapes." The lowest buckets scooped water from the river and then, upon being raised, tilted to pour the water into the corresponding buckets on the opposite column, which did the same at the next alternation, so that the columns, as they stomped up and down, passed the water from each level up to the next. "All sortes of plants, both great and small, may be compared to our first engine of the waterworke att Toledo," Digby observed, "for in them, all the motion we can discerne, is of one part transmitting unto the next to it, the juice which it received from that immediately before it: so that it hath one constant course from the roote (which sucketh it from the earth) unto the toppe of the highest sprigge."[74]

Animals, in contrast, according to Digby, were like the Segovia mint, also built by Philip II and also driven by a huge waterwheel, this one in the Eresma River. The Segovia mint began operations in 1585 using new, state-of-the-art coin-rolling machines donated by Philip's cousin, Archduke Ferdinand of Tirol.[75] In the Segovia mint, unlike in the Toledo waterworks, Digby emphasized, each part performed a different, specialized motion. One formed gold and silver ingots of the requisite size and thickness for stamping, then passed them along to the next part, which stamped them and handed them over to a third apparatus, which cut them into the correct shapes and weights. Finally, the finished coins fell into a reserve in a different room, where an officer received them.

From where he was stationed, the officer could not know the intermediate steps in the process. If he went into the room where the machines were at work, he would see that each of them "considered by it selfe might seem a distinct complete engine." But they were working together toward a single purpose, "to make money," and if any one of them were to be removed, "the whole is maymed and destroyed." Likewise, in "sensible, living creatures," many different little engines worked together cooperatively, some driving or setting others in motion, some handing off tasks to others, thereby forming one big engine and producing a common result. In the Segovia mint, Digby saw an analogy to animal life in the collusion of parts. When all the parts, moving and moved, formed a single whole, Digby concluded, "we call the entire thing *Automatum* or *se movens*; or a living creature."[76]

Even to Descartes himself, his new model of the body as a machine did not immediately seem in itself sharply divergent from the ancient model. The ancient model of living machinery had persisted and been developed through the work of the medieval Scholastics, so that by the mid-seventeenth century, it was as familiar, and as indigenously Catholic, as the automata on clocks and organs in churches and cathedrals. Like those engaging machines, the ancient and medieval animal-machine model of living bodies—and Descartes's new model too, in the first instance—was fluid, responsive, and full of many agencies.

It Was the Human Soul That Drained the Life from the Animal-Machine

Descartes's dangerous idea was not the animal-machine but rather its counterpart: the human soul. What am I, he asked in the second *Méditation,* and the answer he gave was "a thing that thinks," a purely intellectual entity, a disembodied mind, a thinking soul.[77] This was the essence of his being, because it was the one matter on which he could never be mistaken or deluded. In order to be mistaken or deluded, he would still need to be a thinking being. He might be wrong regarding any of his bodily experiences—he could be dreaming or being tricked, he imagined, by an evil demon of some kind—but it would be contradictory to imagine he could be wrong about being a thinking thing, because he would have to be a thinking thing in the first place in order to be wrong about it. Accordingly, to be a "thing that thinks" was not only the essence of his existence; the knowledge of this fact was the bedrock on which all the rest of his knowledge rested.

Descartes located the boundary between these two absolutely distinct entities, body-machine and thinking soul, in what he called "gland H," now known as the pineal gland, where he judged that the soul met the brain.[78] The two sides of this boundary, the body-machine and disembodied intellect, defined one another by the infinite distance between them. It is no coincidence that Descartes, who is frequently credited as the first to put into words a modern sense of subjective selfhood (I am a thing that thinks; this is the essence of my being and the bedrock of all my knowledge)[79] was also the earliest influential proponent of the idea that the living body was a machine. Montaigne, commonly named as Descartes's precursor in the invention of modern subjectivity, also, as we have seen, wrote exhaustive descriptions of machinery, including lifelike machines.

By looking and listening "within my selfe" to discover his self's own "peculiar form," Montaigne gave Descartes a starting point.[80] But Montaigne also looked attentively outside himself, to record the workings of wind chambers, springs, jets, pipes, toothed wheels, levers, and keyboards. For Descartes, and no doubt for Montaigne, the inner and outer views were related. Seeing the world as a pure machine, lifting his thinking soul out of the world, even out of its own bodily interface with the world, Descartes accomplished the distancing of self from world that defines modern subjectivity, the sense of fully autonomous, inner selfhood, and modern objectivity, the sense of regarding the world from a neutral position outside of it.[81] It was in Descartes's philosophy that modern selfhood and modern science created one another.

The distinction between mechanical body and nonmechanical self appeared again and again in the wake of Descartes's articulation of it.[82] It informed even the work of philosophers who radically disagreed with Descartes on other matters. The philosophy of John Locke was in some ways antithetical to Descartes's. Locke argued that all people began life with minds that were tabula rasa, blank slates. These blank slates were then inscribed by people's experiences: everything they knew, therefore, originated in experience. Descartes's "cogito" argument described above ("I think, therefore I am," in Latin, *cogito ergo sum*) says the opposite: experience can offer no certain knowledge because it can be mistaken. Accordingly, the foundation of all my knowledge is not experience but logic, the infallible demonstration that I must exist as a thinking being. Given the sharp contrast between the Lockean tabula rasa (all knowledge comes from physical experiences) and the Cartesian cogito (all knowledge comes from logic, which is an innate capacity of the rational soul), it is striking that in one crucial respect—seeing human selves as purely intellectual entities attached to clockwork bodies—Locke agreed with Descartes.

The body of an animal or a human being, Locke wrote, was like a watch, a system of moving parts. During his travels in France in the 1670s, Locke was as meticulous as Montaigne had been in his descriptions of clocks, fountains, and their automata: the mechanical Annunciation (with rooster) on the Cathédrale Saint Jean in Lyon; the waterworks at Versailles and Fontainebleau; the techniques of a watchmaker on the Quai de Peletier in Paris.[83] Like Descartes, Locke combined with his focus on mechanism an equal and opposite conviction that "personal identity," the quality ensuring that "every one is to himself that which he calls *self*," was something different. Selfhood resided in "that consciousness which is inseparable from thinking." Consciousness, Locke reasoned, could not be a "substance" or arrangement of material parts, for it could flicker in and out of existence, could be interrupted by forgetfulness or deep sleep, without ever being lost.[84]

Although people's selves attached to their watchwork bodies, the self was not made of matter and could never be accounted for by any kind of machinery. This model operated in Locke's political system. His "self" was that creature born into a state of absolute freedom and equality to whom political rights were due.[85] Lockean selves, having no physical attributes, began life utterly equal. No unevenness of physical parts could undermine the essential equality of selves.[86] The bodily machine, meanwhile, was the self's first possession. By its labor, the machine manufactured all the self's other possessions and was therefore the basis of property, which was in turn the basis of civil society.[87]

In Descartes's pairing, the mechanical body served as a foil against whose stark limits Descartes took the measure of his disembodied self's transcendence. A pure machine, for example, could utter words and could even do so responsively, asking what you want upon being touched or crying out that you are hurting it. But Descartes argued that a physical mechanism could never arrange words so as to give meaningful answers to questions. Only a spiritual entity could achieve the limitlessness of interactive language, putting words together in indefinitely many ways. Any material machine must specialize: while a machine might do very well some of the things people do, it would necessarily be unable to do others. Any part or organ needed a particular configuration to achieve each task, and it was impossible to have enough different parts with the requisite configurations in a single machine "to make it act in all the contingencies of life in the same way that our reason makes us act." Only disembodied reason could be "a universal instrument."[88]

While the idea of living bodies as machines was ancient and unthreatening, Descartes's idea of a disembodied human self was more radical.[89] In theological terms, as Arnauld emphasized and subsequent commentators have often noted, Cartesian dualism—the idea that matter and spirit or soul were two absolutely distinct things—was superficially continuous with a much older dualist Christian tradition, which was undergoing a resurgence among both Catholics and Protestants, namely the tradition of Augustinian piety, recalling Saint Augustine's injunction to seek the truth within oneself rather than in the material world.[90] But by extracting soul from matter, Descartes did something different: he gave the human intellect a God's-eye view of the physical world. To have a soul meant to have a rational perspective on the physical world, a position fully outside of it from which one could grasp it rationally.[91]

Descartes turned Augustinian inwardness outward, giving it an Olympian outlook on God's creation and an imperial program. To seek the truth within was not to turn away from the world, but on the contrary, to turn upon it and grasp it.[92] Moreover, the ability to do this, Descartes assured his patroness,

Queen Christina of Sweden, made human beings "in a certain way equal to God, and seems to exempt us from being his subjects."[93] Small wonder that the disembodied human soul seemed more dangerous than the mechanical body. A strictly incorporeal soul was, at any rate, at odds with the Scholastic spectrum of soul and matter, in which the vegetative souls of plants and the sensitive souls of animals occupied intermediate positions between matter and spirit: they were closer to spirit than inanimate matter, but closer to matter than the immortal souls of humans.[94]

Descartes's severing of soul from body ran counter to Christian doctrine by conflicting with the Trinitarian (in relation to the Holy Trinity) understanding of incarnation and with the theology of resurrection, both of which required the embodiment of the soul. The inextricability of soul from body was an urgent concern in medieval and Renaissance Christian practice, provoking worries about what would happen to the hair and fingernail clippings lost during a lifetime and how the bodies of people who had been devoured by wild beasts could ever be resurrected.[95] Aquinas held that after death, and until the resurrection of the body, the separated soul was incomplete: only with the resurrection of the body was human personal identity restored.[96] The Aristotelian framework of "hylomorphism," the doctrine that all substances are composed of matter and form, as Aquinas and his followers applied it, meant that the soul was incomplete without the body it informed: it must resume its very flesh in resurrection.[97]

The relation of soul to body was a fraught problem in late Scholastic discussion.[98] On the one hand, the human soul must be incorporeal in order to be immortal; this was reaffirmed by the Fifth Lateran Council (a council of the Church held at the Lateran Palace in Rome) in 1513. On the other hand, the Gnostic[99] idea of the human soul as a piece of the true divinity, trapped by a fallen God in corrupt matter, had inspired anathemas against the view that the soul was something separate or separable from the body. Church doctrine, therefore, having posited the distinction between soul and body, was otherwise devoted to binding the soul to matter as tightly as possible. The trick was to arrive at a doctrine that emphasized both the incorporeality of soul and its inextricability from body. The body could not be the mere instrument of the soul, but rather the two must constitute one single, complete whole.[100] Aquinas argued that the kind of knowledge appropriate to human beings necessarily required sensitivity and imagination, which he regarded as bodily functions. He, the Coimbran commentators, Jesuit philosophers such as Franciscus Suárez and

Franciscus Toletus, all agreed that "I am not my soul alone, nor my body alone, but the union."[101]

The urgency of tying soul to body imparted a related importance to the indivisibility of the human soul, composed of a vegetative and a sensitive as well as a rational part according to the Scholastics, and to the intermediate positions along the spectrum of matter and spirit. The human soul, though incorporeal, was closer to matter than angels were, because it required a body for its operation.[102] While the rational part of the human soul connected it to the immortal realm of the angels, the vegetative and sensitive parts were what tied it to the corporeal world.[103] On the Scholastic spectrum, the sensitive soul of animals was so closely related to matter as even to acquire extension and divisibility.[104]

Caterus, the Dutch Catholic theologian from whom Descartes solicited responses to his *Méditations*, demurred that Descartes's proof of the distinction between the soul and the body was ill founded: the fact that these could be distinctly conceived did not mean they were essentially different.[105] Even Arnauld, who ultimately espoused an extreme form of dualism, protested in his objections to the *Méditations* that it remained unproven "that the mind can be completely and wholly understood without the body."[106] He warned that Descartes had offered no good defense against those who believed that the human mind was a bodily thing.[107] The Jesuit Pierre Bourdin similarly insisted, "There are many people, and serious philosophers at that," who believed that thought was "certainly not a property that uniquely and necessarily belongs to a mind or spiritual substance!"[108]

The authors of the sixth set of objections to the *Méditations* solicited by Descartes and Mersenne called the idea that animals were automata an "impossible and ridiculous claim," but they agreed with Descartes that animals contained nothing "distinct from their bodies." Animals were purely material things, to be sure, but why must a purely material thing lack a soul? These authors invoked a medieval Christian tradition of construing sensation and other mental functions as essentially corporeal. Several of the Church Fathers, they pointed out, had believed that "Angels were corporeal," and held the same view even of the rational soul, "which some of them thought to be passed from father to son" like other physical attributes. These Church Fathers had nevertheless believed that Angels and souls were capable of thought. They must therefore have believed that thought could be "made of corporeal movements, or that the Angels themselves were but corporeal movements, which they did not at all distinguish from thought."[109]

The separateness of the Cartesian soul provoked the Jesuit historian Gabriel Daniel to write a mischievous parody entitled *Voyage du monde de Descartes* (A Voyage to the World of Descartes) (1690). In Daniel's tale, Descartes, having trouble falling asleep, asks his soul to "go on a little trip." Unfortunately, his doctor then arrives to find Descartes's body-machine speaking and behaving irrationally in the absence of its soul, and decides that Descartes has a brain fever. Subjecting the poor machine to violent remedies, the doctor breaks it. Descartes's soul is thus left to wander in infinite space looking for a place to live. It invites Daniel's soul to join it, and the rest of the story is a travel journal.[110]

Catholic respondents were not the only ones to object on theological grounds to Descartes's disembodiment of the rational soul. The ambivalent Cartesian and Cambridge Platonist Henry More worried that a wholly spiritual soul could never connect with a material body and that even God must have extension in order to have presence in the physical world.[111]

In sum, the severing of soul from body initially provoked more consternation than the accompanying relatively familiar notion of bodily machinery. But Descartes's disembodied self also transformed its mechanical counterpart beyond recognition. The animal-machine, as Descartes described it in the first instance, was warm, mobile, living, responsive, and sentient. The same living machinery, when measured against a disembodied, transcendent self, looked different: confined, rote, passive.

The Cartesian removal of soul from the machinery of the world, like the Reformist removal of God from nature, left behind something starkly different. Recall the Rood of Grace, that went from being at once mechanical and inspirited, a representation of divinity in motion, to being a fraudulent and inert assemblage of parts. Something equivalent took place with the animal-machine model of living beings when Descartes wrested the soul from the living body-machine, and both developments were expressions of a post-Reformation insistence upon the utter distinctness of matter and spirit.[112]

To a Cartesian, a machine, like all the rest of the physical world, was nothing but variously shaped bits of uniform matter. The bits of matter had no qualities except hardness, size, and shape: they were little quality-less chunks. In ancient and medieval Scholastic philosophy, in contrast, a machine did not just have the artificial forms imposed by the builder—which, like Cartesian forms, were essentially restricted to size and shape—but it also had, like all the rest of the physical world, the natural forms that made the parts of the machine what they were. Each kind of natural substance—iron, copper, water—was more than a disposition of homogeneous parts: it had a "substantial form," an essence that rendered it different from all other natural substances.[113] To understand an animal, or indeed a machine, as a system of moving bits of homogeneous matter is very different from considering it as a system of moving bits of informed matter.

Aquinas, for example, although he described living things as a kind of machinery, rejected the popular notion of artificial mechanical statues capable of living behaviors such as speech. That was because the principle of life, Aquinas argued, was a substantial form not transmitted in the forging of a statue. He reasoned that a thing could not receive a new substantial form without giving up its old one, and metal did not give up its substantial form in becoming a statue, but only underwent a change in accidental shape.[114] Descartes's removal of souls and forms from the world-machine left behind a forever altered meaning of machinery. Having signified beauty, complexity, virtuosity, wisdom, and agency in the ancient and Scholastic intellectual universe, machinery acquired a new set of meanings: passivity, limitation, and constraint.

The body-machine, in other words, began to look inanimate, inert, brute. Descartes himself was equivocal regarding its sentience. Although he ascribed sensation to the movements of the animal-machinery, specifically to the animal spirits that were set in motion in the sense organs, which then transmitted their movement through the filaments of the nerves to cause a commotion in the central gland of the brain, he also wrote that it was "the soul that senses, and not the body," since when the soul is lost in reverie, the body remains senseless.[115] On another occasion he subsumed the capacities to understand, will, imagine, remember, and sense under "thinking," which would place all of these, including sensation, under the auspices of the soul and not the body.[116]

Moreover, when Alphonse Pollot, a French Protestant refugee living in the Netherlands and a friend of Descartes's, objected to Descartes's animal-machine idea, writing that animals betrayed "affections and passions" setting them apart from machines,[117] Descartes did not respond by insisting that affections and passions were within the purview of a mechanical creature. Instead he imagined a hypothetical person who had never seen a real animal but had spent his life building automaton animals that behaved like natural ones, including showing signs of passions such as crying when struck or fleeing from loud noises. Such a person, Descartes thought, seeing natural animals for the first time, would assume they were automata "composed by nature," therefore much finer than those he himself had made, but would never judge that these creatures had "any true sentiment, any true passion, as in us."[118]

Descartes maintained that "beasts do not think at all"[119] and dismissed believers in animal thinking as childish and weak-minded.[120] In his last published work, *Les passions de l'âme* (*The Passions of the Soul*) (1649), he returned to the question of where feelings took place, whether in the soul or in the bodily machinery. He noted that what philosophers called a "passion" with regard to the subject to whom it occurred, they called an "action" with regard to that which caused it to occur. Therefore, action and passion were a single thing having two names "according to the subject with which we associate it." That which was a "passion" with regard to the soul was an "action" with regard to the body. Animals, who had only bodies and therefore no reason and "perhaps" no thought either, shared the same hydraulic system of animal spirits. In them, this system gave rise not to "passions, as we have," but instead simply to those "movements of the nerves and muscles that customarily accompany" the passions.[121] In a letter to the French theologian Guillaume Gibieuf, Descartes expressed the same view: although one could observe in animals the movements associated with imagination and sentiments in humans, this did not mean that animals had true imagination and sentiments. They might simply be exhibiting the corresponding movements.[122]

On other hand, Descartes objected when people construed his animal-machine as lifeless and inert. The authors of the sixth set of objections to the *Méditations*, for example, represented Descartes as having said that animals had "neither sensation, nor soul, nor life."[123] In response, Descartes protested that he had not deprived animals of sensation or life and, indeed, that he was even willing to attribute a kind of soul to them as well: "I have never denied them what is commonly called 'life,' a corporeal soul and organic sensation."[124] Since the authors of the sixth objections themselves denied the existence of an incorporeal soul in animals, their position was essentially the same as Descartes's, except for his insistence that true souls were incorporeal.

In two letters to the Dutch doctor Regius (Henri le Roy), Descartes struggled to clarify his view of animal souls by agreeing that animals had what people generally understood as "soul" in a fully corporeal form. But, he explained, he preferred to call this a "vegetative and sensitive force" in order to reserve the word "soul" for the fully incorporeal source of human reason.[125]

In a letter to the Marquis of Newcastle in 1646, Descartes again ascribed a gamut of "passions"—hope, fear, joy—to magpies, dogs, horses, and monkeys. These passions required no thought. That animals had feelings and emotions but no thoughts was clearly demonstrated, Descartes wrote, by the fact that they expressed their passions to us: if they had thoughts, surely they would express these too.[126] Several years later, in a letter to Henry More, Descartes insisted with regard to his refusal of a soul to animals, "I would like to note, however, that I spoke of thought, not life or sense: for no one denies the life of animals." To say animals had no soul only meant they had no reason, and did "not entail refusing life to them." Animal life, Descartes explained, consisted in warmth, mobility, responsiveness: the heat of the heart and the sensibility and arrangement of the organs.[127]

Could a purely material entity—a machine—be *alive* in an animal sense, having sensations, feelings, imagination, passions? Descartes had meant to include living bodies in the realm of material nature, as fully comprehensible to the rational mind. He had therefore disembodied the mind, separating it, he thought, from the living body. But the result seemed instead to evacuate life as well as mind from the natural world, leaving only inanimate machinery behind.

Descartes's respondents almost all took him to mean that animals were essentially inanimate. Locke was a rare exception: when he rejected the Cartesian beast-machine, it was on the grounds of what he took as evidence that certain animals could in fact reason.[128] He responded to Descartes's initial meaning that animals, as machines, lacked only the capacity for rational thought. But virtually everyone else, Cartesians and anti-Cartesians alike, understood Descartes to have meant that animals had no sensations or feelings. The possibility that Descartes labored to introduce, the possibility that matter contained life but not soul, became virtually inconceivable in the philosophical, theological, cultural, and political moment he was struggling to define. Two violently disputed principles crowded out the initial Cartesian possibility of a living yet soulless animal-machine: the older, Scholastic principle that matter contained souls in many forms, and the newer, post-Reformation insistence that matter was passive and inert, utterly distinct from spirit.

The rationalist philosopher Nicolas Malebranche, the leading Cartesian during the 1670s, promulgated a version of animal automatism in which animals lacked all sensations and feeling.[129] The writer and academician Bernard le Bovier de Fontenelle said he had seen the gentle Malebranche kick a pregnant dog, insisting that it felt nothing.[130] (Descartes by all accounts treated his own dog, Monsieur Grat, with kindness and affection.)[131]

This version of a Cartesian picture of animals became notorious. "In Holland," Leibniz reported in 1684, "they are now disputing, loudly and soundly, whether beasts are machines, and the people even get amusement from this, and call the Cartesians ridiculous, since they imagine a dog that is beaten cries in just about the same way as a bagpipe that is pressed."[132] Leibniz himself rejected the view that animals lacked souls, which he construed to mean that animals had neither sensation nor feeling.[133] So did Huygens, who objected emphatically to this version of the idea, writing that animals had "as great a gusto of Bodily Pleasures as we, let the new Philosophers say what they will, who would have them go for nothing but Clocks and Engines of Flesh." Animals regularly demonstrated the strength of their feelings so plainly by "crying and running away from a Stick, and all other Actions, that I wonder how any one could subscribe to so absurd and cruel an Opinion."[134]

By the end of the seventeenth century, virtually everyone took Descartes to have denied animals both sensations and emotions and, moreover, understood this denial to have been the salient feature of his materialist theory of life. Despite Descartes's early and late protestations, to say that animals were fully corporeal had come to mean denying them all sensation and feeling. The removal of the thinking soul had drained the life from the body-machine.

With Descartes's extraction of the human soul from the natural world, the old continuum of matter and soul began to fade from view. But the assumption of that continuum, full of intermediate forms, in which animals had had a kind of soul lower than the human soul but higher than plants, had been Descartes's point of departure. On the older continuum dogs were neither dandelions nor humans: they occupied an intermediate position as dogs. Arnauld put it clearly: "Those who maintain that our mind is corporeal do not on that account suppose that every body is a mind."[135]

Descartes took the old spectrum and made it—except for the human soul, which he removed altogether—a fully material continuum. But he still assumed that there were differences along the material continuum. He himself, after all, was neither pre- nor post-Cartesian, but both. He was working with Scholastic training and intuitions in a mode that was at once conservative and revolutionary. He wanted to preserve the world the Scholastics had described but also to subject it to his new, mechanist philosophical method. To understand living things as machines did not, in the first instance, render them inanimate. Descartes set out to understand animals as alive *and* as machines. He did not mean that a dog was a clock: a dog was alive and a clock was not. In the unstable moment of transition that he inhabited, it seemed briefly possible to see living creatures as machines in the modern sense *and* as alive in the ancient sense.

Many Cartesians labored to bring the animal-machine back to life. Pierre Sylvain Régis, a devoted promulgator of Cartesianism during the second half of the seventeenth century, struggled with this problem in his efforts to present a Cartesian account of animal life that would be at once orthodox and plausible. Régis faithfully reproduced Descartes's analogy of animal-machinery to the hydraulic grottoes of palace gardens. But he then hastened to add that he would never deny life or sentiment to animals. Provided, that is, that one understood by animal life and sentiment simply "the heat of the blood" and the "particular movements of the sense organs," definitions seemingly narrow enough to be irrefutable but also distinctly unsatisfying. In the next breath, Régis confessed himself sympathetic to those who "wanted to say that beasts have a Soul distinct from the Body." He acknowledged that he had no way to disprove such a view and that he himself would even be inclined to share it, if only the possibility of immortal souls in animals were not precluded by Christian faith.[136]

Some tried to rescue the model by sharply limiting its purview. Descartes's original purpose had been ambitious: to render living beings fully intelligible, by subjecting all the functions of life and mind, except for the ability to reason, to mechanical explanation. Later mechanists greatly curtailed the mechanist project by stipulating an immaterial animal soul responsible for life, sensation, and sentiment. The organs of the body could then be "true machines" in the brute sense. "The muscle itself is a dead and inert machine," declared the Italian mathematician and physiologist Giovanni Borelli in the opening passages of his 1680 treatise on the motions of animals. Everybody, he affirmed, now agreed that the source of motion in animals was an immaterial soul.[137]

This soul, Borelli reckoned, was like the suspended weight in a clock, in the sense that it provided the original source of motion to the animal-machine. But, just as the falling weight would cause the clock to spin like a top if not for the regulator, so the soul would "impart a frantic motion" to the body, the blood and spirits racing "frantically and foolishly" about, if not for the regulating mechanism of the lungs.[138] The brute-mechanist division of labor was here fully established: vitality from the immaterial soul; and from the body-machine, constraint.

Perrault, for his part, judged it so crucial to declare right at the outset of his *Mécanique des animaux* (1680) that animals had feelings, and immaterial souls to guarantee these, that he devoted a special "avertissement" before the actual text of the book to this purpose. To forestall the "bad effect" that his title might produce in those weary of hearing the claim that animals were pure machines, Perrault opened his book with the declaration, "I avow that I understand by animal a being who has sentiment, and who is capable of exercising the functions of life by a principle that we call Soul." The machine of the animal body required this soul to move and conduct it just as an organ could do nothing without an organist (Perrault avoided any mention of automatic organs).[139]

The immaterial animal soul as Perrault invoked it was a stipulation, allowing him to declare that animals experienced sensations and feelings, and then to set the matter squarely outside the domain of natural science.[140] "I content myself with explaining the machine of the body of animals," he wrote, without going into that principle that moves and governs it. This limited sort of knowledge was "the only thing we are permitted to know in nature," but no less worthy for that.[141] Free now to restrict his discussion of sensation to the mechanics of the senses and nervous system, Perrault explained that in order to transmit sensation, the sense organs had to be so delicate that the least touch, the subtlest external movement would set them in motion. The nerves were accordingly made of a soft, delicate, mobile substance full of "spirits" that reproduced the movements of objects outside.[142] Just how this mechanical mirroring of the world became a *feeling* lay outside Perrault's carefully circumscribed purview.

Thomas Hobbes took exception to the general rejection of the idea that animals were machines. He himself embraced this idea, but unlike Descartes, he extended it to human beings as well, and he did not have the old model of vital, inspirited living machinery in mind. On the contrary, Hobbes had specifically deflationary intentions. He directed Descartes's mode of mechanist, rationalist philosophical argument against Descartes's notion of limitless selfhood. While it was perfectly true that "I am a thinking thing," Hobbes acknowledged, that did not mean that "I am mind" any more than "I am a walking thing" implied "I am a walk." The thing doing the thinking could still be corporeal and, indeed, Hobbes argued that it must be. He reasoned that all actions required subjects to carry them out and, like Descartes, Hobbes took material entities and their motions to be the only rationally intelligible causes. One could no more separate thought from thinking matter, said Hobbes, than walking from the physical body of a walker.[143]

Accordingly, everything about a person was machinery in the brute sense, with no distinction between mind and body. Thoughts originated in sensations, and sensations were nothing but "motions of matter": motions of the material medium by which external objects pressed against the sense organs, and motions of the sense organs and body in response.[144] Thought itself was a bodily "motion," nothing more. Hobbes took "mechanism" to signify finiteness and constraint: a "machine" was something with boundaries and a limited repertoire. From the premise that humans were machines, Hobbes drew specifically deflationary conclusions, for example, that the human mind could not encompass infinity: "No man can have in his mind an image of infinite magnitude; nor conceive of infinite swiftness, infinite time, infinite force, or infinite power." How could a finite material object, the brain, possibly contain an infinite magnitude of any kind? Therefore, when we say we conceive of something as infinite, we simply mean we cannot conceive of its limits.[145]

Working to puncture all claims to authority originating in divinity, whether royal, clerical, or rational (as in the case of Descartes's immaterial self), Hobbes took mechanism as his instrument. Being machines, humans differed from animals as one device from another. Hobbes agreed with Descartes that humans were distinguished from beasts by their capacity for reason, but he construed this as a difference of degree rather than kind. Reason was among several distinctly human traits, along with the "privilege of absurdity," enjoyed exclusively by humans, especially philosophers. Only humans too, Hobbes thought, had curiosity, that "lust of the mind . . . [that] exceedeth the short vehemence of any carnal pleasure." On the other hand, beasts shared the human capacities for imagination and prudence, as well as appetites, aversions, hopes, and fears, the ability to deliberate, and even will.[146]

The non-infinite, bounded, and material nature of humans supported Hobbes's idea that one could build them into collective persons. The machinery of political representation could make multitudes of men into a single person.[147] The momentous opening paragraph of his great political work, *Leviathan*, uses the material, mechanical nature of humans to explain why the Commonwealth itself is a constructed person, made of material parts just like any individual human, differing in degree but not in kind:

> For seeing life is but a motion of limbs, the beginning whereof is in some principal part within, why may we not say that all automata (engines that move themselves by springs and wheels as doth a watch) have an artificial

> life? For what is the heart, but a spring; and the nerves, but so many strings; and the joints, but so many wheels, giving motion to the whole body, such as was intended by the Artificer? Art goes yet further, imitating that rational and most excellent work of Nature, man. For by art is created that great LEVIATHAN called a COMMONWEALTH, or STATE (in Latin, CIVITAS), which is but an artificial man.[148]

Hobbes's idea of the Commonwealth as an artificial person, along with the theory of representation it supported, allowed him to accomplish two seemingly incompatible tasks simultaneously. He rejected the notion of the divine right of kings: the Commonwealth and its sovereign were human-made, "Artificiall" things. But he also replaced this right with an equally, indeed, a surpassingly absolute, indivisible sovereign power.[149] Hobbes argued that when a multitude of persons constituted a single artificial person through representation, the result was "more than Consent, or Concord; it is a reall Unitie of them all, in one and the same Person." One could no more legitimately claim a grievance against one's sovereign than against oneself.[150]

This was a formidable achievement: a conceptual basis for absolute power without divine transcendence. Not only in being mechanical, but in specifically excluding anything immaterial—soul, spirit, mind—Hobbes's greater person of the state parted company with older, Platonic, and medieval notions of the political entity. Plato and his followers had assumed a permeation of the natural world by mind, which organized both individual life and the functioning of the state.[151] The sort of machinery Hobbes described, in contrast, was something new: machinery defined by its limits, machinery whose very role was to be constrained, rote, brute. Although Hobbes differed from most of his contemporaries not only in embracing the idea of animals as machines but in extending it even to human beings, he shared with them the new model of living machinery, defined by its finitude and lack of spirit.

When the Jesuits finally did include the claim that animals were pure automata on a list of prohibited theses established at their Fifteenth General Congregation in 1706, it was the new version of the view that they described: "that animals are pure automata deprived of all knowledge and sense."[152] By then, the idea of the animal-machine had devolved into an easy mark for satirists and other wits. "Put a male-dog-machine and a female-dog-machine near one another," Fontenelle quipped, and you could get a third little machine, whereas two watches could spend all their lives together without every producing a third little watch. "And so we find by our philosophy, Madame B. and I, that all things which, being two, have the virtue of making themselves three, are of a considerably higher nobility than the machine."[153]

The metamorphosis of Descartes's original idea is captured in the naturalist Georges Buffon's misquotation, toward the end of the following century, of a bit of verse sent by Catherine Descartes, the philosopher's niece, to her friend Madeleine de Scudéry. Writing about the black-headed warbler that returned to Mademoiselle de Scudéry's window each year, Catherine Descartes had written, "with all due respect to my uncle, she has judgment."[154] Mademoiselle Descartes had been accurate: judgment was precisely what her uncle, at least in the first instance, had denied to the warbler. But Buffon, in quoting Catherine, replaced her word "judgment" with "feeling."[155] Sentiment was for Buffon's generation what the Cartesian animal-machine most saliently lacked.

"How pitiful, how poor, to have said that beasts are machines deprived of knowledge and sentiment," wrote the philosopher, historian, writer, and doyen of the Enlightenment, Voltaire, in 1764. It is not because I talk, he chided, that you judge me to have feelings and ideas. Imagine you were to see me enter my house looking worried, search anxiously for a piece of paper, open a desk where I remember putting it, find it and read it with joy. You would judge that I have sentiments of worry and joy, that I have memory and knowledge. So extend the same courtesy to the dog who has lost his master, who searches for him everywhere with piteous cries, who finds him at last with demonstrations of joy. "Barbarians" seize the gentle and loving dog, Voltaire continued, they nail him to a table, they dissect him alive and reveal that he has all the same "organs of sentiment as are in you. Answer me, machinist: has nature arranged all the springs of sentiment in this animal in order for him not to feel? Does he have nerves in order to be impassive?"[156] What had begun, in Descartes's original formulation, as an attempt to explain the mechanisms of sensation, had become, by the time Voltaire described it, a categorical denial of animal sensation, a philosophical absurdity and an act of both moral and physical violence.

Yet, consider the conclusion of Voltaire's story above. The materialist account of animal sensation had become so well established that Voltaire, even as he associated the "machinist" approach with misguided brutality, himself deployed the very same understanding by invoking the "springs of sentiment," the nerves. The idea that animals were machines had become at once axiomatic and absurd as a result of the shifting meaning of animal-machinery. The notion that animals had corporeal souls, as the English experimentalist, theologian, and dissenting minister Joseph Priestley pointed out, went back at least to Plato. That animals were fully material entities, Priestley said, had been "the opinion, I believe of all the christian world till very lately." But the same idea in Descartes's system came to hold radically different implications. "The opinion of Descartes was much more extraordinary," Priestley wrote, "for he made the souls of brutes to be mere *automata*, and his disciples in general denied that they had any perception. Malebranche says that they eat without pleasure, and cry without pain, that they fear nothing, know nothing."[157]

The body-machine, rendered inanimate, its life drained away by the human soul, was just beginning its long career at the heart of modern science.

~ 3 ~

The Passive Telescope or the Restless Clock

The telescope (and lens instruments more generally) and the clock both served as key models of living structures in the new mechanist life sciences of the mid- to late seventeenth century. These devices were themselves undergoing dramatic changes.

To begin with lens instruments, Galileo built the first astronomical telescope in 1609, based upon reports of a recent invention by Dutch spectacle makers, a spyglass that made distant objects appear nearer. By experimenting with different combinations of focal lengths in his lenses, Galileo greatly improved the device and used it to revolutionize his contemporaries' view of the heavens.[1] He was also a pioneer of the microscope; it was for Galileo's instrument, built in 1623 or 1624, that a friend, Giovanni Faber, papal doctor and curator of the Vatican botanical garden, coined the term "microscope."[2] By the 1660s and '70s, at the hands of the Dutch microscopists Jan Swammerdam and Antoni van Leeuwenhoek and the English polymath Robert Hooke, microscopes became sufficiently powerful to reveal a previously unknown microcosm.[3] Lens instruments riveted the attention and transformed the thinking of several generations around the mid-seventeenth century.

Now clocks: another major technical development of the mid-seventeenth century was the addition to timekeeping devices of the balance spring. This

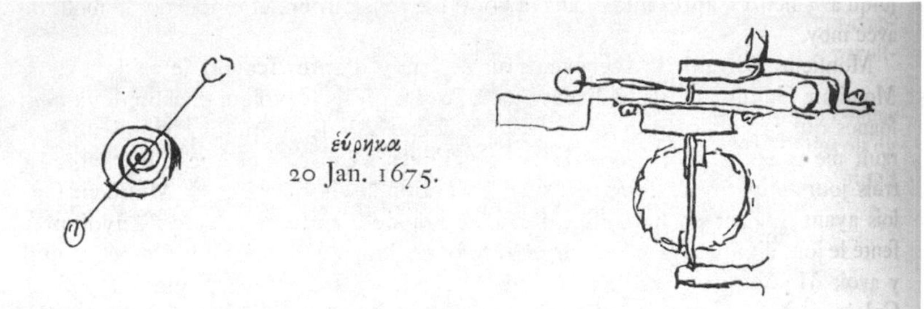

Figure 3.1 Christiaan Huygens's drawing of a spring balance for a watch (1675), courtesy Department of Special Collections, Stanford Libraries.

part transformed the accuracy of mechanical clocks and watches, permitting the addition of minute and even second hands to watches, rendering these accurate enough to be useful. Hooke was the first, in 1658, to add a spring to the balance wheel; and Huygens developed the balance spring during the 1670s principally by using a spiral spring (see figure 3.1). With its spring, the balance became what physicists call a "harmonic oscillator": a physical system that vibrates at a certain resonant frequency and resists disturbing forces such as friction.[4]

The analogy of the eye to a lens instrument was the favorite example, through the seventeenth and eighteenth centuries and well into the nineteenth, in arguments from design. In each case, the analogy confirmed the mechanical perfection of nature: a purely passive, externally endowed perfection. What about clockwork? In most cases, the comparisons of living beings to clocks also supported arguments from design that construed the living world as passive machinery, its order and movement imposed from without. But not quite always. In one instance, the clock, specifically, the balance, served as a model for the inner, restless, resisting agency of the mechanism itself.

The Divine Optician

"Man is a meer piece of Mechanism," wrote the English doctor and polemicist William Coward in 1702, "a curious Frame of Clock-Work, and only a Reasoning Engine." To any potential critic of such a view, Coward added, "I must remind my adversary, that *Man is such a curious piece of Mechanism, as shews only an Almighty Power could be the first and sole Artificer, viz.*, to make a *Reasoning*

Engine out of dead matter, a Lump of Insensible Earth to live, to be able to Discourse, to pry and search into the very nature of Heaven and Earth."[5]

Here was the new animal-machine drained of life (dead matter, a Lump of Insensible Earth) and it brought with it a new, modern science-theology (*only an Almighty Power could be the first and sole Artificer*). I have written science-theology with a hyphen to emphasize their deep connectedness, indeed, their essential oneness. We are accustomed to thinking of science and theology as distinct endeavors, and moreover, historians traditionally trace their separateness to the very moment of this chapter, the middle of the seventeenth century and the founding of the New Science. Before that moment, the traditional story goes, science and theology were essentially the same thing, and people explained natural phenomena in terms of all sorts of spirits and essences, forms and souls. Modern science was born, still according to the traditional story, as written by the seventeenth-century actors themselves, when philosophers such as Descartes, Francis Bacon, Isaac Newton, and others insisted that nature be understood, not in terms of these mysterious properties, but only in terms of material parts and their movements, as brute machinery. However—and this is one reason why the origin myth of modern science is just that, a mythologized version of events—the model of nature as brute machinery carried a powerful new theology of its own.

The new, modern science-theology was devoted to demonstrating God's existence by detailing the mechanical perfection of his artifact, the world-machine. At the core of this modern science-theology figured the transformed model of nature's machinery, now emptied of spirit and agency, and the new figure of the animal-machine: brute, inert, passive. The argument from design was a starkly different way of proving God's existence, and of regarding nature's machinery, than any that had come before. To be sure, natural theology, the search for evidence of the existence of God in nature, was an ancient tradition.[6] Philosophers and physiologists from Aristotle and Galen onward had considered that bodily organs and other natural phenomena represented divine craftsmanship. Aquinas had also inaugurated an empirical tradition of proving God's existence from the appearance of order and purpose in nature.[7] But the idea that God had built the world out of inert material parts was something new under the sun, along with the associated idea of nature as passive machinery.

Descartes himself, having established God's existence, like his own, by rational necessity, died before he could be dismayed by the new tradition of natural theology, invoking his world-machine to prove God's existence empirically. Nothing could have been worse from a Cartesian perspective, given the unreliability of experience, than to make the existence of God a matter of empirical demonstration. Descartes's philosophical achievement was tremendously powerful and yet unstable: his science of life devolved into a science that seemed to explain life out of existence, while his transcendent and logically necessary God collapsed into a contingent feature of His own mundane contraptions.

Henry More was one of the first to inaugurate the new approach to science and theology. More was a great admirer of the eye. In his *An Antidote against Atheism* (1655), More invoked the eye as the first of his "Unavoidable Arguments for divine Providence." The "*number*, the *situation*, the *fabrick*" of the eyes was such, More argued, "that we can excogitate nothing to be added thereto." Leaving the eye's beauty to "Poets and amorous persons," More dwelt instead, in minute detail, on its anatomy and physiology: the eyelids, "fortifi'd with little stiff *bristles* as with *Palisadoes*, against the assault of Flyes and Gnats, and such like bold *Animalcula*";[8] the transparency of the "Humours and Tunicles,"[9] which meant that light and colors could enter "unfoul'd and unsophisticated by any inward tincture;" the convexity of the crystalline lens, serving to focus incoming light rays; the inner surface of the iris, "black'd like the walls of a Tennis court" so that the light would be absorbed there rather than reflected back into the retina; the suspension of the lens, such that the ciliary muscle could thrust it forward or draw it back; the whiteness of the retinal membrane, suited like a piece of white paper to the reception of images; and the six ocular muscles that moved the eye in all directions.[10]

Overall, the body was so artfully formed, More reasoned, that if only we had had the same matter to work with, and could be as clever, we would surely have built ourselves exactly as God built us: "If the wit of Man had been to contrive this Organ [the eye] for himself, what could he have possibly excogitated more accurate?"[11] More's collegial adversary in other matters, Robert Boyle, agreed.[12] "There is no more rashness," Boyle wrote, "to say, that an Eye, than that a Telescope, was made for an Instrument to See with."[13] Boyle, whose commitment to mechanism was matched by his ardor in defending Christianity from "notorious infidels, namely atheists, theists, pagans, Jews and Mahommedans,"[14] often referred to animals as "Living Engines" and "Automata."[15] For example, he extolled "those living automatons, human bodies" and described the world itself as a "great automaton" built by God.[16]

Cribbing extensively from More and Boyle, the English naturalist and theologian John Ray advanced the argument from design in his much translated and reprinted *The Wisdom of God Manifested in the Works of the Creation* (1692). He too dwelled extensively on "the Eye, a Part so artificially composed, and commodiously situate." Ray added refinements to More's and Boyle's lists: the biconvexity of the crystalline lens; the retinal cavity, a convenient "Room" for the "collection of the Rays receiv'd by the pupil"; and the situation of the cornea, which "riseth up, as it were a Hillock" above the white of the eye, a good thing, since otherwise "the Eye could not have admitted a whole Hemisphere at one View."[17]

Natural theologians dwelt especially and increasingly on the eye's properties of accommodation: the pupil's dilations and contractions in response to light and the several ways in which the eye varies its refracting power to see objects at different distances. The Jesuit astronomer Christoph Scheiner had described some aspects of ocular accommodation in his 1619 treatise, *Oculus*.[18] Ray described a simple experiment to demonstrate these phenomena, requiring only a candle, a bead, and a willing child: "Setting a Candle before him, bid him look upon it, and [you] . . . shall observe his Pupil contract itself very much, to exclude the Light, with the Brightness whereof it would otherwise be dazzled and offended." Next, withdraw the candle and watch the child's pupil re-expand. The same contraction will take place when the child looks at a bead held close to his eye, and the same expansion when he looks at the bead from a distance.[19]

A tradition of arguments from design resting heavily on the physiology of vision stretched through the eighteenth century, including the Scottish physician George Cheyne's *Philosophical Principles of Natural Religion* (1705);[20] and the English clergyman William Derham's *Physico-Theology* (1713).[21] But the essentials of the argument were in place by the end of the seventeenth century; subsequent texts included much that was paraphrased or unceremoniously lifted from the earlier ones. While this brute-mechanist, argument-from-design tradition was originally and predominantly Protestant, it came to include Catholic authors as well, notably three Frenchmen, the Cartesian Nicolas Malebranche, about whom more follows presently; the natural historian René Antoine Ferchault de Réaumur; and the abbé Noël-Antoine Pluche, author of a popular mid-eighteenth-century work of natural history and natural theology, *Le spectacle de la nature*.[22]

Throughout the argument-from-design tradition, the mechanism of the eye and its similarities to lens instruments such as the telescope, the microscope, and the camera obscura, an optical device for projecting an image, remained a favorite example (see figure 3.2).[23] The eye-lens-instrument analogy was by no means the only example: every bodily mechanism offered rich material. Discussions of the eye are especially striking, however, in that the writers persisted in avoiding all reference to perception—to *seeing*—and confined themselves to a discussion of the eye's moving parts alone.

The animal-machinery of this tradition was devoid of agency or consciousness. Authors of arguments from design rarely mentioned perception, and when they did, they treated it in purely brute-mechanical terms. Cheyne, for example, marveled, "What can be more amazing, than that the Particles of Matter shou'd be so fram'd, as by their means to shew us the *Shapes, Positions, Distances, Motions,* yea and *Colours* of remote Bodies?" Cheyne shared the idea of Descartes and many others that light was a "fine Fluid" that physically carried the images of objects in the world to the surface of the eye. Here was evidence of the "Wisdom" and "Skill" of the Creator, to so contrive things, since "it seems impossible that any other Composition of the Eye, shou'd be equally fitted" to receiving the particles of light. As an example of this absolute fitness, Cheyne judged that the retina was shaped and positioned just right: with a different curvature, or at a different distance from the crystalline lens, he reckoned, objects would appear distorted, magnified, or miniaturized.[24]

The fact that people perceive objects right-side up despite the inversion of images on the retina, a phenomenon that the German astronomer Johannes Kepler had demonstrated in 1604,[25] does not seem to have troubled Cheyne's assumption that perception was identical to projection. Ray did take up the inversion of the retinal image, but assigned its lack of effect on perception to a material cause: the nerves' ability to convey the situation and position of the objects that pressed upon them. Ray proved this ability by invoking a common tactile illusion. Cross your first two fingers and press them against a round object (such as the tip of your nose—feel free to try this if you are alone or unselfconscious). From the sensation in your fingers, you will have the impression that you are touching two separate objects.

Ray explained this error by saying that when the fingers are crossed, the round object comes into contact with the nerves on their *outside* edges, which are usually distant from one another, and so "signifie to the soul Bodies separate and distant in like manner, two Fingers lying between them. And tho' our Reason, by the help of our Sight, corrects this Error, yet cannot we but fancy it to be

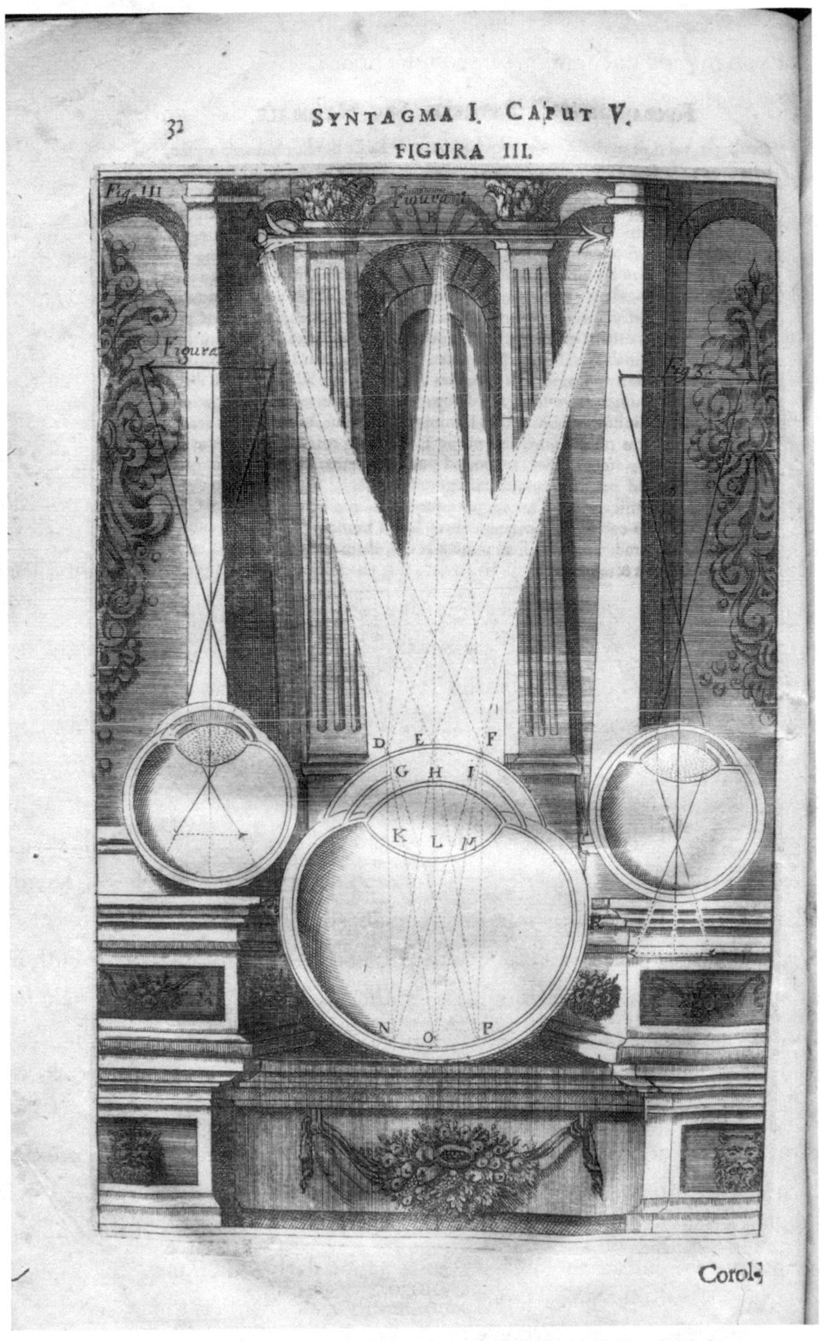

Figure 3.2 Rendition of the eye as camera obscura, from Johann Zahn, *Oculus artificialis* (1685.86), courtesy Department of Special Collections, Stanford Libraries. Photo by Andrew Schupanitz.

so."[26] Here, as elsewhere, Ray took perception to be a literal impact of nervous fluids on the seat of common sense in the brain.[27]

A consistent disregard for perception was built into the tradition of arguments from design from the outset. After all, an eye is nothing like a telescope if one takes into account that a telescope does not see. One might well think this makes a world of difference. But the authors of arguments from design carefully avoided finding evidence for the existence of God in the intractably mysterious fact that living creatures see. Mysteries were not the sorts of proofs they wanted. Rather, they sought material evidence. They cited the physical complexity of nature, the articulation of parts. Surely, More argued, such arrangements of matter could not have happened by accident, "For is it not a wonder that even all our flesh, should be so handsomely contriv'd into distinct pieces," able to move independently of one another?"[28]

Natural theologians of More's and Boyle's tradition were looking for a new source of theological certainty, indeed, a new *kind* of theological certainty. This new kind of theological certainty was meant to come from empirical arguments regarding material evidence. The passive-mechanical artifact world was the terrain of these new arguments with their new kind of certainty.[29] In short, the new mechanism worked at the heart of the new theological program and, reciprocally, the new theological program shaped the development of the new mechanism, propelling it to its most reductive extreme. The world became an entirely passive machine whose job it was to indicate the existence of a divine Designer, whose omnipotence was guaranteed by his monopoly on agency. Classical mechanism was, in the first instance, theological mechanism.

This new combination of mechanist science and theology led the authors of arguments from design to an important intellectual outcome, namely the concept of physiological adaptation or fitness. This was the idea that living creatures were constructed so as to be suited to their particular tasks and environments, and the methodological principle that one must understand their structures in terms of this suitability. The concept of physiological fitness developed centrally within the argument-from-design tradition of natural theology.[30] If God was an engineer, his machinery must be perfectly designed: means ideally adapted to the ends, structures flawlessly suited to the functions.

Thus the eye, Boyle wrote, "(to single out again that Part for an Instance) is *so* little fitted for almost any other Use in the Body, and is *so* exquisitely adapted for the Use of Seeing, and That Use is *so* necessary for the welfare of the Animal, *that* it may well be doubted, whether any Considering Man can really think, that

It was not destinated to that Use.[31] And Ray concurred: "To this Use and Purpose of informing us what is Abroad round about us in this aspectable World, we shall find this Structure and Mechanism of the Eye, and every Part thereof; so well *fitted and adapted*, as not the least Curiosity can be added."[32]

Boyle generalized the argument by taking it beyond human structures and added a new dimension to the notion of fitness: suitability to a particular environment. He sought examples of fitness and adaptation to various circumstances in the structures of the different animals occupying them. Frogs, for example, need a nictitating membrane to protect their eyes from sedges and other plants as they move through the water, and birds to shield their eyes from twigs and leaves. Flies, unable to move their eyes, are compensated by "a multitude of little *protuberant* parts, finely ranged upon the convex of their large and protuberant Eyes," allowing them to see in many directions at once. Grazing animals have a seventh ocular muscle, beyond the six they share with humans, allowing them to look downward for long periods without weariness. These animals also have a horizontal pupil, affording them a wide view of the ground from which they gather their food. In contrast, the vertical pupil in cats gives them a tall view of the world to glimpse their climbing prey. Moles have tiny eyes because, living underground, they do not need to see much. Chameleons can move their eyes independently of one another, the better to catch flies approaching from many directions. The crystalline lens in fishes has a greater curvature to accommodate the different refraction of light in water.[33]

Physiological fitness seemed the appropriate measure of perfection for an Engineer deity. And yet there were difficulties inherent in the project of demonstrating, from the physical mechanisms of the material world, the existence of a single omnipotent and perfect God.

The first difficulty was that these mechanisms are many and varied. Yet early modern natural theologians assumed that human bodies were the most perfect. Why, then, would the all-powerful Author of nature ever diverge from this plan? According to this logic, Boyle pointed out, "we may think he must make no Animals but Men."[34] Boyle's answer added another new level of complexity to the notion of fitness. Since an animal structure needed to adapt the animal to its particular surroundings, Boyle argued that one could not determine perfection or imperfection, where fitness was concerned, in the abstract: "The eye is not to be consider'd abstractedly as an Instrument of Vision, but as an Instrument belonging to an Animal of this or that kind; and who is ordinarily to make use of it in such and such Circumstances."[35]

Moreover, not only did environments vary, but any given structure had many aspects. So fitness must be measured not only in context but along several axes at once. Boyle observed, "Divers Things may be Useful in an Organical Part"; for example, the eyelids have no role in vision, but they protect the eye. In keeping with this principle, Boyle enumerated four separate sorts of utility to be found in the human body: anatomical, chemical, mechanical and, in the eyes, optical.[36] Perfection in engineering, even divine perfection, in other words, was a multifarious and contextual quality.

A further problem will have occurred to the reader: eyes are not only various; they are also decidedly imperfect. Apart from the inversion of images on the retina, mentioned above, which natural theologians seem not to have treated as an imperfection, there was also the matter of chromatic aberration: the distortion resulting from the failure of a lens to focus all the colors to the same point of convergence. Throughout the seventeenth century, people including Huygens and Newton worked to solve this problem in artificial lenses. The trouble was that correcting it introduced other kinds of aberrations.[37]

There was also the startling discovery of the retinal blind spot. Edme Mariotte, a Dijon prior, natural philosopher, and early member of the French Académie royale des sciences, identified the retinal blind spot in 1660. He was trying to test the consequences of a curious anatomical fact, namely that the optic nerve did not join the retina at its center, but "somewhat higher, and on the side towards the Nose." Investigating the consequences of this ugly, asymmetrical arrangement, Mariotte fixed two circles of paper to the wall two feet apart. Standing in front of the leftmost piece of paper with his left eye shut and his right eye fixed upon it, he backed away from the wall. At a distance of ten feet, Mariotte made a remarkable discovery: "The second paper totally disappear'd." He could still see objects beyond it to the right, ruling out the possibility that the angle was too oblique; moreover, he could make the paper reappear by shifting the position of his eye the smallest amount. Having repeated the experiment with his left eye, and using different arrangements of pieces of paper, Mariotte drew a momentous conclusion: at the point where the optic nerve joined the retina, the eye was blind.[38]

No matter: natural theologians inverted the retinal blind spot as neatly as the sensing soul does the retinal image. Two imperfections, they discovered, constituted a perfection. The blind spot provided a reason for the apparently arbitrary lack of symmetry in the eye's structure, that displeasing arrangement of an off-center junction between optic nerve and retina, which Mariotte had been investigating when he discovered the blind spot. That the optic nerve was "not situate directly behind the Eye, but on one side," Ray observed, turned out to be a very good thing, since otherwise the blind spot would be right in the middle of our visual field.[39] Cheyne added that the off-center placement of the optic nerve meant that what was invisible to one eye was visible to the other.[40]

Structural perfection had assumed a new significance. The brute mechanisms of the argument-from-design tradition, being nothing but assemblages of moving parts, must be structurally perfect. What other virtues could they possibly have? How else could they indicate a perfect Designer? Accordingly, the natural theologians of this tradition inaugurated a physiological as well as a theological program: to identify and to describe as minutely as possible the fitness of structures to functions. Fitness replaced both perception and agency in the animal-machine. If animals behaved in ways that seemed to reveal understanding, it was because "God, having made them to preserve them, has formed their bodies so as mechanically to avoid whatever might hurt them."[41] The animal-machine of the New Science was perfect in its passivity and passive in its perfection.

The Restless Clock

Against this passivity, however, there were those who struggled to hold matter, feeling, and will together: to keep the machinery not just alive but active, lifelike. These holdouts accordingly had something very different in mind when they talked about the "animal-machine." William Harvey, whom we have already seen comparing the heart to a pump or other kind of hydraulic machinery, also invoked automata to describe the process of animal generation. Observing the development of a chick embryo, Harvey noted that a great many things happened in a certain order "in the same way as we see one wheel moving another in automata, and other pieces of mechanism." But, Harvey wrote, adopting Aristotle's view, the parts of the mechanism were not moving in the sense of "changing their places," pushing one another like the gears of a clock set in motion by the clockmaker winding the spring. Rather, the parts were remaining in place, but transforming in qualities, "in hardness, softness, colour, &ce."[42] It was a mechanism made of *changing* parts.

This was an idea to which Harvey returned regularly. Animals, he surmised, were like automata whose parts were perpetually transforming: expanding and contracting in response to heat and cold, imagination and sensation and ideas.[43]

These changes took place as a succession of connected developments that were also, somehow, all occurring at once. Similarly, Harvey wrote with regard to the heart that its consecutive action of auricles and ventricles was like "in a piece of machinery, in which, though one wheel gives motion to another, yet all the wheels seem to move simultaneously."[44] Geared mechanisms represented constellations of motions that seemed at once sequential and simultaneous, a congress of mutual causes and effects.[45]

The first appearance of life itself, as Harvey described it, seemed to happen both all at once and as a sequence of events. Harvey wrote of seeing the chick first as a "little cloud," and then, "in the midst of the cloudlet in question," the heart appeared as a tiny bloody point, like the point of a pin, so small that it disappeared altogether during contraction, then reappeared again "so that betwixt the visible and the invisible, betwixt being and not being, as it were, it gave by its pulses a kind of representation of the commencement of life.[46] A gathering cloud and, in its midst, a barely perceptible movement between being and not being: the origin of life. Harvey invoked clockwork and firearm mechanisms to model a defining feature of this cloudy pulse that was the beginning of life: causes and effects happening all at once, together.

Chick embryos inspired another close observer to think in terms of interacting machines. The Italian doctor and physiologist Marcello Malpighi also saw tiny, organic machinery at work in the process of embryonic development and described the development of the embryo as a sequence of little mechanisms giving rise to one another.[47] Malpighi recommended that anatomists reason by analogy from the larger and simpler natural machines, which they were able to see, to understand the invisibly tiniest and most complex machinery.[48] When he called the process mechanical, he meant that it was a rule-governed unfolding of material parts, each causing the next in line.

Malpighi's observations were adopted by a new party in the conversation about embryonic development: those who saw the embryo as preexisting in miniature within the egg, as a tiny homunculus, so that its development took the form of a simple expansion rather than an emergence of parts. Proponents of this idea, which came over the next century to be known as the theory of "preformation," included Swammerdam and van Leeuwenhoek (with some, including van Leeuwenhoek, locating the preexisting embryo in the sperm instead of the egg). Meanwhile, the idea that the preformationists set out to refute, Harvey's notion of the gradual formation of the embryo part by part, came to be known as the theory of "epigenesis," the word by which Harvey himself described the process. The theory of epigenesis had other advocates as well, including Descartes and Borelli.[49]

Although Descartes himself was an epigenesist, many of his principal followers from the 1670s onward were preformationists, including Claude Perrault and Nicolas Malebranche. Preformationism could easily become a version of the argument from design, firmly associating mechanist explanation with divine cause, smoothing away the critical ambiguities in Descartes's original theory of living machinery. Perrault's and Malebranche's was the generation of Cartesians who decisively redefined Descartes's animal-machinery as lifeless and inert, Perrault by attributing animal life and sensation to an immaterial soul, and Malebranche by insisting that animals, lacking souls, were entirely without perception or sensation.

Despite these opposite assessments of animal capacities, Perrault and Malebranche agreed that bodily machinery itself was essentially inanimate: the only proximate as well as the ultimate cause of life, and mind, was God. Likewise, in their versions of the theory of preformation, the tiny machinery of reproduction pointed the way back to its divine source. Malebranche added an important refinement in this regard: the idea that the preformed embryo itself contained a preformed embryo, which in turn contained another, and so on, nested like Russian dolls all the way back to the Creation. (Malebranche also elsewhere recited the watchmaker argument.)[50]

In contrast, Descartes had described a process of sexual reproduction that began with a "confused mixture of the two [seminal] fluids." Each of these served as "leavening" for the other, heating one another such that "certain of their particles acquire the same agitation as fire, dilating, pressing against the others, and by this means disposing them little by little so as to form the members."[51] Far from indicating an original, divine source, Descartes's account ascribed the whole agency of the creation of life to the heat and agitation of the fluids themselves and their active particles. Borelli, meanwhile, referred embryonic development to a kind of "automaton or a clock" formed in the egg by means of a magnetic-like force emanating from the sperm. This device contained even its own "motive faculty," set in motion by air particles that acted as "oscillatory machines" in the blood and other liquids of the body.[52]

The causes named in these epigenetic accounts of reproduction were internal to the animal-machinery itself. The active living machinery contained the forces and causes that set it in motion, as with Harvey's analogy to a geared mechanism, with its congress of mutual causes and effects. The versions of preformation put forth by Cartesians such as Malebranche and Perrault, on the other hand, represented the new, passive-mechanist science, in which any identification of causes moved quickly outside nature's machinery to its divine Engineer. This view would become predominant in the later part of the century alongside the argument from design.[53]

In addition to geared mechanisms and firearms, Harvey invoked another analogy that would become commonplace by the end of the century—we have seen Descartes and his followers invoke it—the analogy between an animal body and a church organ. Muscles, Harvey suggested, worked like "play on the organ, virginals." Under James I, English churches had resumed the use of organs in services, so they were once again a feature of the landscape and available as a source of models for living systems. The organ signified to Harvey something more like what it meant in the ancient and medieval tradition of animal-machinery, rather than the intricate sequence of contrived movements of parts that it later came to signify. Harvey wrote that the muscles performed their actions by "harmony and rhythm," a kind of "silent music." Mind, he said, was the "master of the choir": "mind sets the mass in motion."[54]

The particular ways in which Harvey invoked artificial mechanisms make it difficult to classify him, as historians have been inclined to do, either as a "mechanist" or otherwise,[55] the problem being that the meaning of "mechanism" and related terms was very much in flux. Lecturing at the College of Physicians in London in April 1616, Harvey told his anatomy and surgery students that anatomy was "philosophical, medical and mechanical."[56] But what did he mean, and what did his students understand, by "mechanical"?

In part, he likely meant that there was no need to invoke ethereal or celestial substances in explaining physiological phenomena, because the mundane elements seemed to transcend their own limits when they acted. The "air and water, the winds and the ocean" could "waft navies to either India and round this globe." The terrestrial elements could also "grind, bake, dig, pump, saw timber, sustain fire, support some things, overwhelm others." Fire could cook, heat, soften, harden, melt, sublime, transform, set in motion, and produce iron itself. The compass pointing north, the clock indicating the hours, all were accomplished simply by means of the ordinary elements, each of which "exceeded its own proper powers in action."[57] This was a form of mechanism that was not reductive, but really the reverse: a rising of mechanical parts to new powers, which could conceivably include the power to produce life.

Similarly, Harvey elsewhere defined "mechanics" as "that which overcomes things by which Nature is overcome." His examples were things having "little power of movement" in themselves that were nonetheless able to move great weights, such as a pulley. Mechanics, understood in this way, could include natural phenomena that overcame the usual course of nature, not just artificial ones. Harvey again mentioned the muscles. When he said that the muscles worked *mechanically* in this instance he meant that the muscles, like artificial devices such as a pulley, overcame the usual course of nature and moved great weights without themselves being weighty.[58]

Motion, relatedly, was a term with various meanings, as Harvey himself emphasized. He noted many different kinds of local movement: the movement of a night-blooming tree and that of heliotrope; the movements caused by a magnet and those caused by a rubbed piece of jet.[59] In what were likely some notes for a treatise on the physiology of movement, he jotted down any form of local movement that came to mind, such as the presumably peristaltic and undeniably graphic "shit by degrees not by squirts." He identified too, as a distinct form of movement, a kind of controlled escalation, as "in going forward, mounting up, with the consent of the intellect in a state of emotion."[60]

Harvey drew upon another form of causal motion to resolve another critical mystery in the generation of life: how did the sperm act upon the egg once it was no longer in contact with the egg? Like the apparently simultaneous occurrence of causally connected events, this quandary seemed to pose a problem for a properly "mechanical" anatomy. Invoking Aristotle, Harvey proposed that embryos arose from a kind of contagion, "a vital virus" with which the sperm infected the egg.[61] But after the initial moment of contact, once the contaminating element had disappeared and become "a nonentity," Harvey wondered, how did the process continue? "How, I ask, does a nonentity act?" How could something no longer extant continue to act on a material entity? The process seemed to involve too a kind of action at a distance: "How does a thing which is not in contact fashion another thing like itself?"[62]

Aristotle had invoked "automatic puppets" to explain precisely this seeming mystery. He had surmised that the initial contact at conception set off a succession of linked motions that constituted the development of the embryo.[63] According to this model, as Harvey explained, the seed formed the fetus "by motion" transmitted through a kind of automatic mechanism. Harvey rejected this explanation along with a whole host of other traditional explanations by analogy: to clocks, to kingdoms governed by the mandates of their sovereigns, and to instruments used to produce works of art. All, he thought, were insufficient.[64]

In their place, Harvey proposed a different analogy: one between the uterus and the brain. The two, he observed, were strikingly similar in structure (see figure 3.3) and a mechanical anatomy should correlate structures with physiological functions: "Where the same structure exists," Harvey reasoned, there must be "the same function implanted."[65] The uterus, when ready to conceive, strongly resembled the "ventricles" of the brain and the functions of each were called "conceptions." Perhaps, then, these were essentially the same sort of process.[66]

Harvey taught his anatomy and surgery students that the brain was a kind of workshop, a "manufactory."[67] Brains produced works of art by bringing an immaterial idea or form to matter. Perhaps a uterus produced an embryo in the same way, by means of a "plastic art" capable of bringing an idea or form to flesh. The form of an embryo existed in the uterus of the mother just as the form of a house existed in the brain of the builder. This would solve the apparent problems of action at a distance and nonentities acting upon material entities. The moment of insemination endowed the uterus with an ability to conceive embryos in the same way that education endowed the brain with the ability to conceive ideas. Once the seed disappeared, it no longer needed to act: the uterus itself took over the task of fashioning the embryo.[68]

The idea that the uterus functioned like a brain, actively fashioning an embryo the way a brain fleshes out an idea, was for Harvey not only within the bounds of the "mechanical," but a model that could actually rescue mechanism by eliminating the need for action at a distance.

In keeping with his inclusion of purposeful action—the womb's active fashioning of an embryo—within the mechanisms of life, Harvey disliked arguments from design, which instead referred all order to the rational foresight of an external engineer. This did not discourage Boyle from trying, nevertheless, to represent Harvey as an argument-from-design natural theologian. "I remember that when I asked our famous Harvey," Boyle reminisced, what had ever induced him to think of a circulation of the blood, "he answer'd me, that when he took notice that the Valves in the Veins of so many several Parts of the Body, were so Plac'd that they gave free passage to the Blood Towards the Heart, but oppos'd the passage of the Venal Blood the Contrary way"— this was what guaranteed the one-way circulation of the blood through the body—"He was invited to imagine, that so Provident a Cause as Nature had not so Plac'd so many Valves without Design."[69]

Yet Harvey himself maintained that nature did not work by design, skill,

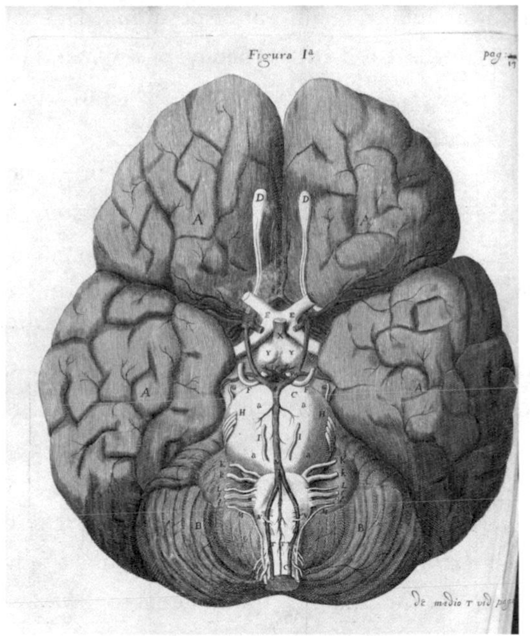

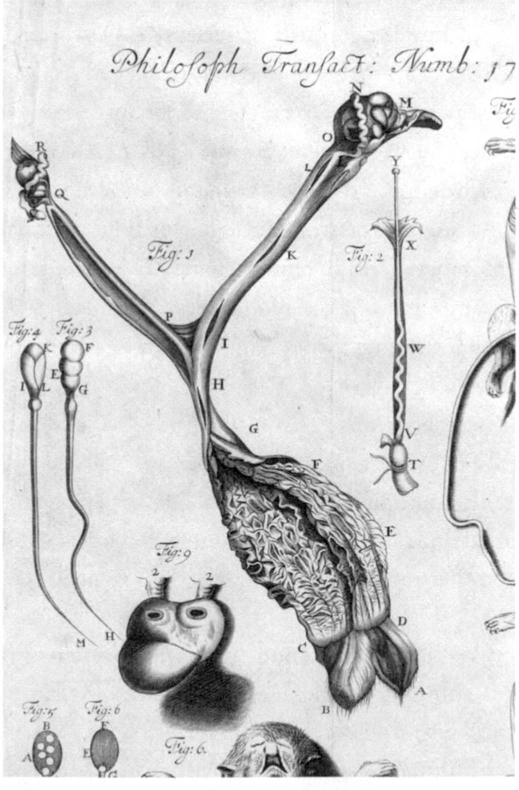

Figure 3.3 Christopher Wren's depiction of a brain from Thomas Willis's *Cerebri Anatome* (1664) and Antoni van Leeuwenhoek's drawing of a dog's uterus (1685), both courtesy Department of Special Collections, Stanford Libraries.

foresight, or reason but rather by an inborn "genius or disposition. " Human beings created things by means of intellect and art, but natural phenomena operated by an "innate" agency. Therefore, those who would "refer all to art or artifice" were but "indifferent judges of nature or natural things."[70]

During the Civil War, taking refuge in Oxford, Harvey lived briefly in Merton Street, where he became a neighbor of a younger doctor and anatomist, Thomas Willis, who was an early cartographer of the brain and nervous system. Like Harvey, Willis resisted the central tenets of the new brute mechanism. He understood Descartes to have meant that animals, as fully corporeal beings, were "meerly passive": they moved only when set in motion by "other Bodies, striking some part of the Soul" so that their actions were nothing but the "artificial Motion of a Mechanical Engine."[71] Willis disliked the passivity of this view of the animal-machine. Why, after all, must matter be incapable of agency and perception? God could surely have endowed matter with these capacities. Willis therefore described a "self-moving" animal-machine possessed of a fully material soul common to both beasts and humans. This material soul was responsible, he reckoned, for life, sensation, and motion.[72]

Willis's account of animals was thus as rigorously materialist as Descartes's, but in Willis's view, the material soul of animals was also "Knowing and Active." It was even capable of learning by means of the "Accidents" it encountered as it stumbled through the world in the course of its day. By means of these contingencies, the soul acquired new knowledge and skills. Its capabilities accordingly increased in complexity. To his knowing, active, educable, yet fully material soul, Willis ascribed physical "Members" and "Organical Parts." He singled out two parts in particular, a vital component in the blood and a sensitive component in the animal spirits, which traveled throughout the "Pipes and other Machines" of the brain and nerves.[73]

The animal soul as Willis described it was a part of the body, made of particles of the same matter, but the choicest among these: the most "subtle and highly active." These most "nimble and Spirituous Particles" played a principal role in the formation of the animal body. Gathering together in dynamic heaps, they jostled, stirred, and steered the other, grosser particles into their proper places.[74]

Artificial machinery offered Willis plenty of models on which to base his idea of a vital, perceptive, active animal-machine. He went beyond the clockmaker's assemblage of wheels and gears, which was, after all, but a narrow domain in the growing expanse of human-made devices. "Mechanical things,"

Willis pointed out, required "Energetical" components: "Fire, Air, and Light." Any smith, chemist, glassmaker, lens grinder, or instrument maker could easily testify to the truth of this. Likewise the "Great Workman," in creating the animal-machine, had made the souls of living creatures out of the most active and energetic particles of matter.

The movements of these particles through the body was like the passage of air through a pneumatic engine: coursing through the animal-machine, the active particles produced sensation and movement.[75] Like Harvey, Willis had church organs in mind: his drawing of the nerves of the trunk, indeed, strikingly resembled the symmetrical geometry of an organ (see figure 3.4). When Willis looked at automatic organs, he saw constraint, to be sure, but not passivity. If the soul of a man was like a musician playing any tune he liked on his instrument, the soul of an animal was more like an automatic water organ: it could play only a limited repertoire of tunes and yet it nevertheless "Institute[d], for Ends necessary for it self, many series of Actions."[76]

Artificial mechanisms provided models not only for agency, but also for indeterminate, variable and responsive activity. Guillaume Lamy, a member of the Paris Medical Faculty,[77] invoked a weather vane, for example, in his rigorously mechanist account of the sensitive soul, to show how a mechanism could behave as unpredictably as a young man's passions. These would turn and shift according to the prevailing winds like the cock on a church tower.[78]

The person who probably went the farthest toward establishing the mechanism of the new philosophy along a different trajectory was the German philosopher, theologian, mathematician, logician, historian, and poet Gottfried Wilhelm Leibniz (see figure 3.5).[79] Like everyone else at the time, Leibniz described animal and human bodies as equivalent to automata.[80] But he meant something very different by it from what was rapidly becoming the dominant meaning. The shared keywords of the seventeenth century— "mechanical," "clockwork"— camouflaged radical differences of opinion. In Leibniz's view, neither animals nor, indeed, machines were passive or brute. Leibniz was among the many who rejected Descartes's claim that animals lacked souls. But in his case, it was part of a more general refutation of Cartesian physics: to Leibniz, nothing really lacked a soul.

To be sure, no sort of brute machinery, no matter how subtle, could by itself account for perception. But neither could it account for any sort of action or motion in nature.[81] Descartes's clockwork, to Leibniz, was implausible even as

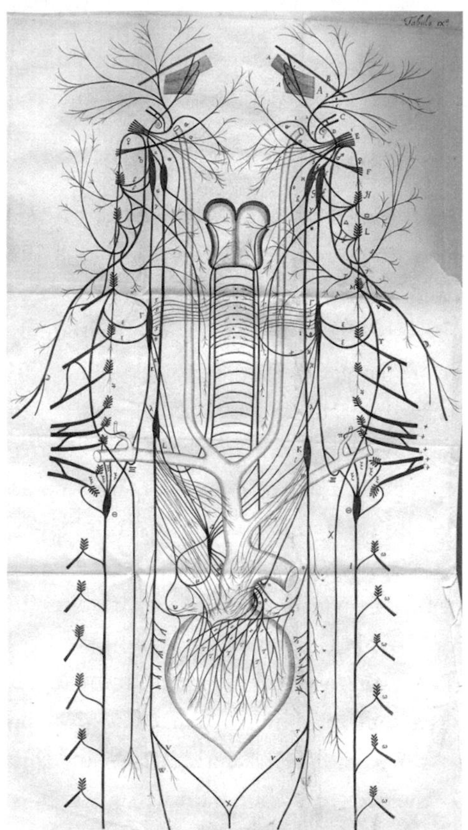

Figure 3.4 The nerves of the trunk, from Thomas Willis's *Two Discourses concerning the Souls of Brutes* (1683), courtesy Department of Special Collections, Stanford Libraries, and the organ at King's College, University of Cambridge, courtesy Hugh Taylor.

Figure 3.5 Portrait of Gottfried Wilhelm Leibniz, after a painting by Bernhard Francke (1729), bpk, Berlin / Kupferstichkabinett, Staatliche Museen, Berlin / Art Resource, New York.

an account of a clock, let alone of a dog. Mere "*extended mass*" could never suffice. Mechanics required something more: force. Leibniz called force a "metaphysical" thing, meaning that it was not a material entity in itself, but rather a principle underlying all material events. Without such a metaphysical principle, nature was unintelligible. Matter drained of spirit or force could explain nothing, he argued, not even inanimate machinery.[82]

Leibniz elaborated his idea of a *vis viva* or "living force" in all mechanical phenomena during the so-called *vis viva* controversy of the 1680s and '90s. Here he defended a view of conservation of *vis viva*, mathematically equivalent to the later notion of kinetic energy (mass multiplied by the square of the velocity), against Cartesians who argued for Descartes's principle of conservation of motion, equivalent to the later notion of momentum (mass multiplied by velocity).[83] Motion, Leibniz argued, was "not something entirely real" because it consisted merely of relations among objects, whereas force, a "force of acting," was "something real," belonging to a given body in itself.[84] Moreover, he said, no one had ever explained force,[85] and he rejected what he saw as the current tendency to instead "summon God *ex machina,* and withdraw all force for acting from things." This was a marionette mechanism with God as the puppeteer, such that "when a person thinks and tries to move his arm, God moves the arm for him," an idea so absurd it "ought to have warned these writers they were depending on a false principle."[86]

On the other hand, Leibniz also disapproved of appeals to an "Archaeus," a term that had been used by medieval and early Renaissance alchemists, such as the fifteenth-century German alchemist Paracelsus, to refer to a vital spirit originating in the astral realm. Leibniz called the idea "unintelligible": "as if not everything in nature can be explained mechanically." The big mistake, he judged, was to assume that a mechanist science must eliminate incorporeal things, when in fact a mechanist science required incorporeal things.[87] Leibniz was after a third way: neither marionette mechanism nor what he saw as alchemical abandonments of mechanism, but a fully mechanist account of nature that included immaterial "active force."[88]

These principles were not purely abstract, but constituted an alternative approach to physics and engineering. Leibniz's notion of *vis viva* informed a tradition of physics and engineering growing through the eighteenth century and into the nineteenth. In 1740, the French marquise, mathematician, philosopher, and lover of Voltaire Emilie du Châtelet, used Leibniz's notion of living force to present a revised version of Newtonian physics. As the translator of Newton's *Principia* into French, and one of a select few able to read and understand that text, du Châtelet was influential in shaping the promulgation and development of physics in the eighteenth century, and she saw Leibnitz's living force as its salvation.

Both Descartes and Newton, she wrote, had tried and failed to describe a force in the universe that would remain constant and undying. If one equated force with the motion of a moving body, proportional to its velocity, as the Cartesians and Newtonians did, one could not avoid the terrible conclusion that the universe was winding down, losing force, and would one day need to be set in motion again. But if one considered the "living force" in a moving body to be proportional to the *square* of its velocity, the problem was solved. In that case one would immediately see that this force, "which is something real, and which endures like matter" would "remain steadfast" and never perish.[89]

Leibniz's active mechanism,[90] with the understanding of force and motion that du Châtelet promoted, was no idle speculation, but had momentous scientific and practical consequences. The Leibnizian tradition in physics and engineering also included such people as Lazare and Sadi Carnot, Gaspard Monge, Jean-Victor Poncelet, and John Smeaton, whose work culminated in the development of an energistic physics and the concepts of energy and work around the middle of the nineteenth century.[91]

To Leibniz and his followers in this tradition, the limits of brute mechanism seemed glaringly evident. Boyle, for example, according to Leibnitz, had failed to distinguish ultimate from proximate causes when he presented his vision of nature as machinery. When Boyle and the other argument-from-design natural theologians exiled perception and agency to a position fully outside their world-machine, it appeared to Leibniz to be a kind of blinkering, a refusal to see the obvious: the machinery could never work without metaphysical forces driving it.[92] Everywhere in material events one found an active agency working from within: "a flare that runs the length of a cord or a fluid that runs in a channel." In the "last analysis of the laws of mechanics, and the nature of substances," Leibniz argued repeatedly, one had ultimately to appeal "to active indivisible principles."[93] In his world-machine, perception and agency were not banished but active within the very works.

In 1676, on the way home from four years' stay in Paris, Leibniz stopped off in the Netherlands. There, he met a controversial figure, the excommunicated Jewish thinker and lens grinder Benedict Spinoza, who was living in exile in the Hague, having been banished from Amsterdam twenty years earlier. Spinoza was in chronically poor health and would die the following year at the age of forty-four, never having left the Netherlands. But he corresponded warmly and extensively with Henry Oldenburg, secretary of the Royal Society of London, and through him came to know the state of conversation in natural philosophy, especially the work of Robert Boyle. Reciprocally, the radical understanding of nature, matter, and mechanism for which Spinoza had been expelled from Amsterdam became a fulcrum of debate in the later seventeenth and eighteenth centuries. Leibniz had already been in correspondence with Spinoza for several years before they met and had read some of his work.[94]

During his visit to The Hague, Leibniz had several long talks with Spinoza. He later described the encounters in a letter to a friend in Paris, the abbé Jean Gallois: "Spinosa died this winter. I saw him when I was passing through Holland, and I spoke with him several times and very long. He has a strange Metaphysics, full of paradoxes."[95] This strange and paradoxical metaphysics marked Leibniz's own thinking, however. He was particularly impressed by Spinoza's distinctive form of naturalism: his idea that nothing existed or operated outside of nature, neither God nor souls nor any spiritual entity or power. Rather, these *were* nature.

Insisting on the inseparability of God from nature, Spinoza argued that with regard to natural phenomena, there could be no external criteria of good or ill. Nature simply *was*. Responding against Boyle's and the other brute mechanists' arguments from design, Spinoza wrote that in natural phenomena, there could be no such thing as either defectiveness or excellence of design, no perfection or imperfection, neither good nor bad. These would require external criteria, but there could be no such thing. Nature had no goals; it did not "act with a purpose," for how could it? A goal is separate from the means that accomplish it, while Spinoza's nature was all-encompassing. Therefore the "eternal and infinite Being, whom we call God or Nature," simply existed.[96]

Nor was the human mind with its "infinite power of thought" something apart, as Descartes had taught, but rather, it was an integral part of the eternal and all-encompassing nature. To account for the oneness of the human mind with nature, Spinoza invoked a common striving (*conatus*) to persist and grow. All finite things in nature, from the motion of a thrown rock to the ambition of a conscious mind, he judged, shared a definitive striving to persist.[97] Here was the inherent force, the source of motion, in Spinoza's all-encompassingly natural world. He and other contemporaries, including Leibniz, borrowed the term *conatus* from classical and medieval theories of motion, and employed it to characterize the tendency of natural entities to persevere, develop, and expand.[98] To characterize the inherent, comprehensive activity of nature, Spinoza also distinguished "*Natura naturans*" (nature naturing), or nature actively conceiving itself, from "*Natura naturata*" (nature natured), nature as the passive product of an external god.[99]

While Leibniz hated one aspect of Spinoza's system, the "blind" and "fatal" necessity he understood Spinoza to be ascribing to natural phenomena, he judged other aspects "excellent," even in keeping with his own views.[100] In particular, Leibniz shared Spinoza's refusal of extra-natural causes, of a marionette mechanism, and his insistence that nature was everything. Whereas Spinoza took this to imply that nature encompassed every possibility, leaving no room for contingency or paths not taken, Leibniz instead construed it to mean something like the opposite: that perceptive agency (and therefore contingency) was integral to the cosmic machinery.[101] They arrived at these opposite poles, however, from a shared principle of unbounded naturalism.

Perhaps Leibniz's search for alternative models for a mechanist natural science helped to motivate his great interest in Chinese philosophy:[102] in Confucianism he found an idea to counter brute mechanism. Leibniz studied written Chinese and acquired a limited knowledge of its structure. He also read the history of China and translations of Chinese classical texts that Jesuit missionaries had produced during the 1670s and '80s, published as *Confucius Sinarum philosophus sive scientia Sinensis latine exposita* (The Knowledge of the Chinese Philosopher Confucius Expounded in Latin) (1687). Shortly afterward, during a trip to Rome, Leibniz met Claudio Grimaldi, a Jesuit who had served as a diplomatic aide at the imperial court in Beijing and been involved in various mathematical and astronomical pursuits there. After their meeting, Leibniz sent Grimaldi a list of thirty questions, largely about Chinese natural sciences, resources, and practical arts.[103] He did find, in Chinese approaches to nature, concepts that he thought crossed the traditional philosophical categories in Europe.

When Leibniz wrote the "Discourse on the natural theology of the Chinese" (1716) during his last year of life, his project was accommodationist in the tradition of the sixteenth-century Italian Jesuit Matteo Ricci, a founder of the Jesuit China Mission. Like Ricci, rather than urging that the Chinese must renounce their ancient beliefs in favor of doctrinal Christian ones, Leibniz meant to demonstrate the reconcilability, indeed the essential sameness, of Christianity and Confucianism.[104] To that end, he argued against the allegations leveled by some Europeans of Chinese materialism, focusing particular attention on the Confucian notion of *Li*, which Jesuit interpreters had rendered as equivalent to the Scholastics' prime matter.[105] One must not interpret *Li* that way, Leibniz cautioned, since the Scholastics' prime matter was "purely passive, without order or form." *Li* in that case could not possibly be "the origin of activity," as Confucianism taught.

"I do not believe [the Chinese] to be so stupid or absurd," Leibniz wrote, as to attribute "the active power, and the perception that regulates this active power" to so passive and inert a thing as the Scholastics' prime matter.[106] *Li*, he emphasized, in Chinese usage denoted not only "the material basis of Heaven and Earth and other material things," but also "the moral basis of virtues, customs, and other spiritual things."[107] In Confucianism (as rendered by European commentators), Leibniz identified a principle that was at once active and mechanical, material and moral: one that violated, in other words, the increasingly entrenched injunction in the sciences to segregate purposeful agency from mechanism.

In reacting against the brute mechanism of his contemporaries, Leibniz began with its insufficiency as an account of life and mind. He appealed to internal experience, the inner consciousness of "this *me*" not explicable by "figures or movements."[108] A sentient, thinking being, he argued in this instance, was not "a mechanical thing like a watch or a mill such that one could conceive of sizes, shapes and movements whose mechanical conjunction could produce something thinking and even feeling."[109]

In a thought experiment designed to show the nonmechanical nature of thought, he instructed his reader to imagine a big machine, the size of a mill, that could think, feel, and perceive. Imagine, he wrote, walking into this great factory of thought and looking around. You would find only "pieces that push each other and never anything to explain a perception." You would understand consciousness no better than before entering the mill of mind. Looking at the machinery, the pushing and pulling, the moving parts, the thing you would be led to understand was that perception and consciousness were not that. Perception and thought, Leibniz wrote, resided not in the operation of the mechanism but in its very substance. Indeed, perception *was* the primary substance, the stuff of the machine.[110]

But if perception and thought could never be explained by the interactions of brute parts, Leibniz believed, neither could clocks. Matter itself could not be made of inert chunks: "*Atoms of matter* are contrary to reason." Any corporeal entity, no matter how "invincibly attached," must still be composed of parts: one could imagine dividing it further. The indivisible atoms that made up the world, therefore, must be something else: "We could call them *metaphysical points*: they have something *vital* and a kind of *perception*."[111] Over the course of his career, Leibniz thus developed an increasingly distinct understanding of mechanism.[112]

Since chunks of inert matter, on their own, in his view explained nothing, he replaced extension (size and shape) with perception, and offered an equal and opposite philosophy to Hobbesian materialism: a reduction of matter to spirit.[113] In place of chunks of matter, he put perceiving subjects. The building blocks of his cosmos were not blocks but little souls.

No difference in substance, therefore, divided material from spiritual things. By the time he wrote the *Monadologie* (1714), Leibniz had decided that both the material and the spiritual were composed of "Monads," elementary spiritual substances whose defining attribute was perception. The difference between material and spiritual things lay only in the laws governing them: laws of force and transfer of motion for material things, laws of justice for spiritual ones. Animals were material in the sense that God governed them as a "machinist." Rational souls, able to think, to recognize God and eternal and necessary truths, were spiritual in the sense that God governed these as a "Prince" or "Legislator."[114]

Leibniz described the cosmos and everything within it, including living animals and humans, as a great nesting of machines within machines within machines, all built out of little perceiving spirits. What he now meant by "machinery" was therefore not just different from, but in fact opposite to, what Hobbes and the Cartesians meant. There was no actual pushing and pulling, no action by impact, but just the appearance of these mechanical causes. Indeed, matter itself was an appearance, a secondary effect of the perceptual substance out of which the world was composed. Monads could not change one another, having no extension, no parts: as Leibniz famously described it, no "windows" through which anything could come in or go out. Rather, each little soul followed its own internally directed sequence of changes that had been set in motion at the beginning of time, and a divinely given, "preestablished harmony" coordinated all these sequences so as to follow the laws of mechanics.[115]

The same eternal, harmonious order also correlated the laws of matter and spirit such that "the springs of bodies are ready to act by themselves, as necessary, at the moment when the soul has a congruent desire or thought," and the mechanical laws of nature would always bear out the moral order of justice.[116] Here, crucially, was how Leibniz's notion of a "preestablished harmony" differed from the idea of "design" in the argument from design: in Leibniz's preestablished harmony, the bodies and substances of nature were active participants rather than passive parts.

This is an important and little-noted distinction. When Voltaire devastatingly lampooned the Leibnizian idea of preestablished harmony in his satirical novel *Candide*, and the associated principle that this harmonious action brought about the best of all possible worlds, he rendered Leibniz's view more as design than harmonization. Dr. Pangloss, the ostensibly Leibniz-esque tutor and professor of "metaphysico-theologico-cosmolonigology," explains to his eager pupil Candide that noses exist to support spectacles, legs to carry shoes, stones to be carved and built into castles, and pigs to be eaten.[117]

This would be an apt parody of the argument from design, according to which each structure in nature exists to perform its given function, but is less so for Leibniz's vision of dynamic, organic machinery, in which the parts actively collaborate in bringing about an overall harmony: a grand collusion of all things in nature including noses and spectacles, legs and shoes, stones and castles, and all of their busy, organic-mechanical parts within parts within parts to infinity. Ironically, Voltaire himself returned regularly to a version of the argument from design, something rather closer to Dr. Pangloss's philosophy, as for example when he wrote: "The universe troubles me and I cannot think that this clock could exist and have no clockmaker."[118]

When Leibniz compared God to a watchmaker and his creations to automata, when he wrote that the bodies of men and animals were "as mechanical" as watches,[119] he meant something profoundly different from the very same statement by Malebranche or Boyle. To see how very different his meaning was, consider the image from which this book and this chapter-section take their title, Leibniz's image of the "restless clock." I have already, in the introduction, discussed the beginning of the passage in which Leibniz presented this image, but I reproduce it here in full.

> In German, the name for the balance of a clock is *Unruhe*—that is to say *disquiet*. One could say that it is the same thing in our body, which can never be perfectly at ease: because if it were, a new impression of objects, a little change in the organs, in the vessels and viscera, would change the balance and make these parts exert some small effort to get back to the best state possible; which produces a perpetual conflict that is, so to speak, the disquiet of our Clock, so that this appellation is rather to my liking.[120]

Nowhere in this passage, among the metaphorical significations of "clock," do we find what had by then become, and still remain, the expected connotations: regularity, imperturbability, precision. Instead we have something like their opposites: disquiet, unease, exertion, conflict. Wherever Leibniz returned to the notion of an animal as a machine his meaning was similarly exotic. He described natural machinery as "entangled," waxing and waning, enfolding and unfolding, "frail" and yet capable of self-maintenance.[121]

In the conversation surrounding Leibniz's writing too machinery held implications strikingly different from those, such as absolute uniformity, that were even then becoming ever more firmly associated with clockwork. The French Calvinist Pierre Bayle, for example, who engaged Leibniz on the subject of human-machinery during the late 1690s, argued that a machine was distinctive in its capacity to act variously, and that it differed in this regard from a simple entity, which was constrained always "to act uniformly if no foreign cause divert[ed] it." Being composed of several pieces, a machine "would act diversely, because the particular activity of each piece could at any moment change the course of the others."[122] To Bayle, a machine was something made of active parts whose actions generated possibilities.

Likewise, when Leibniz called a process "mechanical," he did not mean that it was devoid of spirit, agency, and perception. Rather, such a process followed entirely from its own internal principles, with no appeal to a deus ex machina.[123] The motions of the celestial bodies, the formation of plants and animals, the organism of a living being, all of these were divinely formed mechanisms, their only miracle being at their origin, and "what follows is purely natural and entirely mechanical."[124]

Leibniz's great objection to the Newtonian system, which he voiced steadfastly during the last couple years of his life in an intense epistolary debate with Newton's friend and translator Samuel Clarke, was that Newton described the cosmos as an artifact, a device, of brute mechanism. Like any such device, it required its Maker to step in and adjust it, to rewind it and keep it running. The very running of Newton's cosmic Clock, therefore, rested upon an extra-natural cause, an actor intervening from outside the system.[125]

In Leibniz's view, this was not only to denigrate God's handiwork but also to violate the principles of naturalism and mechanism. It made neither for good theology nor for good science. Several decades later, the Scottish skeptic and empiricist David Hume would make a similar case that the argument from design was the instrument of its own undoing. The real trouble was not those questions about retinal blind spots and other imperfections. The argument's downfall did not lie in imperfections of contrivance, but in contrivance itself as an argument for an omnipotent being. Any contrivance, after all, is a particular and limited affair.[126] It might be a good thing, as Boyle had pointed out, that flies, not being able to move their eyes, have compound eyes to compensate for it. But why can't they move their eyes? Boyle waved the question aside, leaving it between parentheses,[127] whence it stubbornly continued to indicate the same core problem that Hume identified: how to get from a particular and limited contrivance to an absolute power?

So regarding the argument from design, Leibniz disliked it both as theology and as science. A thoroughly naturalist, mechanical theory of nature, he argued, could accommodate no divine interventions from outside. Rather, such a theory would need to include the ultimate causes as well as the proximate ones; it must encompass the metaphysical principles governing force and the laws of motion. According to Leibniz's ideal of science, there should be one single system encompassing all of nature, with nothing outsourced, nothing rendered exceptional or external. The authors of the mechanical philosophy—Galileo, Descartes, Hobbes, Gassendi–had "purged inexplicable chimera from philosophy" but they had left a metaphysical gap and filled it with a meddlesome God who acted supernaturally. "I tried to fill this gap," wrote Leibniz, "and have at last shown that *everything happens mechanically in nature, but that the principles of mechanism are metaphysical*."[128]

Metaphysical principles operated at the crux of Leibniz's mechanical systems. His guiding axiom, for example, that "there is nothing without a reason" was a mechanical as well as a metaphysical principle. This "principle of sufficient reason," Leibniz reckoned, constituted the very connection between mechanical causes and their mechanical effects. The Archimedean principle of equilibrium, in which two equal weights sat in a balance in the same relation to its axis, provided Leibniz with an example: since the situation was perfectly symmetrical, the balance had no reason to tip to one side rather than the other, and therefore it remained in equilibrium.[129] Not only physiology but physics itself required purposeful agency. Moreover, since the "entrails" of nature, the movements of her internal parts, were all hidden from view, understanding could come more easily through a study of her "designs" than of her structural "movements."[130] Metaphysical purposes drove nature's active, restless machinery from within.[131]

Organisms as Restless Clocks

In the first years of the eighteenth century, acting upon the same principle that informed his physics, the essential importance of inner agencies in nature, Leibniz developed a theory of organisms that would shape discussion of living beings over the following century.[132] Living things, he explained, were more thoroughly mechanical than any artificial device created by human beings. That was because God's machines were irreducibly mechanical down to their "smallest distinguishable parts."[133] "I define the Organism, or the natural Machine," Leibniz wrote to the English philosopher Lady Damaris Masham in 1704, as "a machine in which each part is a machine, and consequently the subtlety of its artifice goes to infinity, nothing being small enough to be neglected, whereas the parts of our artificial machines are not machines. That is the difference between Nature and Art."[134] Nature was machinery all the way down.[135] The human body, therefore, Leibniz continued, was "a sort of world full of an infinity of creatures who also deserved to exist."[136]

He later presented this idea in its full form in the *Monadologie*: "Thus each organic body of a living being is a sort of divine machine, or a natural Automaton, that infinitely surpasses all artificial automata." No human-made machine could be a machine in all its parts. For example, the teeth of a brass wheel were made of simple parts having nothing "artificial," mechanical, or contrived about them. "But natural machines, that is, living bodies, remain machines in their least parts to infinity.[137]

This cosmos of machines within machines without end, their ultimate components living souls, was intrinsically active, self-moving, containing its own energies, forces, and powers of action. Leibniz presented this view of an active-mechanical world system in explicit response to Boyle's mechanical cosmos, evacuated as it was, like Descartes's and Newton's cosmic machinery, of all agency.[138]

Leibniz's mechanical cosmos was also brimming with life and sentience in every part. Every possible living form was present in this profusion: Leibniz's "law of continuity" held that "everything goes by degrees in nature, and nothing by jumps," so that every spot in the continuum must be filled.[139] The supreme perfection of God's Creation, the assurance that it was the best of all possible worlds, resided in this completeness, the presence of every form of being.[140]

To inform his descriptions of this profuse, blooming machinery, Leibniz took an active interest in the work of anatomists and physiologists such as Harvey, Willis, and Malpighi.[141] Having met Swammerdam and van Leeuwenhoek in 1676 during the same visit to the Netherlands where he met and talked with Spinoza,[142] Leibniz cited microscopic evidence in support of his claim that the tiniest particle of matter contained whole worlds of living beings.[143] Each speck was a garden of plants, each drop a pond teeming with fish. Every part of every plant or animal and all the air between them was lush with living creatures. Nowhere in the universe was anything actually "uncultivated, sterile or dead," nothing chaotic or confused: creatures lived and moved everywhere according to their harmonious governing laws, could one but see them.[144] "All of nature," Leibniz wrote, "is full of life,"[145] and exactly as full of mechanism.

This lively, mechanical nature was of course permeated to the core with perception too being made of perceiving spirits (the "monads"). In Leibniz's view, the Cartesians had failed to appreciate the "unconscious perception" everywhere in nature, and this was what had led them to divest animals of souls and to detach human souls from bodies. Leibniz argued that "unconscious perception," in its ubiquity, was the connection between soul and body, human and animal.[146] The living and perceiving natural machinery Leibniz described, drawing on the work of observers such as Swammerdam, was also intrinsically active and self-moving (see figure 3.6).

Swammerdam, for example, described how a cuttlefish sperm, upon being removed from the cuttlefish gonad, began to act, and he ascribed full agency to the "minute machine" itself. Moreover, the self-directed movements of this tiny machine were not ticks and tocks but twists and turns, evolutions, unfoldings, emergences: "The extremity begins to evolute and unfold itself," Swammerdam wrote, "and the two slender ligaments, which emerge out of the case, turn and twist themselves in various directions."[147]

Invoking such accounts, Leibniz responded to an increasingly vocal contingent: those who argued that living creatures were more than machines. Leibniz answered them by insisting that *machines* were more than machines. The German doctor and physicist Georg Ernst Stahl, for example, became one of Leibniz's chief interlocutors on this topic. Stahl was the author of "phlogiston," an idea that caught fire during the eighteenth century and burned out by the turn of the nineteenth. Phlogiston was the fiery element that Stahl claimed got released during combustion and respiration: the very stuff of heat, light, and life. A candle or a bird would smother in a closed container, Stahl reasoned, because there was no air to carry away the phlogiston it struggled to release.[148]

Amidst the mayhem of the French Revolution, an experimental physicist, Antoine Lavoisier, who would soon lose his own head to the Terror, turned phlogiston on its head (by showing that breathing and burning were a taking-in rather than a release of a substance) and named it oxygen. Meanwhile, however, phlogiston while still current expressed not just a theory of respiration and combustion but also an understanding of the fundamentally nonmechanical nature of life. It marked what Stahl and his fellow travelers took to be an unbreachable distinction between living creatures and inanimate mechanisms.

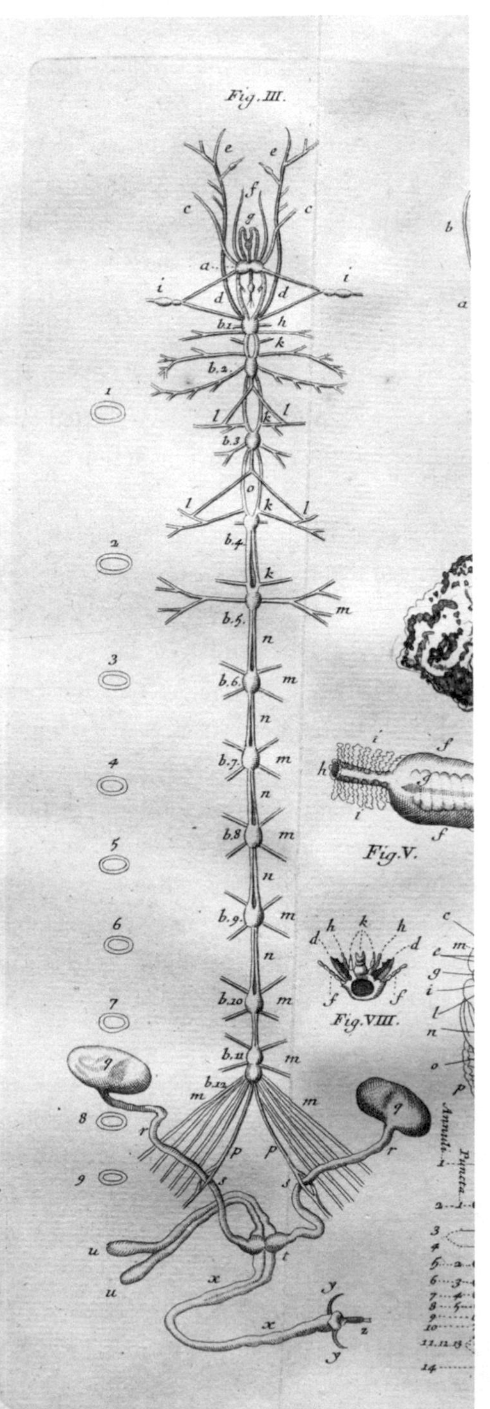

Figure 3.6 The nervous system of a silkworm, from Jan Swammerdam, *Book of Nature* (1658), courtesy Department of Special Collections, Stanford Libraries. Photo by Andrew Schupanitz.

The source of life, phlogiston was definitively nonmechanical, a substance that acted according to its own vital and fiery nature. Within the "machine of the human body," Stahl maintained, their resided a "*truly organic mystery*."[149]

Leibniz objected to this organicism in the same way he objected to brute mechanism. In both cases, he disliked the divorce of agency from matter and the setting of life apart from nature. Life according to Leibniz did not originate in a special life-giving substance but in a "vegetative force" whose mode of action he could deduce "from the very structure of the machine."[150] For example, he proposed that the mechanical force at work in physiological processes might be something like elastic force, allowing things to be "more energetic than according solely to their mass."[151] Living beings were not different in kind from the rest of the world, which also operated not just by matter in motion but also by forces.

In response to Stahl's analogy between the heat of combustion and the warmth of living creatures, Leibniz agreed that bodies were like steam engines: hydraulic-pneumatic machines.[152] But to Leibniz, this vital heat was a part of the cosmic mechanism, not external to it. Stahl acknowledged that organisms *included* machines, Leibniz wrote, but this was not enough: "One must add that an organism is formally nothing other than a mechanism, even if it is more subtle and more divine, because, in nature, everything must happen mechanically."[153] This mechanical naturalism was the very source of order in Leibniz's science. It supplied the limits that gave a shape and pattern to natural phenomena. If creatures acted by means of an immaterial soul rather than mechanically and naturally, there would be no reason for any limits to their actions. For example, Leibniz argued, if it were our soul that allowed us to jump into the air, why could we not jump to any height arbitrarily?[154]

This distinct mode of mechanism had a momentous result. Leibniz's insistence that mechanism encompass purpose and agency set the living world in forward motion. For if God worked exclusively through the purposeful machinery of nature, this meant he had also to work in natural time. As with the argument-from-design natural theologians' God, Leibniz's God remained the source of order in the cosmos, but with a crucial difference. As we have seen, the natural machinery participated actively in bringing about this order, which was therefore not delivered all at once at the beginning of time, but rather it unfolded over all eternity. This sort of order was not "design" but "organization," not a static structure but a patterned process.

"Organization" was initially an Aristotelian idea pertaining to living things and distinguishing them from deliberately designed things. In living things, according to Aristotle, the parts were not put together "for the sake of something" but just happened to be constituted and arranged as if they had been put together for a purpose. Wherever this was the case, the organism survived, "being organized spontaneously in a fitting way; whereas those which grew otherwise perished."[155]

A little over a decade before Leibniz introduced his definition of the "organism," John Locke had picked up the idea of organization and made it the basis of a living thing's identity: that which makes it the same living thing over its lifetime despite the fact that its material parts are continually transforming. An oak tree remains the same oak tree although new particles of matter are always being "virally united" to it, because these preserve the same organization. An animal, likewise, was "a living organized body" whose identity attached to its organization.[156] The identity of a human being consisted in nothing else: a person was a particular, ongoing organization of ever-renewing material parts. While an artificial device such as a watch owed both its organization and its movement to an external source, and could therefore be organized yet motionless, in an animal the organization was the source of motion, including the motion that constituted life itself, all originating "from within."[157]

Leibniz's God worked through organization rather than design. He did not work *upon* matter but *through* it, through "non-intelligent" and "plastic natures devoid of knowledge," endowing these with a plan rather than a design: "God *preformed* things, such that the new organizations are but a mechanical consequence of a preceding organic constitution; as when butterflies come from silkworms."[158] In Leibniz's version of preformation, unlike in Perrault's and Malebranche's, God did not preform the animals themselves, but rather, through preformation, "made nature such that she is capable" of forming them.[159]

Nature worked such marvels "by a certain instinct" and through continual "transformation and augmentation" of organic forms. Accordingly, everything in Leibniz's lively world was in a perpetual state of flux. His cosmos flowed like a river. His natural machines were always developing and changing, just as the "spermatic machine" developed into another machine for producing a human body.[160] Natural machines transformed in this way by means of "enfoldings," "extensions," and "contractions." Moreover, no natural machine ever appeared or disappeared. When it seemed to be gone, it was merely "as if concentrated."[161]

Living machines, that is, creatures, had no true beginnings, apart from the beginning of time, and no real endings: births and deaths were but appearances. Leibniz cited in evidence the transformations revealed by the microscopists whose work he had studied, “Messieurs Swammerdam, Malpighi and Leewenhoeck,” which he took to show that animal substance had no beginning. Its apparent generation was “nothing but a development, a sort of augmentation.”[162] In reality there was only development and growth, envelopment and diminution. And in this continual waxing and waning, certain souls rose “to the degree of reason and to the prerogative of minds.”[163]

Leibniz’s form of purposeful, active mechanism extended even to include the generation, over time, of a thinking mind.

~ 4 ~

The First Androids

In 1677, Salomon Reisel, a Swabian doctor, philosophical mechanist, and personal physician to the Duke of Württemberg, announced that he had built an entire "artificial man" with all the internal bodily functions: circulation, respiration, digestion. Reisel even planned to endow his creation with speech and the ability to move about on its own. In the *Journal des sçavans*, along with reports of a three-hundred-year-old swan, a tree that grew small animals in place of leaves, a woman who vomited a cat, and a monstrous thirteen-footed pig, there appeared the following account:

> *Surprising Machine of the artificial Man of Reyselius, Esquire.* This wise *Physicien*, to demonstrate the circulation of the blood, composed a statue with such similarity & resemblance to man in all internal parts that, except for the operations of the rational soul, one sees in it all that happens in our bodies, & this by the principles of Physics-Hydrostatics. The Author hopes to perfect it to the point of giving it a voice & natural motion. This artificial man has its vessels & viscera in the same shape, structure & size as natural man, such that the water or whichever other liquid one likes, swallowed through the mouth, falling from there through a canal as through the esophagus, soon goes from the stomach to the right ventricle of the heart, after having passed through a marvelous artifice in the intestines, the Pancreas, the portal vein & all the

> other areas that we have explained in our XII. Journal of this year. Nevertheless the same liqueur, descending by the emulgent vessels, & having been filtered by the artificial Kidneys of the Machine, falls into the bladder, whence it discharges itself. One notices in this Machine the natural movement of the lungs the attraction & expulsion of air, in a word all the movements of the pulse, & all the others that are natural in man.[1]

A subsequent issue of the journal included an intriguing addendum: "We did not mention," it said, "that the coarsest part of the liqueur that circulates in this machine . . . goes out the by back of the machine as excrement, and the less coarse part goes out by the front as urine.[2]

Reisel's artificial man seemed to bring the hypothetical living "statues" of Descartes's *Traité de l'homme* from page to flesh. Leibniz was interested; indeed, he had already contemplated the idea. Two years earlier, he had witnessed an android running across the surface of the Seine and returned home with a "funny thought" regarding a new sort of spectacle, a display of "beautiful curiosities and especially machines." His ideas for such a spectacle, which filled several pages, included a race run by artificial horses and a "machine representing the human body," one just such as Reisel subsequently advertised.[3]

When a correspondent wrote to ask, "What man is this Mr. Reyselius, then?" Leibniz responded with the assurance that Reyselius was a doctor and a "man of judgement and experience," but the machine seemed to be only a "mediocre" model of animal movement.[4] Reports of actual animal- and man-machines, built neither for churches nor to amuse but for specifically philosophical purposes, were multiplying. Leibniz's own teacher, Erhard Weigel, a professor of mathematics at the University of Jena, earned a mention in the *Journal des sçavans* of the same year for his spring-driven Bronze Horse, covered with real horse skin, which had "a movement strong and continuous enough to make it go in one autumn day four German miles, that is to say, 8 French leagues, providing it is in a flat Country."[5]

A new breed of machines emerged toward the end of the seventeenth century, developing through the eighteenth and into the early nineteenth. Clockmakers and inventors, and their philosophical audiences, turned theories about animal- and human-machinery into an experimental program: they built androids and automata that reproduced as closely as possible their natural subjects.[6] These machines, like Reisel's artificial man, belonged as much to Leibniz as to Descartes, in the sense that they seemed active, self-moving, and imbued with inner agency. They were musicians, artists, writers, and chess players. They twisted and turned, enfolded and unfolded, as much as they ticked and tocked. And in so doing, they replicated animal and human life from the earthiest bodily functions to the loftiest expressions of mind and spirit, testing the relations of the lofty to the earthy, purpose to mechanism.

These experimental machines arose in close conjunction with a second new breed: industrial automata, notably the automatic loom. The same key people and devices were instrumental in each development. The struggle between competing brute- and active-mechanist sciences of life held momentous economic and social stakes, to accompany the philosophical, theological and scientific ones. We have witnessed this struggle as it took place in cabinets, churches, and libraries; in physical and physiological theory and experimentation; and in religious conflict and iconoclastic violence. In this chapter we will see the same struggle enter workshops and factories, economic policy making, and industrial reform.

The First Androids

The word "android," derived from Greek roots meaning "manlike," was the coinage of Gabriel Naudé, French physician and librarian, personal doctor to Louis XIII, and later architect of the forty-thousand-volume library of Cardinal Jules Mazarin. Naudé was a rationalist and an enemy of superstition. In 1625 he published a defense of Scholastic philosophers to whom tradition had ascribed works of magic. He included the thirteenth-century Dominican friar, theologian, and philosopher Albertus Magnus (Albert the Great), who, according to legend, had built an artificial man made of bronze.[7]

This story seems to have originated long after Albert's death with Alfonso de Madrigal (also known as El Tostado), a voluminous commentator of the fifteenth century, who adapted and embellished the tales of moving statues and talking brazen heads in medieval lore.[8] El Tostado said that Albert had worked for thirty years to compose a whole man out of metal. The automaton supplied Albert with the answers to all of his most vexing questions and problems and even, in some versions of the tale, obligingly dictated a large part of Albert's voluminous writings. The android had met its fate, according to El Tostado, when Albert's student, Thomas Aquinas, smashed it to bits in frustration, having grown tired of "its great babbling and chattering."[9]

Naudé did not believe in Albert's talkative statue. He rejected it and other tales of talking automaton heads as "false, absurd and erroneous."[10] The reason Naudé cited was the statues' lack of equipment: being altogether without "muscles, lungs, epiglottis, and all that is necessary for a perfect articulation of the voice," they simply did not have the necessary "parts and instruments" to speak reasonably.[11] Naudé concluded, in light of all the reports, that Albert the Great probably had built an automaton, but never one that could give him intelligible and articulate responses to questions. Instead, Albert's machine must have been similar to the Egyptian statue of Memnon, much discussed by ancient authors, which murmured agreeably when the sun shone upon it: the heat caused the air inside the statue to "rarefy" so that it was forced out through little pipes, making a murmuring sound.[12]

Despite disbelieving in Albert the Great's talking head, Naudé gave it a powerful new name, referring to it as the "android."[13] Thus deftly, he smuggled a new term into the language, for according to the 1695 dictionary by the French philosopher and writer Pierre Bayle, "android" had been "an absolutely unknown word, & purely an invention of Naudé, who used it boldly as though it were established."[14] It was a propitious moment for neologisms: Naudé's term quickly infiltrated the emerging genre of dictionaries and encyclopedias. Bayle repeated it in the article on "Albert le Grand" in his dictionary.[15] Thence, "android" secured its immortality as the headword of an article—citing Naudé and Bayle—in the first volume of the supplement to the English encyclopedist Ephraim Chambers's *Cyclopaedia*.[16] In denying the existence of Albert's android, Naudé had given life to the android as a category of machine.

But the first actual android of the new, experimental-philosphical variety for which the historical record contains rich information—"android" in Naudé's root sense, a working human-shaped assemblage of "necessary parts" and instruments—went on display on February 3, 1738. The venue was the opening of the annual Saint-Germain fair on Paris's Left Bank. This android differed crucially from earlier musical automata, the figures on hydraulic organs and musical clocks, in that it really performed the complex task it appeared to perform, in this case, playing a flute, rather than merely making some suggestive motions. The device was, in this sense, a novelty, but it must have looked familiar to many of the fairgoers, being modeled on a well-known statue that stood in the entrance to the Tuileries Gardens and that is now at the Louvre Museum: Antoine Coysevox's *Shepherd Playing the Flute* (see figure 4.1).

Like the statue, the android represented a faun, half man and half goat. The mechanical faun, like the marble one at the Tuileries, held a flute. The second

Figure 4.1 *Faune jouant de la flûte*, statue by Antoine Coysevox, © Musée du Louvre, Dist. RMN-Grand Palais / Philippe Fuzeau / Art Resource, New York.

faun, though, became suddenly animate and began to play its instrument, executing twelve tunes in succession. At first, skeptical spectators were persuaded this must be a music box, with an autonomous mechanism inside to produce the sound, while the external figure merely pretended to play. But no, the android actually did play a real flute, blowing air from its lungs (three sets of bellows), and exercising flexible lips, a supple tongue, and soft, padded fingers with a skin of leather. It was even reported that one could bring one's own flute, and the machine would oblige by playing that one too.[17]

The flute-playing android was the work of an ambitious young engineer named Jacques Vaucanson. The last of ten children of a Grenoble glove maker, Vaucanson had been born in the bitterly cold winter of 1709, at the waning of Louis XIV's long reign, in the midst of a terrible famine and the bloodiest year of a war that France was losing. Emerging from this dark moment, Vaucanson's life and the Enlightenment would take shape in tandem, and his work would become a point of reference for the world of letters.

As a child, he had liked to build clocks and repair watches. While a schoolboy, he had begun designing automata. After a brief stint as a novice in Lyon, ending when a church dignitary ordered Vaucanson's workshop destroyed, he had come to Paris at the age of nineteen to seek his fortune. Thinking he might train as a doctor, he had attended some courses in anatomy and medicine, but had soon decided to apply these studies to a new area of research: re-creating living processes in machinery. The Flutist was the result of five years' labor.[18] When it was finished, Vaucanson submitted a memoir explaining its mechanism to the Paris Academy of Sciences. This memoir contains the first known experimental and theoretical study of the acoustics of the flute.[19]

Following an eight-day debut at the Saint-Germain fair, Vaucanson moved his android to the Hôtel de Longueville, a gilded hall in a grand sixteenth-century mansion at the center of the city. There it attracted about seventy-five people a day, each paying a hefty entrance fee of three livres (roughly an average week's wages for a Parisian worker). Among its audience were the members of the Paris Academy of Sciences, who traveled as a body to the Hôtel de Longueville at the behest of their president, the Cardinal Fleury, to witness the android Flutist.[20] Greeting his public in groups of ten or fifteen, Vaucanson explained the Flutist's mechanism and then set it to play its concert.

The reviews were effusive. "All of Paris is going to admire . . . the most singular and agreeable mechanical phenomenon perhaps ever seen," wrote one reviewer, emphasizing that the android "really and physically plays the flute."[21] The music-making statue, another agreed, was "the most marvelous piece of mechanics" that had ever been.[22] The abbé Pierre Desfontaines, a journalist and popular writer, advertising Vaucanson's show to readers of his literary journal, described the insides of the Flutist as containing "an infinity of wires and steel chains . . . [which] form the movement of the fingers, in the same way as in living man, by the dilation and contraction of the muscles. It is doubtless the knowledge of the anatomy of man . . . that guided the author in his mechanics."[23]

In the article "Androïde" in the monumental *Encylopédie*, a universal compilation of knowledge edited by the philosopher and writer Denis Diderot and the mathematician and philosopher Jean d'Alembert, Vaucanson's mechanical Flutist became the paradigm of an android. The article, written by d'Alembert, defines an android as a human figure performing human functions, and virtually the whole piece is devoted to the Flutist.[24]

Soon after the Academy of Science members came to the Hôtel de Longueville, Vaucanson returned the visit to read a memoir on the design and function of his Flutist.[25] The android's mechanism was moved by weights attached to two sets of gears. The bottom set turned an axle with cranks that powered three sets of bellows, leading into three windpipes, giving the Flutist's lungs three different blowing pressures. The upper set of gears turned a cylinder with cams, triggering a frame of levers that controlled the Flutist's fingers, windpipe, tongue, and lips. To design a machine that played a flute, Vaucanson had studied human flute players in minute detail. He had devised various ways of transmitting aspects of their playing into the design of his android. For example, to mark out measures he had had a flutist play a tune while another person beat time with a sharp stylus onto the rotating cylinder.[26]

Vaucanson's memoir on the Flutist begins with a theory of the physics of sound production in the flute, the first known such theory. Vaucanson's idea was that the pitch of a note depended upon three parameters: blowing pressure, the shape of the aperture, and the sounding length of the flute damping the vibrations, which was determined by the player's finger positions. Vaucanson wanted to test the influence of these three parameters on pitch: his Flutist was an acoustical experiment. He told the members of the Academy of Sciences that he had investigated the "Physical Causes" of the modification of sound in the flute "by imitating the same Mechanism in an Automaton."[27]

Vaucanson explained that he had chosen the flute because it was unique among wind instruments in having an "undetermined" aperture, which depended upon the position of the player's lips and their situation with respect to the flute's hole. This made flute playing subject to an "infinity" of variations, which he claimed to approximate using only four parameters. The lips could open, close, draw back from the flute's hole (to approximate tilting the flute outward), and advance toward the hole (to approximate tilting the flute inward).[28]

Vaucanson was able to produce the lowest note by using the weakest blowing pressure, further attenuated by passing through a large aperture and damped by the flute's full sounding length. The higher notes and octaves resulted from stronger blowing pressures, smaller apertures, and shorter sounding lengths. These results confirmed his hypothesis that the three parameters together—blowing pressure, aperture, and sounding length—governed pitch.[29]

The following winter, Vaucanson added two more machines to the show (see figure 4.2). One was a second android musician, a life-size Provençal shepherd that played twenty minuets and other dance tunes on a pipe grasped in its left hand, while accompanying itself with its right on a drum slung over its shoulder.[30] The Piper was also an acoustical experiment, and Vaucanson chose the pipe too because it posed a challenge to mechanical imitation. The pipe, unlike the flute, had a fixed aperture, but it had only three holes, which meant that the notes were produced almost entirely by the player's variations of blowing pressure and tongue stops. Working to reproduce these subtleties in his automaton, Vaucanson found that human pipers employed a much greater range of blowing pressures than they themselves realized and emphasized the enormous labor involved in producing each one by an arrangement of levers and springs.

The Piper yielded a surprising discovery. Vaucanson had assumed that each note would be the product of a given finger position combined with a particular blowing pressure, but he discovered that the blowing pressure for a given note depended upon the preceding note, so that, for example, it required more pressure to produce a D after an E than after a C, obliging him to have twice as many blowing pressures as notes.[31] The higher overtones of the higher note resonate more strongly in the pipe than the lower overtones of the lower note; but pipers themselves were not aware of compensating for this effect, and the physics of overtones was explained only in the 1860s by Hermann von Helmholtz.[32] Like the Flutist, the Piper was a successful experiment: it yielded a new result, in this case unpredicted by theory.

These were lifelike machines in a new sense. The android musicians did not just make music, a feat that music boxes had achieved for more than two centuries, but they did so using flexible lips, moving tongues, soft fingers, and swelling lungs. The machines also did well by their maker. They won praise from the dean of the Enlightenment: Voltaire celebrated Vaucanson as "Prometheus's rival" and later persuaded his young patron, the newly ascended Frederick II of Prussia, to invite Vaucanson to join his court in Berlin (Vaucanson declined).[33] He

Figure 4.2 The Flutist, the Piper, and the Duck from Vaucanson's *Mécanisme du fluteur automate* (1738), courtesy Department of Special Collections, Stanford Libraries.

became a regular at the salon of Madame de La Poupelinière, whose husband, a wealthy and philandering tax farmer,[34] was one of the century's leading patrons of music and the arts. There, Vaucanson rubbed shoulders with the likes not only of his admirer, Voltaire, but of Jean-Philippe Rameau, Georges Buffon, the Baron von Grimm, and the duc de Richelieu among others.[35]

That Vaucanson was an intimate in the La Poupelinière household is evident from the role that he and his mechanical genius played in the scene that ultimately tore that household asunder, at least according to the wag and litterateur, Jean-François Marmontel. Madame de La Poupelinière, at the time, was carrying on a much-noticed affair with the duc de Richelieu, mysteriously spiriting him in and out of her bedchamber. One day while she was out, her husband called Vaucanson in to investigate her apartment. The crack *mécanicien* noticed that the back of the fireplace in the lady's harpsichord room was mounted on cunningly hidden hinges. Behind it was an opening in the common wall shared with the adjoining house, covered on the other side by a mirror. An associate of the duke, of course, rented the neighboring house. As Marmontel tells the story, Vaucanson's engineering skills far outpaced his tact in this instance. In his admiration of the clever arrangement, he clean forgot its purpose:

> Oh! Monsieur, cried he, turning toward La Poplinière [sic], what a beautiful piece of work I see there! and what an excellent workman was he who made it! This plate is moveable, it opens; but its hinges are of such delicacy!.....no snuff-box could be better worked. What a clever man this one is! –What! Sir, said la Poplinière, turning pale, you are sure that this plate opens? – Truly! I'm sure of it, I see it, said Vaucanson, ravished with admiration; nothing is more wonderful. – what good is your wonder to me? this is not a matter of admiration!– Ah! Sir, such workmen are very rare! I have good ones, assuredly; but I have not one who. . . .– Leave off with your workmen, interrupted la Poplinière, and let one be sent for who can force this plate.—It's a shame, said Vaucanson, to destroy a masterpiece as perfect as that.

"As for Vaucanson," was Marmontel's verdict, "all his mind was in his genius; and outside of mechanics, no one could be more ignorant or limited than he."[36] It is a measure of the cultural importance of Vaucanson's androids that, among the salon regulars, they could trump even a total lack of wit.

Marmontel himself had occasion to make use of Vaucanson's talents when he needed some special effects for his new play, *Cléopatre*. "Vaucanson agreed to make me an automaton asp who, at the moment when Cleopatra pressed him to her bosom to provoke his bite, imitated almost to perfection the natural movement of a living asp." But the asp was cut from the cast after the first performance. According to Marmontel, it diverted the audience's attention from the dramatic interest of this moment in the play.[37] Others told a different story. The asp, in its realism, slithered and even hissed. In the foyer after the opening performance, the marquis de Louvois, upon being asked how he had liked the play, devastatingly replied, "I, sirs, am of the opinion of the asp."[38]

Vaucanson's androids did not only open the houses (cabinets, closets, chimneys, and stages) of the rich and powerful to him. They also helped him to secure a much-coveted appointment to the Paris Academy of Sciences as "associated mechanician" in 1757 (a contest in which he beat out the philosophe and master planner of the Enlightenment, Denis Diderot).[39] And the famous machines inspired imitations, generating a venture that quickly seized the imaginations of engineers, philosophers, noble patrons, and paying audiences: to reproduce as closely as possible the functions and behaviors of living beings in machinery. Over the course of the century, this venture would yield machines that (actually or apparently) ate, shat, bled, breathed, cavorted, walked, talked, swam, made music, drew, wrote, and played an almost unbeatable game of chess.

The last was the achievement of a Hungarian engineer named Wolfgang von Kempelen who had been hired at the age of twenty-one by the Empress Maria Theresa to serve at the court of the Holy Roman Empire in Vienna. In 1769, for the amusement of his patroness, Kempelen built an android Turk that played an expert game of chess (see figure 4.3). The Turk was exhibited across Europe and America by Kempelen himself and then by others through 1840.[40]

The android did not merely play human opponents, but it also generously corrected their mistakes, and in the course of its long career it bested Frederick the Great, Benjamin Franklin, Napoleon, and Charles Babbage. Napoleon reportedly made several deliberately false moves to test the automaton's reactions. The first time, as was its habit, it corrected the emperor's mistake and then made its own move in return. When Napoleon again made a false move, the Turk removed the wrongly moved piece from the board altogether. The third time, having apparently lost its patience, it swept all the pieces to the floor and refused to continue playing.[41]

In addition to playing chess, Kempelen's android could perform a Knight's Tour[42] and respond to questions from the audience, spelling out its answers by pointing to letters on a board.[43] As it happened, the Turk's motions were

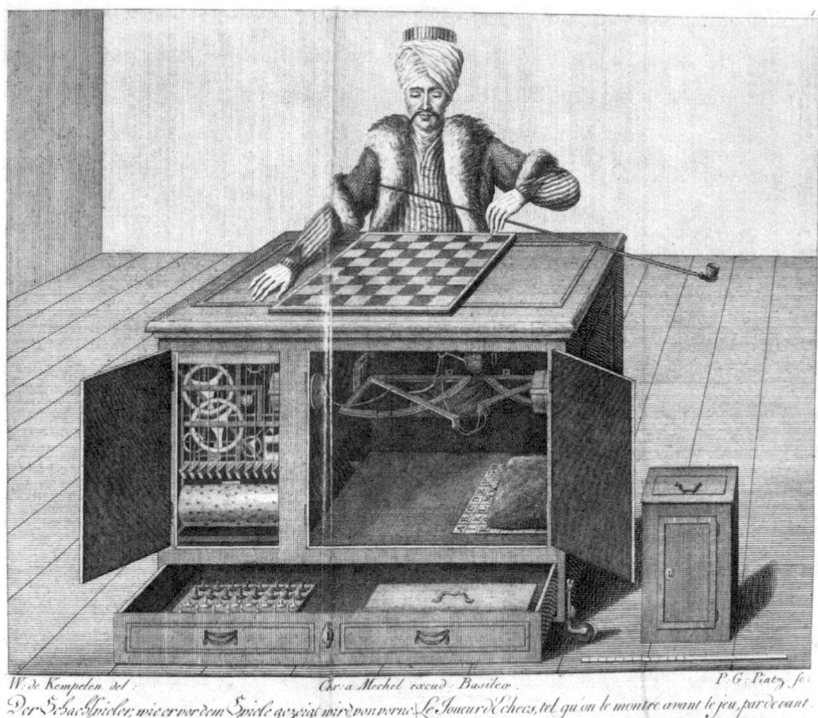

Figure 4.3 Wolfgang von Kempelen's Chess-Playing Turk, from Windisch, *Lettres sur le joueur d'echecs de M. de Kempelen* (1783), courtesy Department of Special Collections, Stanford Libraries. Photo by Andrew Schupanitz.

secretly directed by human chess players ingeniously concealed in its pedestal. Although this fraud was not established until the middle of the following century,[44] Kempelen himself spoke deprecatingly of the Turk as a mere "bagatelle" and even insisted that his major achievement in it had been to create an "illusion."[45] On one occasion, he jokingly apologized to a spectator on behalf of his automaton, explaining that it had been badly jostled in transit and consequently played a very poor game. Kempelen promised to have the automaton working properly again within eight or ten days. "I fear only that to eyes such as yours, this little invention will not have the effect it had upon plenty of people who, out of friendship or indulgence, have found more of the marvelous in it than is actually there. In a word, you will see a little bit of mechanics and a little bit of charlatanry, just between us."[46] Yet these caveats did not detract from the machine's fascination, which was fueled by a growing interest in the question whether a machine could cross the Cartesian divide, that is, whether intelligent mental processes could be reproduced in artificial machinery.

The Turk, in its fraudulence, dramatized this question and provided a context for arguing both sides. Arriving in London in the winter of 1783, it riveted the attention of the city, rivaling Edmund Burke's attempt to reform the East India Company, the first Montgolfiers, and the star dancer of the Paris Opera: "The winter is not dull or disagreeable," Horace Walpole wrote to his friend, the Earl of Strafford, in 1783. "The India Bill, air-balloons, Vestris and the automaton, share all the attention."[47]

In 1784, a friend of Kempelen's, Karl Gottlieb von Windisch, published an account of the Turk that rehearsed the arguments both for and against a thinking machine. In his account, which he entitled *Inanimate Reason,* Windisch wrote that the Turk did for "the understanding" what Vaucanson's Flutist did for the ear. At the same time, Windisch believed that the Turk was "a deception" and identified two separate "powers," a visible "*vis motrix*" (motive power) and a hidden "*vis directrix*" (directing power). Kempelen's ability to unite these two powers, Windisch deemed "the boldest idea that ever entered the brain of a mechanic."[48] Windisch's analysis of the Turk was picked up by later commentators and remained influential well into the following century.

Only one eighteenth-century critic rejected the idea of a chess-playing automaton on the grounds that such a thing would be impossible. He was the writer, polemicist, and army officer Philip Thicknesse, known as "Dr. Viper" because of his reputation as a blackmailer.[49] Thicknesse was outraged by "Foreigner[s]" who charged "HALF A CROWN a piece admittance" and twice that sum to witness talking heads and automaton chess players. He fumed:

> That the human voice may be imitated, and many, or most words, articulated by valves, and bellows, like the barrel organ, there is no doubt; but that a mechanical figure can be made to answer all, or any such questions, which are put to it, or even to put a question, is UTTERLY IMPOSSIBLE. That an AUTOMATON may be made to move its hand, its head, and its eyes in certain and regular motions, is past all doubt: but that an AUTOMATON can be made to move the Chessmen properly, as a sagacious Player, in consequence of the preceding move of a stranger, who undertakes to play against it, is also UTTERLY IMPOSSIBLE.[50]

Another critic doubted that "simple springs" could accomplish the work of a "well-organized brain."[51] A third thought a responsive machine would have taken more than six months (the time Kempelen claimed to have taken) to build.[52] But for the most part even the critics were willing to entertain the possibility of a chess-playing machine.

In 1819, the British mathematician and engineer Charles Babbage encountered von Kempelen's Chess Player at Spring Gardens in London, where it was being exhibited. Babbage watched the android play a game, which it won, and took careful notes on its performance in the margins of a copy of Windisch's pamphlet. Babbage noted that the automaton's hand and arm movements were inelegant, but that it played "very well and had a very excellent game in the opening." He returned the following year to play the Turk, which beat him, as mentioned, in about an hour. On this occasion, Babbage found that the android played "very cautiously."[53] A reviewer of the Turk's performance wrote that the machine displayed "a power of invention as bold and original, as any that has ever been exhibited to the world," but also thought it was directed by "some human agent."[54]

Although Babbage did not believe his opponent was fully mechanical, he was nevertheless extremely interested in the possibility that a machine could perform complex calculations and play games of skill—in short, could think. Other animal-machines and androids of the preceding century had prepared his interest in this idea during his childhood. Around 1800, as a child of eight, Babbage had gone with his mother to visit the Mechanical Museum in Hanover Square, London, run by the inventor John Joseph Merlin.

There, for three shillings, Babbage had seen a sampling of the artificial creatures of the late Enlightenment: an automaton bat, an android Turk "who will chew and swallow an artificial stone; which any of the Company may put in his mouth," and above all, a foot-high silver lady who danced and "attitudinized in the most fascinating manner," her eyes "full of imagination, and irresistible," while holding a fluttering bird on her right forefinger. Many years later, Babbage would buy the silver dancer, restore her, commission a new dress for her, and install her in his drawing room.[55]

After his match against Kempelen's chess-playing Turk, Babbage thought about designing a game-playing machine. He was sure the Turk was a trick, but he believed it *could* have been genuine, and indeed that "every game of skill is susceptible of being played by an automaton." For a time, he toyed with the idea of building one. He imagined a machine consisting of two figures of children playing a game, accompanied by a lamb and a rooster. The winning child would clap his hands while the cock crowed, after which the losing child would cry and wring his hands while the lamb bleated. Babbage abandoned the project.[56] But sometime later, he devised a party trick to show how machinery might demonstrate what looked like willful, responsive, or unpredictable behavior. He set his mechanical calculator prototype (about which more follows presently) to print a series of integers from one to a million, with occasional discontinuous leaps forward by ten thousand, surprising spectators at each discontinuous leap. Willful behavior, the trick implied, was all in the eye of the spectator and was nothing but unforeseen behavior.[57]

Johann Maelzel, an inventor and instrument maker, collaborator of Beethoven, and best known to posterity as the author of the metronome, was the owner and demonstrator of the Turk when Babbage played it, having purchased it for 10,000 francs from von Kempelen's son after the engineer's death. Maelzel discontinued the question-and-answer part of the performance, which he thought would be unbelievable to contemporary audiences.[58]

Now, indeed, the critical commentaries all focused on the Turk's ability to respond. In 1821, Robert Willis, the twenty-year-old son and grandson of royal physicians, later to be a professor of applied mechanics at Cambridge, wrote a commentary on the Turk ruling out the possibility of a responsive machine. The Chess Player could not be a mere mechanism, he wrote, "however great and surprising the powers of mechanism may be," because the physical parts of a mechanism and their movements were "necessarily limited and uniform," whereas the circumstances of a game of chess were endlessly variable. Like Descartes, Willis judged that to be able to confront an indefinitely variable situation was "the province of intellect alone."[59] In the following decades, Willis's assessment was frequently cited.[60] It pursued the Turk across the ocean, where an American commentator agreed that "the interposition of some intelligence, at each seperate [*sic*] move, must be admitted."[61]

In America too the Turk received a further critique from the pen of Edgar Allan Poe, who encountered the android in Richmond, Virginia.[62] In 1839, Poe published a refutation of the Turk in which he reproduced Willis's argument that a machine could never play chess, since no machine could react to varying external conditions. The key point for Poe, as for Willis, was that "no one move [in chess] . . . necessarily follows upon any other." Even if the android's own moves were scripted, he had continually to confront "the indeterminate will of his antagonist." The machine's ability to respond to these made it "quite certain that the operations of the Automaton are regulated by *mind*, and by nothing else."[63]

In 1844, the French magician and inventor Jean-Eugène Robert-Houdin took his turn against Kempelen's chess-playing Turk. In his memoirs, Robert-Houdin spun a fictitious and catchy story featuring a mutinous Polish officer in the Russian army named Worousky, who had lost his legs during a rebellion. Worousky's skill at chess and diminished stature, the story went, had given Kempelen the idea of building a fake chess "automaton" and concealing Worousky inside.[64] The tale, though entirely false, had the legs its protagonist lacked. It inspired plays and novels throughout the nineteenth and early twentieth centuries, including Henry Dupuy-Mazuel's *Le joueur d'échecs* (1926), adapted the following year in the film by the same title starring Pierre Blanchard and Charles Dullin, and is reproduced in the 1911 edition of the *Encyclopaedia Britannica*.[65]

By the mid-nineteenth century, the consensus was that if a machine exhibited responsive agency of the sort needed to play chess, there must be a Worousky in the works. But over the previous century, Kempelen's chess-playing Turk, like Vaucanson's musician androids, had challenged audiences to consider whether a purely mechanical entity could achieve rational mental functions. The machines performed Descartes's thought experiment and extended it to include the functions of the rational soul.

Today, one can visit the best surviving representatives of this tradition, three of the most extraordinary androids of the eighteenth century. They hold court in a small theater in the Musée d'art et d'histoire in Neuchâtel, Switzerland.[66] Listening to the curator give his greeting, one is distracted by the conviction that the almost 250-year-old figure behind him is breathing. Otherwise she seems quite still: her eyes stare straight ahead, her fingers remained poised on the clavichord before her, the folds of her dress and the curls of her hair lie undisturbed. But surely, one thinks, her chest is rising and falling, isn't it? The gentle, barely perceptible, rhythmic motion makes her stillness seem eerily animate.

In fact, the lady was designed to breathe for an hour before and after each of her performances. For two and a half centuries, spectators filing into her presence have done the same uneasy double take. As the Musicienne (see figure 4.4 and plate 5) was being built, in 1772, a skeptic observed that although Vaucanson had demonstrated the motions of music making to be mechanically reproducible, "I defy M. de Vocanson [*sic*] and all the machinists on earth to make an

Figure 4.4 The Jaquet-Droz Musicienne, courtesy Musée d'art et d'histoire de Neuchâtel.

artificial face that expresses the passions, because to express the passions of the soul, one must have a soul."[67] Yet the Musicienne, bending her head with grace, following her fingers with her eyes as she played, her breast heaving in time to the music with apparent emotion, seemed to suggest just the opposite.[68]

The unsettling musician is flanked on either side by two young brothers:

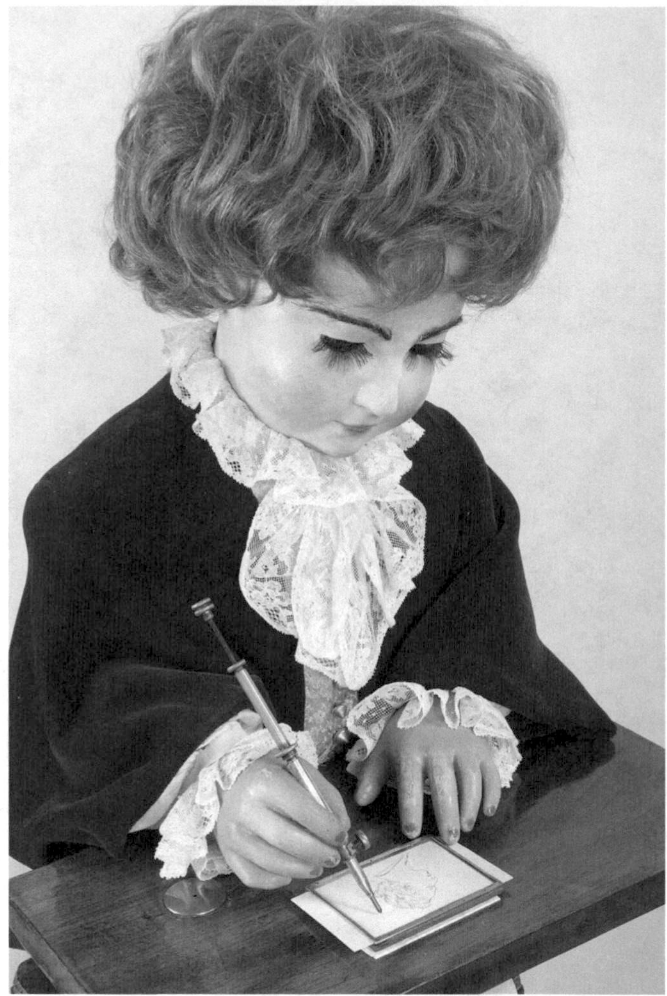

Figure 4.5 The Jaquet-Droz Dessinateur, courtesy Musée d'art et d'histoire de Neuchâtel.

rosy, toddler-size boys in fetching velvet suits, seated at desks with small tablets of paper(see figure 4.5 and plate 6). One of the young boys leans over his table and uses a bit of charcoal to sketch several pictures. Like the Musicienne, the Draughtsman also breathes, occasionally blowing the charcoal dust from his paper to survey his work. The other child dips his pen in a bottle of ink and writes a message of several lines in an elegant, eighteenth-century script. The Writer is the first known programmable machine in modern terms: it can be arranged to write any message of up to forty characters.[69]

Watching the letters form is unsettling in a different way from seeing the rising and falling chest of the musician, like hearing a voice echo down the centuries. For, although the ink is fresh, the handwriting is that of a man who has been dead for over two hundred years, a clockmaker named Pierre Jaquet-Droz. With the help of his son, Henri-Louis, and his adopted son, Pierre Leschot, Jaquet-Droz built the three machines in his workshop in La Chaux-de-Fonds in 1774, using springs to drive them and gears and cams to program them.[70] They used lifelike materials such as leather, cork, and papier-mâché to give their machines the softness, lightness, and pliancy of living things. They also likely designed the machines' hands with the help of the village surgeon, modeling their skeletal structures on real human hands.[71]

Current visitors to the androids join a two-and-a-half-century tradition. When they first went on display in the Jaquet-Droz workshop, people flocked "from everywhere," reported one witness, "as on a pilgrimage. The garden and the main road were each day filled with carriages, and the rains deterred only a few." Beginning at six in the morning and continuing until eight in the evening, the Jaquet-Droz family, assisted by two workers, demonstrated their automata for the crowds of spectators. All the local nobles of the surrounding areas came to see the show, as well as the "bailiffs of the cantons with their families" and even "the ambassador of France himself came incognito."[72]

The Jaquet-Droz automata, like their predecessors, went on to tour the cities and courts of Europe; they had an audience with Louis XVI and Marie Antoinette, whose portrait the Draughtsman obligingly sketched.[73] (According to another version of the story, the Draughtsman was supposed to sketch the profile of the late Louis XV, but Leschot in his nervousness set up the machine wrong, so that it sketched "mon toutou," a little dog, instead.)[74] Vaucanson himself went to see and admire the Jaquet-Droz androids when they were in Paris.[75]

While in Paris, Henri-Louis Jaquet-Droz received a commission from a wealthy tax farmer that gave the family an occasion to extend their idea that an artificial machine could have an organic, lifelike texture. The commission was to design two artificial hands for the man's son, Alexandre Balthazar Laurent Grimod de La Reynière, soon to achieve fame as a gastronome and notorious eccentric (who would plan his own ostensible funeral dinner and then startle his guests by arising mid-meal from his coffin).[76] Grimod de La Reynière's own hands were congenitally deformed, a fact that his parents had creatively hidden. They had had their son baptized in secret (hence his birth certificate lacked the noble cachet, a stroke of luck that likely saved him from the guillotine during the Revolution) and cloistered him away in childhood.

To those who did glimpse his clawlike hands, the parents had offered various explanations, such as that the boy had fallen into a pigpen and been chewed by its greedy occupants. To Jaquet-Droz, Grimod de La Reynière senior claimed less sensationally that his son had lost his hands in a hunting accident. The result was a pair of prostheses made from the same materials Jaquet-Droz had used in his automata: leather, cork, parchment, and papier-mâché on a steel frame. The mechanical hands, which Grimod de La Reynière always clothed in white gloves, were very light, about 480 grams, and reportedly quite versatile, enabling their wearer to write and draw. The Jaquet-Droz operation continued to design prosthetic hands and arms of this sort through the 1790s. Vaucanson, upon seeing Grimod de La Reynière's pair, reportedly remarked to Jaquet-Droz fils "Young man, you begin where I would have liked to finish."[77]

Vaucanson himself was by then a celebrity. In the early 1770s, when Diderot and his fellow materialist, the polemical Claude-Adrien Helvétius, got into a dispute over just how great men achieve greatness, their first example of greatness was Vaucanson.[78] Meanwhile, well-prepared audiences awaited the next mechanical-philosophical creature. When the Jaquet-Droz machines were in Paris in 1775, the writer and *salonnière* Julie de Lespinasse, then in the final throes of a fatal passion for the comte de Guibert (she did die soon afterward), wrote to the author of her suffering, "Midnight, 1775 . . . Before dinner I will go see rue de Cléry some automata, which are prodigious, they say." Weary of going about in society, she wrote, she nevertheless thought the androids worth a visit since "they act and don't speak."[79] Very soon, however, there would be chattering androids too.

Defecating Ducks and Talking Heads

The first androids, then, seemed anything but clocklike in the traditional sense of the term. They seemed clocklike, rather, in Leibniz's sense: unquiet, restless, visceral, active, responsive. A certain kind of responsiveness to its environment, or at least the appearance of a certain kind of responsiveness, characterized Vaucanson's third automaton, which, unlike its confrères, was not a musician, and did not exercise any skill. In the frontispiece to the brochure for Vaucanson's 1738 exhibit, this automaton stands modestly on its pedestal with the baroque figures of the two musicians posed ornately on either side.

It was a duck, and what the Duck did, though unremarkable in a duck, was so extraordinary in a machine that it immediately seized center stage. Like Reisel's artificial man and certain other machines had been purported to do—but this time in live performance—the Duck shat (see figure 4.6). It did so, appropriately, in response to being presented with a meal. First it gobbled up and gulped down some bits of corn and grain; then it took a pregnant pause; and at last it relieved itself, through its tail end, of an authentic-looking burden. The Duck also did other duckish things—flapping, splashing, ruffling—but its main attraction, drawing people from all over Europe to see it, was its final malodorous delivery.[80]

Vaucanson advertised his Duck as transparent: its gilded copper feathers were perforated to allow an inside view. Wittily, he observed that although some "Ladies, or some People, who only like the Outside of Animals, had rather have seen . . . the Duck with feathers," his "Design [had been] rather to demonstrate the Manner of the Actions, than to shew a Machine." The Duck was powered by a weight wrapped around a lower cylinder, which drove a larger cylinder above it. Cams in the upper cylinder activated a frame of about thirty levers. These were connected with different parts of the Duck's skeletal system to determine its repertoire of movements, which included drinking, playing "in the water with his Bill, and mak[ing] a gurgling Noise like a real living Duck" as well as rising up on its feet, lying down, stretching and bending its neck, and moving its wings, tail, and even its larger feathers.[81] Vaucanson carried out exhaustive studies of natural ducks and each wing on his artificial Duck contained over four hundred articulated pieces, imitating every bump on every bone of a natural wing.[82]

On the subject of the Duck's digestive and excretory processes, Vaucanson was suspiciously equivocal. On the one hand, he protested, "I don't pretend to give this as a perfect Digestion. . . . I hope no body would be so unkind as to upbraid me with pretending to any such Thing." On the other hand, he maintained that these processes were "copied from Nature," the food digested "as in real Animals, by Dissolution," this taking place in a "Chymical Elaboratory" in the Duck's stomach. These things, Vaucanson promised to demonstrate on "another occasion."[83] This postponement of further explanations aroused doubts. In 1755 a critic accused the Duck of being "nothing more than a coffee-grinder."[84] At the time of Vaucanson's death in 1782, the marquis de Condorcet, as perpetual secretary of the Académie des sciences, was charged with writing his eulogy and delicately intimated that he did not believe the Duck really shat either: he imagined nature was inimitable in this regard.[85]

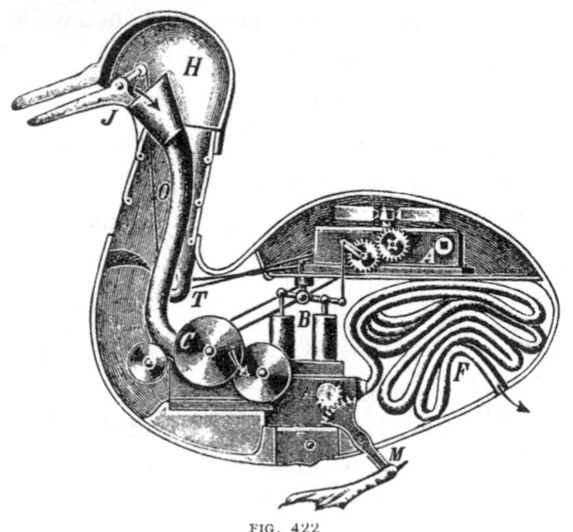

Figure 4.6 The top image is a much-reproduced rendition of Vaucanson's Duck that originated in the *Scientific American* issue for January 21, 1899. The accompanying description bears as little resemblance to Vaucanson's actual Duck as does the illustration. However, an arrow helpfully indicates where the main action takes place. The bottom image is one of a mysterious set of photographs discovered around 1950 by the curator of the Musée des arts et métiers in Paris. The photographs were in a folder left by his predecessor, labeled "Pictures of Vaucanson's Duck received from Dresden." From Chapuis and Droz, *Automata* (1958), 233–38.

The following year, a close observer of the Duck's swallowing mechanism found that the food did not continue down the neck and into the stomach but rather stayed at the base of the mouth tube. Reasoning that digesting the food by dissolution would take longer than the brief pause that the Duck took between swallowing and expulsion, this observer concluded that the grain input and excrement output were entirely unrelated and that the tail end of the Duck must be loaded before each act with phony excrement.[86]

Even though fraudulent, the Duck enacted the possibility that a machine could eat, digest, and defecate. It posed a question people were eager to think about in 1738: whether natural phenomena worked in essentially the same way as artificial ones. If a machine could defecate, a separation between life and art had been truly bridged. Five years before the Duck, in 1733, a mechanician named Maillard had presented to the Paris Academy of Sciences an "artificial swan" that paddled through the water on a paddle wheel while a set of gears swept its head slowly from side to side(see figure 4.7).[87] Vaucanson's Duck was a horse of an entirely different color. Not content to paddle, a feat any paddle wheel could achieve, the Duck pretended to perform an act that epitomized its animal nature. Its musical companions used fingers, lungs, lips, and tongue to play their instruments. Vaucanson took his project with a new literalness. Even when he cheated, his dishonesty was in the service of verisimilitude, not virtuosity: making the machine seem *lifelike* in the earthiest sense.

Vaucanson had twice before designed what he called "moving anatomies." He described the first as a machine containing "several automata, and in which the natural functions of several animals are imitated by the movement of fire, air and water." Little information remains regarding this first machine except that Vaucanson took it on a successful tour of France.[88] After the Flutist, Duck, and Piper, Vaucanson returned to his moving anatomy project, and in 1741 he presented to the Académie de Lyon a plan to build an automaton "whose motions will be an imitation of all animal operations, such as the circulation of the blood, respiration, digestion, the movement of the muscles, tendons, nerves and so forth." He meant to use the automaton to "carry out experiments on animal functions" in order to better understand "the different states of human health."[89]

This experimental machine was never finished, but more than twenty years later, Louis XV asked Vaucanson if he could build a mechanical model of the circulation of the blood.[90] Vaucanson made plans for a hydraulic model of the circulatory system, and the king approved his request to have the machine built

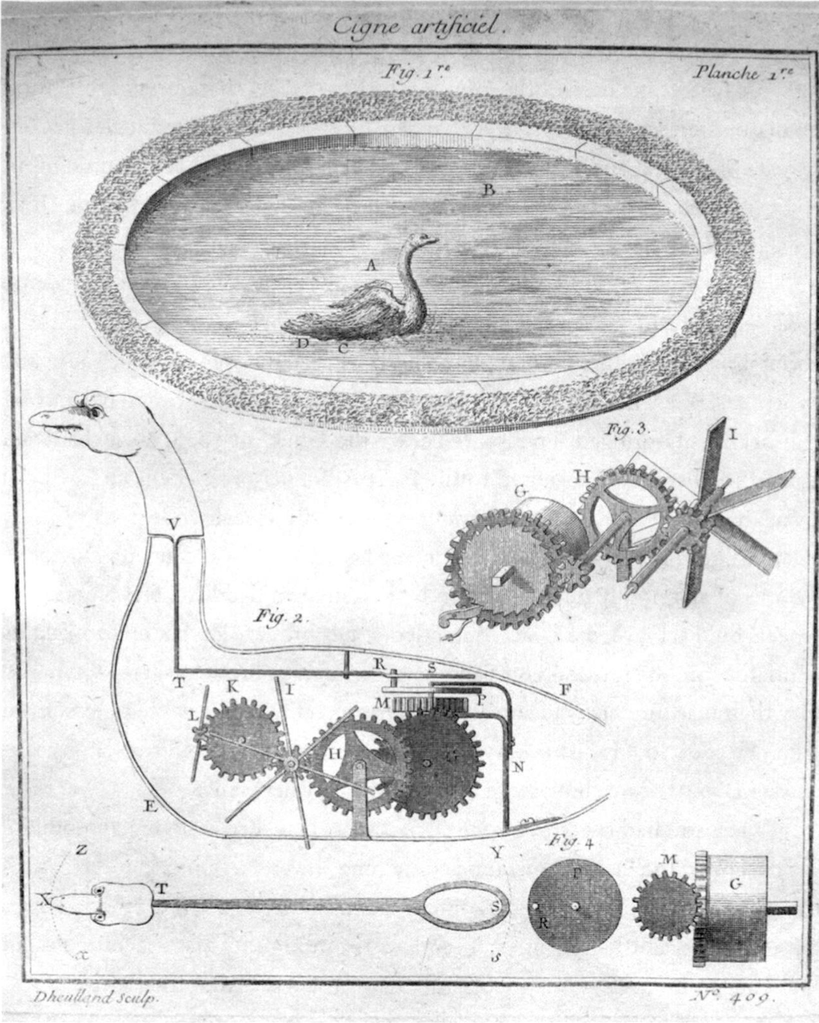

Figure 4.7 Maillard's artificial swan, courtesy Department of Special Collections, Stanford Libraries. Photo by Andrew Schupanitz.

in Guyana, where he proposed to use "elastic gum" (rubber) to make the veins. This project too was never completed, but the veins would have been the first flexible rubber tubes.[91] Vaucanson's moving anatomies were warm, heated by fire, their fluids coursing and airs rushing through their rubbery passageways.

Artificial machines that breathed and bled became commonplace.[92] During the 1730s and '40s, moving anatomies were a fad. The French surgeon and economic reformer François Quesnay designed one upon which to test medical therapies, in particular the therapy of bleeding. Quesnay believed, as had Descartes, that the source of motion in animal bodies was an extremely volatile fluid, variously called "animal spirits" or the "vital principle," distributed through the nerves. Physics could not account for the action of this fluid, Quesnay thought, without recourse to its particular, intrinsic property of activeness.[93] But he also appealed to the active properties of ordinary fluids, in particular their striking tendency to seek equilibria. He emphasized that bleeding did not diminish the amount of blood in a given vessel, because when a surgeon depleted the blood in one of a patient's blood vessels, an equal amount of blood came from branching vessels to replace it, and he persuaded himself of the truth of this purposeful activity of blood and other fluids by building a mechanical model of the circulatory system.[94]

In 1739, a competitor of Quesnay's, a fellow surgeon named Claude-Nicolas le Cat, published a description, now lost, of an "automaton man in which one sees executed the principal functions of the animal economy," circulation, respiration, and "the secretions."[95] It is not clear what became of this early project, but Le Cat returned to the idea in 1744 when, according to the proceedings of the Académie de Rouen, he read a sensational memoir there. A great crowd was assembled to hear it, and one witness reported, "Monsieur Le Cat told us of his plan for an artificial man. . . . His automaton will have respiration, circulation, quasi-digestion, secretion and chyle, heart, lungs, liver and bladder, and God forgive us, all that follows from it." Le Cat's idea, like Vaucanson's and Quesnay's, was that one could experiment on this automaton to test the effects of medical therapies. "Let him have a fever, we will bleed him, we will purge him, and he will but too much resemble a man.[96]

Le Cat's automaton man was to have "all the operations of a living man," including not only "the circulation of the blood, the movement of the heart, the play of the lungs, the swallowing of food, its digestion, the evacuations, the filling of the blood vessels and their depletion by bleeding," but also—apparently crossing the Cartesian boundary between mechanical body and rational soul—"even speech and the articulation of words."[97]

This idea, the possibility of simulating articulate speech, had generated a tradition of philosophical discussion over the preceding century. If some continued to find it a quixotic notion, it was in fact literally so: when Don Quixote himself encounters a talking bronze head (connected to a hidden human being), he is fully captivated by it, though his less suggestible squire, Sancho Panza, is unimpressed by its conversation.[98] Cervantes's contemporary, the

Spanish writer on magic, Martín del Río, also found it unreasonable to suppose "that an inanimate thing should produce the human voice and give answers to questions. For this requires life, and breath, and a perfect cooperation of the vital organs, and some discursive ability in the speaker."[99]

Several decades later, some if not all the items on del Río's list seemed possibly achievable in an artificial machine. Athanasius Kircher wrote in 1673, with regard to the legends of Albert the Great's talking head and the ancient Egyptian speaking statues, that while certain skeptics believed these devices must have been "either non-existent or fraudulent or constructed with the help of the devil," many others believed it was possible to build such a statue having throat, tongue and other organs of speech that would emit an articulated voice when it was activated by wind. Kircher included a sketch of a design for a talking figure (see figure 4.8).[100] His student, Gaspar Schott, also a prolific natural philosopher and engineer, adopted the same attitude, even alluding to a question-answering statue that Kircher was building for Queen Christina of Sweden.[101] No doubt the queen's previous philosophy teacher, Descartes, had interested her in the relations between rational speech and a mechanical body.

Although the idea of simulated speech was not new, around the middle of the eighteenth century, experimental philosophers and mechanicians took a renewed interest in it. They assumed that speech was a bodily function akin to respiration or digestion—they did not explicitly distinguish the rational from the physiological aspects of speaking—and even the skeptics expressed their skepticism in connection with physiological details rather than principled objections. In his effusive review of Vaucanson's Flutist in 1738, for example, the abbé Desfontaines predicted that articulate speech could never be produced in artificial machinery because the bodily process of speaking would remain impenetrably mysterious: one could never know precisely "what goes on in the larynx and glottis . . . [and] the action of the tongue, its folds, its movements, its varied and imperceptible rubbings, all the modifications of the jaw and the lips."[102] Speaking was an essentially organic process, Desfontaines reckoned, and could only take place in a living throat.

Desfontaines was not alone in this belief: in this period, skeptics about the possibility of artificial speech generally argued that the human larynx, vocal tract, and mouth were too soft, supple, and malleable to be simulated mechanically. Around 1700, Denys Dodart, personal physician to Louis XIV, presented several memoirs to the Paris Academy of Sciences on the subject of the human voice, in which he argued that the voice and its modulations were caused by

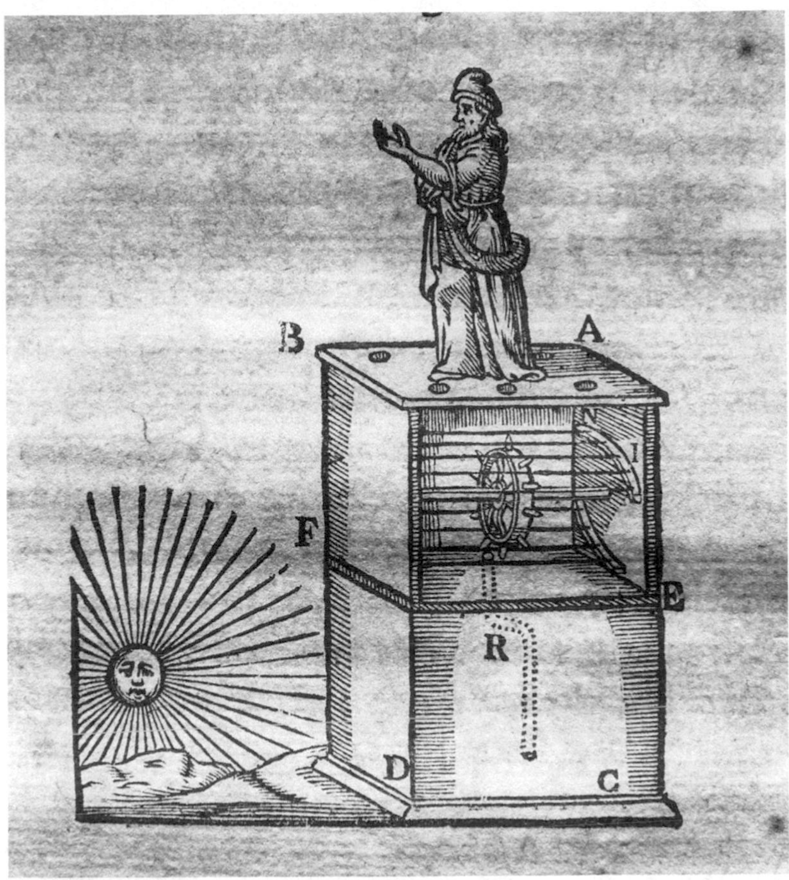

Figure 4.8 Athanasius Kircher's design for a speaking figure, from *Oedipus Aegyptiacus* (1652), courtesy Department of Special Collections, Stanford Libraries. Photo by Andrew Schupanitz.

constrictions of the glottis, and that these were "inimitable by art."[103] Fontenelle, who was then Perpetual Secretary of the Academy, commented that no wind instrument produced its sound by such a mechanism (the variation of a single opening) and that it seemed "altogether outside the realm of imitation. . . . Nature can use materials that are not at all at our disposal, and she knows how to use them in ways that we are not at all permitted to know."[104]

A last skeptic citing material difficulties was the philosopher and writer Antoine Court de Gébelin, who observed that "the trembling that spreads to all the parts of the glottis, the jigging of its muscles, their shock against the hyoid bone that raises and lowers itself, the repercussions that the air undergoes against the sides of the mouth . . . these phenomena" could only take place in living bodies.[105] On the other hand, there were plenty who disagreed. For example, the polemical materialist Julien Offray de La Mettrie took a look at Vaucanson's Flutist and concluded that a speaking machine "could no longer be regarded as impossible."[106]

During the last three decades of the century, several people took up the project of artificial speech. All of them assumed that the sounds of spoken language required a structure as similar as possible to the throat and mouth. This assumption, that a talking machine required simulated speaking organs, had not always dominated thinking about artificial speech. In 1648, John Wilkins, the first secretary of the Royal Society of London, had described plans for a speaking statue that would synthesize, rather than simulate, speech by making use of "inarticulate sounds." He wrote, "We may note the trembling of water to be like the letter L, the quenching of hot things to the letter Z, the sound of strings, to the letter Ng [*sic*], the jirking of a switch to the letter Q, etc."[107] But in the 1770s and '80s, builders of speaking machines mostly assumed that it would be impossible to create artificial speech without building a talking head: reproducing the speech organs and simulating the process of speaking.

The first to attempt such a machine was the English poet and naturalist Erasmus Darwin (grandfather of Charles Darwin) who in 1771 reported that he had "contrived a wooden mouth with lips of soft leather, and with a valve over the back part of it for nostrils." Darwin's talking head had a larynx made of "a silk ribbon . . . stretched between two bits of smooth wood a little hollowed." It said "mama, papa, map and pam" in "a most plaintive tone."[108]

The next to simulate speech was a Frenchman, the abbé Mical, who presented a pair of talking heads to the Paris Academy of Sciences in 1778 (see figure 4.9). The heads contained "several artificial glottises of different forms [arranged] over taut membranes." By means of these glottises, the heads performed a dialogue in praise of Louis XVI: "The King gives peace to Europe," intoned the first head; "Peace crowns the King with Glory," replied the second; "and Peace makes the Happiness of the People," added the first; "O King Adorable Father of your People their Happiness shows Europe the Glory of your Throne," concluded the second head.[109]

The Paris gossip and memoirist Louis Petit de Bachaumont noted that the heads were life-size, but covered tastelessly in gold. They mumbled some words and swallowed certain letters; moreover, their voices were hoarse and their diction slow (and their conversation, he might have added, uninspiring).

I.re Tête
Le Roi donne la Paix à l'Europe
II.e Tête
La Paix couronne le Roi de Gloire
I.re Tête
Et la Paix fait le bonheur des peuples
I.re Tête
O Roi adorable Pere de vos peuples
leur bonheur fait voir à l'Europe
la gloire de votre trône
TÊTES PARLANTES
Probleme résolu en Mécanique qui jusqu'à ce jour avoit été regardé comme insoluble ou du moins comme très difficile.
L'Accadémie des Sciences a dit dans son raport que ces têtes parlantes peuvent jetter le plus grand jour sur le Mecanisme de l'Organe Vocal, et sur le mistere de la parole : Elle ajoute que cet Ouvrage est digne de son approbation par sa nouvauté par son importance, et par son execution.

Figure 4.9 The abbé Mical's pair of discoursing heads. Bibliothèque nationale de France.

Yet despite all this, they undeniably had "the gift of speech." The academicians appointed to examine Mical's talking heads agreed that their enunciation was "very imperfect" but granted their approval to the work anyhow because it was done in imitation of nature and contained "the same results that we admire in dissecting . . . the organ of the voice." Bachaumont recorded that the academicians were so impressed with the abbé Mical that, on the occasion of the Montgolfière test at Versailles on September 19, 1783, in which a sheep, a rooster, and a duck became the world's first aviation passengers, the six delegates from the Académie des sciences invited Mical to accompany their delegation and presented him to the king as the author of the celebrated talking heads.[110]

The following year, probably at the instigation of the mathematician Leonhard Euler, the Saint Petersburg Academy of Sciences sponsored a prize competition to determine the nature of the vowels and to construct an instrument like *vox humana* organ pipes to express them. C. G. Kratzenstein, a member of the Academy, won the prize. He used an artificial glottis (a reed) and organ pipes shaped according to the situation of the tongue, lips, and mouth in the pronunciation of the vowels.[111]

Several more people built talking heads before the turn of the century. Among them was Kempelen, working this time not to amuse a queen with clever tricks but for scientific purposes: to uncover the secret of articulate speech. In 1791, he published "a description of a speaking machine"[112] in which he reported having attached bellows and resonators to musical instruments that resembled the human voice, such as oboes and clarinets; he had also tried, like Kratzenstein, modifying *vox humana* organ pipes. Through twenty years of such attempts, he had been sustained, he said, by the conviction that "*speech must be imitable*." The resulting apparatus had bellows for lungs, a glottis of ivory, a leather vocal tract with a hinged tongue, a rubber oral cavity, a mouth whose resonance could be altered by opening and closing valves, and a nose with two little pipes as nostrils. Two levers on the device connected with whistles and a third with a wire that could be dropped onto the reed. These enabled the machine to pronounce liquids and fricatives: Ss, Zs, and Rs.[113]

This machine produced an empirical finding reminiscent of Vaucanson's discovery that the blowing pressure for a given note depended upon the preceding note. Kempelen reported that he had first tried to produce each sound in a given word or phrase independently but failed because the successive sounds needed to take their shape from one another: "The sounds of speech become distinct only by the proportion that exists among them, and in the linking of whole words and phrases." Listening to his machine's blurred speech, Kempelen perceived a further constraint upon the mechanization of language: the reliance of comprehension upon context.[114]

Kempelen's machine was only moderately successful. It reportedly prattled in a childish voice, reciting vowels and consonants. It pronounced words such as "Mama" and "Papa," and uttered some phrases, such as "you are my friend—I love you with all my heart,"[115] "my wife is my friend,"[116] and "come with me to Paris,"[117] but indistinctly. Today the machine resides at the Deutsches Museum in Munich, Germany. Kempelen and his supporters emphasized that the device was imperfect and explained that it was not so much a speaking machine in itself as a machine that demonstrated the possibility of constructing a speaking machine.[118]

After this flurry of activity in the 1770s, '80s, and '90s, there was a decline in interest in speech simulation. A few people over the course of the nineteenth century, including the inventors Charles Wheatstone and Alexander Graham Bell, built their own versions of Kempelen's and Mical's speaking machines and of other talking heads from an earlier period.[119] But for the most part, designers of artificial speech turned their attention once again to speech synthesis rather than simulation: reproducing the sounds of human speech by other means rather than trying to reproduce the actual organs and physiological processes of speech.[120]

In 1828, Robert Willis—the same who had rejected the possibility of the Turk's intelligence, and who had by now assumed his professorship at Cambridge—wrote disparagingly that most people who had investigated the nature of the vowel sounds "appear never to have looked beyond the vocal organs for their origin," apparently assuming that the vowel sounds could not exist without being produced by the vocal organs. In other words, they had treated the vowels as "physiological functions of the human body" rather than as "a branch of acoustics." In fact, Willis argued, vowel sounds could perfectly well be produced by other means.[121] Whether or not the vocal organs themselves could be simulated artificially became a separate question from whether the sounds of speech could be reproduced. As late as 1850, the French physiologist Claude Bernard wrote in his notebook: "The larynx is a larynx and the crystalline lens is a crystalline lens, that is to say their mechanical or physical conditions are realized nowhere but in the living organism."[122]

Disenchantment with speech simulation was so deep that when a German immigrant to America named Joseph Faber designed quite an impressive talking head in the late 1840s, he could not get anyone to take any notice of it. Faber's talking head was modeled on Kempelen's and Mical's, but was far more elaborate (see figure 4.10). It had the head and torso of a man dressed like a Turk, and inside were bellows, an ivory glottis and tongue, a variable resonance

Figure 4.10 Joseph Faber's talking head advertised as Barnum's "Euphonia." Circus World Museum, Baraboo, Wisconsin.

chamber, and a mouth cavity with a rubber palate, lower jaw, and cheeks. The machine could pronounce all the vowels and consonants, and was connected by way of levers to a keyboard of seventeen keys, so that Faber could play it like a piano. He first exhibited the machine in New York City in 1844, where it aroused very little interest. He then took it to Philadelphia where he had no better luck. P. T. Barnum found Faber and his talking head there, renamed the machine the "Euphonia," and took them on tour to London, but even Barnum could not make a success of it. Finally the Euphonia was exhibited in Paris in the late 1870s, where it was mostly ignored, and soon thereafter all traces of it disappear.[123]

The moment for talking heads had passed. In the early part of the twentieth century, designers of artificial speech moved on from mechanical to electrical speech synthesis.[124] The simulation of the organs and process of speaking—of the trembling glottis, the malleable vocal tract, the supple tongue and mouth—was specific to the last decades of the eighteenth century, when philosophers and mechanicians and paying audiences were briefly preoccupied with the idea that articulate language was a bodily function: that Descartes's divide between mind and body might be bridged in the organs of speech.

The Division of Labor

Which animal and human functions could be accomplished by artificial machinery? The experimental automata of the previous section—the androids, moving anatomies, and talking heads—addressed this question for philosophical (and entertainment) purposes. Their close cousins, an emerging breed of industrial automata that were often designed by the same people and that employed the same key devices, addressed the same question in its social and economic dimension. Follow the android flutists and mechanical birds and they will soon lead you to an automated factory.

For example, almost half a century before Vaucanson's Duck made its first deposit, an iron and ivory peacock had already staked the same claim to fame. The peacock swallowed food and, according to reports, digested it by means of a process of fermentation, then displayed the results. The peacock was the work of a Frenchman named Jean-Baptiste de Gennes, a count and mechanical tinkerer with an abiding interest in mechanizing the actions of living beings. He had already presented to the Académie des sciences in Paris a design for a partially automatic loom, though both the loom and the peacock appear to have lacked some crucial parts.[125]

The peacock inventor would go on to a flamboyant career as an explorer, ship's captain, slave trader, and, finally, governor of the colony of Saint-Christophe in the Antilles.[126] But his reputation as the creator of the artificial bird followed him through all these exploits. Witness the conversation about him that Jean-Baptiste Labat, a Dominican priest and missionary, naturalist, explorer, soldier, engineer, plantation-owner, and slaveholder on Martinique, recorded having had with Christopher Codrington, the Barbados-born captain-general of the Leeward Islands, during a visit in 1700. Labat recalled:

> I noticed in their conversation how vain [the English] are, & the little esteem in which they hold other Nations, & above all the Irish. For when someone said that the French Colony was very weak, M. de Codrington instantly responded that it was up to M. de Gennes to augment it with Irishmen, if he couldn't do it with Frenchmen. I asked him to tell me the secret of this, so I could let M. de Gennes know it. Gladly, he said, do you know that M. de Gennes made a Peacock that walks, eats, digests. I answered that I did know it. Well, he continued, let him make five or six regiments of Irishmen. He will have far less trouble making these sorts of gross brutes than a Peacock. Since

> he is so clever, he will find the way to give them the necessary movements for firing, & for fighting, & by this means he will grow the Colony as much as he likes.[127]

Automata were a preoccupation among this set of modernizing proprietors, imperialists, and slaveholders. The equivalence of machines to lowly people of various sorts—slaves, conscripts, workers—was an attractive supposition and a theme that would recur throughout the development of automatic machinery, industrial as well as experimental. In such conversations, machines unsurprisingly had a decidedly Cartesian meaning: they signified the lack of a rational soul, of a capacity for reason and intellect.

Automatic textile machinery and mechanical defecating fowl, surprising though it may seem, shared an author in more than one instance. Like De Gennes, Vaucanson also built an automatic loom, having been recruited in 1741 by the king's finance minister, Philibert Orry, to become inspector of silk manufactures.[128] This invitation was in fact an effort to retain a rising star: one of the first acts of the young Frederick II of Prussia upon ascending to the throne had been to invite Vaucanson to join his court. But Vaucanson remained in France and went to work for his own monarch. This minor scuffle between Frederick and Louis XV is an indication of the seriousness with which android building (not to say defecating-mechanical-bird-making) was taken in the halls of power.

The automatic loom of 1745, now at the Musée des arts et métiers in Paris, was a close cousin of Vaucanson's three automata: it was built by the same Parisian artisans, and it worked similarly. A rotating cylinder was perforated according to the pattern to be woven. It turned against a frame of horizontal needles connected to vertical cords coming up from the warp threads. The spaces in the cylinder pushed the corresponding needles forward, while the holes allowed them to remain in place. The needles remaining in place, attached to the corresponding cords, were then raised by the bar, raising the selected warp threads.[129]

Vaucanson boasted that with his machine a "horse, an ox, an ass makes fabrics much more beautiful and much more perfect than the most clever workers of silk." He imagined a ghostly factory in which "one sees the fabric weave itself on the loom without human intervention. . . . The warp opens, the shuttle propels itself through, the reed pounds the cloth, the cloth rolls itself onto the cylinder." These claims were quoted in an enthusiastic review of the loom in November 1745 in the *Mercure de France*.[130] The "reading of designs," Vaucanson noted, was "the operation that demands the most intelligence" in silk production. "It is so difficult that it requires three or four years to learn." But, on the automatic loom, this operation became "so simple that . . . the only science required is to know how to count to ten." Thus the "most limited people," even "girls," could be substituted for those who were more intelligent and so would demand a higher salary.[131]

The inventor had not invented these divisions of workers: contemporary political economy relied upon them. The French school of economic theorists and reformers known as the Physiocrats, for example, led by Quesnay, rested their economic program on a different yet analogous distinction between "productive" and "sterile" classes of workers. The only productive workers according to the Physiocrats were the agricultural laborers, who produced all the value in the economy literally by growing it from the land; the rest of the economy—manufacturing and commerce—consisted merely of "sterile" exchanges of equal values, producing no new value. In fact, even the productive agricultural workers were less the producers of value than its midwives, delivering it from the land. Viewing human productions and economic activities as essentially barren, the Physiocrats also advocated replacing human workers wherever possible with animals and machines.[132]

The particular discrimination that Vaucanson invoked between intelligent and unintelligent work was central to the social hierarchy of the Old Regime. Diderot and d'Alembert's *Encyclopédie*, for example, defined *artist* as the name given to workers in the mechanical arts whose work required the most intelligence, while the work of *artisans* required the least intelligence.[133] But the automatic loom shifted the categories of intelligent and unintelligent work. It demoted the reading of designs, which had been the most intelligent job in textile production, to the very bottom of the hierarchy. While weaving became unintelligent work, fabric design remained a matter of human art; indeed, Vaucanson promised that his automatic loom would open "vast new fields . . . to the genius of fabric-designers."[134]

The comparatively lowly task of silk reeling also remained a skilled job. In fact, the biggest problem confronting French textiles centered upon this basic task of drawing the long fibers from cocoons and reeling them into thread. Silk thread available on the domestic market was so poor that French manufacturers often had to import their thread from Piedmont. Orry, the finance minister who had recruited Vaucanson, was especially worried about Italian competition in silk, so Vaucanson's first effort as silk inspector was directed at improving domestic primary material.[135] His diagnosis was that silk reeling was a delicate and skilled job, but the peasants who raised silkworms generally took the cocoons to market and sold or traded them to merchants and artisans of all sorts. These people would then reel the silk themselves or hire peasant women to do it; none had any training.

To remedy this situation, Vaucanson proposed to educate a population of expert *tireuses,* women trained in silk reeling, and to establish standards. He meant to accomplish both by creating a company of silk merchant-manufacturers, who would in turn establish seven factories, constituting a Royal Manufacture guaranteed by the Royal Treasury, where silk would be reeled under ideal conditions.[136] Established by a regulation of the city of Lyon in 1744, the Royal Manufacture was instantly embroiled in a fierce struggle between the roughly 250 silk merchant-manufacturers of Lyon and the roughly 3,000 master workers who ran their shops and who sometimes succeeded in setting up their own (there were about 160 independent shops in 1744). The workers had recently won a repeal of certain merchant-manufacturer monopolies, increasing their chances of becoming independent.

Vaucanson wanted the cooperation of the merchant-manufacturers, so he restored their monopolies and provoked a silk workers' strike accompanied by some of the worst pre-Revolutionary rioting of the century. He was forced to flee Lyon in the dead of night, disguised as a monk, and the regulation was annulled. Neither the automatic loom nor the Royal Manufacture succeeded in the first instance, but both systems, the mechanical and the social, would be established during the decades of and following the Revolution. By then, mechanical criteria for social arrangements had become commonplace, and mechanical and social engineering were inextricable.[137]

Mechanical calculators had an analogous effect to the automatic loom, demoting calculation from a paradigm of intelligence to the antithesis of intelligence.[138] If a machine could calculate, then something else—say, decision making or language—must be emblematic of human intelligence. Blaise Pascal, inventor of one of the first calculating machines, located the line between memory, which he said a machine could supply, and judgment, which he assigned to the human operator of his mechanical calculator.[139] Even Leibniz, who designed another early mechanical calculator, seemed to attribute no restlessness or capacity for perception to his device. Instead he wrote that it would rescue "excellent men" from losing "hours like slaves in the labor of calculation."[140] And when Charles Babbage designed and partially accomplished his computation machines, the "analytical engine" and the "difference engine," during the 1830s, he based his thinking crucially on a hierarchical taxonomy of human workers.

Babbage modeled the analytical engine on Vaucanson's automatic loom (later modified by a weaver named Joseph Jacquard, whose name it still bears). The loom's punch cards represented essentially the same system as the android's camshaft, with holes rather than pegs directing the movement of a system of levers and pulleys. The loom, Babbage later explained, could weave "any design which the imagination of man may conceive" by means of "a set of pasteboard cards" punched with holes according to the pattern to be woven.[141] Likewise, an analytical engine of the same design ought to be able to perform any algebraic computation. Ada Byron, Countess of Lovelace, Lord Byron's illegitimate daughter and Babbage's collaborator, observed, "We may say most aptly that the Analytical Engine *weaves algebraic patterns* just as the Jacquard loom weaves flowers and leaves."[142]

Babbage traced his idea for engines of calculation to an evening in 1812 or 1813, when he had been a student at Cambridge University and had been lounging at the Analytical Society, a mathematical club he had formed with some fellow undergraduates. Babbage had been sitting with his head "leaning forward on the table in a kind of dreamy mood." He later recalled, "A table of logarithms [was] lying open before me. Another member, coming into the room, and seeing me half asleep, called out, 'Well, Babbage, what are you dreaming about?' to which I replied, 'I am thinking that all these tables (pointing to the logarithms) might be calculated by machinery.'"[143] The idea stayed with him. Shortly after his chess match with the automaton Turk, in the summer of 1821, Babbage sat down with a friend, the astronomer John Herschel, to check over some mathematical tables for the Astronomical Society. Finding that the tables were strewn with errors, Babbage exclaimed in frustration, "I wish to God these calculations had been executed by steam!"[144] By then, he was already engaged in designing engines that could perform such a task.

Explaining his scheme for a mechanical computer, Babbage placed computation at the bottom of a tripartite hierarchy into which he divided the making of mathematical tables. The top of the hierarchy, establishing the formulas, had to be the work of "eminent mathematicians." The second level, working out how to apply the formulas to a given calculation, required "considerable skill." And the third, carrying out the actual calculations, required so little ability that Babbage believed it could be done by his calculating engines. He attributed this "division of mental labor" to the French engineer Gaspard Riche de Prony, who in turn said he had been inspired by Adam Smith's famous description of pin making as an example of the division of labor:

> One man draws out the wire, another straights it, a third cuts it, a fourth points it, a fifth grinds it at the top for receiving the head; to make the head requires two or three distinct operations; to put it on, is a peculiar business, to whiten the pins is another; it is even a trade by itself to put them into the paper; and the important business of making a pin is, in this manner, divided into about eighteen distinct operations, which, in some manufactories, are all performed by distinct hands.[145]

This had inspired de Prony to try reducing table making to operations simple enough that they could be performed by unskilled workers. As it happened, de Prony hired hairdressers left unemployed by the simplified hairstyle of the post-Revolutionary era. Their ability to do the job implied for Babbage that a machine could do it too.[146]

During the eight decades separating Vaucanson's Flutist and Babbage's calculating engines, inventors, philosophers, and social and economic reformers contemplated the possible relations among body, feelings and intellect, mechanism and mind. As they did so, the meaning of the word "machine" underwent a gradual transformation. Diderot's *Encyclopédie* (1765) defines "machine" neutrally as "that which serves to augment or direct moving forces."[147] The definition of "machine" in the first five editions of the *Dictionnaire de l'Académie française* is essentially the same: "Engine, instrument suited to move, pull, lift, drag or throw something," with figurative uses such as "man is an admirable machine."[148]

It is only in the sixth edition (1832–35) of the *Dictionnaire* that the sense of "machine" comes to include "Prov. and fig., *It is nothing but a machine, it is a pure machine, a walking machine*, a person without spirit, without energy."[149] The new "figurative" meaning of "machine" reflected a newly configured landscape of machines and people. A brute-mechanist distinction between mind and mechanism informed the process of industrialization and that process made manifest an ultimately brute-mechanist division of the social and economic world into parts, mind on one side and mechanism on the other. The new class of automaton workers included humans and animals as well as machines. The industrial reformers and inventors of automated machinery alike understood their task as the ingenious and lucrative division of intelligence from labor, design from execution, agency from mechanism.

~ 5 ~

The Adventures of Mr. Machine

With the chattering, sighing, and singing of the androids in their ears, and with the growing spectacle of automaton labor in view, a rising generation of writers took up the notion of living machinery. The result of their efforts was the Enlightenment man-machine,[1] a hypothetical figure that oriented his century's philosophical, moral, social, and political discussions. The Enlightenment man-machine was a great thought experiment, an attempt to reintegrate the human self into the material world, the soul into the machinery. For the first time, people tried to imagine whether a human being could actually be a machine through and through, a purely material set of moving parts, and what that might mean.

One thing it meant was a new basis for categorizing and ranking varieties of human-machine. Men and women, weavers and poets, Hottentots and Hurons and Laplanders could now be classified and ranked by the organization of their machinery. The Enlightenment man-machine thought experiment also raised another problem that was at once moral, philosophical, and scientific. It was, once again, the familiar problem of agency: what sort of agency could a purely material creature have? An answer arose within the conversation about this question. It was a momentous new possibility, a form of agency for a rigorously material being, namely, the capacity to create oneself over time, the power of "self-organization."

The man-machine thought experiment had a lead author(see plate 7), and with his checkered tale our story begins.

Mr. Machine Goes to Berlin

In the autumn of 1744, in the throes of the War of Austrian Succession, having entered into a secret alliance with Frederick II of Prussia, Louis XV laid siege to the city of Freiburg. During the siege, a young medical officer named Julien Offray de La Mettrie, who was attached to a regiment of the Gardes françaises, contracted a raging fever. As Frederick II, soon to become La Mettrie's protector and patron, later told the story, "For a philosopher an illness is a school of physiology." Being of a philosophical bent, the delirious doctor seized the occasion to observe the effects of fever on mental function. "He believed he could see clearly that thought is but a consequence of the organization of the machine, and that the disturbance of the springs has considerable influence on that part of us which the metaphysicians call the soul."[2] Frederick related, with or without irony I leave to the reader to judge, that La Mettrie had taken his intellectual direction from this early moment of delirium.[3]

Even in the pink of health, La Mettrie was apparently hotheaded. "Tumultuous and open-mouthed" is how Thomas Carlyle later described him, with a "*minimum* quantity" of discretion.[4] Born and raised in the port city of Saint-Malo, Brittany, the son of a wealthy textile merchant, La Mettrie was a polemicist. During his convalescence and after, he pursued the idea that machinery was the basis of thought and "found only mechanism where others had supposed an essence superior to matter."[5] Persuaded that thought must be a bodily function, La Mettrie said so in his first philosophical work, *L'histoire naturelle de l'âme* (Natural History of the Soul) (1745). When the Paris Parlement condemned this book to be burned by the public executioner, La Mettrie left for Leiden. Once there, however, he went back to work on developing his idea. Illness was not the only example of the influence of physical upon mental states. La Mettrie compiled a long list that included the mind-altering power of opium, wine, coffee, sleep, pregnancy (with its "frightful schemes"), age, climate, weather, hunger ("The power of a meal!"), and of course lust, "that other frenzy of Man or Woman . . . hounded by continence and health."[6]

He also considered the effects of mental upon physical states. Why, for example, "does the sight or the mere idea of a beautiful woman cause such singular movements?" The response of "certain organs" to the simple thought of womanly beauty demonstrated beyond a doubt, La Mettrie reckoned, the intimate connection between the imagination and the muscles. The imagination had the capacity to excite a sequence of springs in the body, he supposed, and "how can this be, if not by the disorder and tumult of the blood and spirits that gallop with extraordinary promptitude and swell the hollow tubes?"[7] Seized thus by the idea of human-machinery, La Mettrie produced his most important work, *L'homme machine* (Man a Machine) (1747). Though no longer ill, he had a feverish style, as even his champion acknowledged: "He wrote his *Man a Machine* or rather," amended Frederick, "he put on paper some vigorous thoughts about materialism, which he doubtless planned to rewrite."[8] (Privately, in a letter to his sister following La Mettrie's death, Frederick observed that he was "regretted by all those who knew him. He was gay, a good fellow, a good doctor, and a very bad writer; but, if one didn't read his books, one could be very happy with him.")[9]

Among the vigorous thoughts that La Mettrie put on paper was the announcement that soul was but "a vain word" signifying "that part that thinks," namely the brain. This organ, in turn, had "muscles for thinking as do the legs for walking."[10] With his customary mixture of serious and mischievous intentions, and to the consternation of his beneficiary, La Mettrie dedicated this philosophical hot potato to the Swiss physiologist, poet, novelist, political theorist, and theologian Albrecht von Haller. The two were exact contemporaries and both had studied with Herman Boerhaave, botanist, doctor ,and mechanist philosopher at the University of Leiden.[11] But, in contrast with La Mettrie's materialism, Haller was a Calvinist and a temperamental as well as a doctrinal moderate.[12] The two intellectual offspring struggled over the legacy of their carefully ambiguous father: Haller construed Boerhaave as a devout dualist while La Mettrie made him an unflinching materialist.

At the time, Haller was poised to unleash a controversy by presenting the core idea of his physiology, first in lectures and then in print, which was an identification of two capacities of animal tissues. The first, "irritability," was specific to muscle tissues: it was the capacity to contract in response to stimulus, and Haller viewed it as the basis of animal motion. The second capacity, "sensibility," resided in the nerves.[13]

Irritability and sensibility were versatile ideas that lent themselves to exploitation both by those who saw evidence of a special, animating force in living tissue, and to materialists, who saw support for the explanatory reach of ordinary matter alone.[14] Vehemently, Haller rejected both ideas. As a strict mechanist, who saw physiology as simply the description of the movements of the "animated machine,"[15] he disliked special, vital forces. But he was also a self-styled scourge of atheists and materialists. La Mettrie, whose philosophy was at once materialist and filled with vital fluids and forces (he did not much trouble himself over the details of just what sort of matter composed these), invoked irritability and sensibility in the service of both; indeed, he anticipated Haller in his discussion of muscular irritability in *L'homme machine*.[16] In short, while Haller was fending off vital fluids on one flank and materialism on the other, La Mettrie, with his dedication of *L'homme machine*, neatly planted a thorn in each.

La Mettrie's manifesto was too hot even for Holland and went the way of its elder sibling, to a ceremonial burning in the city square by the public hangman.[17] It rose from the ashes, however, provoking one of the defining controversies of the accelerating Enlightenment and becoming a fulcrum of philosophical debate.[18] The banished author too landed on his feet, at Frederick's court in Berlin. There he played a mixed role, officially the monarch's reader but equal parts gadfly and court fool.[19] He affected a great familiarity with the king, not hesitating to "throw himself down and stretch out on the couches. When it was hot, he opened his collar, unbuttoned his vest, and threw his wig on the floor."[20] Frederick hated to be parted from La Mettrie, who played with him and made him laugh,[21] inspiring much envy. Carlyle's unflattering characterization, indeed, relies on the accounts of jealous rivals for Frederick's attention, especially Voltaire, and should be taken with a grain of salt, not to say arsenic.

La Mettrie discovered the limit of the Voltairean sense of humor with a bit of well-aimed wit. He told Voltaire that Frederick had remarked of him, "I will want him another year, at most; you squeeze the orange, you throw away the skin." The orange-skin affair tormented Voltaire, as he self-mockingly reported: "Should I believe it? is this possible? What! after sixteen years of kindnesses . . . I sacrifice everything to serve him . . . a king . . . who told me that he loved me . . . it is beyond me."[22] Although he struggled not to believe it, "I am still dreaming of the *orange-skin*," Voltaire lamented, ". . . I am afraid I am like those cuckolds, who force themselves to think that their wives are very faithful."[23] Voltaire's orange-skin panic is representative: by the same quality of outrageous ineptitude or malign cleverness, or likely a potent alchemy of the two, La Mettrie penetrated as deeply under the Enlightenment's collective skin.

In his three short years at Frederick's Berlin court, La Mettrie was a whirlwind of activity. He elaborated his polemical philosophy in eight or ten more

Figure 5.1 Adolf Von Menzel, *Die Tafelrunde* (1848), a dinner hosted by Frederick II at Sanssouci Palace with members of the Prussian Royal Academy of Sciences including Voltaire, leaning forward on the left in a purple coat, and La Mettrie with his back turned. bpk, Berlin / Nationalgalerie, Staatliche Museen, Berlin / Art Resource, New York.

works and, at Frederick's behest and to the chagrin of many of its members, secured election to the Prussian Royal Academy of Sciences (see figure 5.1).[24] Meanwhile he cheerfully stoked the fires of controversy, joining in the fun incognito on the side of his detractors. A philosophy professor at Göttingen gave La Mettrie an opening by suggesting that the author of *L'homme machine*, if he were correct in his claims, would himself be a machine and therefore not responsible for the gibberish he produced.[25] La Mettrie delightedly accepted the soubriquet "Mr. Machine" and dashed out an anonymous, self-satirizing pamphlet describing the life and, more saliently, the death of this personage.

Having decided that opium was the secret to a machine's happiness, Mr. Machine had met an untimely end by indulging in a good dose of rat poison. But one must not blame a creature equivalent to "the ducks of Mr. V*aucanson* in *Paris*": "Remind yourself, if you would, that this is *Mr. Machine*. A machine does not act as it likes, but rather as it must."[26] This mischievous piece of mystification contained a certain poignant clairvoyance and perhaps also an undercurrent of sorrow.

Despite his official jollity, the philosophical jester appears to have felt his exile keenly. "Reader to the king of Prussia though he may be," Voltaire reported, "he burns to return to France. This man so gay, who passes for someone who laughs at everything, sometimes cries like a baby to be here."[27] Voltaire, who evidently would not have been opposed to La Mettrie's repatriation, enclosed a letter from La Mettrie to Richelieu, asking that Richelieu obtain a pardon for him to return to France. But before the matter could be pursued, La Mettrie made an abrupt and suitably dramatic exit from history's stage. "Our crazy La Métrie [*sic*]," Voltaire wrote with ill-disguised satisfaction, "has just made up his mind to die. . . . I cannot get over my astonishment."[28]

La Mettrie had gone to visit the Irish Jacobite Lord Tyrconnel, who acted as Louis XV's ambassador in Berlin. Tyrconnel was unwell and had requested the presence of the monarch's reader-qua-fool to cheer him. Arriving just as Madame Tyrconnel was sitting down to eat, La Mettrie, according to Voltaire, "eats and drinks, and talks, and laughs more than all the others; when he is full to the gills, they bring a pâté of eagle disguised as pheasant, sent from the North, and well mixed with bad lard, chopped pork and ginger. My man eats the whole pâté and dies the next day."[29] Contemporaries called it "indigestion"; historians have agreed upon "food poisoning"; but if Voltaire's account is accurate, then poisoning, tout court, seems as good a name as any. "Voilà, my hero, one of our farces carried out," was Voltaire's punning verdict to Richelieu, a *farce* being both a form of comic theater and a stuffing.[30] (Farcical too is that Voltaire was even still thinking of peeled fruit: "I would have liked to ask La Métrie, at death's door, for news of the *orange skin*. This good soul, on the verge of appearing before God, would not have been able to lie.")[31]

The manner of the materialist's death instantly became a test of his principles. "There is now a great dispute," reported Voltaire, "to know whether he died as a christian or as a doctor." La Mettrie had been a good atheist to the end, according to Voltaire, begging to be buried in Tyrconnel's garden, but had been denied this final courtesy: "His body, swollen and big as a barrel, was carried, willy-nilly, into the catholic church, where it was astonished to find itself."[32] La Mettrie's death by pâté allowed his delighted enemies to equate materialism with gluttony.[33] On the other side, his supporters too were keen to attach their own moral to the parable. Frederick, in particular, scrupulously verified La Mettrie's steadfastness in extremis before undertaking to write his eulogy. "The king inquired very exactingly about the manner of his death," reported Voltaire, "whether he had gone through all the catholic rites, if he had had some sort of edification; finally he satisfied himself that the gourmand had died as a philosopher: *I am much relieved*, the king told us, *for the peace of his soul*; we started laughing, and he did too.[34]

Another chronicler of Frederick's court, the bookseller and writer Christoph Friedrich Nicolai, told the story of La Mettrie's fabled demise more poignantly. According to Nicolai, Tyrconnel's chaplain, urged on by some enemies of La Mettrie who wanted to render him "contemptible" in Frederick's eyes, had "pushed into the sickroom." In this version of the story too the materialist moralist held fast and achieved, indeed, a form of heroism:

> La Mettrie would have nothing to do with this Priest and his talk; who, however, still sat and waited. La Mettrie, in a twinge of agony, cried out "JESUS MARIE!" "AH, VOUS VOILA ENFIN RETOURNE A CES NOMS CONSOLATEURS!" exclaimed the Irishman. To which La Mettrie answered (in polite language, to the effect), "Bother you!" and expired a few minutes later.[35]

The stance to which La Mettrie apparently remained committed even to his death was not merely anti-Christian but also constituted a positive moral program. The core idea of this overtly Epicurean program (La Mettrie titled a book after the ancient Greek moralist Epicurus, ca. 300 BCE, who called pleasure the greatest good)[36] was that the moral as well as the physical order of nature, and the individual human self to which it applied, were built right into the material machinery of the world.

An admirer of Vaucanson, La Mettrie described the body as "a clock," and a "machine that winds itself." He lingered over the "springs of the human Machine": the spring of the entire body backward in terror from the edge of a cliff; the blink of an eye at the threat of a blow; the expansions and contractions of the pupils, the pores of the skin, the heart, the lungs, the sphincters of the bladder and rectum; and the heaving of the stomach when poisoned. He also considered admiringly how "the erector muscles raise up the Rod in man." La Mettrie marveled, "There is a singular spring in this member," whose study had been shamefully neglected even in the present age of enlightened anatomy.[37]

The living machinery that La Mettrie described was actually not very clock-like, despite all the springs. It was filled, for one thing, with fluids and fermentations. But more importantly, La Mettrie's living machinery differed from the clockwork of the argument-from-design natural theologians because the "clockmaker" was no external Designer, but, La Mettrie specified, it was "chyle." Chyle was neither a rational designer nor any sort of external force. It is a bodily fluid (composed of lymph and emulsified fats), and according to La Mettrie, it animated the clock by heating it, through a sort of "fever" or "fermentation." This machine was no passive artifact of an external builder. It contained its own "motive principle," an internal source of both movement and sentiment.[38] Denis Diderot, whose philosophy imbibed much of the tone and substance of *L'homme machine*,[39] affirmed in his *Eléments de physiologie* (1784) that a special, internal force of "sensibility" drove animal-machinery according to its own laws of motion.[40]

La Mettrie's man-machine was an erotic and passionate creature, coursing with sensations and emotions. "To be a machine," La Mettrie wrote: "to feel, think, know how to distinguish good from evil like blue from yellow." Emotions, moral instincts, an aesthetic sense: the man-machine had all of these and also, by the same token, a sex life (for the convergence of mechanism and eroticism, see plate 8): "Who would ever have divined *a priori* that a drop of the liqueur that shoots forth during coupling would make one feel divine pleasures?"[41] Even a rudimentary living machine could experience this last universal boon: La Mettrie extended the joys of sex right down to plants.[42]

His libertinism was not amoral: on the contrary, it constituted a moral scheme of its own. According to La Mettrie, "Natural Law" was a feature of the machinery, a "feeling" for integrity, humanity, and virtue. To treat others as one would want to be treated was not a principle but a sensation built right into the machinery.[43] Diderot would later make a similar argument for another traditional virtue, industry: "Idleness is always contrary to a living machine!"[44]

The greatest vice in La Mettrie's moral universe was therefore rational reflection: the doomed attempt to transcend one's bodily mechanism. La Mettrie's overriding project was to deflate the imperial self of Descartes and his fellow rationalists. Study was "a catalepsy, or immobility of the Mind, so deliciously inebriated . . . that it seems detached by abstraction from its own body." Learning was an "abuse of our faculties." Philosophers trying to understand the world a priori using "the wings of the Mind" were doomed to failure. Worse, they were "Do-nothings" and "vain Pedants" whose "Balloon" brains were swollen with heaps of words and figures—which were, after all, only so much stuff, physical imprints on the "medullar canvas."[45]

The finiteness and materiality of the human-machine constituted the central moral truth in La Mettrie's philosophy: "Man in his first Principle is nothing but a Worm." Viewing the material world from within rather than above or beyond, the man-machine enjoyed the perspective of a "Mole." Therefore, La Mettrie urged, "Let us not lose ourselves in the infinite, we are not made to have the least idea of it; it is absolutely impossible for us to go back to the origins of things." With human knowledge necessarily limited and provisional, the worst pretenders were those "proud and vain beings" who claimed access to a larger, more transcendent kind of truth: rationalist philosophers and, of course, clerics and theologians. "However much they want to elevate themselves, [they] are at bottom nothing but Animals, and perpendicularly crawling Machines."[46]

While the negative component of La Mettrie's moral program was an extended rebuke to both rationalists and theologians, its positive element was a proto-Romantic celebration of feeling, emotion, experience, and mystery. Thought itself, the materialist moralist insisted, was a kind of feeling, a property of living matter comparable to electricity. Moreover, neither the mechanism of thought (i.e., feeling) nor any other mechanism could ever be fully explained. This was because the essential nature of matter and motion was ultimately "an impenetrable mystery." La Mettrie therefore announced himself reconciled to the "incomprehensible Marvels of Nature." To embrace the ultimate ignorance of an essentially material creature—a worm, a mole, a crawling machine—was to live a good, just, and happy life.

"What do we know of our destiny," La Mettrie wrote, "any more than of our origin?" Human beings, being machines, must accept their "invincible ignorance." He who did so would be "wise, just, and tranquil about his fate, and consequently, happy. He will await death neither fearing nor desiring it." Such a creature would cherish life, be "full of respect for Nature; full of gratitude, attachment and tenderness." He would be grateful to partake in the "charming Spectacle of the Universe." He would "pity the vicious without hating them; in his eyes they will be but deformed Men."[47] A materialist would regard all human failings with the tolerant comprehension of a "Physician": "Do you know why I still make something of Men? It is because I seriously believe them to be *Machines*. Under the contrary hypothesis, I know few whose society I would value. Materialism is the antidote to Misanthropy."[48] To realize that humans were nothing but more or less imperfect machines, Diderot later affirmed, was to embrace the following credo: "There is only one virtue, justice; one duty, to be happy; one corollary, not to overrate life, and not to fear death."[49]

The Sentimental Education of Mr. Machine

Reacting against the disembodiment of the human self, the Enlightenment materialist-moralists, led by La Mettrie, went to the opposite extreme. They snatched the soul from the heavens and drove it into the very earth, making it a "*soul of mud*."[50] Their man-machine was a rebuke to the rationalists: a denial of the transcendent intellectual self that rationalists ascribed to humans, above all, to themselves. "What I call *myself*," wrote Joseph Priestley, "is an organized system of matter."[51]

It may sound surprising to suggest that those who most urgently pressed the idea of the human-machine in the eighteenth century were driven by a moral purpose, these polemicists who have appeared in historical writing not only as anticlerical, but as "immoralist" and "anarchic" hedonists who liked to celebrate in print every sexual possibility from masturbation to bestiality.[52] But neither anticlericalism nor lasciviousness actually constitutes moral anarchism. Moreover, neither stance was specific to the materialists; deists and other theological moderates shared both.[53]

A different moral purpose did distinguish the materialist man-machine advocates, however, and that was antirationalism. They, more decisively than any of their contemporaries, were bent on humbling the Cartesian imperial self, returning it to the continuum of nature, whose eternal mysteries they meant to restore by curtailing the authority of reason. In this sense, partisans of the human-machine model in the Enlightenment were not only moralists but also proto-Romantics.

Despite their determination to humble the Cartesian self by returning it to the continuum of nature, La Mettrie and his fellow man-machine moral theorists nevertheless shared their generation's Cartesian intuitions regarding human selfhood. A human self, according to these intuitions, was an autonomous agent whose defining task was to know, first itself, and then the world outside itself, each in relation to the other. Rendering such a being back into matter, post-Descartes, created difficulties. For example, it raised the problem of fragmentation: a dissolution of the human self into the material continuum of nature. If we and the world are all made of a single kind of matter diversely configured, how do we know where one ends and the next begins? Perhaps everything in the material world constitutes just "one great individual." In any machine or animal one can give separate names to the separate parts, Diderot suggested, but that does not make each of these an individual entity. It may make no more sense to identify a person than a bird's wing as an individual.[54]

A material self is in danger of dissolving into its material surroundings. How can physical parts come together to make a single, indivisible, and discrete whole? How can a being have agency if it cannot distinguish itself from the world around it? Diderot canvassed the possibilities in a mischievous bit of writing entitled the *Rêve de d'Alembert* (D'Alembert's Dream). Here, Diderot presented his worldview in the traditional form of a dialogue, casting his estranged friend and former collaborator on the *Encyclopédie,* Jean d'Alembert, in the starring role. Differences in philosophical temperament and doctrine, as well as the usual sorts of vanity and rivalry, had led to d'Alembert's departure from the *Encyclopédie* and a rift between the friends. The long-suffering mathematician was accustomed to serving as Diderot's rationalist foil.[55] The "D'Alembert" character in the dialogues composing the *Rêve de d'Alembert* lapses in and out of a feverish philosophic dream during which, his misguided rational faculty for once suspended, he discerns and announces the truths of the cosmos as Diderot deemed them.

An earlier conversation with "Diderot" (he is himself a character in the work) prepares the momentous night of dreaming. During the conversation, "Diderot" introduces "d'Alembert" to such principles as the idea that there is no essential difference between a canary and a bird-automaton, other than complexity of organization and degree of sensitivity. Indeed, the nerves of a human being, even those of a philosopher, are but "sensitive vibrating strings," so that the difference between "the philosopher-instrument and the clavichord-instrument" is just the greater sensitivity of the philosopher-instrument and its ability to play itself. A philosopher is essentially a keenly sensitive, self-playing clavichord. Imagine, "Diderot" exhorts "d'Alembert," that such a clavichord "organized like you and me" really existed and then grant it the power to reproduce itself, say, with the help of a female clavichord. The result would be little clavichords, no less "alive and resonant" than baby philosophers produced in the equivalent way.[56]

For the character of d'Alembert's doctor and irreverent materialist sage, Diderot thought of La Mettrie, who costars in a manuscript version of the piece,[57] although in the posthumously published version, he assigned the role to the Montpellier doctor Théophile Bordeu, investigator of the nervous and glandular systems and author of the article on medical "crisis" for Diderot's *Encyclopédie*. The fictional d'Alembert is also attended by a Julie de Lespinasse modeled on the actual one with whom the actual d'Alembert actually lived, and who had recently nursed him through an actual illness.[58]

In a passage of the dialogue, Diderot considers the various ways in which physical parts might form a single, integral whole such as a human self, or at least, such as a human self feels itself to be: drops of mercury merging; a swarm of bees; conjoined twins; a monastery whose corporate identity remains intact as long as its membership changes only little by little. Are any of these wholes truly indivisible? A materialist has no choice, lectures the Bordeu character, but to regard a human being as a juxtaposition of "sensitive molecules" each of which had its own "me" before being combined with the others. How did each part "lose this 'me' to produce a single consciousness?" The actual Bordeu had in fact argued that man was a "decentralized" being, a "federation of organs," and had even drawn the comparison to a swarm of bees.[59] Patting himself all over, the sleeping pseudo-d'Alembert mutters: "I am certainly one, I couldn't doubt that, but how is this unity made? . . . I can understand an aggregate, a tissue of little sensible beings, but an animal . . . a whole! A system, a him, having consciousness of his unity! I do not see that, no I do not see it."[60]

In addition to the danger of dissolving into its surroundings, a fully material self also ran the symmetrical problem of having its surroundings dissolve into it. In other words, it posed the problem of solipsism: a creature that could not recognize that there existed a world beyond its own borders. How could it, when all it had to go by were sensations taking place within itself? "There is a moment of madness," Diderot warned, "in which the sensible clavichord thinks that it is the only clavichord there has ever been in the world, and all the harmony of the universe takes place in it."[61]

Diderot's "Julie de Lespinasse" character imagines a spider at the center of a web to represent these twin hazards of dissolution and solipsism: losing oneself in the world and losing the world in oneself. The spider feels each tug on every thread of its web. In the same way, Julie's self sits in her brain at the center of her network of nerves, which in turn connects with the rest of the world at her borders. Her inner spider can feel itself to be connected outward to the wide world through its web, or it can concentrate itself inward on its own sensations and experiences. For example, she says, "[While dreaming] I exist as in a point; I almost cease to be material, I feel only my thought; there is no more place, nor movement, nor body, nor distance, nor space for me: the universe is reduced to nothing for me, and I am nothing to it." Pseudo-Bordeu responds by invoking the complementary danger: "Voilà the extreme limit of concentration of your existence," he allows, but "its dilation could be without limits." The spider might feel all its sensations as internal, in which case it would lose its grasp on the world outside (and without a world it can have no meaningful understanding of its self). Or it might open itself outward, extending itself through its senses to the whole universe, in which case it would lose itself (but without a self it cannot understand that there is a world outside).[62]

A rigorously material creature, a man-machine, seemed ironically in danger of lacking precisely the capacity its proponents cared most about: the capacity to know its place in the material continuum of nature. A sensitive, self-playing clavichord occupied a pitifully contradictory position. It felt itself both infinite and vanishingly small, coextensive with the world and trapped in a single point, lost in the cosmos and solitary in its tiny and transient existence, in constant danger of losing either itself or the wider world, which came ultimately to the same thing. Diderot did not try to resolve these contradictions. Rather, he announced them. They were his destination in the *Rêve de d'Alembert*: a dethronement of the all-knowing rational self by a poignant paradox. Like La Mettrie, Diderot was steadfast in his materialism. That materialism posed problems for human agency, and for knowing, rational selfhood, strengthened rather than shook his resolve. "The peasant who sees a watch move itself," he wrote, "and not knowing the mechanism, imagines a spirit in the needle, is neither more nor less foolish than our spiritualists."[63]

The rigorously materialist view of the human self had a powerful patron in Frederick II. "If the soul of your nerves is in a state of quietude," he wrote teasingly to Voltaire, "I would be charmed to see you this evening," and "if not, I believe it will take vengeance on your body for the wrong your mind is doing it." Joking aside, Frederick continued that he did not believe in a disembodied mind or soul. In fact, he thought the materialist position made more sense with respect to a unified self. "I do not believe I or anyone could be double [body and soul]. Great people, in speaking of themselves, say we; but this does not make them multiplied. Let us put our hand on our conscience and speak frankly; we will admit in good faith that thought and movement, of which our body has the faculty, are attributes of the animated machine, formed and organized as man."[64] An animate machine, perhaps, but no less an imperial self.

Other writers on living machinery thought that a rigorously materialist view of human beings, by erasing any divide between human beings and the material world around them, would require sacrificing the unitary, rational human agent, and they declined to do it. The Genevan Calvinist naturalist Charles Bonnet, for example, inferred from the subjective feeling of unity—"the sentiment I have of my *Me* is always one, simple, indivisible"—that he was "not all Matter."[65]

Georges Buffon, the director of the Jardin du Roi in Paris and the author of the bestselling *Histoire naturelle, générale et particulière,* after carefully imagining how a human being might develop in its absence, also reintroduced an immaterial soul. To carry out this thought experiment, Buffon conjured up an adult version of John Locke's tabula rasa, a newly created yet fully formed human creature with no experiences or ideas. Such a creature would have only its machinery, specifically the "organic machine" comprised of sense organs, brain, spinal cord, and nerves. The senses were the parts of the machine upon which external forces exerted themselves, the brain the fulcrum by which these forces were transmitted to the rest of the mechanism, and the nerves the parts of the machinery set in motion.[66]

Imagining himself as such a machine first beginning to operate, Buffon thought that upon opening his eyes for the first time, he would be utterly solipsistic. He would experience the whole wide and glorious world as contained within his own being: "the light, the celestial vault, the green of the earth, the crystal of the waters." He would believe these things were all within him, parts of himself. Hearing sounds, "the song of the birds, the murmur of the breezes," he would be persuaded "that this harmony was me." Everything he saw, heard, or felt he would experience as "contained in a part of my being." When the breeze brought him "perfumes that caused in me an intimate blossoming," these inspired "a sentiment of love for myself." Agitated by these feelings, the Buffon-machine began to move and to touch things. At first, it touched only itself, and "all that I touched . . . seemed to return sentiment for sentiment," but finally it touched something that did not respond. "I turned away with a sort of horror, and knew for the first time that there was something outside of myself."

A stormy period followed in which the Buffon-machine, like Descartes in his second *Méditation,* was "profoundly occupied with myself, with what I might be." At last he found humility: "My knees bent and I found myself in a situation of rest." He soon made a further discovery: "What was my surprise to find by my side a form similar to my own! . . . It was not me, but it was more than me, better than me, I felt my existence change places and pass entirely into this second half of myself." Buffon's parable ends with the creature-machine's arrival at love. Contemplating the other one, "I wanted to give it all my being; this vivid desire took over my existence."[67] The defining capacities of a human being, then, according to Buffon's fable, were first self-love and second love of others, both residing in the physical mechanism of the sense of touch: the mechanical impact of the creature's parts against one another and against an alien world.

Buffon's thought experiment was a kind of translation of Descartes's second *Méditation*, in which Descartes found certainty in the knowledge of his own existence as a thinking being, and then used that certainty as a basis for knowing the world outside himself. Buffon translated Descartes's process from rational, epistemological terms into sentimental, materialist ones. In his own parable, Buffon replaced Descartes's source of knowledge, rational reflection, with its opposite, the most primitive kind of sensation, the sense of touch. He then replaced Descartes's rational knowledge of his own existence with the feeling of self-love. Finally, Buffon replaced Descartes's arrival at knowledge of the outside world with a different kind of journey outside oneself: falling in love.

Despite all that reversing of Descartes, however, soon afterward Buffon accorded humans a Cartesian soul, a purely "spiritual substance" that brought the capacity for rational thought.[68] Apparently, he had not persuaded himself of the reducibility of human selfhood to material parts.

The struggle for dominion between "me" and "my machine" became a central drama of the age. It occupied the baron de Montesquieu, philosopher, jurist, and legal reformer, throughout his career. For example, we are "ceaselessly terrorized" by our bodies, observed Usbek, the sensitive protagonist of Montesquieu's *Persian Letters*, for if the blood moves too slowly, if "the spirits are not purified enough; if they are not present in sufficient quantities: we fall into sorrow and sadness." But then, have a drink and "our soul becomes again capable of receiving impressions that cheer it, and it feels a secret pleasure in seeing its machine recover, so to speak, its movement and its life."[69] To contemplate the human body, "this machine so simple in its action and so complicated in its springs," Montesquieu told the Academy of Bordeaux, was to witness a tempestuous relationship: "these animal spirits so imperious and so obedient, these movements so submissive and sometimes so free, this will that commands like a queen and obeys like a slave."[70]

A similar struggle takes place in the writings of Jean-Jacques Rousseau, perhaps his age's leading bard of "the *moi*."[71] The primordial truth of his own self is revealed in the most notorious passage of his scandalous *Confessions*: a mechanical response to a mechanical act, a spanking secretly relished. "Who would believe that this childhood punishment received at eight years of age from a woman of thirty determined my tastes, my desires, my passions, myself for the rest of my life." From that origin moment forth, the author's ardent susceptibility to "my machine"—its sensations, its weaknesses, its languishing illnesses, its every passion and foible— provided the *Confessions* with a central story line.[72]

On the other hand, Rousseau, the son of a watchmaker and trained in his father's craft, also thought that the self was something specifically antithetical to clockwork. Physics might explain the mechanism of the senses, but it was helpless to account for human agency, the capacity to act freely, "the power of willing": a "purely spiritual" action inexplicable "by the Laws of Mechanics."[73] The deistic Savoyard vicar who steps into Rousseau's philosophical novel *Émile* (1762) to present the author's own religious credo announces: "A machine does not think at all." Neither "movement nor shape" can produce thought. There must be something else in a person, an immaterial and transcendent something. "Something in you seeks to break the bonds that confine it. Space is not your measure, the whole universe is not big enough for you; your sentiments, your desires, your unease, even your vanity have another principle than the narrow body in which you feel imprisoned."[74] Bounded like Hamlet in his nutshell, Rousseau, through his Savoyard vicar, yearned to count himself the king of infinite space.

On the third hand, still according to Rousseau, the self did not exist when the body's machinery began its functioning. It was absent from a newborn, whose movements and cries were "purely mechanical."[75] (Rousseau, as is often noted, left each of his five children at the foundling hospital).[76] Even a grown man, were he to emerge all at once in his fully developed state "like Pallas from the brain of Jupiter," would be as selfless as a newborn, "an automaton, a motionless and almost insensible statue."[77] In other words, like Buffon and many others in this period, Rousseau was ambivalent and inconsistent on the subject of human selfhood and bodily machinery. Sometimes, he defined a human self by its contrast with mechanism, and at other times, he made selfhood a product of the functioning of the body-machine. Ambivalence on this score, more than decisiveness in one direction or the other, was a defining feature of the Enlightenment conversation about human beings and their place in nature.

The hypothetical android as a thought experiment about the nature of human selfhood and agency was an idea so popular during the 1750s that it provoked rancorous priority disputes.[78] The chief dispute swirled around the author of the best-known hypothetical android of the age, Etienne Bonnot de Condillac. The fight broke out at the Paris home of Claude-Adrien Helvétius, who was an enthusiastic admirer of Vaucanson and another materialist, atheist, moral radical, and protégé of Frederick II. At a gathering there following the publication of Condillac's *Traité des sensations* (1754), featuring the awakening-to-the-world of an imaginary human-automaton, Buffon observed within Condillac's hearing that it was "amusing" that Condillac "could make two volumes" from what Buffon himself had "contained in ten pages."[79]

The writer and literary commentator Baron von Grimm on another occasion lamented that Condillac had "drowned M. de Buffon's statue in a barrel of cold water"[80] (he was referring to Buffon's hypothetical tabula rasa machine-man), and the abbé Raynal observed dryly that Condillac did not "have many ideas of his own."[81] Fending off accusations that he had stolen the idea of the hypothetical android from both Diderot and Buffon, Condillac then saw his own thought experiment repeated essentially verbatim by Bonnet.[82] In other words, the android thought experiment was an idea whose time had come.

Like Vaucanson, Condillac was a native of Grenoble; he was the son of a magistrate and member of the robe nobility, Gabriel Bonnot, vicomte de Mably. Condillac, as a young and still unpublished writer, had had the good fortune to interest Rousseau. Rousseau had worked for a time in the Lyon home of Condillac's brother, Jean Bonnot de Mably, as a tutor to Condillac's two young nephews. The author of *Émile* (which describes an imaginary tutor endowing an idealized pupil with Rousseau's vision of an ideal education) was, by his own account, a disastrous tutor: "When my students did not understand me I raved, and when they showed mischievousness I would have killed them." He made eyes at the mistress and stole bottles from the wine cellar and, when he still was not fired, grew bored after a year and quit.

Unaccountably, he remained welcome at the Mably household where, at a dinner several years later, he met Condillac, who had come there on a visit to his brother. Rousseau felt a sense of discovery: "I am the first perhaps who saw his promise and who estimated his true value." The two struck up a working friendship. "He seemed also to be pleased with me," Rousseau recalled, "and while I was shut up in my room, rue Jean-Saint-Denis near the Opera, writing my Hesiod act, he sometimes came to dine with me tête-à-tête. He was working at that time on his *Essay on the Origin of Human Knowledge*."[83]

When Condillac had trouble finding a printer for this, his first book, Rousseau solicited the help of the best-connected person in the world of Enlightenment publishing: "I spoke to Diderot about Condillac and his work; I had them become acquainted. They were made to get along with each other, they did get along." Diderot, in turn, set Condillac up with his own bookseller, Laurent Durand, one of the leading figures in publishing, especially of the controversial or clandestine variety. For a time the three, Rousseau, Diderot, and Condillac, met weekly at the Palais-Royal and went to dine nearby at the Hôtel du Panier Fleuri. "It must be that these weekly dinners pleased Diderot extremely"; Rousseau reflected with satisfaction, "for he, who missed almost all of his appointments, never missed a single one of these."[84]

The three shared an interest in human-machinery. Around the time of his weekly dinners with Rousseau and Condillac, Diderot conjured a pair of hypothetical androids in an essay on the nature of virtue.[85] But the particular hypothetical machine-man that gossips most often accused Condillac of stealing, apart from Buffon's, appeared in another work by Diderot, one that he was also writing around the time of the Hôtel du Panier Fleuri dinners: his *Letter on the Deaf and Mute* (1751). "Consider man an automaton like a walking clock," Diderot proposed in this essay: "Let the heart represent the main spring. . . . Imagine in the head a bell equipped with little hammers from which an infinite multitude of strings extend to every part of the case." A decorative figure on this bell, "its ear bent, like a musician who listens to see if his instrument is well tuned" could then represent the soul. Would that be a material soul, or an immaterial one? Diderot did not say.

This hypothetical clock-person was meant to illustrate a point about the understanding, namely that the mind experiences many sensations at once. If several of the strings attached to the hammers over the bell in the head were pulled together, the little figure at the top would hear several notes sounding simultaneously. Some of the strings, indeed, Diderot supposed, were always being pulled, and these provided the background noise for all the creature's sensations and thoughts. The little figure on the bell would be unaware of these background tones, just as one never notices the background noise of Paris until it is silenced at night. One such continuous, background sensation, Diderot proposed, would be the creature's sensation of its own existence.[86]

How do we *feel* our own existence? This physiological and moral transposition of Descartes's original question, how do we *know* our own existence, occupied the authors of the hypothetical androids. To answer it, Condillac imagined a statue that, like Descartes's automaton statues in the *Traité de l'homme*, had the internal structure of a human being. Like Locke's newborn baby, Condillac's statue began life with an utterly blank mind. And to these basic conditions, Condillac added a third: his statue, having a skin of marble, would be initially incapable of sensation. The experimenter could then endow it at will with each of the five senses in turn. Condillac began the experiment modestly by granting his statue a sense of smell.

Imagine, then, a statue that can only smell. Waft a rose under its nose. To an observer, it will be a statue that smells a rose. But "to itself, it will be simply the odor of this flower itself. It will be thus the odor of rose, of carnation, of jasmine, of violet, according to the objects that stimulate its organ." The odors the statue smells will seem to it not as properties of an external object, but rather as its own "manners of being." Think now that the statue can only hear. Again, it cannot suspect that there exists anything outside itself. It can only feel itself to change, to vary in tone and intensity. Thus "we transform it, as we like, into a noise, a sound, a symphony." A seeing statue similarly "cannot judge that there is something outside itself." It feels itself to be all light and color.

Condillac's statue would feel itself *become* whatever it sensed. Even if it had four of the five senses, he imagined—smell, hearing, taste, and sight—it would still have no way of knowing that its manners of being had external causes. By reflecting upon its experiences through these four senses, Condillac reckoned, the statue could remember, compare, judge, discern, imagine; it would have abstract ideas of number and duration; it would know general and particular truths; it would have desires, passions, loves, and hates; it would be capable of hope, fear, and surprise. But all of these mental operations and feelings would regard the statue itself and itself alone.[87]

The sense of touch, however, was different. Like Buffon, Condillac imagined that with a sense of touch, the statue would begin to discover its own boundaries and the existence of a world beyond. Moving "mechanically," it would gain all at once the sentiment "of the action of the parts of its body against one another." To the feeling of this purely mechanical phenomenon, the brute impact of matter against matter, Condillac gave the name "*fundamental sentiment* because it is with this play of the machine that the life of the animal begins." This feeling of the collision of solid parts would bring about the statue's realization of "*me*."[88]

In touching its own body, the statue would recognize itself in all its parts. But it could only really fully discover its "*me*" when it touched "a foreign body." At that point, "the *me*, that it feels modified in its hands, does not feel modified in this body. If the hand says *me*, it does not receive the same response." The feeling of solidity in an external object would represent to the statue "two things that exclude each other. . . . Voilà therefore a sensation by which the soul travels outside itself." This journey of the soul would bring fearsome emotions. Condillac imagined his statue "quite astonished not to find itself in all that it touches," and from this astonishment would be "born the anxiety of knowing where it is, and if I dare express it thus, to what extent it is." The statue's disturbing discovery of its own borders would produce a new, outward focus for its thoughts and feelings. Its "love, hate, desire, hope, fear no longer have only its own manners of being for their sole object: there are palpable things that it loves, hates, hopes, fears, and wants. It is thus not limited to loving only itself."[89]

The automaton statue, like Buffon's adult tabula rasa, would acquire a self in the impact of its bodily parts against an alien world. Selfhood and all it entailed, including above all the loving knowledge of a world *outside* one's self, would emerge from this original collision.

Except that maybe it wouldn't, or at least Condillac could not quite make up his mind whether it would. Like Buffon and Rousseau, he was ambivalent, and he backed away from a rigorously materialist denial of an immaterial human self. At the very end of his parable, Condillac makes an abrupt and dramatic reversal: the statue's journey, he decides, has not been a discovery of its "*me*" after all. The statue has arrived only at an *illusion* of "*me*." The actual "*me*" remains elusive: "I see myself, I touch myself, in a word, I feel myself, but I do not know what I am; . . . I no longer know what to believe myself."[90] Condillac decided, at least for the time it took him to write the conclusion to his treatise, that it was only an illusion that sensations take place throughout one's body. In reality, they take place in a purely spiritual "*me*," and the illusion of bodily sensations was simply nature's way of connecting this immaterial soul to a set of physical parts. If the human self were actually a material thing, it would be as divisible as any chunk of marble and could shatter to pieces.[91]

Hottentots, Hurons, Negroes, Laplanders, Southerners, Women, Weavers, Poets, and Other Kinds of Human-Machine

The problem of human agency in relation to human-machinery was not just a subject for thought experiments and speculative philosophy, but had urgent social meaning. This was an age greatly preoccupied by questions about race, in particular, but also by other categorizations of human beings including by sex and class. The model of the human-machine became the conceptual framework within which people worked on these taxonomies, and in particular, worked on the question of whether and how varieties of human-machinery might correspond with varieties of human agency.

The human-machine model did not dictate any particular answer or set of answers to these questions. On the contrary, it underlay the whole conversation about race and other human taxonomies, encompassing radically opposing positions. One could, and people did, draw egalitarian implications from the idea that we are all just made of stuff, material bits and pieces. On the other hand, one could also, and people did that too, argue that the world of human-machinery was arrayed along an infinitely graduated spectrum of imperfection and perfection. Often, the same people argued both positions. A model such as the human-machine is truly powerful, not when it compels a certain way of thinking, but rather when it is ubiquitous, supporting all sides of every dispute, underlying the whole conversation of a given age. Such was the case with the human-machine and the Enlightenment conversation about human agency, universalism, and difference.

La Mettrie, for example, at times emphasized the egalitarian implications of the principle that we are all just stuff. The "Ploughman whose Mind and knowledge extend no farther than the edges of his furrow" is made of the very same stuff as "the greatest Genius, as dissections of the brains of Descartes and Newton would have proven."[92] Montesquieu, for his part, a thinker of a different temperament and politics, yet drew a similarly egalitarian lesson from the idea of human-machinery. He reckoned that if the native people of the New World had known that all human beings were but machines, things would have turned out differently. If only Descartes had gone to Mexico and Peru "a hundred years before Cortez and Pizarro, and had taught these people that men, composed as they are, cannot be immortal; that the springs of their machine wear out, like those of all machines; that the effects of nature result only from laws and the communication of motion, Cortez, with a handful of men, could never have destroyed the empire of Mexico nor Pizarro that of Peru."

How could ignorance of a philosophical principle, that human beings are machines, subject to physical causes, produce the greatest defeat in history? The Mexicans and Peruvians had had everything going for them, Montesquieu mused, including great bravery. But they had taken the European invaders for representatives of "an invisible force" against which they had no hope.[93]

On the other hand, La Mettrie also reasoned that if we are all just moving parts and our every capacity depends upon our machinery, then differences in machinery must create significant differences in merit: "One People have a heavy and stupid mind, another a lively, light and penetrating one." These divergences among peoples, he considered, must derive from "the food they eat, their Fathers' semen, and that Chaos of diverse elements that swim in the immensity of the air."[94] Climate was especially important in conversations about the scale of perfection of human-machinery. "Such is the empire of Climate," La Mettrie observed, "that a Man who changes climates feels the change despite himself." Social climate was just as important: La Mettrie observed that people took on the manners, gestures, and accents of the people around them, just as spectators at a pantomime would "mechanically" imitate the artist. "What I have just said proves that the best company for an intelligent Man is his own, if he cannot find a fellow. The Mind rusts with those who don't have any for lack of exercise."[95]

The role of the environment in the development and functioning of the human-machine received great emphasis starting around midcentury. Montesquieu developed the idea in a pivotal chapter of his monumental study of law, *On the Spirit of the Laws*. He reasoned that different climates required different laws because the human-machines differed correspondingly. In the southern countries, a "delicate, weak machine, but sensitive, inclines to a love that either rises and calms itself ceaselessly in a seraglio, or else, leaving women in a state of greater independence, is exposed to a thousand troubles." Far better luck to be born in the north and enjoy a "healthy and well-constituted machine, but heavy," not so easily set in motion. The northern regions, populated by such machinery, boasted people of "few vices, plenty of virtues, much sincerity and frankness."[96] It seems, Rousseau later amended, that the "organization of the brain is less perfect at the two extremes. Neither the negroes nor the laplanders have the sense of europeans. If therefore I want my student to be able to inhabit the whole earth, I would take him to a temperate zone," such as, let's see now, well, "France, for example."[97]

Given this great diversity in perfection of human-machinery, it was perfectly okay, La Mettrie thought, for superior machines to be proud of themselves. It was a mere acknowledgment of the obvious: "Those whom Nature has filled with her most precious gifts, should pity those to whom these have been refused; but they can feel their superiority without arrogance." All qualities were physical qualities on the man-machine model, even intellectual and moral qualities. To deny them would be as silly as to deny that one is tall or has brown eyes. "A beautiful woman would be as ridiculous to find herself ugly, as a Wise

Man to believe himself a Fool." Moral superiority, in particular, was a physical feature consisting of having sharper, more "exquisite feelings." These well-tuned types were the "Imaginative Men," the "great Poets," those "ravished by a well-expressed sentiment . . . transported by an exquisite taste, by the charms of nature." People whom nature had favored with so felicitous a mechanism, La Mettrie urged, "merit more regard, than those whom she has treated as a Stepmother."[98]

Diderot offered a careful correlation between nervous mechanisms and qualities of character. An extreme mobility of certain threads of the nerves, he explained, produced an overly acute sensibility. If the trunk of the network of nerves was "too vigorous in relation to the branches," the result would be "poets, artists, people of imagination, cowards, enthusiasts, madmen." When such a disposition afflicted great men—Voltaire, for example—they worked tirelessly to dominate it. Artistic greatness arose from the epic struggle of the machine against its own organization. Meanwhile, an energetic, well-regulated, well-ordered system would produce "great thinkers, philosophers, sages." Mediocrity, on the other hand, derived principally from an "extreme mobility of certain threads of the network." If the central component of the nervous apparatus was too weak, it would produce "what we call brutes or ferocious beasts." A whole network that was "slack, limp, lacking in energy" would result in an imbecile.[99]

The differences between male and female machinery, and their moral implications, commanded the interest of the man-machine theorists. For example, Buffon observed that once one understood the "play of the machine," one could clearly recognize a correspondence in women between "the womb, the teats and the head."[100] Diderot likewise deemed the action of the "womb" in the female machine to be "too strong" for the rest.[101] The Swiss physiologist and "physiognomist" Johann Kaspar Lavater notoriously made mechanical organization the central parameter of his physiognomic science, in which he professed to derive all important facts about a person's character from his or her physical structure, especially the face. In his exposition of his methods, for example, Lavater displayed "a silhouette of a youth of most happy organization," one of a "woman very happily organized," and best of all, the author's own silhouette, "of an organization infinitely delicate."[102]

Social class differences too could be explained by variations in human-machinery. Those with the crudest machinery, Rousseau wrote, were best suited to menial tasks: "Weavers, stocking-makers, stone-cutters; what would be the point of employing sensitive men in these jobs?"[103] The equivalence of character and organization of machinery pervaded moral and political writing during the second half of the century from Montesquieu to Adam Smith. The "prudent, the equitable, the active, resolute and sober character," observed Smith in his *Theory of the Moral Sentiments* (1759), had "all the beauty which can belong to the most perfect machine."[104]

The man-machine model also provided a basis for theories of racial difference. On a material continuum of complexity and perfection, the threshold of humanity could go blurry. "There are still people on earth," mused Montesquieu, "among whom a passably educated monkey could live with honor; he would find himself just about at the level of the other inhabitants; they would not find his mind to be singular or his character bizarre; he would pass as one of them, and would even be distinguished by his kindness."[105] The bodies of men and monkeys, Buffon judged, were "machines" of the same sort. Going purely on the basis of machinery, he found no great difference between an orangutan and a Hottentot:

> The head covered with bristly hair, or a frizzy wool; the face veiled by a longer beard, surmounted by two crescents of even thicker hairs, which by their size and projection shrink the forehead, and make it lose its august character, and not only cast shadows over the eyes, but deepen them and round them out like those of animals; the thick and protruding lips; the flattened nose; the stupid or fierce look; the hairy ears, body and members; the skin hard like black or tanned leather; and for sexual attributes, long and soft teats, the skin of the belly hanging down to the knees; children wallowing in the filth and crawling on all fours; the father and mother squatting on their heels, all hideous, all covered with a pestilential grime.[106]

Having canvassed these similarities in Hottentot and orangutan bodily machinery, Buffon abruptly recalled that Hottentots had language, which orangutans lacked. This capacity, which created an "immense" interval after all, could only originate in something nonmechanical, Buffon thought: a God-given soul.[107]

La Mettrie, who believed in nothing nonmechanical, observed that a "well-constructed Animal" might even be more human than a poorly organized human. A "Mindful Monkey," La Mettrie thought, should be able to learn astronomy and predict an eclipse as well as any human astronomer. Conversely "the imbecile, or the stupid, are Beasts in Human form." Human-machinery generally included the capacity for education, he explained, which elevated humans above the rest of the animal world, but the distinction did not apply to badly constructed, hence ineducable human-machines: "To the Deaf, to the Blind-from-birth, to Imbeciles, to Lunatics, to Savages, or those who have been raised in the Woods with the Beasts . . . finally, to all these Beasts in human form. . . . No, all these Men in body & not in mind merit no particular classification."[108]

Bonnet too described a spectrum of human-machinery and a corresponding scale of moral and intellectual perfection among humans. Although he did on occasion invoke an immaterial soul, as we will see, in this context he was persuaded that bodily machinery was the determining factor. The gulf separating a Montesquieu from a Huron, Bonnet reckoned, originated in the brain: if the divine power had lodged the soul of a Huron in the brain of a Montesquieu, "would not this brain, so highly organized, and so richly furnished, have been to this soul a kind of optical machine, by which she would have viewed the universe, in the same manner as the sublime author of the Spirit of Laws viewed it?"[109]

The descent by intervals from a Montesquieu to a Huron to an orangutan was a material continuum, and Bonnet was firmly committed to its continuity. The continuity of nature, with no gaps or breaks, was a core principle of contemporary natural history. Leibniz, we have seen, embraced this originally ancient view and was influential in transmitting it to his contemporaries and followers.[110] Bonnet, for his part, was an avowed Leibnizian: "I drew from this great Man most of my Ideas about the Past and Future of Animals."[111] However, Bonnet departed from Leibniz in his application of the law of continuity to his understanding of the races of humanity. Whereas Leibniz named reason as the defining and universal feature of human beings, and explicitly rejected the idea of "interior," "essential specific" differences among human races,[112] Bonnet arrived at a different conclusion.

In a chapter entitled "Gradations of Humanity," Bonnet explained that if one compared the most perfect man to a monkey, the scale of nature might seem "to suffer a great interruption," but if one took into consideration the whole scale of humanity, one would realize "humanity has its gradations like all the productions of the globe":

> Between the most perfect man and the monkey, there is a prodigious number of continuous links. Glance over the nations of the earth; consider the inhabitants of a single kingdom, of a single province, of a single city, of a single town; what am I saying! Look at the members of a single family and you will believe

> you see as many species of men as you see individuals. The dwarf of Lapland succeeds the giant of the Lands of Magellan. The African with a flat face, black tint and hair of wool gives way to the European in whom the regular features are heightened by the whiteness of his tint and the beauty of his hair. To the uncleanliness of the Hottentot contrast the cleanliness of the Dutch. From the cruel cannibal pass rapidly to the humane Frenchman. Place the stupid Huron next to the profound Englishman. Climb from the Scottish peasant to the great Newton. Descend from the harmony of Rameau to the rustic songs of the shepherd. Balance the ironsmith who makes a roasting spit against Vaucanson creating his automata. Count how many steps there are from the blacksmith who works the anvil to Réaumur analysing iron.

These differences, Bonnet affirmed, in no way depended upon "a real difference between human souls"—returning perhaps to a more Leibnizian principle of the essential sameness of human beings—but rather upon the organization of the machinery.[113]

In this connection it seems important to note that Leibniz's later commitment to the essential unity of the human race did not prevent him, as an ambitious twenty-five-year-old, from extensively recommending the conquest of Egypt to Louis XIV, along with the colonization of the Americas, nor, to that end, from sketching a plan for establishing an army of slave-warriors.[114] However, the differences that supported enslavement and conquest in Leibniz's imperial schemes were not primarily racial nor did they bear any relation to bodily organization or machinery; rather, they were sectarian and linguistic.[115] On the contrary, Leibniz's conception of organic machinery helped to provide the basis for his understanding of human sameness. On the principle that all organic bodies, plants and animals, were "machines" for "perpetuating" certain duties, he identified the human body in all its variations as a machine for "perpetual contemplation."[116]

On the other hand, J.-B.-C. Delisles de Sales, the widely read author of *De la philosophie de la nature* (1769), agreed with Bonnet that the organization of the machinery was the key to human categories, though he dissented on the matter of the Huron brain. A Huron, he thought, was an excellent machine: "Nothing impedes the action of his organs; the animal spirits circulate freely in his fibers; . . . he has a lively mind because he is well organized." Imagine, he urged, that such a "savage . . . born perfectly organized" traveled to Paris. Reading Montesquieu, seeing Racine performed at the theatre, strolling the Tuileries amid Le

Nôtre's landscaping and Girardon's sculpture, the savage would soon be a man of taste. Despite his commitment to Huron educability, however, Delisles de Sales was no believer in universal human equality: to a "feebly organized" machine, he was sure, such an attainment would be forever out of reach.[117]

In contrast, Pierre-Louis Moreau de Maupertuis argued for a different theory of the human races on the basis of his own understanding of the mechanics of generation as a collaboration of innumerable sentient parts. In this process, Maupertuis suggested, "chance" always played a role, resulting in the occasional happenstance of offspring who did not resemble their parents. Such a chance occurrence, he surmised, had produced the first black-skinned people. These people had originated as random variations, but they had then been forced toward the "torrid zone" of the earth when "vanity or fear turned the greater part of the human race against them."[118]

At that point, a process equivalent to animal breeding would have taken over: by reproducing solely with one another, over many generations, any people sharing a distinguishing feature such as skin color would "confirm the race." As another example, Maupertuis invoked the tiny feet of Chinese women. He acknowledged that these were partly the result of the Chinese practice of foot binding but, he thought, not entirely: Chinese women seemed to be born with especially small feet, a result of selective cultivation just like the features of fancy pigeons and dogs. Finally, although Maupertuis was sure that the essential cause of human racial differences was happenstance in the combinations of the tiniest generative particles, in conjunction with selective reproduction analogous to animal breeding, he did "not exclude the influence that climate and nourishment might have," especially over the course of many centuries.[119]

Like Maupertuis, Helvétius also argued for a fundamental sameness of human beings on the basis of his understanding of human-machinery. He insisted that all human beings were machines of essentially equal organization and rejected the idea that organization or climate created the evident differences among countries in "their characters, their genius and their taste." Rather, he insisted, it was the form of government that made all the difference. Citizens of a republic, for example, valued eloquence because it could bring them riches and power, as could politics, jurisprudence, morality, poetry, and philosophy. People living under despotic rule, in contrast, had no use for these sciences.[120]

Helvétius was consistent and exceptional in his refusal to associate human differences with human-machinery.[121] He adopted the same stance with regard to intellectual differences between men and women, arguing that the key factor in creating the inferiority of women was not their mechanical organization but their inferior education.[122] His friend, ally, and collaborator Diderot vigorously disputed this view, insisting that organization was the original source of all variations in temperament, ability, virtue, and passion. If all individuals were different sorts of machines, Helvétius asked, how could divine or human justice treat them the same? Diderot responded that justice must reflect the diversity of human-machinery.[123] Moreover, if Helvétius's books were really true, Diderot observed sardonically, they could just as well have been written by his dog keeper.[124]

The French naturalist Jean-Baptiste Lamarck, writing several decades later, maintained like Helvétius that all humans had the same machinery, the same organs at the same "degree of composition." Differences in intelligence, Lamarck wrote, were due to the fact that the brain, like any organ, differed according to the uses and exercise it got. The "brain of a man of labor, who spends his life building walls or carrying burdens," was not "inferior in composition or perfection to that of Montaigne, Bacon, Montesquieu, Fénelon, Voltaire, etc," but merely lacked exercise.[125]

We are all just stuff. But the world of stuff is infinitely divisible. The human-machine model and the conversation about the relations between human-machinery and human agency underlay a rapid alternation between universalist ideals on the one hand and the whole panoply of relegations on the other: Hottentots are mechanically indistinguishable from orangutans; female machines are driven by the womb; faulty machines such as the deaf and blind are not really human and so on. This dizzying alternation was itself a hallmark of the Enlightenment discussion of human nature. We are all the same, a single type of living machine. We are arranged on a graduated scale of mechanical perfection from Montesquieu down to a monkey, from Vaucanson to his own automata.

Organized Rather Than Designed, Mr. Machine Evolves

Mr. Machine, that moral, material creature, was no passive, brute mechanism. On the contrary, he was an active, self-moving, self-constituting mechanism. No divine Clockmaker assigned him his structure, function, or source of movement. These developed from within himself. In other words, Mr. Machine was not a *designed* machine, but an *organized* one.[126] La Mettrie adopted the notion of "organization" from Leibniz, though he regretted that Leibniz had "spiritualized matter" (recall Leibniz's tiny, perceiving "monads," spiritual entities that made up the world) and had therefore produced an "unintelligible" system.

When La Mettrie adopted core Leibnizian concepts such as "moving force" and organization, he translated these into purely material phenomena.[127]

La Mettrie's mentor, Boerhaave, had defined an organic or organized body as one "consisting of different parts, which jointly concur to the exercise of the same function."[128] An organized machine was a concurrence of active parts, unlike the rigidly deterministic, designed clockwork described by natural theologians.[129] It derived order and agency from its own workings. "Organization," La Mettrie wrote, "is the first merit of Man" and the source of all the others. Every faculty that had been attributed to mind or rational soul came down in the end to "Organization itself; voilà a well-enlightened Machine!"[130]

Joseph Priestley, admirer of La Mettrie, agreed that sensation and thought resulted purely from "the organization of the brain" in the same way that an attractive power resulted from magnetized iron, another example of an apparently intrinsic agency in matter.[131] Priestley ascribed the idea of a disembodied, rational soul to "the vain imaginations of men, flattering themselves with a higher origin than they had any proper claim to." He also called it a modern perversion, the work of Descartes and especially his followers, driven by an unreasonable "dread of materialism."[132]

The fresh-water polyps that the Genevan naturalist Abraham Trembley had studied exhaustively several years earlier provided evidence for La Mettrie's claim that living matter organized itself. The polyps, when cut into pieces, could regenerate themselves into full animals from each piece. "Look at Trembley's polyp," La Mettrie exclaimed. "Does it not contain its own regenerative causes?" The polyp's self-regenerative ability represented an inherent vital force in nature, the study of which could not fail to "make unbelievers," since it indicated an internal rather than an external source of life's order.[133]

With regard to arguments from design, La Mettrie ridiculed the endless demonstrations of God's existence from mechanical details of physiology. The eye might well work like a telescope, he admonished, but that did not mean someone had constructed it expressly to do so. "Nature no more thought of making an eye to see than water to serve as a mirror for the simple Shepherd." Water just so happened to reflect images, as other substances happened to reflect sound, and likewise the eye "sees only because it happens to be organized and placed as it is." Eyes and ears required "no greater artifice" than the "the fabrication of an echo."[134] The optics of vision, so prominent in the argument-from-design tradition that it would later shake the resolve of Darwin himself, was to La Mettrie just happenstance.[135]

At any rate, the celestial Engineer featured in arguments from design constituted a poor image of the deity, according to Buffon. "Who in fact has the greater idea of the Supreme Being," he asked: the one who perceives him as the source of natural law, or the one who "finds him attentively conducting a republic of flies, and greatly occupied with how a beetle's wing should fold?[136] The sneer was directed at Buffon's rival, Réaumur, who had offered his voluminous and minute descriptions of insects as so many demonstrations of God's skill. Buffon judged that one did greater service to God by seeing the order in his Creation as emerging from its own inner organization rather than his tinkering designs.

David Hume, we have seen, made the same argument: design was a fatally contradictory notion because no limited contraption could indicate divine omnipotence. Hume too answered "design" with "organization": a kind of order that did not arise from "reason or contrivance."[137] A tree "bestows order and organization on that tree, which springs from it, without knowing the order: an animal, in the same manner, on its offspring: a bird, on its nest." To insist that all the order evident in plants and animals proceeded "ultimately from design," one would first need to show that order itself was "inseparably attached to thought, and that it can never, of itself, or from original unknown principles, belong to matter," two laws that Hume believed could never be demonstrated.[138]

In the wake of La Mettrie's best seller, *L'homme machine*, the word "organization" proliferated in natural history, first in France but soon in the writings of Swiss, German, and English naturalists.[139] In literature and the arts, in moralist and aesthetic writing as well, "organization" became a keyword after 1750, preserving the sense of an animate order that was internally generated rather than externally imposed: a result of the organized being's own agency.[140]

Buffon picked up the idea of organization in his own best seller. The opening volumes of *Histoire naturelle* identify nature's chief business as the production of life by "organization": "the most ordinary work of Nature is the production of the organic."[141] Organization also provided the key to Buffon's understanding of an individual being, which he defined as "a whole uniformly organized in all its interior parts." These parts, in turn, he supposed to be made up of "an infinity of little organized beings," or "organic molecules." How did these little beings manage to sort themselves into the proper organization? Buffon surmised that an "internal mold" brought them together in generation and nutrition, and they came apart again after death.[142]

Haller, a partisan of design over organization, objected to Buffon's idea of an "internal mold," insisting that it could never accomplish the task: it required a "building master" to ensure that "never could an eye become attached to a knee, or an ear to a forehead."[143] To Buffon, however, as Leibniz had said, a living creature was an organized conglomeration of organized beings: organization all the way down, with an inner self-organizing agency acting at every level.

Like La Mettrie, Buffon emphasized that organized bodies worked differently from artificially built contraptions. The "organic machine" of the sense organs and brain, for example, differed from artificial machines in that it was not only capable of "resistance and reaction," but "itself active."[144] The original "springs" of animal motion were not the visible muscles, veins, arteries, and nerves, but rather the "interior forces" that clearly resided within organized bodies. These forces did not follow the "laws of gross mechanics . . . to which we would like to reduce everything." After all, why must the very few properties of matter admitted by Descartes and other classical mechanists—extension, hardness, shape, movement—be the only ones?[145]

Charles Bonnet also cited "organization" as the reason why traditional mechanical principles could never explain the workings of "animal machines" although they contained wondrous arrays of "levers, counterweights, diversely calibrated tubes, curves and bypasses."[146] Organized bodies, or "organic machines," contained a "secret mechanics" of their own that allowed them to proliferate, grow, and heal.[147] Like his sage Leibniz, Bonnet subscribed to the doctrine of preformation and he also shared Leibniz's view that the development of living beings from preexisting forms happened through nature's inherent power of organization. "Of all the modifications of nature," Bonnet mused, "the most excellent is *organization*."[148]

In his theory of nutrition, Buffon invoked a sort of "active power": the "penetrating" tendency of living, organic matter to organize itself. Like gravity, this self-organizing agency bore no relation to the external features of an object, but only to its "interior," acting upon the "most intimate parts and penetrating them at every point." Such forces would remain forever essentially mysterious: "In a word they escape our eyes" and "we will never reach them by reasoning."[149] His theory of the human constitution as an organic rather than a designed machine, like La Mettrie's, was a rebuke to rationalists, with their hubristic arguments from design, who set humans, and especially themselves, apart from the rest of nature with the idea of the disembodied rational soul. Buffon promised his readers the moral equivalent of a cold shower: they must learn the "humiliating" truth that man was an animal. Worse yet, he occupied the same continuum as "the most unformed matter" and "the most brute mineral": never mind moles and worms, no sharp discontinuity separated humans even from rocks.[150]

La Mettrie's friend and protector, the French mathematician and philosopher Maupertuis, also took up the Leibnizian concept of "organization." Maupertuis, like La Mettrie, was a native of Saint-Malo and a protégé of Frederick II, for whom the monarch had several years earlier secured the presidency of the Prussian Royal Academy of Sciences. It was Maupertuis, in fact, who had devised the rescue plan of bringing La Mettrie to Berlin in 1748 after the publication of *L'homme machine.*[151]

Like Leibniz, Maupertuis understood the physical world—including the inanimate physical world—to regulate itself by means of intrinsic, active principles, by "preferences" and choices. For example, one active physical principle that Leibniz had developed was that light, in its propagation, reflections, and refractions, would always followed the path that allowed it to travel the fastest. But this principle was inconsistent with Descartes's view, shared by Hooke and Newton, that light traveled faster in denser media. According to this view, the refracted path of light moving from air to water was neither the shortest nor the quickest. To rescue the Leibnizian idea of an active principle by which light governed its movement, Maupertuis devised his "principle of least action": he proposed that light would always follow the path "by which the quantity of action is the least," where "quantity of action" was given by a combination of the speed and the distance traversed.[152]

Maupertuis also shared with Leibniz the conviction that organized, living bodies operated by a particular sort of inner agency. Whereas Buffon compared the self-organizing power of living things to gravity, Maupertuis thought, on the contrary, that a gravity-like "uniform and blind attraction spread through all the .parts of matter" could not account for how the elements of a living creature came together: how some parts formed an eye while others made an ear. The very elements of matter must contain, not just a force, but "some principle of intelligence . . . something similar to what we call *desire, aversion, memory*."[153] Elsewhere, Maupertuis called it an "instinct" belonging to "the smallest parts of which an animal is formed," and constituting the true "essence of the animal."[154]

Perception, sentience, in other words, must be an elementary property of matter. In which case the elements themselves, having perception, according to Maupertuis, might be considered "animals" themselves. Maupertuis thought it implausible that living beings had come together randomly out of brute and unintelligent parts, and equally implausible that God had used such parts to create them the way an architect builds with stones. Rather, "the elements themselves, endowed with intelligence, arranged and united themselves to carry out the vision of the Creator": no externally imposed construction of blocks or stones but a confluence of sentient participants.[155]

Authors of the concept of "organization" generally avoided building metaphors, blocks and stones, in favor of another sort of image: weaving. A living being was not like a wall built of bricks, but rather like a self-moving loom or a self-weaving fabric. Bonnet described organized bodies both as "*looms*" that assimilated and incorporated materials into themselves and also as "cloths, networks, sorts of fabrics in which the warp itself forms the woof." Each fiber, each "fibrilla," of these loom/fabrics was itself a miniature machine and the "entire machine is in a sense nothing but a repetition of all the *machinelets*."[156]

In fact, an organized body was at once the loom and the fabric, weaving itself such that its parts were themselves self-weaving looms, the whole in constant flux, never remaining the same for a single instant.[157] An animal, wrote Diderot, is a "machine that is born from a point, from a churned fluid," whose development depends upon "a bundle of thin, separate and flexible threads, a sort of skein in which the least strand cannot be broken, worn out, [or] displaced."[158]

The organic self-weaving agency of living beings extended beyond the individual into its living environment. Bonnet emphasizes that organisms wove themselves from the world around them. Each organized body was a burgeoning mix of all the others. Each was "a little earth, where I perceive in miniature all the species of plants and animals." An oak tree was "composed of *plants, insects, shells, reptiles, fishes, birds, quadrupeds,* even *men.*" The earth itself, with its air and water and soil, appeared to Bonnet as "but a mass of seeds, a vast organic whole."[159]

The organizing agency was in fact all-pervasive, extending indefinitely outward to encompass the world itself and indefinitely inward to the microscopic workings of organic bodies: "We do not know where organization finishes, what is its smallest term.[160] Perhaps it would be an exaggeration "to *organize* or to *animalize* everything," Bonnet allowed, but on the other hand, what appeared unorganized or inanimate might not really be so. There was "no philosophical reason to limit the scale of animality to this or that production": the material world might well be fully permeated "with life and sentiment."[161]

These theorists of "organization," rather than design, as the basis of life described a physical world imbued with perception, feeling, and self-organizing agency. Diderot agreed that sensitivity was a universal and primitive property of matter, extending to every stone or speck in nature, corresponding with its level of organization.[162] A living, sentient cosmos thrumming with feeling traveled arm in arm with Mr. Machine, a further effect of eliminating Descartes's disjuncture between mechanism and self, self and world.

Through the pump-like lungs of the passionate, sensitive, moral, organized man-machine breathed the first approximations of a modern theory of evolution.[163] If man was a worm and a mole, La Mettrie observed, no sharp discontinuity separated him from animals. "Go, proud reasoner," likewise proclaimed Erasmus Darwin, "and call the worm thy sister!"[164] Darwin was the leader of the Birmingham philosophical dinner club, the Lunar Society,[165] attended by Priestley, and as we have seen, shared an interest in living and artificial mechanisms, and the relations between these.

Like La Mettrie, Darwin believed that all animals including human beings shared the same basic mechanical "organization,"[166] and that sensation and sentiment were spread throughout living matter (he too rhapsodized about the sex lives of plants).[167] From comparative anatomy, La Mettrie drew examples of structural similarities between the organs of human beings and their equivalents in other animals. The great ape resembled human beings so strongly, he thought, that it should be able to learn a language.[168] Regarding the orangutan at the Jardin du Roi in Paris, Diderot wrote similarly, "Cardinal Polignac said to him one day, 'speak, and I will baptise you.'"[169]

In general, La Mettrie saw no problem in imagining that "an intelligent Being could come from a blind Cause." After all, he pointed out, it required no genius on the part of parents to produce intelligent children. As a woman's womb, "from one drop of liqueur, makes a child," so a part of the machinery of human beings had simply turned out to be suited to retaining and producing ideas. "Having made, without seeing, eyes that see, [nature] made, without thinking, a machine that thinks."[170] The resulting human-machine, La Mettrie mused, was like a vessel constructed to sail on its own, but also constantly pulled this way and that by the wind and the currents: "a ship without a pilot in the middle of the sea."[171]

If human-machinery had not been designed all at once from brute parts, but had instead organized itself from active, sentient parts, this suggested a very different kind of process. Like Leibniz, La Mettrie thought it must have taken place over a long time. Unlike Leibniz, however, La Mettrie assumed that the self-organization of living matter had a kind of randomness at its core. Acting as a "blind Cause," nature must have produced people and animals only "little by little" from the smallest and humblest beginnings. Matter would have had to pass through an "infinity of combinations" before arriving at the one that produced "a perfect Animal."

In a striking anticipation of Darwinian natural selection, La Mettrie imagined that less perfect animals would die before reproducing, while more perfect animals would survive for longer and reproduce more, bringing about an overall change in the population. The first animals must have been very flawed: "Here the Esophagus would have been missing; there the Stomach, the Vulva, the Intestines, etc." Only those animals with the necessary equipment, not only for survival but also for reproduction, could have perpetuated their species. The others, "deprived of some absolutely necessary part, would be dead either soon after their birth, or at least without having reproduced. Perfection has no more been the work of a day for Nature than it is for Art."[172]

Maupertuis picked up the same idea several years later. "Chance, let us say," he wrote, "produced an innumerable multitude of Individuals," but only a small minority of these had "turned out to be constructed such that the parts of the Animal could satisfy its needs." The others had perished, leaving only "the smallest fraction of what a blind fate had produced," those lucky species "in whom there was order and fitness."[173] In such a process of trial and error, La Mettrie found an alternative to both "Chance" and "God," namely "Nature."[174]

Here were two principal ingredients of the evolutionary theories that were to emerge over the next century: the idea that living organisms including humans might be the result of a gradual process and the possibility that nature could be orderly without being designed. Both arose through La Mettrie's and others' attempts to describe a kind of machinery that was, in keeping with Leibniz's organic machines-all-the-way-down, not rationally designed but self-organizing. The materialist moralist impulse that motivated these attempts, and the accompanying notion of "organization," found expression in the work of a growing class of people who renounced the doctrine of special creation—the idea that God had designed each form of life and placed it in its niche in the natural world—for the conviction that species changed over time: what the French anatomist and anthropologist Paul Broca would retrospectively, in 1870, call "transformism."[175] I will use Broca's term to avoid reading aspects of later "evolutionary" theories back into these early ideas about species-change.

When Lamarck, whom Broca, and subsequent tradition, designated as the original "transformist," scrutinized the many varieties and gradations in the ongoing composition of the "animal organization," he decided that reason itself was no peculiar capacity of humans, but a function of the nervous system and therefore common in lesser or greater degrees to all animals.[176] In Lamarck's view, God was only indirectly the creator of the observable world, acting through the intermediary of nature itself. Nature, Lamarck judged, was "certainly not a reasonable being," but rather a "blind power, everywhere limited and constrained." He objected to arguments from design on the grounds that it was a mistake to attribute intentions or goals to such a power.[177]

But Lamarck was not really the first transformist, although as we will see in the next chapter, he was the first to work out a fully elaborated modern theory of the transformation of living forms.[178] From Leibniz onward, those who tried to build the agency responsible for structuring living machines into the machines themselves arrived at self-organizing beings that changed themselves over time. La Mettrie described his man-machine as emerging "little by little" from the "smallest beginnings," and this inspired a host of other self-transforming organic machines.

Buffon, perceiving similarities joining even the most divergent models of "animal machine," observed that one might even believe all the animals had originated from just a single one by means of a continual "mixing," a "successive variation," and the gradual perfection or degeneration of the resulting forms.[179] Maupertuis suggested along similar lines that the intelligent, sentient action of the parts of matter in forming animal bodies could explain the proliferation of species. If the elementary parts combined themselves too readily or forgot the order of the parent animals, each mistake would produce a new species, and "by means of these repeated divergences would have come the infinite diversity of animals that we see today." This diversity would likely continue to increase with time, but very slowly, so that "the succession of centuries would bring only imperceptible increases."[180]

Transformism was an idea that traveled through the eighteenth century in tandem with rejections of the argument from design, with its passive-mechanist picture of living beings. For example, Hume reasoned that the world was much more like an animal or a vegetable than it was like any "artificial machine," such as a watch or a loom. "The cause, therefore, of the world, we may infer to be something similar or analogous to generation or vegetation," a growth or emergence or development over time.[181] "If faith had not taught us," Diderot asked, "that animals left the hands of the creator just as we see them," would we not begin to suspect that the "elements" of "animality" had come together differently? Could we not imagine that these elements had started out "separate and confused in the mass of matter"? And that they had come together simply "because it was possible for that to happen"? The embryo they had formed must have had to pass through "an infinity of organizations and developments." It had gradually gained "movement, sensation, ideas, thought, reflection, sentiments, passions, signs, gestures, sounds, articulate sounds, a language, laws, sciences, and arts."

Each of these developments took "millions of years." And the process was ongoing, with unknown developments still under way. Perhaps life would "disappear forever from nature, or rather it will continue to exist, but in a different form, and with faculties altogether different from those we remark at this instant in time."[182] Regarding the idea that animals had always been what they were at present, Diderot wrote, "How silly! We no more know what they have been than what they will become."[183] He imagined race upon race of animals coming into and passing out of existence in unending succession. The present moment represented but an instant "in the succession of these animal generations."[184] The Jesuit abbé Augustin Barruel, scandalized by this suggestion of Diderot's, and intending a devastating satire, restated its point with eloquent clarity despite himself. Could we not conclude, Barruel wrote, that the embryo described by Diderot "was at first a simple machine, an automaton, next a fly, a mouse, a dog, a fox, a horse, a parakeet, an eagle, an elephant, a man directed by laws, and finally author of the sciences and arts?"[185]

Diderot's eternally transforming organic machines constituted a strikingly nonprogressive view of the transformation of living forms. This is a critical way in which the early transformism of La Mettrie, Diderot, Buffon, and Maupertuis differed from later evolutionary ideas, as well as from Leibniz's earlier one. "The imperceptible worm that wriggles in the mire" might be on its way to becoming a great and fearsome beast; but likewise, today's enormous and terrifying animals were likely tending into worms.[186] Diderot's vision of natural history was also startlingly indeterminist. Remote and indifferent, the sun was the cause of all. Extinguish it and everything would perish; relight it and the resulting "infinity of new generations" might never include our own plants and animals.[187] Let "the current race of existing animals pass; let the great, inert sediment act several million centuries," and there was no telling what sort of beings might result.[188]

Above all, humans represented only a brief and haphazard moment, and no kind of culmination. Indeed, we might not even be the protagonists of our own moment. Being made up of infinite "animacules," themselves in constant flux, might we not be but the "breeding-ground of a second generation of beings, separated from this one by an inconceivable interval of centuries and successive developments?"[189]

The power of self-creation was the new form of agency suitable to a rigorously material being, a man-machine. But this power was matched by a commensurate impotence: with self-organization came the profound uncertainty of an un-Designed creature forever in danger of losing itself in an un-Designed world.

~ 6 ~

Dilemmas of a Self-Organizing Machine

Consider Frankenstein's monster, cobbled together from dead matter yet driven by an inexorable living agency. Born of a nightmare in the summer of 1816, the monster personified the predicament of the man-machine, Designer-less, orphaned and contradictory, torn between active- and brute-mechanist principles, its living agency at terrible odds with its inanimate material parts. The man-machine's quandary was a chief preoccupation of the Romantic movement, as this period is traditionally known in the history of literature and the arts, and more recently, in the history of science too.[1] The Romantics struggled mightily with the idea whose beginnings we witnessed in the previous chapter, that living beings might be self-organizing and self-transforming machines, striving to constitute and reconstitute themselves in the dynamic, living machinery of nature. From this period and its struggles there emerged both the disciplinary science of biology and the first full-fledged theory of the transformation of species, Lamarck's theory of species-change (see figure 6.1).

The condition of a self-organizing living machine inhabiting an un-designed world inspired an effusion of poetry and science—or to put it better, an effusive *fusion* of poetry and science—around the turn of the eighteenth to the nineteenth centuries. A remarkable intimacy between poetry and science characterized the Romantic movement.[2] While poets did electrical experiments, attended chemistry lectures, flocked to physics demonstrations and pored over

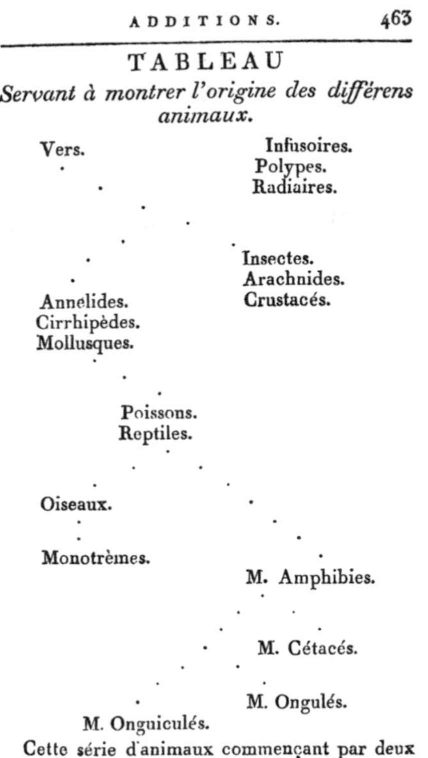

Figure 6.1 Lamarck's chart of species-transformation from *Philosophie zoologique* (1809) showing more complex animals arising from simpler ones.

the latest results in physiology, the physicists, chemists, and physiologists wrote poetry, often presenting their experimental results and theories in poetic form.

Immanuel Kant proposed formal grounds for this intimacy of science and poetry in his last major work, the *Critique of Judgement* (1790). He argued that living nature must be regarded as full of agency, inherently "purposive," and that people were only able to apprehend it as such, in the first instance, by means of an act of aesthetic judgment, a feeling of pleasure or displeasure. Judgments of natural objects, therefore, all began with an aesthetic response.[3] Like Descartes,

Kant is often identified as the first modern philosopher, and like Descartes, he worked at the fault line between the intellectual self and the mechanical world, which might suggest that this fault line was itself the starting-point for modern philosophy. The German philosopher, writer, and poet Johann Wolfgang von Goethe, whose discovery of Kant's *Critique of Judgement* inaugurated "a wonderful period in my life," rejoiced in the union of science and poetry he found there.[4]

Goethe feared that his own efforts to combine science and poetry constituted a lone act of rebellion, remarking in frustration that "nowhere would anyone grant" the essential oneness of scientific and poetic knowledge.[5] He was mistaken: virtually everyone granted it, though they did tend, like Goethe, to suggest that their poetic approach to nature constituted a rebellion. If so, it was a large rebellion, and one that defined itself by opposition to an ideal of nonpoetic science: the rational account of a passive mechanical cosmos.

The Romantic love affair between science and poetry constituted a search for an alternative approach to science, and in particular to the science of life. From this search came an understanding of the mechanisms of life as self-organizing, energetic, and operating historically: not all at once, by fiat, at the beginning of time, but gradually, contingently, and over all eternity.

Striving Machines

The German Romantic poet Heinrich Heine thought that Kant was himself an automaton: "He had neither life nor history." This is ironic, given that Kant reached for a historical approach to understanding living beings and argued that automata were rote, inert, passive things that could never serve as models of life. Heine continued,

> He led a mechanically ordered, almost abstract bachelor existence in a quiet, remote little street in Koenigsberg, an old town on the northeastern border of Germany. I do not believe that the great clock of the cathedral there performed more dispassionately and methodically its outward routine of the day than did its fellow countryman Immanuel Kant. Getting up in the morning, drinking coffee, writing, giving lectures, eating, walking, everything had its appointed time, and the neighbors knew for certain that it was half-past three when Immanuel Kant, in his gray frock-coat, his Spanish cane in his hand, stepped out of his house and strolled to the little linden avenue called after

> him to this day the "Philosopher's Path." Eight times he walked up and down it, in every season of the year, and when the sky was overcast, or gray clouds announced a rain coming, old Lampe, his servant, was seen walking anxiously behind him with a big umbrella under his arm, like an image of Providence.

If the citizens of Koenigsberg, Heine went on, had had an inkling of the "destructive, world-crushing" nature of Kant's ideas, they would have felt dread at his approach. "But the good folk saw in him nothing but a professor of philosophy, and as he passed by at his customary hour, they gave him a friendly greeting and perhaps set their watches by him."[6]

Heine's portrait of Kant's personal life was reproduced in generation after generation of biographies. But Heine's topic was only superficially Kant's lifestyle; his true target was Kant's style of philosophy. Either way, it is ironic that Kant should have been immortalized in this image. He did espouse the mechanist ban on ascriptions of agency to nature. But he also labored to reconcile this principle with what one might think its exact opposite, the view that teleology, purposeful agency, was ineradicable from natural science.[7] He was as drawn toward the rejection of rationalist, mechanist science as toward its apotheosis in Newtonian physics. The sole picture in Kant's house, biographers have liked to note,[8] hanging above the writing desk, was an engraved portrait of Rousseau, passionate scourge of rational and mechanical systems.[9]

The struggle between a mechanist rejection and a naturalist embrace of intrinsic purposefulness in nature shaped Kant's philosophy from beginning to end and, in important part through Kant, defined the preoccupations of a generation of Romantic poets and philosophers. Kant's first publication, for example, a work that his admirer and promulgator, the English poet Samuel Taylor Coleridge, later warmly recommended, was a defense of Leibniz's *vis viva*, the living force that animated Leibniz's world-machine. Kant thought that rational science went against such a notion but he maintained that fact and experience were both in its favor.[10] Later, in his *Critique of Pure Reason* (1781), Kant was sharply critical of Leibniz, censuring in particular Leibniz's notion of preestablished harmony and the way in which it connected the order in nature with a divine will.[11]

In his last major work, Kant took on the seemingly insurmountable problem of how to accommodate both the mechanist ban on ascribing purposefulness to natural phenomena and also contemporary accounts of living beings for which purposefulness was abidingly, in fact, increasingly crucial. The result was a work so tortuous it has moved some to allege that the philosopher was senile by the time he wrote it.[12] Here I want to advance a more charitable view: not senility but the same lifelong commitment to both mechanist science and an understanding of living nature as intrinsically purposeful can explain the convolutions of Kant's third *Critique* (it came after two others, the Critiques of Pure and of Practical Reason). Here, at the end of his career, Kant took on the most seismically active fault line in science and philosophy: the trench between understandings of life and of science.

Consider the contortionist's view of living things that Kant presented in this final work.[13] He wrote that organisms demanded to be understood in terms of purposive action. They simply could not be conceived in any other way. Their parts must be seen as having agency, as working individually and collectively to produce the whole organism, each part "existing *for the sake of the others* and of the whole." But this did not mean they necessarily were the results of purposeful action by their parts, merely that they could only be understood as such. A living creature thus appeared, though it was not necessarily so in fact, "*both cause and effect of itself*," an "*organized* and *self-organized* being."[14]

Kant referred to the by then standard comparison of a living organism to an artificial device such as a watch. The resemblance, he said, was only superficial: an organism's parts, like a watch's, all worked together toward a purpose. But a living organism was different from any artificial device, because the agency that produced an artificial device, and the purpose it served, were always external to it, whereas the agency and purpose of an organism were internal. An organism was its own end, and its parts worked together by their own internal agency toward that end, which was simply the organism itself. The analogy between organisms and artificial machines was misleading, therefore, since it suggested "an artist—a rational being—working from without. But nature, on the contrary, organizes itself."[15]

An organism had "a self-propagating formative power" specific to living mechanisms. Matter in motion alone could not explain this power, Kant thought. For as we have seen, since the advent of classical mechanism in the seventeenth century, the dominant view of matter held that it was definitively inert. "The possibility of a living matter is quite inconceivable," Kant wrote. "The very conception of it involves a self-contradiction, since lifelessness, *inertia*, constitutes the essential characteristic of matter."[16] This characterization of matter as fundamentally lifeless and inert in contrast with living beings as essentially active, would become the constant theme of Romantic authors.

Naturalists were caught on the horns of a dilemma, as Kant described it, "quite as unable to free themselves" from what he called "teleological" approaches to living beings —by which he meant an internal teleology, a purposeful agency within the natural forms themselves—as from the abiding principle of "general physical science" that precluded internal purposeful agency in nature. "Each mode of explanation excludes the other," and yet naturalists could renounce neither.[17] For Kant accepted the classical mechanist model of natural science, and yet he rejected the argument from design on which it rested. He thought it was contradictory to seek proofs of God's existence through natural science. Science could never uncover the purpose of nature itself, because such a purpose must lie outside of nature, beyond the purview of science.[18] In order to maintain the integrity of natural science, one had to treat the agency in nature as operating within nature itself, not externally imposed as in the argument from design. Otherwise "there will be no more nature. There will be only a God in the machine who produces the world's changes."[19]

Kant's dilemma was how to reconcile the mechanist ban on agency in nature, which he embraced, with his own conviction that the argument from design was unintelligible: that one could only meaningfully understand the apparent purposefulness of living forms as intrinsic to themselves. His solution was to distinguish between how people understood organisms and how organisms actually were in themselves (which was anyway, he said, unknowable). The apparent purposefulness of organisms did not mean that these were in fact the results of an internal agency, but rather that our reason or "cognitive faculty" had to understand them as such.[20] Naturalists should therefore retain their methodological principle that all living forms originate in purposeful agency, but they should also regard this principle as "reflective" and "regulative" rather than "determining" and "constitutive." In this way, they might carry on seeking natural purposes while assuming no "underlying end," no actual "teleology."[21]

Organisms viewed from this as-if perspective, Kant thought, lent natural science a legitimate basis for a teleology that would otherwise be "absolutely unjustifiable."[22] Mechanism and teleology, "thesis" and "antithesis," could coexist in science without contradiction as long as people treated these as "maxims of reflective judgement" or "regulative principles of investigation" rather than "constitutive principles." One could speak "with perfect justice of the wisdom, the economy, the forethought, the beneficence of nature," but in doing so, one must never represent nature as an intelligent being, "for that would be absurd." On the other hand, neither must one portray nature as a lifeless, passive artifact.

This would amount to placing an intelligent being "above nature as its architect," which Kant deemed "presumptuous" and contradictory.[23]

It was a quandary, but a productive one: if Kant's struggle over thesis and antithesis did not produce a clear synthesis, it did produce some considerable new possibilities. One was the principle that teleological judgments must be local and relative, operating only within the system of nature, rather than absolute. Kant offered the Arctic region as an example. The people living in the Arctic had found that snow protected their seeds from frost; reindeer served their needs for travel and communication; other Arctic animals provided sources of fuel, food, and clothing; and the sea brought them driftwood to build houses. It was a "truly marvelous confluence" of means to ends, the ends being "Greenlanders, Laplanders, Samoyedes, Jakutes, and the like." However, why should human beings live in the Arctic at all? What larger purpose or benefit could there be? If all this natural utility were absent, it would never be missed.[24] A naturalist must note all this local suitedness of means to ends, and yet must leave a deliberate and necessary blank in the spot where an ultimate end might make sense of it all, or not.

Besides the subordination of teleology to a greater contingency, another powerful possibility to emerge from Kant's struggles to reconcile mechanism with teleology was the notion that the two approaches might one day come together in a single, genetic account of living beings ("genetic" in the root sense, pertaining to the "genesis" of a thing). That is, Kant considered the possibility that mechanical causes might explain the purposefulness displayed by living forms, not by acting in a single moment of construction, but through a process of "descent from a common parent." The fact that an enormous variety of species shared a single, underlying structural plan suggested to Kant a "kinship" among them, like a great family whose "genesis" one could trace, the forms flowing from one another "by the shortening of one part and the lengthening of another, by the involution of this part and the evolution of that." Perhaps the apparent purposeful agency of living beings had come about gradually through a process of genesis: "The womb of mother earth as it first emerged, like a huge animal, from its chaotic state, gave birth to creatures whose form displayed less purposiveness," but these gave birth to new, more perfectly "adapted" forms.[25]

With this idea of a genesis of new organisms over time, Kant wrote, "there gleams upon the mind a ray of hope, however faint that the principle of the mechanism of nature, apart from which there can be no natural science at all, may yet enable us to arrive at some explanation of organic life." In the next paragraph, Kant quenched this "ray of hope" for a purely mechanical, non-purposeful account of living nature, finding that it merely pushed the purposeful agency back a stage to the primordial mother earth.[26] Nevertheless, he had introduced the notion of time as a possible force in the production of purposeful, natural, living mechanisms, and he would return to it frequently.

For example, in connection with his rejection of the argument from design, Kant considered the apparently beneficent lay of the land for collecting water. He observed that however "wisely the configuration, elevation and slope of the land may now seem to be adapted" for the collection of rainwater, for the springs that well up from beneath the earth and for the courses of rivers, we can see upon inspection that all of this has arisen from causes such as volcanic eruptions, floods, and movements of the ocean. "If now the abode for all these forms of life—the lap of the land and the bosom of the deep—points to none but a wholly undesigned mechanical generation, how can we, or what right have we to ask for or to maintain a different origin?"[27]

For the sake of describing how this "undesigned mechanical generation" acted over time, Kant regretted that the phrase "*natural history*" was commonly employed to mean "the description of nature" rather than the changes in nature over time.[28] This was indeed the meaning of "natural history" in seventeenth- and eighteenth-century usage.[29] If it was to continue that way, Kant proposed, "we may give the name of *archaeology of nature*" to what might otherwise have been called "history," that is, the "account of the bygone or ancient state of the earth." Such a history, in Kant's archaeological sense, would, for example, study fossil remains.[30]

Throughout this chapter, I intend the word "history" and its forms in the sense that Kant was reaching for, the sense that the word was in fact gradually acquiring during the eighteenth and early nineteenth centuries, to mean a secular, material transformation driven from within by internal agencies of various kinds. This new sense of "history" was coming into usage in relation to the development of human society as well as of natural living forms. Voltaire, for example, who transformed the writing of history in the middle decades of the eighteenth century, carefully distinguished profane from sacred history (the latter, he said, was a respectable endeavor but not his subject), ruling out gods, fables, anything counter to natural law, and in general, "all that violates the ordinary course of nature." Not only the categorical distinction of profane from sacred history but also the attachment of profane history to the "course of nature" moved the meaning of history in a naturalist, empirical and rule-governed direction. Voltaire observed that the invention of the printing press and the rise of the sciences at the end of the fifteenth century had "at last" allowed for the appearance of some correct and reliable histories.[31]

"One requires from modern historians," Voltaire instructed, "more details, facts that are better established, precise dates, authorities, more attention to customs, laws, mores, commerce, finance, agriculture, population. It is for history as it is for Mathematics and Physics," a matter of empirical and rational rigor. These standards dictated, in part, that one consider the element of time to be necessary to the emergence of complex phenomena. For example, the astronomical achievements of the Babylonians, Voltaire thought, indicated that they had existed as a people for a great many centuries. "Arts are but the product of time, and the natural laziness of men leaves them for thousands of years without any knowledge or talent except to nourish themselves, to defend themselves and to kill each other." Similarly, the first Egyptian cities must have existed "a prodigious time" before the pyramids, in order for the ancient Egyptians to have developed the skills and tools needed to build them. Having restricted historical explanations to processes within "the ordinary course of nature," Voltaire considered time not only as chronology, but also as a necessary dimension in the production of human history.[32]

The idea that a purely material, mechanical entity might overcome its physical finitude by operating over indefinitely long periods of time would achieve an important currency in the early nineteenth century. When Charles Babbage designed his mathematical engines during the 1820's and '30s, for example, the crucial ingredient that he added to Vaucanson's automatic loom was an awareness of the dimension of time. Babbage pointed out a hitherto unremarked aspect of punch cards, namely their indefinitely great temporal capacity: one could string together any number of them, and could indeed feed the result of an operation back into the machine, a process Babbage described as "the Engine eating its tail."[33] The machine could therefore have an "*unlimited*" capacity despite the fact that "no *finite* machine can include infinity." One could never "construct machinery occupying unlimited space," but Babbage reckoned he could substitute "*infinity of time* for the *infinity of space*."[34] No mechanism could transcend its own spatial limits. But perhaps, particularly if it transmitted its mechanical action from one generation to the next, it might achieve a form of transcendence through the dimension of time.

A scientific mode of explanation that was at once mechanist and historical, operating over time and by means of various intrinsic, natural agencies: this was not, in fact, a new idea for Kant. Already in 1755, he had described the cosmos itself as arising "from the mechanical laws of matter striving to develop."[35] Here already an inner agency was the means by which natural mechanisms transformed and created their own history. This sort of genetic account seemed to Kant to hold out the best hope of uniting mechanical and final causes with the "least possible expenditure of the supernatural"[36] in comparison with the leading contenders. First among these contenders was *occasionalism*, in which God acted directly to shape each organized being on the occasion of its birth. Second was the theory of preformation, which we have seen in chapter 3, and which went in contemporary natural history by the name of *evolution* (not to be confused with the later theory of "evolution," which has an entirely different meaning): the Creator had acted in advance, preforming every organism and nesting the homunculi generation within generation.[37]

In preference to both of these, Kant chose epigenesis: the Supreme Cause had endowed organized beings themselves with the ability to reproduce. He commended the doctor and physiologist Johann Friedrich Blumenbach, whose theory of epigenesis provided Kant's model,[38] for having acknowledged both the purposeful agency of life and the mechanist dictates of science in framing his theory. In Blumenbach's view, each generation of organisms following the first, propelled by a *Bildungstrieb* or "formative impulse" in matter itself, mechanically constituted the next. A living being, Blumenbach wrote, was the result of unorganized matter taking on "a particular action." This action, which constituted the life of the organism, was distinct from other animal capacities such as sensibility. He also called it a "*nisus*": a striving.[39]

Blumenbach sent Kant a copy of his essay on generation, featuring the "formative impulse" or "striving" (*Bildungstrieb* or *nisus*), to which Kant responded warmly. The essay, he wrote back, treated a subject that had lately preoccupied him: "the union of two principles that people have believed to be irreconcilable, namely the physical-mechanistic and the merely teleological way of explaining organized nature." Kant promised that Blumenbach would find an acknowledgment of the debt in his latest book, which was even then on its way to Blumenbach direct from the book merchant.[40]

Sure enough, in the third *Critique*, Kant offers Blumenbach's epigenetic reconciliation of mechanism and purpose as his example of right thinking about the problem: "It entrusts to nature the explanation of all steps subsequent to the original beginning."[41] His hesitation between mechanism and teleology led Kant to favor an epigenetic account of living beings, an account that operated in terms of self-organization rather than externally imposed design, and made the "purposiveness" of living structures central to every explanation, but in which teleology was also always subordinate to a greater contingency. It was a process driven by ubiquitous but limited agency. Struggling to reconcile mechanism and teleology, Kant arrived at a view of life that was fundamentally historical.

Several years later, Goethe announced his own embrace of a genetic approach to living forms, regarding the stages of the creation of life as a "progression of uninterrupted activity."[42] His *Metamorphosis of Plants* (1790) had already exemplified the genre. It presented a minutely detailed account of the transformations of the material parts of plants in growth. The driving force in this account was each plant's ability to "express its vitality" by continually making and remaking its parts.[43] Elsewhere, Goethe described this as a force of "intensification . . . a state of ever-striving ascent."[44]

A mechanical striving propelled the emergence and development of living beings also according to Jean-Baptiste Lamarck (see figure 6.2), author of the discipline of biology and professor of natural history at the Muséum national d'histoire naturelle in Paris. Although he appears only briefly in this chapter, his appearance is crucial, for he will be a key figure throughout the rest of this book. When he coined the term *biologie* in 1802, Lamarck defined the distinctive subject area of his new field in terms of such a vital, mechanical striving.[45] An intrinsic "*pouvoir de la vie*" (force of life) he observed, drove "animate machines," plants and animals, not only to compose themselves, but to elaborate and complicate their organization over time. The process took place over an "incalculable series of centuries," and began with the most rudimentary form of life, an "animated point" to which Lamarck gave the Leibnizian name *monade*. Organisms developed, grew, and transformed purely as a result of their own movements, specifically of the movements of fluids within them. Plants and animals were the sole beings on the planet to form this way, using materials of their own composition.[46]

According to Lamarck, in addition to this rudimentary upward-striving force of composition and complexification, in higher animals another kind of agency acted as well. Animals engaged with their environments by exercising their will, forming "habits" and "ways of life" in response to their circumstances. Through this agency they gradually transformed their bodies. "When the will determines an animal to perform a given action," Lamarck wrote, "the organs that must execute this action are immediately provoked by the affluence of sub-

Figure 6.2 Portrait of Jean-Baptiste Lamarck, lithograph by Louis-Léopold Boilly © Bibliothèque de l'Académie nationale de Médecine, Paris.

tle fluids" to carry it out. Many repetitions of these "acts of organization" would then "fortify, extend, develop and even create the necessary organs."[47]

Regarding the standard comparison of living bodies to watches and clocks, Lamarck noted that it would work only if one took into account the original cause of motion. He wrote, "The comparison people have made between life and a watch in active motion is imperfect, to say the least," unless one took the spring of the watch to be equivalent to the "exciting cause of the vital movements." Without the spring, the mechanism would remain inactive, while without a suitable arrangement in the parts of the mechanism, the movement of the spring would be useless. Lamarck insisted that the "essential motor" of the living mechanism was an integral part of it. A living thing was like a watch only insofar as a watch could be considered to contain the agency that set it in motion.[48]

Lamarck's living machinery formed and transformed itself by two different sorts of internal agency, a rudimentary, primitive force of life and a higher force of will. Both sorts were internal to the works, constitutive of the machinery from within. Lamarck was convinced that such a process was the only way to account for sentient life. If each creature owed its organization to a "force entirely exterior and foreign" to it, then instead of being animate machines, animals would have been "totally passive machines." They would never have had "sensibility or the intimate sentiment of existence that follows from it," nor the power to act, nor ideas, nor thought, nor intelligence. In short, they would not have been alive.[49]

The notion that living beings produced themselves by their own agency was controversial. Lamarck's fellow naturalist and critic, the zoologist Georges Cuvier, was prominent among those who rejected the idea.[50] Moreover, he rejected it on the grounds that ascribing agency to natural phenomena might make good poetry but never good science. Alas, poor Lamarck! It was Cuvier who wrote his eulogy, which he read to the Academy of Sciences in November 1832, three years after Lamarck's death. Rarely has a eulogy offered fainter praise. Cuvier observed that no one had found Lamarck's theory of life "dangerous enough to merit attacking." It rested upon the "arbitrary" supposition "that desires, efforts, can engender organs," an idea that might "amuse the imagination of a poet" but could never persuade a true anatomist.[51] And yet Cuvier himself defined life as an activity: the faculty of "enduring" through give and take, assimilating substance from one's surroundings and rendering substance back.[52] Even Cuvier, who dismissed as "poetry" the idea of ascribing agency to natural phenomena, understood life as a form of activity.

By the turn of the nineteenth century, a living being in scientific, philosophical, and literary understanding had become, in essence, an agent. An agent, in turn, was a thing in constant, self-generated motion and transformation of material parts. Living agency took the form of a responsive "striving," a capacity to bring about developmental change over time and in response to external circumstances: in short, to produce a history. It therefore demanded a kind of understanding that was at once mechanist, in the sense that it concerned interactions of material parts, and diachronic, operating over time. Lamarck invoked the "history of the facts of organization"[53] in support of his argument that species changed inexorably—albeit "with extreme slowness"—and that these changes were essential to the "history of our globe."[54]

Here Lamarck departed from the traditional meaning of "history" as in the phrase "natural history," meaning a complete description of nature, and adopted instead "history" in something more like Kant's archaeological or genetic sense, to describe a continual state of transformation in time. "Nature is a line in constant and continuous evolution," wrote Coleridge.[55] The science of living nature must take the form of history.

The Striving Machinery of Life Dramatized

Kant's genetic approach to the phenomena of life, torn between a mechanist model of science and a "purposive" model of living beings, became a touchstone for the generation of poet-philosophers working in the first decades of the nineteenth century. "The writings of the illustrious sage of Koenigsberg," wrote Coleridge, "took possession of me as with a giant's hand." Kant's works, he judged, had shaped his thinking more than any other, had "invigorated" and "disciplined" his understanding, and he returned to Kant over the years with "undiminished delight and increasing admiration."[56]

Coleridge made it his mission to bring Kant to his compatriots.[57] William Wordsworth, Thomas Carlyle, and in America, Ralph Waldo Emerson and Edgar Allan Poe encountered the sage of Koenigsberg through Coleridge.[58] His proselytizing focused upon the third *Critique*. At a gathering at the home of the essayist Charles Lamb, Coleridge told the lawyer and diarist Henry Crabb Robinson, according to Robinson's diary, that Kant's *Critique of Judgement* was the "most astonishing" of all his books.[59] The dilemma that Kant had struggled to describe and overcome was the focus of Coleridge's writing. Rather than working to resolve it, he dramatized it. Descartes's mechanist system, Coleridge wrote, was "a lifeless Machine whirled about by the dust of its own Grinding," a reduction of "the living fountain of Life" to "Death." On the contrary, in Coleridge's judgment, living organs were different from artificial machines in that, rather than being composed of parts, they actively assimilated foreign matter into themselves: "As the unseen Agency weaves its magic eddies, the foliage [eaten by an Ox or Elephant] becomes indifferently the Bone and its Marrow, the pulpy Brain, or the solid Ivory."[60]

Agency was the key word in Coleridge's understanding of living nature. He rejected the Cartesian-mechanist approach for having excluded "life and immanent activity from the visible Universe." In Newtonian physics, Coleridge perceived "the necessity of an active power, of positive forces present in the Material Universe," and he lamented the natural-theological identification of these forces with God.[61]

The argument from design left nature inert and God indistinguishable from gravitational force. Coleridge preferred the "dynamic spirit" he saw at work in contemporary physical science, such as the electromagnetic chemistry of the English chemist Humphry Davy and the Danish physicist Hans Christian Oersted, which found active forces and tendencies at work throughout matter. Surely, Coleridge thought, these dynamic sciences had dealt a "mortal blow" to scientific mechanism. They had shown the way to the conclusion that a living being was not a particular material structure but a "distinct and individualized Agency" expressing itself through successive combinations of particles.[62]

As a child, Mary Shelley, the future creator of Frankenstein's monster, the leading hypothetical man-machine of the Romantic period, inhaled these principles and preoccupations with the air she breathed. Coleridge was a good friend of her father, the novelist and radical social theorist William Godwin, and Shelley knew him well from childhood. As an adult, she remembered having hidden beneath the parlor sofa with her half sister, during a gathering one summer evening shortly before her ninth birthday, to hear Coleridge croon out the *Rime of the Ancient Mariner*.[63]

The evening conversations that gave rise to *Frankenstein* took place at the Geneva villa of George Gordon, Lord Byron during the rainy summer of 1816, when the author was nineteen, and they touched upon the latest theories of life.[64] Percy Shelley, with whom Mary Shelley had eloped two years earlier, was interested in the work of the French doctor and physiologist Pierre-Jean-Georges Cabanis, one of the leading contemporary proponents of machinery as a model of animal and human life.[65] Cabanis wrote that the brain was an organ that produced thought just as the stomach digested and the liver filtered bile.[66] His model of living machinery was however intrinsically active and sentient rather than brute and inert: sensation spread throughout the "living machine" by the constriction and relaxation of the nerves.[67] Living machinery needed "to feel and to act: and life is that much more whole, when all the organs feel and act strongly."[68]

One evening at Byron's villa, the group discussed some experiments that had been performed by Erasmus Darwin. In addition to being the creator of a mechanical talking head, a believer in the mechanical and sentient oneness of all living things, and a rhapsodist of the sex lives of plants, Darwin was also a prolific and popular naturalist who discussed zoology and natural history in poetry as well as prose. He was interested in the artificial production of lifelike entities and, indeed, of life itself. The experiments in question were reported, like Darwin's talking head, in the "philosophical notes" to a poem entitled *The Temple of Nature,* published posthumously in 1802.[69]

Darwin claimed to have produced life—or rather, triggered life's production of itself—by spontaneous generation. Shelley, in a later discussion of the origins of her story, referred to the spontaneous generation experiments: Darwin, she said, had locked a noodle in a glass case until it began to move on its own. The actual experiments involved a paste of flour and water left to putrefy in a closed container, yielding "animalcules called eels, vibrio anguillula" that displayed "wonderful strength and activity."[70]

Alongside Erasmus Darwin, another advocate of an intrinsic living agency in matter informed Shelley's story. He was Sir Humphry Davy (mentioned above in connection with Coleridge), the son of a Cornish wood-carver who had risen to become one of Britain's most prominent natural philosophers. Davy, who was also an acquaintance of Shelley's father, was the chemical lecturer at the Royal Institution and, like everyone else, a poet. His approach to the subject of "Life's warm fountains,"[71] equal parts lyrical and experimental, included research into "the conversion of dead matter into living matter."[72]

Darwin and Davy together contributed a crucial element of the story of *Frankenstein*: they were leading proponents of a widely held theory that the "living principle, or spirit of animation, which resides throughout the body" was a sort of "electric fluid."[73] The idea of "animal electricity," also known as "galvanism" after the man who had proposed it just over a decade earlier, was another topic of conversation at the villa on the rainy night in question. The Bolognese anatomist Luigi Galvani had reported in 1791 that he could make the leg of a dissected frog jump by applying electric sparks to its nerves. He had surmised that animal tissue contained a vital force, which he named "animal electricity" by analogy with the "natural electricity" generated by lightning and the "artificial electricity" produced by friction in an electrostatic generator. Galvani believed animal electricity was secreted by the brain and conducted by the nerves and acted as the medium of both sensation and muscular motion.[74]

The torpedo fish and electric eel, which produced shocking electrical discharge, provided additional support for Galvani's view, and commanded much attention, again at once poetic and philosophical. “The tropic eel, electric in his ire,” Darwin rhymed, “alarms the waves with unextinguish'd fire.”[75] When the Italian physicist Alessandro Volta (also the author of a scientific poem featuring electricity), while trying to reproduce Galvani's experiments, invented the first electric battery and discovered current electricity, he regarded his contraption as a sort of artificial torpedo fish.[76]

As a student at Oxford, Percy Shelley had kept electrical equipment in his rooms and had liked to demonstrate his electrostatic generator to guests by having them turn the crank while he drew off the fire, perched on a glass-footed stool, “so that his long, wild locks bristled and stood on end.”[77] During the evening conversations at Byron's villa, he and the others wondered aloud whether artificial electricity might be used to reanimate a corpse.[78]

They were able to cite experimental data. In a London anatomy theater in 1803, Galvani's nephew, Giovanni Aldini, had used an electrical battery to cause the corpse of a hanged criminal to grimace and twitch its muscles (see figure 6.3).[79] (An Edinburgh doctor, Andrew Ure, repeated the experiment in 1818 with “truly appalling” results. “The scene was hideous,” recalled a witness, including an episode of apparent “laborious breathing” on the part of the corpse.[80] Ure wrote that “several of the spectators were forced to leave the apartment from terror or sickness, and one gentleman fainted.” In both cases, the experimenters were persuaded that “vitality might, perhaps, have been restored” if circumstances had not intervened.[81])

In the night following these discussions, Mary Shelley had a horrible vision: she saw Dr. Frankenstein, the monster's creator, kneeling beside his creature as it comes hideously to life, then fleeing in terror, falling into a troubled sleep from which he awakes to find the monster gazing upon him “with yellow, watery, but speculative eyes.”[82] In the midnight reverie that was the germ of *Frankenstein*, the defining feature of the finished novel was already in place: the conflict between dead matter and living agency.

Matter had become antithetical to life: passive, inert. The surgeon and physiologist Richard Saumarez described matter this way in his *New System of Physiology* (1799), a book Coleridge greatly admired. The “Principle of Life” according to Saumarez was a power of organization that, passing from a state of dormancy into one of “energy and action,” could “overcome” the “passivity of common matter.” Deprived of this living principle of organization, Saumarez wrote, the matter of an organism was “as imbecile and inert as the shoe without the foot.”[83]

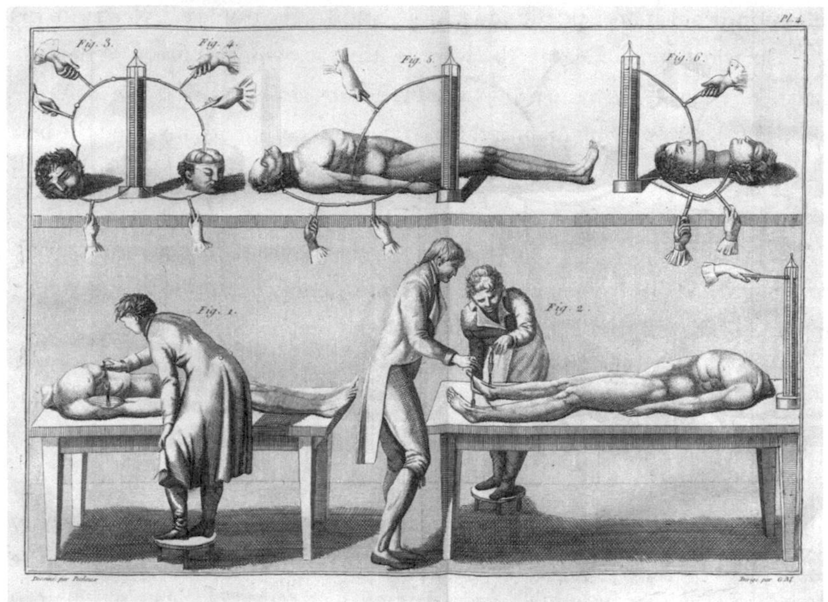

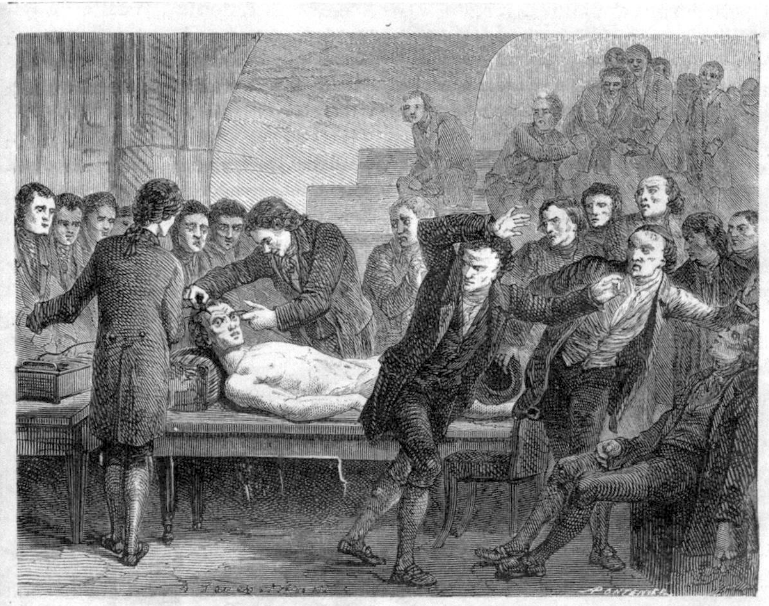

Fig. 333. — Le docteur Ure galvanisant le corps de l'assassin Clydsdale.

Figure 6.3 Giovanni Aldini's electrical antics with corpses, from *Essai théorique*, courtesy Cornell University Rare Book and Manuscript Collections, and Andrew Ure galvanizing the corpse of an executed murderer ("le docteur Ure galvanisant le corps de l'assassin") from Figuier, *Merveilles de la science* (1867).

The Romantic poet-philosophers who took up the question of living machinery in the first years of the nineteenth century carried the reasoning a step further. In their hands, inanimate matter, which Saumarez had described as the shoe without the foot, became the dead and severed foot itself. Inert matter became not just inert, but dead, and a horrified fascination with the juxtaposition of "dead" matter and living agency worked away at the heart of the Romantic understanding of the nature of life, which was inseparably poetic, moral, and scientific.

In his popular 1802 "Discourse on Chemistry," read not only by Mary Shelley but by many other Romantic writers including Coleridge and William Wordsworth, Humphry Davy used the phrase "dead matter" and similar ones—"dead nature," the "dead state"—on almost every page.[84] This was not Davy's idiosyncrasy but a standard way of speaking. "The matter that surrounds us," the London doctor William Lawrence, Percy Shelly's own physician, told audiences at his much-discussed public lectures on the nature of life in 1814 and 1815, "is divided into two great classes, living and dead."[85] Rather than contrasting life with nonlife—the inanimate—the Romantics set life up against death. What was not alive was dead.

Frankenstein's monster represented the central dilemma of contemporary science according to which all living beings were constituted by an inherent agency and yet made out of dead matter. *Frankenstein* was the best-known dramatization of this dilemma but not the only one. A great many others involved the androids of the preceding century, which positively haunted the Romantics. For example, Lewis, the hero of E. T. A. Hoffmann's 1814 story "Automata," gets a "gruesome feeling" from androids, which he finds to be "mere images of death or inanimate life." Having gone to see a collection of android musicians play, with their human owner accompanying them on the piano, Lewis flies into a rage. He finds it dreadful to see a human engaging with an automaton in any way and imagines an even worse scenario, a human being dancing with an android: "a living man putting his arms about a lifeless partner of wood. . . . Could you look at such a sight, for an instant, without horror?[86]

In 1805, Goethe made a pilgrimage to see Vaucanson's automata. He traveled with his son August, then fifteen, and Goethe's good friend, Friedrich August Wolf, a philologist at the University of Halle. Their journey was southward to the village of Helmstädt in Franconia, not far from de Caus's Palatine Gardens at Heidelberg, where they would pay a visit to the current owner of the automata, one Gottfried Christoph Beireis, a privy councilor and polymath, professor of physics, chemistry, medicine, and surgery at the University of Helmstädt, and personal physician to the Duke of Braunschweig. Beireis was a shadowy and eccentric figure known for his "remarkable possessions, strange behavior, and . . . secret brooding." The threesome had a rowdy voyage. Wolf was a "humorous fellow-traveler" who was "constantly plying the boy with railleries," to which August responded in kind, so that the trip was spent in "nudges and boisterous frolics, not altogether convenient in a carriage."[87]

Arriving in Helmstädt, the party found Beireis gallant and hospitable, with an "incredibly high and arched forehead, out of all relation to the lower parts of the face," indicating "peculiar intellectual powers." Beireis gave the travelers free rein to explore his collection of "extraordinary and scarcely conceivable treasures." First among these were Vaucanson's automata, now venerable creatures nearing seventy. In the hands of a series of exhibitors, the automata had traveled the length and breadth of Europe. Beireis had acquired them in 1782 upon their return from Russia, where they had been exhibited in Saint Petersburg. They were now in "the most lamentable state." Languishing in "an old garden house," the two androids were "utterly paralyzed," the Flutist without its camshaft fallen "mute." The Duck, having lost its feathers, was a mere skeleton, and "still devoured the oats briskly enough, but had lost its powers of digestion."[88] In such a state, the automata were neither impressive nor scary. Several years earlier, Goethe had met the touted talking machine of Wolfgang von Kempelen, and been similarly unimpressed. Bored by the machine's conversation, the great poet pronounced it "not very loquacious."[89]

Kant found Vaucanson's automata more ominous and invoked them as a counterexample to argue for one of the central principles of his philosophy: that the nature of the human self was fundamentally unknowable—unthinkable—because (as Descartes had said first) it existed outside the mechanical world of time and space. This had to be so, Kant wrote, or else a human being would be nothing but "a Vaucansonian automaton." It might even be a thinking, self-conscious automaton, but still, its subjective sense of freedom would be a delusion: "nothing better than the freedom of a turnspit, which, once it is wound up, accomplishes its motions of itself."[90]

That matter was dead, the opposite of life, and that life was a form of activity, a continual effort to constitute oneself from and against dead matter, were the Romantic principles that informed the founding of biology as a discipline. Life was the struggle against extinction, "the sum of the functions that resist death," as the French surgeon and physiologist Marie-François-Xavier Bichat defined it.[91] Erasmus Darwin put it in verse, "Life clings trembling on her tottering throne."[92] Lamarck, in a manuscript sketch of the book he planned to write to inaugurate his new science entitled "Biology, or Considerations on the Nature, the Faculties, the Developments and the Origins of Living Bodies," began by dividing nature into two parts: "brute bodies" and "living bodies." This division, Lamarck wrote, was "infinitely distinct," with no "intermediary" between the brute and the living. A living body was "a natural body limited in its duration, organized in its parts . . . possessing what we call life, and subject necessarily to lose it, that is, to succumb to death," at which point it must be re-classed among the brute bodies.[93] In addition to these two sorts of bodies, Lamarck announced two opposing forces to govern them: one a force of composition and life, the other a force of destruction and death.[94]

The effortful activity of life, the striving agency, Lamarck declared, was ultimately doomed.[95] His colleague and fellow traveler, Étienne Géoffroy Saint-Hilaire, described the situation of life in similarly combative terms. Saint-Hilaire was the author of a theory broadly sympathetic to Lamarckism, in which species transformed by the direct impingement of the environment. He described this process as an "engagement of actions and reactions." The "battle," Saint-Hilaire wrote, must always favor the forces of organization over disorganization as long as "the machine is not definitively totally disorganized."[96]

Life could triumph on a grand scale, if not in any individual case. Erasmus Darwin made the point in verse again: "Organic forms with chemic changes strive, / Live but to die, and die but to revive."[97] The living, organizing power by which organisms produced themselves from "senses, feelings and conceptions," wrote the Prussian philosopher and poet Johann Gottfried Herder, did not perish with an individual's death. Even in the "dissolved machine" of a dead flower the power remained "active."[98] Erasmus Darwin agreed: "The births and deaths contend with equal strife, / And every pore of Nature teems with Life."[99] Redefined in terms of their duel with dead matter, the vital powers and striving impulses of the eighteenth century took on heroic capital letters and became the "Vital Power,"[100] the "World Spirit."[101]

The institutional establishment of the discipline of physiology, like biology, occurred amid these philosophical-literary-scientific-aesthetic developments. The notion of a generalized, circulating living agency, larger than the life of any given organism, connected the new science of physiology with the tradition in physics that had arisen from Leibniz's *vis viva,* or "living force." With *vis viva,* we have seen, Leibniz had built the source of action into his world-machine rather than attributing it to an external source, distinguishing his intrinsically active form of mechanism from what he saw as the passive machinery of both Descartes's and Newton's cosmoses.[102]

Leibniz's idea of innate "living force" became, during the late eighteenth and especially the early nineteenth century, a primary inspiration behind the German Romantic movement in philosophy and science known as *Naturphilosophie* (nature-philosophy), which received its name and much of its direction from Goethe's friend, Friedrich Wilhelm Joseph Schelling.[103] A core idea of *Naturphilosophie* was that a single, inherent formative power pervaded all of nature, manifesting itself in the various natural forces and phenomena including light, gravity, electricity, magnetism, chemical forces, and also the animate forces of living beings, sensibility, growth, and movement. One had to understand all of these, according to the proponents of *Naturphilosophie,* as expressions of the same underlying and intrinsic power. Schelling and the other nature-philosophers took it as crucial that this power in its many manifestations was internal to nature, not externally imposed.[104] Primarily through the developments of German Romantic science, Leibniz's "living force" informed the development of a new physics of energy.[105]

Physiologists, starting with the German physiologist Johannes Müller and later his students, Emil du Bois-Reymond and Hermann von Helmholtz, began to understand living agency and sentience in terms of energies and, later, a single overarching energy.[106] Quoting Kant, Müller wrote that living matter contained "a principle constantly in action," which "adapted" the parts of an organism to one another, perhaps a "vital energy." Müller explained sensation itself, and in particular its differentiation, in terms of energies specific to each nerve.[107] Cabanis similarly supposed that the brain exerted its control over the rest of the "living machine" by means of its "energy and activity."[108]

Like the older ideas of a striving impulse in living beings—Blumenbach's *nisus*, Lamarck's *pouvoir de la vie*—the idea of vital energy incorporated action and purpose into the machinery. But while the earlier forms of vital agency had constituted a distinction between living beings and inert matter, the new energy reconnected animate with inanimate nature. This energy was not a primitive feature of living substance but the currency of its engagement with dead matter, and moreover, this engagement was not specific to organisms: "The animal machine resembles every other machine, the action of which necessitates the destruction of some material," wrote Müller.[109]

Plate 1 Sketch of an automaton she-devil by Giovanni Fontana, fifteenth century, Bayerische Staatsbibliothek München, Cod.icon. 240, fol. 63v.

Plate 2 Automaton devil from the collection of Ludovico Settala, fifteenth–sixteenth century, Raccolte d'Arte Applicata, Castello Sforzesco, Milano,

Plate 3 Männleinlaufen (Parade of little men) on the Frauenkirche clock, Nuremberg, © Birgit Fuder / Stadt Nürnberg.

Plate 4 Automaton Franciscan monk attributed to Juanelo Turriano, ca. 1560. Division of Work and Industry, Archives Center, National Museum of American History, Smithsonian Institution.

Plate 5 Close-up of the Jaquet-Droz Musicienne's hands, courtesy Musée d'art et d'histoire de Neuchâtel.

Plate 6 The Jaquet-Droz Ecrivain, courtesy Musée d'art et d'histoire de Neuchâtel.

Plate 7 Portrait of Julien Offray de La Mettrie, oil on canvas by Georg Friedrich Schmidt, ca. 1750. bpk, Berlin / Gemaeldegalerie, Staatliche Museen, Berlin / Volker-H. Schneider / Art Resource, New York.

Plate 8 Erotic pocket watch with automaton birds and musicians on the dial plate and concealed erotic automaton scene, by Henry Capt, Geneva, ca. 1810. Images courtesy Antiquorum Auctioneers.

Plate 9 Cog the robot. Credit: Sam Ogden / Photo Researchers.

Helmholtz invoked the previous century's android makers to explain the difference. Vaucanson's Duck, his Flutist, and the Jaquet-Droz trio had been the marvels of their age, he said, but the efforts to create life in artificial machinery and the related attempts at perpetual motion machines had been based upon a fundamental misconception about living beings: "The development of force out of itself seemed to be the essential peculiarity, the real quintessence of organic life." By construing life as the capacity to produce force from nothing, philosophers and mechanicians had set organisms apart from the rest of nature. "To the builders of automata of the last century, men and animals appeared as clockwork which was never wound up, and created the force which they exerted out of nothing."[110] This was a misconception. Living machines could not create force out of nothing any more than inanimate machines could.

Moving forces acted throughout inorganic and organic nature alike, according to Helmholtz, who rejected his teacher Müller's notion of a special life force, distinct from chemical and physical forces.[111] All the forces ultimately originated in the sun. Despite his sharp criticisms of Romantic science, Helmholtz was an heir (as he himself affirmed) to some of the core principles developed by Romantic-era natural philosophers, notably those involving force and energy.[112] In keeping with these, Helmholtz regarded all moving forces as interconvertible forms of the same essential "activity," and in this context, he stated one of the first principles of energy conservation: "Whenever the capacity for work of one natural force is destroyed, it is transformed into another kind of activity."[113] There was a cosmic circuit: sunlight enabled plants to grow and produce the fuel and nutriment for animals to "burn" in their lungs, while the products of this combustion became the plants' nutriments taken from air, water, and soil: carbon, hydrogen, nitrogen. The whole process originated in the sun, and therefore "all force, by means of which our bodies live and move, finds its source in the purest sunlight."[114]

The living agency of organisms was integral to a more general living agency of nature itself. The poets had arrived there before the physiologists. "O! the one Life within us and abroad," Coleridge rhapsodized,[115] while Erasmus Darwin explained, "With finer links the vital chain extends / And the long line of Being never ends."[116] To live was now to transcend, not from matter into spirit, but from the individual into the universal. It was a form of "transcendental materialism," in Anson Rabinbach's aptly oxymoronic phrase: a materialism that assumed, not a reductive sameness in nature, but instead a dynamic "unity of all material being" in the form of interconvertible forces and, ultimately, of energy.[117] "Be it," wrote Herder, "that we know nothing of our soul as pure spirit: we desire not to know it as such. Be it, that it acts only as an organic power: it was not intended to act otherwise."[118] Electricity, magnetism, heat, mechanical and chemical forces, and life were all forms of a single entity, namely energy. They were modes of one another, convertible into one another, and so living creatures were participants in the life of nature itself. "That what you see *is* blood, *is* flesh, is itself the work . . . of the invisible Energy."[119]

Not only life but also consciousness and sentience were forms of dynamic participation in the greater life and sentience of nature itself: "A motion and a spirit, that impels / All thinking things, all objects of all thought, / And rolls through all things."[120] Our own "machine," wrote Herder, was "a growing, flourishing tree," and so it "feels even with trees; and there are men, who cannot bear to see a young tree cut down or destroyed."[121] According to Erasmus Darwin, Mind likewise characterized all Life: even flowers could entertain "ideas" and "passions."[122] Looking into one's own mind, one looked "into the Mind of Man,"[123] and also into the mind of all plants and animals too: "And what if all of animated nature / Be but organic Harps diversely framed, / That tremble into thought, as o'er them sweeps / Plastic and vast, one intellectual breeze."[124]

How to revoke the monopoly on agency that mechanist science had assigned to God? How to bring the inanimate, clockwork cosmos of classical mechanist science back to life while remaining as faithful as possible to the core principles of the scientific tradition? A whole movement of poets, physiologists, novelists, chemists, philosophers and experimental physicists—roles often combined in the same person—struggled with this question. Their struggles brought the natural machinery of contemporary science from inanimate to dead to alive once more. The dead matter of the Romantics became animate, not at the hands of an external Designer, but through the action of a vital agency, an organic power, an all-embracing energy intrinsic to nature's machinery. Along with this new understanding of matter and force, organic life and vital agency, came two major developments.

Both developments represented ways in which living machines might transcend the limits of their own machinery without leaving the realm of material nature. The first was a genetic (again, genesis-related) approach to life whose proponents thought that by tracing the developments of limited agents working in specific contexts over periods of time, they might reconcile the demands of mechanism with the appearance of living purpose. In other words, they tried out the idea of transcendence through the dimension of *time*, placing their hope in a form of explanation that was coming to be called "historical" in Kant's and Lamarck's new sense. "Historical" in this sense described a gradual, open, undetermined, self-driven process of transformation.

The second development was the idea of a kind of transcendence through energy: the idea that organisms, understood as living machines, were connected through a great web of energy exchange to the cosmos itself. This was an attempt to rescue Frankenstein's monster, made of dead matter, from his terrible isolation without granting him a soul. The Romantics had arrived, through a dramatic collaboration of science and poetry, at the possibility of a rigorously, exhaustively natural science: one in which the divine Clockmaker ceded his monopoly to inherent agencies operating over time, through matter and energy, across space, and always within the continuum of nature.

~ 7 ~

Darwin between the Machines

Charles Darwin did not love his grandfather's poetry, which he found "artificial," though he did make allowances for changes of taste. "I have myself," he observed, "met with old men who spoke with a degree of enthusiasm about his poetry, quite incomprehensible at the present day. . . . No one of the present generation reads, as it appears, a single line of it."[1] Erasmus Darwin's Romantic fusion of poetry, natural philosophy, physiology, and medicine was a source of as much embarrassment as pride to his Victorian grandson.

Still, when Charles Darwin began a secret notebook on the transmutation of species in the summer of 1837, he entitled it *Zoonomia,* a private reference to his grandfather's work of that title, in which Erasmus Darwin had presented his own theory of the transformation of species. This theory had operated by means of a "power" of "animality": of "acquiring new parts, attended with new propensities, directed by irritations, sensations, volitions and associations; and thus possessing the faculty of continuing to improve by its own inherent activity."[2] Although Charles Darwin professed to rid his own science of his grandfather's Romantic idea of inner agency, he also incorporated this idea, through the genetic approach to life that it informed, at the very core of his doctrine.

Meanwhile, from the argument-from-design tradition, which he studied at Cambridge, Darwin adopted certain other core principles: the notion of mechanical adaptation or "fitness" of parts, and the related requirement that parts be passive, that a properly scientific account of living phenomena ascribe no agency to the phenomena themselves. Yet, Darwin eliminated the invocations of the Designer that had made sense of these principles in classical mechanist natural theology from the seventeenth century onward.

In other words, officially, Darwin rejected both internal (striving) and external (divine) agency as elements of scientific explanations. But he adopted the modes of explanation that each had informed: the genetic or historical mode that went with the notion of an internal, striving agency, according to which living beings actively transformed themselves over time; and the fitness mode that went with the assumption of divine agency, according to which living beings were static, passive, designed devices.

Despite his distaste for his grandfather's poetry, the culture of Romanticism had an unmistakable influence on Darwin. Apart from his own grandfather, another important figure was the German Romantic explorer, naturalist, and poet Alexander von Humboldt, whose writings Darwin treasured, carried about with him, read aloud to his friends, and tried hard to imitate.[3] Like the Romantics, Darwin was deeply though fruitfully torn between the mechanist dictate to banish agency from nature and the organicist impulse to naturalize agency, to make agency synonymous with life. His conflicting intellectual inheritances and allegiances put him in a state of profound and extraordinarily productive ambivalence.

Action without Agency

Notebook B, as the secret transmutation notebook of 1837 is known, has several references to Lamarck's "monads," that Leibnizian name that Lamarck assigned to animal life in its most elementary form: as we saw, the "animate point" that, through its upward-striving power of life, provided the source and basis for all higher organisms.[4] Together with Erasmus Darwin's power of animality, Lamarck's rising monads constituted the context in which Darwin began to think his way into the problem of species change. "Each species changes," he wrote in the notebook. "does it progress? Man gains ideas. the simplest cannot help—becoming more complicated; & if we look to first origin there must be progress. If we suppose monads are constantly formed, would they not be pretty similar over whole world under similar climates & as far as world has been uniform at former epoch." A bit later he added "Man is derived from Monad each fresh—" but he crossed that out (see figure 7.1). Later yet, he decided, a plague on both their houses: species changed, but not by means of an internal tendency toward higher forms and greater complexity, nor by animals' efforts to satisfy changing needs.

> Lamarck's "willing" doctrine absurd (as equally are arguments against it—namely how did otter live before being modern otter—why to be sure there are a thousand intermediate forms.—Opponents will say. show them me. I will answer yes, if you will show me every step between bull Dog & Greyhound) I should say the changes were effects of external causes, of which we are as ignorant as why millet seed turns a Bullfinch black.[5]

The wording of this private reflection is telling. The idea that species changed from within struck Darwin not just as incorrect but as "absurd," and he characterized it to himself as "Lamarck's 'willing' doctrine." This was not just a rejection of Lamarck's notion, but a principled one. It was a resolute espousal of the classical mechanist ban on ascriptions of agency to natural phenomena. To say that species change themselves would be a scientific absurdity, thought Darwin; one must instead maintain that they *are changed* by "external causes."

The geologist Charles Lyell, Darwin's close friend and intellectual role model, contributed to Darwin's fluctuating attitude toward Lamarck's complexifying power of life and upward-tending monads. In his *Principles of Geology* (1830–33), Lyell offered a meticulous but scathing rendition of Lamarck's theory in which he found fault with it along the very same lines.[6] The first volume of Lyell's monumental work had recently been published when Darwin, at the age of twenty-two, set out on a five-year voyage aboard the *HMS Beagle* as naturalist-without-portfolio and traveling companion to Robert Fitzroy, the ship's captain. Fitzroy gave Lyell's new book to Darwin as a gift; he received the second volume, whose opening chapter contained the refutation of Lamarck, at Montevideo in 1832, and the third before reaching Valparaiso in 1834.[7] In Lyell's book, Darwin read that Lamarck showed an "unpardonable" disregard for evidence and that his "efforts of internal sentiment" and "acts of organization" were no better than so many "fictions . . . and other phantoms of the middle ages."[8]

Twenty-five years later, however, when Darwin sent Lyell the proofs of the *Origin of Species*, Lyell had changed his tune about Lamarck. Responding warmly to the proofs, Lyell congratulated himself for having urged Darwin to publish his majestic idea "without waiting for a time which probably could never have arrived, though you lived till the age of a hundred, when you had

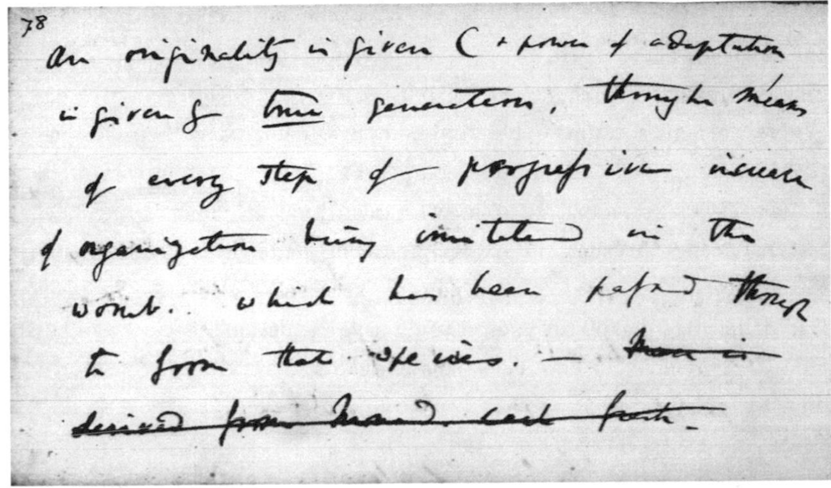

78
an originality is given (& power of adaptation
is given by true generation, through means
of every step of progressive increase
of organization being initiated in the
womb. which has been passed through
to form that species. — ~~Man is~~
~~derived from Monad each fork~~

Figure 7.1 Page from Darwin's secret transmutation notebook of 1837, Notebook B.

prepared all your facts on which you ground so many grand generalizations." However, Lyell had rather a weighty query relating to Lamarck's monads and their upward agency: where was the creative power in Darwin's theory? Natural selection, after all, could only work if it had choices among which to select. "Lamarck's monads coming daily into being supply a perpetual crop of the simplest forms," Lyell wrote. In Darwin's system, where did variety come from? What produced "new powers, attributes & forces"?[9]

Lyell explained that he would welcome "any hints" of an answer for the sake of his own ongoing "protest against the finality men," advocates of teleological appeals to a designer God in natural history. He meant in particular the Swiss paleontologist and geologist Louis Agassiz, who was then founding the Museum of Comparative Zoology at Harvard, and the English anatomist and paleontologist Richard Owen, who was establishing a national natural history museum under the auspices of the British Museum. It seemed to Lyell that these adversaries, who held positions of the greatest influence in the public promulgation of natural history, willfully misinterpreted his own rejection of teleology as a rejection of progress. On the contrary, Lyell protested, he was inclined to believe in a slow, internally driven progress over ages of time. What did Darwin take to be the active, driving force in nature, if any?[10]

Darwin's response was sharply dismissive: "On the continued Creation of Monads.—This doctrine is superfluous (& groundless) on the theory of

Natural Selection, which implies no *necessary* tendency to progression." As far as Darwin was concerned, a monad could stay the same for age upon age as long as it underwent no beneficial (or, presumably, harmful) structural changes. Whereas genetic accounts of life from Leibniz to Lamarck had replaced an external designer God with an inner, burgeoning force operating within living matter, Darwin now appeared to side with the natural-theological brute mechanists in rejecting any such internal agency. Living matter, he said, did not strive. Regarding the "primeval creative power" that Lyell had thought necessary, Darwin retorted that no such power could have any part in his science and that, indeed, he preferred to leave an explanatory gap at its very core, as Newton had done for gravity:

> I am not sure I understand your remarks which follow the above. We must under present knowledge assume the creation of one or of a few forms,—in same manner as philosophers assume the existence of a power of attraction, without any explanation. But I entirely reject as in my judgement quite unnecessary any subsequent addition "of new powers, & attributes & forces"; or of any "principle of improvement," except in so far as every character which is naturally selected or preserved is in some way an advantage or an improvement, otherwise it would not have been selected.

If Lyell actually thought such powers as he proposed were necessary to natural selection, Darwin concluded, "I would reject it as rubbish."[11]

These remarks were more adamant than explanatory. Indeed, in their vehemence about the inadmissibility of natural forces, powers, or tendencies, they specifically renounced explanation, and right at the spot where, even a sympathetic Lyell could not help but feel, one needed it most. Lyell wanted a source of creativity and progress *within* nature as a defense against the "finality men" with their meddlesome, supernatural Designer, who stepped in with new and improved models of life at each stage of progressive development. Between Lamarck's creative power of life and the "finality men," when push came to shove, Lyell chose Lamarck.

He tried to get Darwin on that side too. A few years after the publication of the *Origin of Species*, Lyell wrote a sort of apology to Lamarck's memory in which he credited Lamarck with having been the first to "introduce the element of time into the definition of a species," indeed, with having had a clearer appreciation than Lyell himself had had for the great depth of deep time, the slowness of change, and the "insignificance of thirty or forty centuries." Lyell here, to Darwin's annoyance, referred to the *Origin of Species* as a modification of the Lamarckian theory of progressive development.[12]

In his next letter to Lyell, Darwin wrote testily, "If this is your deliberate opinion there is nothing to be said," but he did not see that the *Origin* had anything more in common with Lamarck's "wretched book," from which he had "gained nothing," than with "Plato, Buffon or my grandfather," all of whom had "propounded the obvious view that if species were not created separately, they must have descended from other species."[13] Lyell held his ground, writing back, "When I came to the conclusion that after all Lamarck was going to be shown to be right, that we must 'go the whole orang,' I reread his book, and remembering when it was written, I felt I had done him an injustice." Many of Lamarck's ideas now seemed, Lyell wrote with sly anachronism, "very Darwinian." Even more troubling, Lyell found that the "substitution of the variety-making power for 'volition,' 'muscular action,' &c. (and in plants even volition was not called in) is in some respects only a change of name,"[14] and he now endorsed Lamarck's idea of a "renovating force" in living nature.[15]

Similarly, Darwin's greatest German disciple, Ernst Haeckel, made a place for Lamarck in his Darwinism. Haeckel was the most successful promulgator of Darwinism in the period before World War I;[16] and in his best-selling popularizations, he represented Darwin's work as the culmination of a transformist theory of life inaugurated by Goethe and Lamarck.[17] From these two originators of transformism, Haeckel adopted an array of materialist generative principles. Goethe, he wrote, had discovered "the active causes of organic formations," namely, the "two organic formative tendencies—on the one hand the conservative, centripetal, and internal formative tendency of inheritance or specification; and on the other hand the progressive, centrifugal, and external tendency of Adaptation, or metamorphosis."[18]

Meanwhile Lamarck, Haeckel said, had pioneered the first "monistic (mechanical) system of nature," which already contained all the core principles of modern biology. First among these was "the unity of the active causes in organic and inorganic nature." Like Goethe, Lamarck had recognized the action of the "two organic formative tendencies of Adaptation and Inheritance." Haeckel's version of Darwinism accordingly included an important generative aspect counterbalancing natural selection: an "inner constructive force, or inner formative tendency" that acted throughout organic and inorganic nature.[19]

All of Darwin's main proselytizers returned to this key question, and their queries not only pestered and worried Darwin, but also substantially directed the course of his theory. Just what agency drove the modification of species? Was it an internal or an external form of agency? The philosopher and social theorist Herbert Spencer is an important example, since he shaped the development and presentation of Darwin's theory by supplying its key words and phrases. Spencer is the source of the phrase "survival of the fittest," which appeared in the fifth (1869) edition of the *Origin of Species*,[20] and of the word "evolution," which is absent from all but the very last, sixth edition in 1872.[21]

These terms became the slogans by which people grasped, and continue to grasp, Darwin's theory of what he had first called "descent with modification." Both added a powerful teleological agency to a theory whose author had originally, deliberately excluded such a thing. "Survival of the fittest" implied, as Darwin expressed it, "a power incessantly ready for action, . . . as immeasurably superior to man's feeble efforts as the works of Nature to those of Art."[22] "Evolution," as we know, had hitherto referred to the preformationist theory of embryology according to which organisms developed from preexisting homunculi, each generation nested inside the previous one like Russian dolls. When Spencer attached the word "evolution" to Darwin's theory of the modification of species, he brought with it a sense of the unfolding of a preordained order, something profoundly at odds with Darwin's original conception. Spencer used "evolution" promiscuously to mean a growth from simple to complex, from lower to higher, from "barbarous" to "civilized," in a social as well as a natural context, indeed, expressly to blur the line between the two.[23]

Spencer construed Darwin's theory as describing progress from the primitive toward the advanced, in a sense that was simultaneously moral and physical. Although he rejected Lamarck's and Erasmus Darwin's ideas about internal forces or tendencies driving the transformation of species, calling these "unphilosophical,"[24] he endorsed Lamarck's claim that in higher animals, habits and ways of life brought about heritable, bodily changes. It was Spencer who attached the phrase "inheritance of acquired modifications" to this idea (a formula whose influence almost matched "evolution" and "survival of the fittest"), finding that although Lamarck and Darwin had both described such a process, neither emphasized it enough. Spencer, characteristically, seized the idea by turning it into a slogan, and ran with it, arguing that for the more complicated, higher forms of life, their actions and habits actually overtook natural selection as the main source of transformation.[25]

Darwin had to overcome strong distaste before incorporating Spencer's ideas and terms into his work. At the beginning, he found Spencer's style to be "detestable," obscure, unedited,[26] clever but empty,[27] a lot of "dreadful hypothetical rubbish,"[28] and a disappointing tissue of "words and generalities."[29] Sometime in 1866, Darwin began to come around to Spencer, just around the time, in fact, that Spencer began to have considerable success at promoting his own version of "Darwinism." His books, promulgating renditions and applications of Darwin's theory, became worldwide best sellers in the 1860's and '70s.[30] Darwin's tune changed, though only partially: Spencer now sometimes seemed to him to show an astonishing "prodigality of original thought," and if he had sacrificed some of that cleverness for sharper powers of observation, he might have been "a wonderful man."[31]

Darwin met Spencer for the first time that same year, 1866, at the home of a neighbor, the baronet and banker Sir John Lubbock, who was a promoter of the natural sciences, especially evolutionary theory. Darwin reported afterward that he had much enjoyed his talk with Spencer "though he does use awesomely long words"; Lady Lubbock, in Darwin's opinion, had found Spencer a dreadful bore.[32] Even in 1872, as he incorporated Spencer's terminology into his final edition of the *Origin*, Darwin still maintained that he disagreed with Spencer on many points,[33] took a "malicious" pleasure in criticisms of him,[34] and found the phrase "natural selection" more "convenient" than "survival of the fittest."[35]

That Darwin nevertheless incorporated Spencer's vocabulary into his final presentation of his theory in the sixth edition of the *Origin* is a measure of his eagerness to gain followers and of Spencer's extreme popularity, of which Darwin was keenly aware. Regarding his own distaste, he wrote perhaps disingenuously, "I suppose this is all my stupidity; as so many think so highly of his work."[36] The word "evolution" and its various forms appear ten times in the sixth edition of the *Origin of Species*, and five of these are contained in a single new chapter, added after the fifth edition, chapter 7, entitled "Miscellaneous Objections to the Theory of Natural Selection." The other five all occur also in connection with people's adherence, or not, to Darwin's theory.[37] In other words, "evolution" entered the *Origin of Species* only at the last moment and as a polemical term, to define the position that one must either adopt or reject. Darwin included it in the end because Spencer had succeeded in consolidating the "Darwinian" position around it.

While Darwin was ambiguous in presentations of his system regarding the question of internal forces and agencies, much to the confusion of his supporters, he continued, under direct questioning, to deny that there was any such thing. The matter came up with another good friend and particular ally against Agassiz, the Harvard botanist Asa Gray. Already in September 1857, Darwin had sent Gray a preliminary abstract of his theory of natural selection. Gray apparently responded that Darwin seemed to be attributing a form of agency to natural selection.[38] The sketch began, as would the *Origin* itself, with the effects of artificial selection in breeding (see figure 7.2). Then Darwin wrote,

> Suppose there was a being, who did not judge by mere external appearance, but could study the whole internal organization—who never was capricious,—who should go on selecting for one end during millions of generations, who will say what he might not effect! . . . I think it can be shown that there is such an unerring power at work, or *Natural Selection* (the title of my Book), which selects exclusively for the good of each organic being.[39]

Gray, finding that Darwin seemed to be ascribing a kind of agency to natural selection, was certainly not alone in this reaction. Reading such passages, Lyell objected that Darwin was "deifying" natural selection.[40] But to Gray's comment about agency, Darwin responded, surprisingly, with surprise: "I had not thought of your objection of my using the term 'natural Selection' as an agent; I use it much as a geologist does the word Denudation, for an agent, expressing the result of several combined actions."[41]

If it is possible to make "nature" sound even less like the kind of agent a geologist might consider "denudation," and more like the sort with actual agency, Darwin managed it in the subsequent, published version:

> Man can act only on external and visible characters: nature cares nothing for appearances, except in so far as they may be useful to any being. She can act on every internal organ, on every shade of constitutional difference, on the whole machinery of life. Man selects only for his own good; Nature only for that of the being which she tends. Every selected character is fully exercised by her; and the being is placed under well-suited conditions of life.[42]

This question of nature's agency was deeply fraught for Darwin. To see how it worried him, one need only follow the backing and forthing he did with this key passage from one edition of the *Origin* to the next. The passage remains the same in the second edition except for a tiny but significant change: nature becomes Nature.[43] In the third edition, as if to compensate for that capital let-

Figure 7.2 The results of pigeon-fancying, an example to Darwin of the malleability of living forms and the shaping power of selection, from *Variation of Animals and Plants* (1868).

ter, the passage has acquired a long parenthetical apology: "Nature (if I may be allowed thus to personify the natural preservation of varying and favoured individuals during the struggle for existence) cares nothing for appearances."[44] In the fourth edition, the parenthetical apology remains unchanged,[45] but in the fifth, it loses its parentheses and incorporates "survival of the fittest," a phrase appearing for the first time in that edition, as we have seen, borrowed from Spencer: "Nature, if I may be allowed to personify the natural preservation or survival of the fittest, cares nothing for appearances."[46] If the capital letter had intensified the suggestion of agency, and the parenthetical apology had seemed to qualify it by suggesting it was just a figure of speech, "survival of the fittest" surreptitiously insinuated not just a discriminating power, but one whose decisions were driven by a moral impetus.

Darwin's personifications of natural selection were not simply figures of speech. Or to put it better, figures of speech are not simple things. These images provided the terms in which Darwin thought his way into the idea of natural selection.[47] Consider the thought experiment that Darwin proposed in a preliminary, unpublished sketch of his theory. It involved an imaginary Being:

> Let us now suppose a Being with penetration sufficient to perceive differences in the outer and innermost organization quite imperceptible to man, and with forethought extending over future centuries to watch with unerring care and select for any object the offspring of an organism produced under the foregoing circumstances; I can see no conceivable reason why he could not form a new race (or several were he to separate the stock of the original organism and work on several islands) adapted to new ends. As we assume his discrimination, and his forethought, and his steadiness of object, to be incomparably greater that [sic] those qualities in man, so we may suppose the beauty and complications of the adaptations of the new races and their differences from the original stock to be greater than in the domestic races produced by man's agency. . . . With time enough, such a Being might rationally (without some unknown law opposed him) aim at almost any result.[48]

This wise and perceptive Being never reappears onstage in subsequent versions of the theory. But having played so dramatic a role in the mock-up, it seems hard to believe he fully vanished. Rather, he went underground, whence he rumbled his presence like the thunder machine in a Renaissance theater.

What, meanwhile, of the ostensible antecedent agency to natural selection, Lamarck's "power of life," Lyell's "renovating force," the source of creativity and change? All of those who had formerly believed in species-change had ascribed it to some such inner activity, distinguishing their views from the theological mechanists who called upon a supernatural God instead. Even T. H. Huxley, the English naturalist and polemical Darwinist who referred to himself as "Darwin's bulldog,"[49] and who was adamantly opposed to Lamarckian inner powers, wondered about the source of variation in Darwin's theory. Upon first reading the *Origin of Species*, Huxley later recalled that he had exclaimed to himself, "How extremely stupid of me not to have thought of that," and immediately wrote to Darwin professing himself "ready to go to the Stake" for the doctrine. Yet in the same letter, he did also wonder why "if external physical conditions are of so little moment as you suppose variation should occur at all—However, I must read the book two or three times more before I presume to begin picking holes."[50]

Huxley's good opinion, Darwin responded, had made him feel like "a good Catholic, who has received extreme unction": he could now die happy. Except, that is, for the cause of variation: "You have most cleverly hit on one point, which has greatly troubled me; if, as I must think external conditions produce little *direct* effect, what the devil determines each particular variation. . . .—I shall much like to talk this over with you."[51] In championing natural selection, Huxley adopted Darwin's Newton-inspired strategy of confessing ignorance about the cause of variation, as Newton had ostentatiously done regarding the cause of gravity. The "important point," Huxley affirmed, was that, granting the existence of variability, natural selection could then explain the subsequent development of living forms.[52]

Then, also like Darwin, Huxley wavered in his resolve. At one point, he tried to produce a source of variability without recourse to inner forces or powers by iterating natural selection downward, suggesting that molecules too engaged in a struggle for survival, and that the outcomes of their competition explained both heredity and variation. In this case, Huxley wrote, the causes of variation would not be "actively productive," but "passively permissive; they do not cause variation in any given direction, but permit and favour a tendency in that direction which already exists." Huxley himself must have realized that this was simply shifting the problem down a level. "I think it must be admitted," he added, "that the existence of an internal metamorphic tendency must be as distinctly recognized as that of an internal conservative tendency."[53]

Blumenbach's striving, "formative power," the *nisus formativus*, made an important appearance in Darwin's 1868 treatise, *Variation of Animals and Plants under Domestication*, where it received a section of its own. Darwin described the *nisus* as "that co-ordinating and reparative power which is common, in a higher or lower degree, to all organic beings."[54] This concession was partly in response to Spencer, who had been pressing for a greater role for creatures' "use and disuse" of their parts in directing the course of their transformations. How else, Spencer asked, to account for cases in which whole constellations of coordinated changes must occur simultaneously, for example, those changes necessary to support the great antlers of a moose or an Irish elk?[55] In a letter to his good friend, the botanist J. D. Hooker, Darwin admitted that Spencer had hit upon "a blot in the 'Origin,'" but said he reckoned he had solved the problem in *Variation of Animals and Plants*, which was then in manuscript, by adding the coordinating *nisus formativus* into his system (in a chapter in which he also emphasized the importance of use and disuse among the "laws governing Variability").[56]

Darwin also introduced another form of inner agency in *Variation of Animals and Plants,* acting at the crux of his tentative theory of inheritance. His idea was that inheritance took place by a process he called "pangenesis," in which every part of an organism threw off particles he named "gemmules." These had a "mutual affinity" for one another that led them to gather in the reproductive organs, whence they were transmitted to the offspring, directing their development. In addition to their mutual affinity, gemmules were also modified by the organism's use and disuse of the corresponding organs.[57] This was in keeping with Lamarck's view that in higher animals, their force of will or habit could transform their bodies gradually and heritably. Darwin throughout his career, in each successive edition of the *Origin* and every one of his later writings, maintained the importance of the Lamarckian principle of use and disuse of parts. Darwin believed, that is, that individual organisms, through the heritable effects of their habits and behaviors, shaped the course of evolution.[58]

Indeed, higher organisms even directed evolutionary development by exercising their tastes, desires, and aesthetic preferences. Darwin called the process by which they did so "sexual selection": "the advantage which certain individuals have over others of the same sex and species solely in respect of reproduction." Sexual selection took place partly in the competition among males for a female, and partly in the selective agency of the female.[59] A female bird, by choosing the most melodious and beautiful of her suitors, acted as a human breeder with equivalent effects. If humans find the resulting bright plumage and lilting song delightful, Darwin observed, we may infer "that a nearly similar taste for beautiful colours and musical sounds runs through a large part of the animal kingdom." Sexual selection could accordingly help explain "how it is that there is so much beauty throughout nature: for this may largely be attributed to the agency of selection."[60] Animals (especially *female* animals) were continually shaping and reshaping the world of living things in accordance with their own standards of beauty and worth:

> He who admits the principle of sexual selection will be led to the remarkable conclusion that the nervous system not only regulates most of the existing functions of the body, but has indirectly influenced the progressive develop-

> ment of various bodily structures and of certain mental qualities. Courage, pugnacity, perseverance, strength and size of body, weapons of all kinds, musical organs, both vocal and instrumental, bright colours and ornamental appendages, have all been indirectly gained by the one sex or the other, through the exertion of choice, the influence of love and jealousy, and the appreciation of the beautiful in sound, colour or form.[61]

Animals governed the transformation of species not only by their habits and behaviors but by a very particular sort of act of will, "the exertion of choice." Clearly, of the two kinds of inner agency at work in Darwin's theory of variation and inheritance—an inherent tendency of the parts, such as the *nisus formativus* or the affinities of the gemmules, and the behavior of the overall organism—the second posed no problem for Darwin. He believed in it unreservedly and dedicated a major work, *The Descent of Man and Selection in Relation to Sex,* to examining how it might have shaped human evolution.[62]

He was as abidingly ambivalent about the first sort of agency as he was unwavering with regard to the second. In the *Origin of Species,* Darwin wanted to allow no such inner tendency or force, or at least, so he said and repeated. In the third edition of the *Origin,* he added a historical sketch of his Romantic predecessors, chiefly Lamarck and Saint-Hilaire, relegating his grandfather to a footnote. Here, he tried to give credit where credit was due while still distancing himself from the Romantic approach to transmutation of species represented by Erasmus Darwin, Goethe, and Saint-Hilaire, and especially from the assumption that the simplest forms of life arose through a manifestation of upward-striving material tendencies in spontaneous generation:

> It is curious how largely my grandfather, Dr. Erasmus Darwin, anticipated the erroneous grounds of opinion, and the views of Lamarck, in his "Zoonomia" (vol. i. p. 500–510), published in 1794. According to Isid. Geoffroy there is no doubt that Goethe was an extreme partisan of similar views [i.e., spontaneous generation], as shown in the Introduction to a work written in 1794 and 1795, but not published till long afterwards. It is rather a singular instance of the manner in which similar views arise at about the same period, that Goethe in Germany, Dr. Darwin in England, and Geoffroy Saint Hilaire (as we shall immediately see) in France, came to the same conclusion on the origin of species, in the years 1794–5.[63]

The idea of spontaneous generation was now, Darwin thought, quite out of the question: "I need hardly say that Science in her present state does not countenance the belief that living creatures are now ever produced from inorganic matter."[64] A decade later, however, Darwin wrote to Alfred Russel Wallace that he thought spontaneous generation probably did take place.[65] Wallace, a naturalist, explorer, and friend of Darwin's, was a sort of parallel author of the theory of natural selection, having arrived independently at a theory of transmutation of species that was essentially similar.[66]

Between Wallace and Darwin, the question of what might account for the generative side of transmutation became an important source of divergence. Having argued steadfastly alongside Darwin for more than a decade that natural selection was the source of every plant, animal, or human trait, in 1870, Wallace announced a change of heart. He had come to believe that the human intellect—the power of abstract thought, the moral and aesthetic faculties—served no purpose related to survival and reproduction and therefore could not be accounted for by natural selection.[67]

The salient aspect of Wallace's change of heart here is that he framed it also in terms of the failure of matter to explain any sort of action or movement. Atoms of matter themselves, Wallace wrote, were inert, and could not conceivably cause the forces that ruled their movements. It was therefore "logical" to eliminate atoms of matter from one's physical theory leaving only "centres of force to represent them." Wallace concluded that matter itself was "essentially force, and nothing but force; that matter, as popularly understood, does not exist, and is, in fact, philosophically inconceivable." Moreover, force, in turn, Wallace argued, must be "WILL." He reasoned thus: if human agency were purely illusory, and all our movements were the results of external causes, then how and why would consciousness or the illusion of agency ever have arisen? Whereas, if one assumed the appearance of human agency to be genuine, and surmised that other natural forces resembled it, then all natural forces would be forms of agency: "the WILL of higher intelligences or of one Supreme Intelligence."[68] Wallace emphasized that this "Mind," as he elsewhere called it, worked "by and through the primal forces of nature," such as gravitation, electrical repulsion, and "by means of Natural Selection in the world of life."[69]

Although Wallace arrived at a sort of divine presence, this was by no means an argument from design, but really the reverse, an argument from agency. Wallace did not claim that living beings, due to their intricate and perfect contrivance, must be the artifacts of a divine Engineer. He argued the opposite: brute machinery could not account for any kind of action, movement, or agency on its own, and yet these phenomena were everywhere in nature. It therefore made better epistemological sense to assume that nature was made, not of matter, but of force: action, agency. Here once again is the badly strained seam whose line we have been tracing through the heart of Darwin's theory of natural selection: on the one hand, the genetic approach to development, founded in an ascription of agency to living forms; and on the other, the notion of mechanical adaptation, originating in the brute-mechanist tradition of the argument from design, which strictly banished agency from nature. For Wallace, the seam finally came apart, and he chose agency over matter, nature over a supernatural God.

Darwin was dismayed. He wrote Wallace a letter full of kindness and compliments, saving his real message for the end, where he delivered it with gentle humor: "But I groan over Man—you write like a metamorphosed (in retrograde direction) naturalist, & you the author of the best paper that ever appeared in Anth. Review! Eheu Eheu Eheu / Your miserable friend / C. Darwin."[70]

During the mid-1870s, the problem of the generative aspect of species-change also drove a painful wedge between Darwin and another onetime supporter, the writer and polemicist Samuel Butler.[71] He had been an early convert to Darwinism upon first reading the *Origin* soon after its publication. At the time, Butler was in his midtwenties and farming sheep in New Zealand. He read Darwin's book by candlelight on his antipodean sheep farm and immediately became "one of Dr. Darwin's many enthusiastic admirers."[72] But within a couple of years, he had grown ambivalent. He published a broad spoof of Darwin's theory entitled "Darwin among the Machines," which later became "The Book of Machines," a section of his novel *Erewhon* (1872).

The spoof purported to be a manifesto that had, four centuries before the story's present, brought about a revolution in the land of Erewhon, persuading its readers to destroy all machines invented in the last three centuries. Erewhon at that earlier time had been undergoing rapid industrialization, and the manifesto had argued that machines, in their current state of advance, would take on a life and consciousness of their own, supplanting the human race. Who, after all, could say "that a vapour engine has not a kind of consciousness?" And even if "mechanical consciousness" did not yet exist, it might evolve.[73]

Whether Butler intended this as philosophy or parody was never entirely clear; no doubt his intentions were a mixture. Soon after *Erewhon*'s publication, Butler apologized to Darwin: "I am sincerely sorry that some of the critics should have thought that I was laughing at your theory, a thing which I never meant to do, and should be shocked at having done."[74] But in the next few years, he would turn dramatically against not only natural selection, but Darwin himself. Butler loudly and repeatedly charged that Darwin had failed to acknowledge his debts to his predecessors, particularly his own grandfather, Erasmus Darwin, as well as to Buffon, Lamarck, and Saint-Hilaire.[75] Also, Butler himself began to champion what he described as an older form of evolutionary theory, one that posited the principal source of transformation within the transforming organisms themselves. Darwin's theory, Butler said, lacked a "motive power" such as Lamarck had described, a "power to vary," a force "to originate and direct the variations" that natural selection accumulated.[76]

The innate power to vary, moreover, in Butler's view, was essentially the power of thought. Heredity was a form of memory; variability arose from an "intelligent sense of need."[77] Thus Butler did believe in an intelligent designer, but not an external one:

> The designer of all organisms is so incorporate with the organisms themselves—so lives, moves, and has its being in those organisms, and is so one with them—they in it, and it in them—that it is more consistent with reason and the common use of words to see the designer of each living form in the living form itself, than to look for its designer in some other place or person. Thus we have a third alternative presented to us. Mr. Charles Darwin and his followers deny design, as having any appreciable share in the formation of organism at all. Paley and the theologians insist on design, but upon a designer outside the universe and the organism. The third opinion is that suggested in the first instance, and carried out to a very high degree of development by Buffon. It was improved and, indeed, made almost perfect by Dr. Erasmus Darwin, but too much neglected by him after he had put it forward. It was borrowed, as I think we may say with some confidence, from Dr. Darwin by Lamarck, and was followed up by him ardently thenceforth, during the remainder of his life, though somewhat less perfectly comprehended by him than it had been by Dr. Darwin. It is that the design which has designed organisms, has resided within, and been embodied by, the organisms themselves.[78]

Butler proposed to "see God henceforth as embodied in all living forms." Invoking Lamarck and Saint-Hilaire, he espoused a system that combined descent with design, mechanism with teleology, by making "the organism design itself."[79]

Butler was a thorn in Darwin's side. "The [Butler] affair has annoyed and pained me to a silly extent," he wrote to Huxley. Darwin felt the loss of an ally: "Until quite recently, he expressed great friendship for me . . . and I have no idea what has made him so bitter against me."[80] Furthermore, Butler's idea that organisms transformed by their own powers of mind made little sense to Darwin. Even if a cell had memory and the power to wish for improved parts, he wrote in another exasperated letter to the German biologist Ernst Krause, who had also been caught up in the "Butler Affair," Darwin did not see how these would enable it to change itself chemically or structurally.[81]

But privately, Darwin continued to wrestle with the idea of a continual increase in complexity of living forms. He agreed with Lamarck that such an increase did take place. But he labored to characterize this increase as a trend, not an innate tendency or a force. "The enormous number of animals in the world," he jotted in one of his notebooks, "depends on their varied structure & complexity.—hence as the forms became complicated, they opened fresh means of adding to their complexity.—but yet there is no necessary tendency in the simple animals to become complicated." At any rate, he thought, it should be immaterial to his theory whether an inner force of some kind caused the variations or not. All that should matter is that the changes were transmitted. Whether a body altered as a result of "the Nisus formativus" or due to a mere "tendency," in either case, "the effects are equally handed to the offspring.—"[82] Still, he prized what he took to be evidence against the idea of an inherent, transforming force, commenting in his notebook: "The identity of (or only closeness) of some species—(especially of mammifers) in old beds & existing species is valuable because it shows no innate power of change."[83]

Yet, precisely an innate power of change recurs in key passages of the *Origin*. The unnamed something responsible for variability constitutes a kind of subterranean, dynamic presence, periodically bursting up through the limpid surface of Darwin's prose. Here it suddenly appears, for example, along with another innate tendency—the tendency to revert to an earlier form—in Darwin's discussion of how difficult it can be for breeders to maintain desired traits in a population. He writes, "There may be truly said to be a constant struggle going on between, on the one hand, the tendency to reversion to a less modified state, as well as an innate tendency to further variability of all kinds, and, on the other hand, the power of steady selection to keep the breed true." A bit further on, he proposes naming this tendency "the *generative variability*."[84]

Generative forces are even more in evidence in Darwin's preliminary sketches of his theory. A species that transforms advantageously, he wrote in the first of these, might "go on beating out forms" forever, might even "people the earth."[85] But just what was it that was beating out those forms and filling the world with new versions of life? In his initial outlines, Darwin depicted living nature as positively unruly with tendencies—a "tendency to vary," a "hereditary tendency," a "tendency to revert to ancestral forms"—which breeders and, by analogy, natural selection had continually to battle and tame.[86] These tendencies are somewhat pruned back and pinned down by the first edition of the *Origin*. The "hereditary tendency," for example, gets a whole section in Darwin's preliminary sketch of 1842, but a brief mention in the equivalent spot in the first edition of the *Origin*.[87] Still, although Darwin toned down the manner in which he described them, the "tendencies" of living organisms remained pervasive all the way through to the definitive sixth edition.[88]

In fact, between the fourth, fifth, and sixth editions, worrying away at a key passage in chapter 5 on the Laws of Variation, Darwin progressively strengthened the role of an inherent tendency to vary. Initially, he had said that he thought the direct action of an organism's environmental conditions played a relatively small role in inducing it to change.[89] In the fourth edition, he appeared to contradict himself, adding a paragraph in which he then said that the conditions of life *were* the cause of variability.[90] The contradiction remained in the fifth edition,[91] but in the sixth, Darwin finally allowed the generative cause of variability to burst through. In light of the instances in which similar varieties arose under dramatically different conditions, he wrote, or in which dissimilar varieties arose in the same conditions, external circumstances must be less important than "a tendency to vary, due to causes of which we are quite ignorant."[92]

Proponents of the argument from design were quick to exploit Darwin's hesitation regarding an innate tendency to vary.[93] One such antagonist was George Douglas Campbell, Duke of Argyll, who was skillful at deploying the terms of transformist theories against them. Argyll noted that Darwin's theory of descent with modification did not explain how new forms appeared, but only how some survived and proliferated while others disappeared. "Natural selection," wrote Argyll, "can originate nothing." Argyll pointed out that contemporary theories of life were full of vital forces and tendencies, and invoked these to argue that natural science was tending, not toward materialism, but toward "transcendentalism."[94]

In fact, we have seen that these ideas of inner vital force, locating the source of life's agency within material nature, were anti-transcendental, in Argyll's terms; that is, they had arisen precisely in opposition to arguments from design, which attributed natural arrangements to an external agency. Moreover, Argyll himself certainly did not advocate a theory of the development of life by innate forces or tendencies. Rather, he concluded that life must be due to an external "Creative Will" working analogously to the way human intelligence produces its "contrivances": the steam engine, Babbage's calculating machine, the electric telegraph, and so on.[95] Clocks had given way to newer technologies, but the argument was essentially the same as it had been when Robert Boyle had stated it two centuries earlier: a designed artifact implies an Artificer.[96]

Part of Darwin's ambiguity on the subject of innate tendencies arose from a conflation of various different sorts of tendency: the tendency to vary, to complexify, to progress, to revert. Darwin remained adamantly opposed to the Lamarckian idea of an innate tendency to progress up the scale of life, or to increase in complexity, although he believed that living forms did in fact advance in organization, but always as the result of externally imposed natural selections rather than an inner force or power.[97] A tendency simply to vary, in contrast, played an ongoing and increasingly explicit role in his theory, at least as presented in successive editions of the *Origin*. And yet in other contexts, Darwin rebelled against even this idea of a tendency to vary. In his 1868 treatise on *Variation of Animals and Plants*, he argued that rather than look at variation as an "innate tendency" or "ultimate fact," one should instead consider it to be an aggregate effect: "each modification must have its own distinct cause." Darwin here described these causes as purely external, such that if one could breed all the individuals of a species over many generations in identical conditions, they would never vary at all. Change came only and exclusively from without. Living forms in themselves were essentially inert.[98]

In the sixth edition of the *Origin*, however, Darwin at last introduced a clear distinction between a tendency to vary and a tendency to progress, and endorsed the first of these. He was provoked to do so by criticism from the Swiss botanist Karl Wilhelm von Nägeli, who had argued that since plants often differ in morphological characteristics with no bearing on their survival, they must have an innate tendency to transform.[99] Darwin had read Nägeli's pamphlet just too late for the fourth edition of the *Origin*, but he noted in the margin that it was a "very good objection"[100] and wrote an appreciative letter to Nägeli saying that his criticisms were "the best which I have met with":

> The remark which has struck me most is that on the position of the leaves not having been acquired thro' natural selection from not being of any special

> importance to the plant. I well remember being formerly troubled by an analogous difficulty. . . . It was owing to forgetfulness that I did not notice this difficulty in the Origin. Although I can offer no explanation of such facts, & only hope to see that they may be explained, yet I hardly see how they support the doctrine of some law of necessary development for it is not clear to me that a plant with its leaves placed at some particular angle or with its ovules in some particular position, thus stands higher than another plant.[101]

Preparing his fifth edition, Darwin appealed to his friend and staunch ally, the explorer and botanist Joseph Hooker, for advice about how to respond to Nägeli's argument. "I find old notes about this difficulty," Darwin wrote, "but I have hitherto slurred it over." He wanted, he told Hooker, to show that morphological differences such as those Nägeli invoked "follow in some unexplained manner" from adaptive changes. "Anyhow, I want to show that these differences do not support the idea of progressive development. . . . Will you turn the subject in your mind, and tell me any more facts."[102]

Hooker responded in a series of letters with some observations about the uses of morphological structures in orchids and other advice which Darwin pronounced "invaluable" and "splenditious."[103] He likewise rejoiced in news from Thomas Henry Farrer, then permanent secretary of the Board of Trade, later to be Darwin's in-law by the marriage of their children, and a botanical researcher who had recently suggested that the staminal tube in the diadelphous Leguminosae performed the task of holding nectar.

> It is delightful to me to see what appears a mere morphological character found to be of use. It pleases me the more as Carl Nägeli has lately been pitching into me on this head. Hooker, with whom I discussed the subject, maintained that uses would be found for lots more structures, and cheered me by throwing my own orchids into my teeth.[104]

Having gathered his counterevidence, Darwin responded to Nägeli's objection in the fifth edition of the *Origin* by arguing that many morphological differences in plants, though apparently useless, might in fact have uses, or might earlier have had uses that were presently imperceptible. Moreover, Darwin insisted, in no case were these apparently useless differences "evidence of an innate tendency towards perfectibility or progressive development.[105]

However, by the sixth edition of the *Origin*, Darwin had fully distinguished a tendency to vary from a tendency to progress and accepted the tendency to vary. "In earlier editions of this work," he wrote, "I under-rated, as it now seems probable, the frequency and importance of modifications due to spontaneous variability." With plants, in particular, he acknowledged, "many morphological changes may be attributed to the laws of growth and the inter-action of parts, independently of natural selection."[106] At last, Darwin had reconciled himself to something that had been actively present, informing his theory from the very first, a kind of change generated from within living forms.

Or had he? That very same year, 1876, he also published the results from a series of experiments on plants designed to *refute* the idea "that there is an innate tendency in all beings to vary and to advance in organisation, independently of external agencies." An important piece of evidence in favor of an innate tendency to vary was that no two individuals of a species were ever exactly alike, even when they were raised in the same conditions. Darwin's experiments showed that what seemed like the same conditions were not really identical: most significantly, even plants growing right near one another would be visited differentially by insects.[107] Variability, in this case, came in on the wing.

It was a deep, an *elemental* ambivalence, as Butler mercilessly pointed out.[108] Moreover, Darwin was drawn to the work of other people torn between the same two impulses: to banish agency from living nature's machinery and to make agency its very crux. The French physiologist Claude Bernard struggled mightily with the same problem, also without solving it, and also extremely productively. Bernard is generally regarded as the author of the concept of homeostasis, for having argued that the essential capacity of living organisms was their ability to maintain a stable internal environment in response to a varying external environment.[109] But how did they manage it? What, indeed, *was* an organism, that it could perform such a feat?

"An organism is nothing but a living machine," was Bernard's constant refrain.[110] Living machines, he insisted, worked by the same forces as the rest of nature; to deny this would be "antiscientific."[111] Bernard described this basic mechanist principle in distinctly Romantic terms: a living organism, he wrote, was not "an exception to the grand natural harmony which makes things adapt to one another." He continued:

> It does not break any accord; it is neither in contradiction nor at war with the general cosmic forces; far from that, it takes part in the universal concert of

> things, and the life of the animal, for example, is only a fragment of the total life of the universe."[112]

Moreover, living machines did not differ from "brute machines" in these respects. They maintained themselves by a continual internal movement of adjustment and renewal; so too did a steam engine. In both cases, the living machine and the steam engine, this action of self-maintenance looked "independent" from the outside, leading observers into "a sort of personification." But if one looked deeply into the internal mechanism responsible, one would find an "absolute determinism" in the living machine as in the steam engine.[113]

Yet, at the same time, Bernard did also believe that living machines had a certain distinguishing feature, unique to them, which he described as their "directive idea": "a creating idea that develops and manifests itself through organization." Throughout a life, it was "always this same vital idea that conserves the being." The vital idea, Bernard said, was the essence of life.[114] In his efforts to articulate this idea specific to living machinery, Bernard also referred to such things as a "vital creation," an "organizing synthesis," and a "creative force." Indeed, he continually used the term "vital force," all the while insisting that there was fundamentally no such thing.[115]

A special vital force or tendency would be "antiscientific," while a reduction of living beings to "external physico-chemical conditions" would be wrong. Accordingly, Bernard rejected both what he called "vitalism," the appeal to a special, vital force, and mechanist materialism.[116] He labored to define a credible third possibility. He proposed "physical vitalism," resting on a "principle of the specificity of vital mechanisms." But his definition of physical vitalism was none too clear: "It is not the action that is vital and of a particular essence, it is the mechanism that is specific, particular, without being of a distinct order."[117] Perhaps Darwin recognized a kindred spirit: he praised Bernard's work, which he took to show that each organ in a living being had "its proper life, its autonomy," and could develop and reproduce itself "independently."[118]

Etienne-Jules Marey, who succeeded Bernard in the chair in physiology at the Paris Academy of Sciences, described living beings in similar terms, as both strictly mechanical and essentially active. Marey was a Darwinian and a mechanist who believed that "the laws of mechanics apply as well to animate motors as to other machines," and he decided to place his studies of the "animal machine at the service of Darwinism. Marey's idea was that, although one could not experimentally demonstrate the transformation of a species, which happened too slowly and gradually, one could perhaps reveal the tiny individual changes of which such a transformation would be constituted.[119] Rather than looking at big changes happening in geologic deep time, therefore, Marey set out to capture the opposite end of the spectrum, the tiniest increments of change taking place in infinitely small increments of time.

Life glimpsed at this level, Marey thought, was an "incessant mobility," a "chaos" of "movements, electric currents, variations of gravity and temperature." He designed inscription devices to register these tiny, fleeting changes, and it was in this context that he conducted his well-known photographic studies of animal movement using a technique that became an important precursor to cinematography.[120] Living machinery at the timescale Marey studied it was a blooming, buzzing confusion. By showing the myriad, minuscule, mutual adaptations that constituted life, Marey believed he could confirm Darwin's theory of evolution. He argued that one could see adaptation itself in progress: the muscular system transforming the skeletal system and, in turn, being transformed by the commands of the nervous system.[121] For his part, Darwin much appreciated the confirmation, citing Marey's work on "co-adaptation," the mutual adaptation of all the parts of an organism, with grateful admiration.[122]

Huxley, in contrast, also in the service of evolutionary theory, proposed the symmetrically opposite idea: that consciousness and agency were not primitive features of the machinery, but evolutionary side effects of its functioning: "There is as much propriety in saying that the brain evolves sensation, as there is in saying that an iron rod, when hammered, evolves heat." Huxley argued that Descartes's idea of living beings as machinery should be extended to include even the human mind: "We are conscious automata." He lavished praise on Descartes, who he said had not only been a great philosopher and mathematician, but was "certainly entitled to the rank of a great and original physiologist," since he had "opened up that road to the mechanical theory of these processes, which has been followed by all his successors."[123]

Regarding the extension of Descartes's hypothesis that animals were automata, Huxley proposed that human states of consciousness were produced by the bodily machinery but were not active in the works. Rather, a state of consciousness, in a human being as in other animals, was an epiphenomenon, a "collateral product" like heat from a hammered rod, the steam whistle on a locomotive, the bell on a clock. Human feelings of volition were therefore not the causes of human actions; they were merely secondary consequences of the physical states of the machinery in the brain that caused the actions.[124]

Darwinist mechanist accounts of life thus encompassed both extremes: a kind of machinery in which agency was elemental, constitutive of the works, and a kind of machinery in which agency was illusory, reducible to brute parts.

Design without the Designer

While Darwin was fending off the inner agency of innate forces, powers, and tendencies on one flank, he was doing battle on the other against the outer agency of a designer God. Just as the striving power of life was built into the structure of his theory through the genetic approach to transmutation that it had informed, so the designer God was also built in, through the notion of physiological adaptation or "fitness" and of living structures as complex but passive, designed artifacts. Therefore, while struggling with the evidence of apparently useless variations in the morphological features of plants, Darwin was also, symmetrically, confronting the formidable intricacies of the eye.

The eye was a great worry to Darwin, giving rise to certain uneasy passages in his writings that are now favorites among proponents of the "Intelligent Design" and "Creation Science" movement. One such passage is in the *Origin of Species* (1859), in the chapter where Darwin discusses difficulties for the theory of natural selection. The eye comes under "organs of extreme perfection and complication":

> To suppose that the eye with all its inimitable contrivances for adjusting the focus to different distances, for admitting different amounts of light, and for the correction of spherical and chromatic aberration, could have been formed by natural selection, seems, I freely confess, absurd in the highest degree.

In the continuation of the passage, Darwin then goes on to explain how "a perfect and complex eye could be formed by natural selection."[125] Lyell judged this a tactical error. The fact that natural selection would require the eye to have formed over time by means of small, random variations seemed such a "startling objection" that it gave "the adversary an advantage" to mention it so prominently.[126]

Despite Lyell's advice, Darwin had the confidence to include the eye and his answer to it in the *Origin*. Still, it continued to trouble him. He confessed to Asa Gray in 1860: "The eye to this day gives me a cold shudder."[127] Perfection was the heart of the problem: Could so complexly *perfect* a device truly be the result of natural selection alone? Or mustn't it require a divine Designer? A few months later, in another letter to Gray, Darwin confessed that intricate structures of other kinds caused him similar anguish: "The sight of a feather in a peacock's tail, whenever I gaze at it, makes me sick!"[128]

The qualms that Darwin felt when contemplating the eye gained force from the currency of the argument from design, which, as we have seen, found its chief exemplar in the complex and perfect eye.[129] This tradition, which we followed through the seventeenth and eighteenth centuries (chapter 3), had, at the beginning of the nineteenth, received its most powerful, definitive statement from the philosopher and Christian apologist William Paley, whose *Natural Theology, or Evidences of the Existence and Attributes of the Deity Collected from the Appearances of Nature,* first published in 1802, went through twenty editions by 1820.[130] Paley gave the argument from design its canonical image of the watch on the heath. "In crossing a heath," he wrote,

> suppose I pitched my foot against a *stone,* and were asked how the stone came to be there, I might possibly answer, that, for any thing I knew to the contrary, it had lain there for ever; nor would it, perhaps, be very easy to show the absurdity of this answer. But suppose I had found a *watch* upon the ground, and it should be inquired how the watch happened to be in that place, I should hardly think of the answer which I had before given—that, for any thing I knew, the watch might always have been there.[131]

A watch implies a watchmaker: it is a perfect encapsulation of the argument from design. What people do not generally notice is the part played by the watch here: it lies upon the heath like a stone about to be kicked. The watch is perhaps unlike a stone in its mechanical complexity, but in Paley's famous passage, it resembles the stone in its passivity and inertness, something one might kick or trip over. It represents the clockwork universe of early modern science not only in being an intricate mechanism, but also in being a profoundly passive object of externally imposed design.

Darwin was deeply marked by Paley's work, which he studied as a student at Cambridge. "In order to pass the B.A. examination," he recalled in his autobiography,

> it was . . . necessary to get up Paley's *Evidences of Christianity,* and his *Moral Philosophy.* This was done in a thorough manner, and I am convinced I could

> have written out the whole of the *Evidences* with perfect correctness, but not of course in the clear language of Paley. The logic of this book and as I may add of his *Natural Theology* gave me as much delight as did Euclid.[132]

The argument from design found its supreme illustrations, as we saw, in analogies between the eye and lens instruments. These too Paley picked up and polished. "Were there no example in the world, of contrivance, except that of the *eye*," Paley wrote,

> it would be alone sufficient to support the conclusion which we draw from it, as to the necessity of an intelligent Creator. It could never be got rid of; because it could not be accounted for by any other supposition. . . . [It is] an apparatus, a system of parts, a preparation of means, so manifest in their design, so exquisite in their contrivance, so successful in their issue, so precious, and so infinitely beneficial in their use, as, in my opinion, to bear down all doubt that can be raised upon the subject.[133]

To Darwin's contemporaries, raised on Paley and the tradition his *Natural Theology* culminated, it seemed evident that the eye was designed. It seemed so to Lyell, and to Darwin himself. To suggest otherwise, even for those who did so, felt equivalent to suggesting that a telescope had shaped and formed itself out of brute matter. This left an uneasy choice. On the one hand, there was divine design, which Darwin rejected. There was, he wrote to Asa Gray, "too much misery in the world":

> I cannot persuade myself that a beneficent and omnipotent God would have designedly created the Ichneumonidæ with the express intention of their feeding within the living bodies of caterpillars, or that a cat should play with mice. Not believing this, I see no necessity in the belief that the eye was expressly designed. On the other hand I cannot anyhow be contented to view this wonderful universe & especially the nature of man, & to conclude that everything is the result of brute force.

Darwin concluded, "But the more I think, the more bewildered I become, as indeed I have probably shown by this letter."[134]

A few weeks later, continuing the theological conversation with Gray, Darwin repeated that he could not believe the world to be "brute": "as I said before, I cannot persuade myself that electricity acts, that the tree grows, that man aspires to loftiest conceptions all from blind, brute force. Your muddled & affectionate friend CH. Darwin."[135] Having rejected, at least in principle, both Lamarck's inner agency and Paley's outer agency, Darwin found himself with a "brute" world in which he did not after all believe.

Paley's arguments for design were rigorously detailed. He devoted several pages, for example, to distance accommodation, which he reckoned gave the eye indisputable superiority over the telescope. In a telescope, after all, the operator, rather than the instrument itself, performed the accommodations by switching lenses or even instruments. In contrast, the eye, whenever it was directed at something nearby, underwent three simultaneous changes in order to adjust itself: the cornea grew rounder and more prominent, the crystalline lens pushed itself forward, and the whole eye elongated itself in depth. "Can anything," Paley demanded, "be more decisive of contrivance than this?" To build such a device, one needed to know the "most secret laws of optics." The fact that an eye can adjust itself, while a telescope cannot, might have destroyed Paley's analogy, but he found another device for that feature. In its capacity for self-regulation, the eye reminded him of John Harrison's self-regulating marine chronometer, the clock that had won the Longitude Prize in 1773.[136]

Other authors of arguments from design, in their determination to treat the eye as a lens instrument, had excluded perception, the fact that an eye can *see*, from their discussion. They treated the eye as a brute mechanism, not a perceiving one. This exclusion, like other aspects of the argument from design, reached its pinnacle in Paley's rendition. Unlike previous writers, Paley acknowledged the potential objection that the eye, unlike the telescope, could see. "To some," he wrote, "it may appear a difference sufficient to destroy all similitude between the eye and the telescope, that the one is a perceiving organ, and the other an unperceiving instrument." But Paley declared this difference to be irrelevant to the matter at hand: "The fact is, they are both instruments." For a creature to see, "an image or picture of the object [must] be formed at the bottom of the eye." How this image connected with sensation, with the experience of seeing, was, Paley argued, immaterial. Having "trace[d] mechanical contrivance a certain way," one might well arrive at "something which is not mechanical, or which is inscrutable." But that was all beyond the scope of Paley's investigation. To him, all that mattered was that the eye projected images by means of an apparatus constructed on the same principles as the telescope and the camera obscura.

"The perception arising from the image may be laid out of the question; for the production of the image, these are instruments of the same kind."[137]

Paley's principal arguments for the artificiality of the eye, like Boyle's and all intervening versions, rested upon fitness: showing that different sorts of eyes were structurally adapted to the circumstances in which they had to operate. Paley extended Boyle's earlier list of adaptations. Eels, he pointed out, had a "transparent, horny, convex case" over their eyes to protect them from sand and gravel. Different species exhibited different degrees of ocular accommodation, corresponding to the range over which they needed to see things. For example, fish eyes in their relaxed state were accommodated to nearby objects but had muscles that could flatten them to view farther objects. Birds needed to see things from the tips of their beaks to a distance of miles; accordingly their eyes had a bony rim

> which, confining the action of the muscles to that part, increases their lateral pressure upon the orb, for the purpose of looking at very near objects . . . . [and] an additional muscle, called the marsupium, to draw . . . the chrystalline lens *back*, and to fit the same eye for the viewing of very distant objects.

"Thus," Paley concluded (one hears in this passage how Darwin would echo the meter and phrasing as well as the content of Paley's writing) "in comparing the eyes of different animals, we see in their resemblances and distinctions, one general plan laid down, and that plan varied with the varying exigencies to which it is to be applied."[138]

By the time Paley repeated these arguments, they had the feel of truisms.[139] The eye's perfection was by this time so well established that Darwin, as we have seen, felt unshakably daunted by it, the more so because Paley's demonstrations of fitness were deeply ingrained, through the concept of physiological adaptation, in his own theory. Darwin imported the concept from the argument-from-design tradition, most immediately Paley, and it carried an unmistakable whiff of theology. "My Dear Darwin," Asa Gray teased in response to some notes from Darwin on the pollen and anthers of the melastomaceous plant *Heterocentron roseum*, "Why you are coming out so strong in final causes that they should make a D.D. of you at Cambridge!"[140]

That the notion of adaptation had originated in the argument from design especially preoccupied Huxley, since he had made it his project to defend Darwinism against all comers, but particularly against Owen and the other "Teleologists," to whom Darwin referred as "Paley and Co."[141] "Some of your friends, at any rate," Huxley had assured Darwin, "are endowed with an amount of combativeness which (though you have often and justly rebuked it) may stand you in good stead. I am sharpening up my claws & beak in readiness."[142] Teleology had received its "deathblow" from Darwinism, Huxley argued, at least that sort of teleology "which supposes that the eye, such as we see it in man, or one of the higher vertebra, was made with the precise structure it exhibits, for the purpose of enabling the animal which possesses it to see."[143]

Yet the adaptation of living structures to purposes, he also admitted, was as crucial to the theory of natural selection as it was to the argument from design: "Every species which exists," he wrote, "exists in virtue of adaptation, and whatever accounts for that adaptation accounts for the existence of the species."[144] With natural selection, Huxley decided, Darwin had found a way to reconcile the "teleological & the mechanical views of nature." In fact, as we know, these needed no reconciling, since the "mechanical view," as Huxley meant it, had relied upon a teleology, indeed a theology, from the start: the divine Clockmaker. Huxley thought Darwin had found a way to detach the teleology that was implicit in classical brute mechanism from its theological implication, to explain the pervasive adaptations to purpose throughout nature, "which Teleology has done good service in keeping before our minds, without being false to the fundamental principles of a scientific conception of the universe."[145]

But the concept of fitness retained a ghostly aura of the divine Engineer. Though they had evacuated God from his machinery, mechanists and authors of arguments from design had left behind a device world, an artifact world: a world utterly dependent upon an external source of purpose and action. If one removed this external source of agency, how could the remnants stand on their own? This was the puzzle at the core of the seven "transcendent world riddles" listed by the German physiologist Emil du Bois-Reymond in 1886. Given an ontology of passive brute matter in motion, du Bois-Reymond said, certain things must be necessarily and permanently inexplicable, such as the nature of matter and force, and the original cause of motion. Fourth on his list was the appearance of purpose everywhere in nature.[146]

However, if the evacuation of agency and purpose from nature left the artifact world hobbled, so too was its counterpart, the ostensibly omnipotent Engineer. The defeat of the argument from design, which Darwin said he had delivered with the law of natural selection,[147] had been prepared by the natural theologians themselves. As Buffon and Hume had explained, the argument from design ought to have been its own undoing. The questions about how to explain blind spots and other imperfections were a red herring. The real trouble was not imperfections of contrivance, but *contrivance itself* as an argument for an omnipotent being. And Paley himself acknowledged it, addressing in a more general form the question that Boyle had waved aside: it may be a good thing that flies have compound eyes to compensate for not being able to move them, but why not instead just give them the ability to move their eyes? ("The reason whereof I shall not now stay to consider," Boyle had written parenthetically.)[148]

"One question," Paley wrote, "may possibly have dwelt in the reader's mind during the perusal of these observations, namely, Why should not the Deity have given to the animal the faculty of vision all at once?"

> Why this circuitous perception, the ministry of so many means; an element provided for the purpose, reflected from opaque substances, refracted through transparent ones; and both according to precise laws, to produce an image upon a membrane communicating with the brain? Wherefore all this? Why make the difficulty in order to surmount it? If to perceive objects by some other mode than that of touch, or objects which lay out of the reach of that sense, were the thing proposed; could not a simple volition of the Creator have communicated the capacity? Why resort to contrivance, where power is omnipotent? Contrivance, by its very definition and nature, is the refuge of imperfection. To have recourse to expedients, implies difficulty, impediment, restraint, or defect of power.

Paley's best answer to this difficulty was to suggest that only by means of contrivances could the Deity make known to his rational creatures his existence, agency, and wisdom. Thus, although "Whatever is done, God could have done without intervention of instruments or means," nevertheless God "has been pleased to prescribe to his own power, and to work his ends within those limits":

> For then, i.e. such laws and limitations being laid down, it is as though one Being should have fixed certain rules; and if we may so speak, provided certain materials; and, afterwards, have committed to another Being, out of these materials, and in subordination to these rules, the task of drawing forth a creation: a supposition which evidently leaves room, and induces indeed a necessity for contrivance. Nay, there may be many such agents, and many ranks of these.[149]

Look where we have suddenly arrived, and how we got here! Paley's Engineer was hoist by his own petard. It was the logic not of natural selection, but of the argument from design, that led to this startling turn of events, in which the single, omnipotent Engineer whose existence had been being irrefutably demonstrated abruptly gave way to legions of anonymous and limited agents.

Meanwhile Darwin, who renounced the idea of a good and all-powerful God, conjured, in his very rejection of the argument from design, a Creator infinitely more awesome than Paley's Optician. "It is scarcely possible to avoid comparing the eye to a telescope," Darwin confessed.

> We know that this instrument has been perfected by the long-continued efforts of the highest human intellects; and we naturally infer that the eye has been formed by a somewhat analogous process. But may not this inference be presumptuous? Have we any right to assume that the Creator works by intellectual powers like those of man? If we must compare the eye to an optical instrument, we ought in imagination to take a thick layer of transparent tissue, with a nerve sensitive to light beneath, and then suppose every part of this layer to be continually changing slowly in density, so as to separate into layers of different densities and thicknesses, placed at different distances from each other, and with the surfaces of each layer slowly changing in form. Further we must suppose that there is a power always intently watching each slight accidental alteration in the transparent layers; and carefully selecting each alteration which, under varied circumstances, may in any way, or in any degree, tend to produce a distincter image. We must suppose each new state of the instrument to be multiplied by the million, and each to be preserved till a better be produced, and then the old ones to be destroyed. . . . Let this process go on for millions on millions of years; and during each year on millions of individuals of many kinds; and may we not believe that a living optical instrument might thus be formed as superior to one of glass, as the works of the Creator are to those of man?[150]

Consider the contrast between Paley's utterly accessible, knowable, and limited Engineer and the inscrutable, boundless, eternal power that Darwin describes in this passage.

Yet the eye still retained its transcendent superiority. It was not Darwin, but Hermann von Helmholtz, who gave the eye its decisive demotion. In his monumental study of the physiology of vision, the *Handbuch der Physiologischen*

Optik (1867), and in his philosophical and popular writing, Helmholtz insisted that the eye was anything but superior. It was, in fact, quite a lousy instrument, riddled with aberrations and blind spots.

To begin with, our vision is accurate only over a very small part of the visual field, corresponding to the fovea or pit of the retina where the cones are packed most tightly, "so that the image we receive by the eye is like a picture, minutely and elaborately finished in the centre, but only roughly sketched in at the borders." Moreover, the point of clearest vision is less sensitive to weak light than the other parts, a fact with which astronomers were familiar since they found they could perceive faint stars better in their peripheral vision (where, of course, they could see them least distinctly). Next, look at a street lamp through a violet glass, Helmholtz advised, and you will see a red flame surrounded by a bluish halo, the "dispersive image of the flame thrown by its blue and violet light." Such halos constitute "simple and complete proof of the fact of chromatic aberration in the eye."[151]

The cornea is not a perfectly symmetrical curve, leading to astigmatisms of varying degrees. Stars look star-shaped because the fibers of the crystalline lens are arranged around six diverging axes, "so that the rays we see around stars and other distant lights are images of the radiated structure of our lens." Dark objects are hard to see next to bright ones because the cornea and crystalline lens are not clear but rather a dingy whitish color, endowing bright objects with a halo that obscures darker ones nearby. The lens is also full of "entoptic objects"—fibers and spots—which you can see when you look at a bright surface. There are also fibers, corpuscles, and folds of membrane that move about with the motions of the eye and become visible when they come close over the retina. In addition to the retinal blind spot, there are plenty of other smaller gaps in the visual field caused by the blood vessels of the retina. These are small, to be sure, but still big enough for an astronomer to lose a fixed star in one.

In short, the eye "has every possible defect that can be found in an optical instrument, and even some which are peculiar to itself." Helmholtz concluded cheerfully, "Now it is not too much to say that if an optician wanted to sell me an instrument which had all these defects, I should think myself quite justified in blaming his carelessness in the strongest terms, and giving him back his instrument."[152]

The marvel, Helmholtz argued, is that creatures are able to see smoothly despite such a faulty apparatus. In this fact—that the eye is as good as it needs to be and no better—Helmholtz found support for Darwin. The eye was useful not because of its perfection, but despite its imperfections. "From this point of view," Helmholtz wrote, "the study of the eye gives us deep insight into the true character of organic adaptation generally. And this consideration becomes still more interesting when brought into relation with the great and daring conceptions which Darwin has introduced into science, as to the means by which the progressive perfection of the races of animals and plants has been carried on."[153]

Helmholtz also thought one needed to bring perception back into the conversation. Vision transcends the limits of the eye, he argued, not thanks to any divine artificer, but because of the mental processes that accompany and crucially help to constitute vision: the very processes of perception that natural theologians had ruled out of bounds, having fully outsourced all that was not brute mechanical in nature and physiology. The "extraordinary value [of the eye] depends upon the way in which we use it: its perfection is practical, not absolute." Directing our attention here and there, we overcome every impediment—an inverted retinal image, aberrations, blind spots, entoptic objects, poor peripheral vision—to form a smooth picture of the world. And we do this, Helmholtz argued, by a continual, unconscious process of inductive inference. We see by thinking.[154]

In the sixth edition of the *Origin*, Darwin added this argument, citing Helmholtz's observations regarding the "inexactness and imperfection in the optical machine." One could even believe, Helmholtz had written, that nature had "taken delight in accumulating contradictions" precisely in order to undermine arguments from design.[155] Even as examples of physiological fitness constituted a core category of evidence for natural selection, Darwin had, in fact, long cited examples of *un*fitness in favor of his theory as well. "What a book a Devil's chaplain might write," he had written to Hooker in 1856, "on the clumsy, wasteful, blundering low & horridly cruel works of nature!"[156]

The eye, then, troubled Darwin *when he regarded it as a brute mechanism.* Viewed thus, like any brute mechanism, it irresistibly implied a Maker. But when he restored to the picture that crucial part of the rudimentary, nascent eye that already differentiated it from any telescope, the "nerve sensitive to light" waiting beneath the thick layer of transparent tissue, he was back on firm ground, able to explain the eye as a product of natural selection. Herein lies a lesson regarding the earlier development of machinelike models of life during the seventeenth and eighteenth centuries, and also regarding Darwin's relation to these models. The most insistently reductive, device-like models of living creatures, models that left their unmistakable mark on Darwinism, relied upon a theology, a supernatural power to provide meaning and purpose. Those who most reductively rendered creatures as brute machines were the ones who believed in a divine Mechanic, while the rigorous naturalizers built perception and agency right into nature's machinery.

A central irony characterized brute-mechanist theories of life and the artifactual models of living beings to which they gave rise. The authors of these models, proponents of arguments from design, insisted upon the distinction between God and his handiwork. They saw this insistence as important on both sides of the matter: crucial to the autonomy of the New Science as well as the source of a new, powerful kind of argument for God's existence. But precisely by removing God from his creation, these new philosophers permeated their ostensibly autonomous science with a pervasive appeal to a supernatural power and also hobbled their omnipotent God by attaching him to particular, limited contrivances. Helmholtz, in contrast, worked in the alternative tradition whose development we have also been following, a tradition devoted to naturalizing agency and order rather than outsourcing these to an extra-natural Engineer. Accordingly, Helmholtz did away with the perfect eye, replacing it with the active process of perception.

Yet Paley's artisan deity did not quite disappear. Instead the ghost of the divine Engineer haunted the age of evolutionism: too connected with particular contrivances to be the full-fledged God of old, yet too ingrained in the development of the passive-mechanist physiological model from which Darwinism departed to be well and truly gone.

The famous final paragraph of the *Origin* is a plea for the interest and beauty of the natural world Darwin has described.

> It is interesting to contemplate a tangled bank, clothed with many plants of many kinds, with birds singing on the bushes, with various insects flitting about, and with worms crawling through the damp earth, and to reflect that these elaborately constructed forms, so different from each other, and dependent upon each other in so complex a manner, have all been produced by laws acting around us. These laws, taken in the largest sense, being Growth with Reproduction; Inheritance which is almost implied by reproduction; Variability from the indirect and direct action of the conditions of life, and from use and disuse: a Ratio of Increase so high as to lead to a Struggle for Life, and as a consequence to Natural Selection, entailing Divergence of Character and the Extinction of less-improved forms. Thus, from the war of nature, from

famine and death, the most exalted object which we are capable of conceiving, namely, the production of the higher animals, directly follows. There is grandeur in this view of life, with its several powers, having been originally breathed by the Creator into a few forms or into one; and that, whilst this planet has gone cycling on according to the fixed law of gravity, from so simple a beginning endless forms most beautiful and most wonderful have been, and are being evolved.[157]

Compare this paragraph from one edition to the next, and you will note two changes. First, the Creator appears in the second edition: "There is grandeur in this view of life, with its several powers, having been originally breathed *by the Creator* into a few forms or into one."[158] The Creator had been absent from that spot in the first edition. Continuing the comparison and arriving at the sixth edition, you will discover that "Variability" has acquired the hint of an internal source: "These laws, taken in the largest sense, being Growth with Reproduction; Inheritance which is almost implied by reproduction; Variability from the indirect and direct action of the [*external* removed here] conditions of life, and from use and disuse."[159]

These were the only two changes that Darwin ever made to his beautiful parting image of a Darwinian natural world: he added a nod to the external, divine agency of the natural theologians; and a nod to the internal, striving agency of the transformists. His theory remained suspended between the two.

~ 8 ~

The Mechanical Egg and the Intelligent Egg

> The bird's egg . . . is a highly complicated historical product.
> —Ernst Haeckel, *The Evolution of Man* (1905)

> The egg is considered as a definite machine which if once wound up will do its work in a certain direction.
> —Jacques Loeb, *The Mechanistic Conception of Life* (1912)

In the competition to interpret Darwinism that took place over the last decades of the nineteenth century and the turn of the twentieth, the passive-mechanists triumphed over the active historical-mechanists. The passive-mechanists established the theory they called "Darwinism," which acquired the name "neo-Darwinism" during the 1890s,[1] as antithetical to a science of inner agencies, formative impulses, and striving forces, and especially as antithetical to "Lamarckism" (I use the quotation marks to distinguish the construal of Lamarckism in these later struggles from what Lamarck himself wrote). These neo-Darwinists described living beings as passive mechanisms, never acting but only acted upon, as passive as Paley's watch on the heath. How and why the passive-mechanists won the battle is the story of this chapter. It takes place mostly in Germany, in the new German research universities of the late nineteenth century, and it begins with an egg.

Protestant *Wissenschaft* and the Redefinition of Darwinism

An egg was a machine, according to an influential generation of mostly German-speaking biologists of the late nineteenth to early twentieth centuries. An egg was a machine and moreover, as such, it was subject to mechanical and not "historical" explanation. By "historical" explanation, these biologists had in mind the Lamarckian-inflected version of Darwinism according to which living forms developed gradually and contingently, driven and directed by their own inner agencies. This antihistorical generation of biologists rejected the idea that intrinsic agencies operate in nature, and they also repudiated something else that this idea brought to scientific explanation: contingency. Contingency is neither randomness nor determinism, but something altogether different. It is the product of limited agencies working in particular changing situations.

Biologists around the turn of the twentieth century accepted that change could take place through a combination of random and mechanically determined events, but not that change could be contingent, directed by tendencies or actions within the living organisms themselves. Neither random nor deterministic processes violated the principled exclusion of agency from scientific explanations. Therefore, it became the habitual stance of the new generation of biologists to reason in terms of a combination of random and deterministic processes, but never of contingent, historical ones.

In keeping with their rejection of historical explanations in science, these biologists asserted a sharp distinction between the natural sciences and the humanities (*Naturwissenschaften* and *Geisteswissenschaften*). Whereas the humanities offered sweeping descriptions and impressions of the world, the natural sciences established causal accounts of discrete phenomena. Brute-mechanist science took on a new twist, becoming antihistorical and distinct from humanist disciplines. A science of living beings must describe their machinery not only as passive and rote but, by the same token, as timeless, without history and devoid of historical meaning.

An embryologist and evolutionary biologist named August Weismann was the principal author of this new version of Darwinism that was purged of all Lamarckian influences. Weismann was a doctor and zoology professor at the University of Freiburg.[2] It would be difficult to overstate the transformative importance of Weismann's interpretation of Darwin's theory: it became the standard by which biologists distinguished legitimate from counterfeit Darwinism and has remained so right through the twentieth century and into the twenty-first. Weismann is this chapter's main protagonist.

But Weismann did not act alone: the new Darwinism was the collaborative work of a couple of generations of biologists, philosophers, and historians. So solid and great was their consensus that it became all but impossible for later generations to see around it. According to this Everest of a consensus, Darwinism was unequivocally "mechanist," notwithstanding Darwin's own ambivalence about the mechanical models of life on which he drew. Darwinism became "mechanist," and "mechanist," in turn, became the opposite of "vitalist," a word often applied to sciences of the eighteenth century and even earlier, but which was in fact an innovation of the nineteenth.[3] "Vitalists" were those who set living beings apart from the rest of the natural world, as requiring a distinct set of explanatory principles. Eclipsed from view was the path that Leibniz, Lamarck, and others had struggled to defend: that of a single, active-mechanist and historical philosophy embracing all of nature.

Under this new regime, the notion that one could give a historical explanation of the functioning of a mechanism began to sound contradictory and strange. To describe the way a clock works, must one consider the clock's biography and ancestry? Yet a tradition of people, going back to the earliest ones to describe species as changing in time—Leibniz, La Mettrie, Buffon, Diderot, Lamarck, Helmholtz—understood their approach as at once mechanist and historical. They took living entities to be machines, by which they meant rationally comprehensible systems of moving parts that developed and changed thanks to a kind of historical agency: an ability throughout the parts of the machinery to constitute and transform themselves and their relations to one another purposefully and responsively. One could not understand the machinery, these active-mechanists believed, except in historical terms: in terms of its self-transforming actions over time.

In contrast, while the new Darwinists described living things as changing over time, they rejected the idea that the results must be understood in historical terms. The living machinery was continually being transformed by a combination of random accidents and external forces, but these transformations were of a "physico-chemical" nature and must be understood as such.[4]

Lamarck played a starring role in this new construal of Darwinist-mechanist biology: he was the foil. Though he had described living organisms as "animate machines," Lamarck, posthumously, became a "vitalist." Though he had closely informed the thinking of Darwin and his followers, Lamarck now came to represent the antithesis of Darwinism. The establishment of a Darwinist biology for the twentieth century rested upon the eradication of any whiff of an active-mechanist, historicist approach now associated with "Lamarckism." Lamarck ultimately became not just wrong but absurd, laughable, beyond the pale. To be sure, in Weismann's generation this was not yet the case. There is evidence that Weismann himself respected and even admired Lamarck: he had a copy of Jacques-Louis David's portrait of Lamarck hanging over his desk.[5] However, in the longer term, Weismann's work to eradicate Lamarckism from evolutionary biology had a powerful effect. By 1920, the word "Lamarckism," according to the Harvard entomologist William Morton Wheeler, designated "the ninth mortal sin" in biology.[6] Why and how did this happen?

The intellectual and national politics of late nineteenth- and early twentieth-century German-speaking Europe, and more specifically, the emergence of the modern research university, provide the original context for this development. The research university as we know it today is an originally German, and predominantly Prussian, Protestant institution.[7] Over the course of just a few decades, following the founding of the University of Berlin in 1810 by Wilhelm von Humboldt, German universities developed from medieval theological seminaries to juggernauts of scientific and industrial research and key organs of the state. By the establishment of the imperial German state in 1871, the research university had taken on a pivotal importance in German society and politics. Beginning in 1876 with the founding of Johns Hopkins on the template of the University of Berlin, the United States began to import the model, which subsequently became the primary locus of scientific research worldwide.[8]

The new German research universities, however, retained their theology faculties, indeed until the present. The theology faculties modernized alongside the burgeoning scientific disciplines, developing a "scientific theology" that, in the Protestant universities, explicitly associated an empirical, mechanist, positivist approach to natural science with Protestantism and against Catholicism. "The professors meet daily within and without the university," wrote a contemporary observer. "The theologian associates with the philosopher, the philologian, the historian, the student of nature; these sciences appear to him, as it were, in personal form, so that he cannot pass them by without notice; he is forced to settle his mental accounts with them. This has, undoubtedly, had the greatest influence on the character of Protestant theology, whose principal tendency is to bring religion and science to an inward reconciliation." In contrast, the writer continued, Roman Catholic theology isolated itself, which allowed for a greater uniformity of doctrine but "also decreases its power of influencing the science and culture of the age. As Protestant theology is affected by all branches of science, it in turn affects them all."[9]

Protestant scholars claimed as their own the positivist ideal of objective neutrality (*Voraussetzungslosigkeit*, or "presuppositionlessness"), a distinguishing feature of "Protestant *Wissenschaft*," setting them apart from the doctrine-laden Catholics.[10] "The devout Catholic," observed Max Weber, participant observer of these developments as well as theorist of the relations between Protestant theology and science, technology, and industrialization, "will never accept a view put forward by a teacher who does not share his dogmatic assumptions."[11] The comment appears in an essay entitled "Science as a Calling" (Wissenschaft als Beruf), first delivered about a generation after the culmination of these developments as a speech at the University of Munich in 1918 in the throes of World War I. By then, the modern research university with its ideal of scientific practice not only was well established but represented the hope for salvation. Weber meant his title literally. Modern empirical science and its technical applications, he claimed elsewhere (as part of his argument that Protestantism had been the source of capitalist society in the West), had grown from the Lutheran idea of a "calling": the notion that worldly activity, as opposed to scholastic reflection, represented the highest expression of religious duty.[12]

The institutional culture of Protestant *Wissenschaft*, in important part a reaction against the older holistic approach of *Naturphilosophie*, dominated the new world of academic research.[13] To dominate, of course, is not to monopolize entirely: people continued to promote other approaches, including, as we shall see, Hermann von Helmholtz and Ernst Haeckel. But they did so in resistance to an increasingly prevalent methodology. A measure of this methodology's prevalence is that, despite its connection with Protestant theology and culture, it was by no means limited to Protestant institutions. Even in the minority of traditionally Catholic universities in the German empire, as we shall see, the same scientific ideals came to hold sway, and the same culture of research grew strong.[14] Across the university faculties in the natural sciences, this new methodology became a moral imperative: not only specialized but anti-generalist, not only empirical but antitheoretical, not only mechanist but brute mechanist, shunning any explanation that reached beyond what was immediately adjacent.

The history faculties at the universities embraced an equivalent principle. The study of history likewise established itself as an autonomous academic discipline in this period and in the transformed universities. Those who inaugurated the modern discipline of history in the new German research universities, led by Leopold von Ranke, criticized the great, progressive, idealist philosophies of history of the late eighteenth and early nineteenth centuries. The authors of these earlier philosophies of history, beginning with G. W. F. Hegel, had construed history as the rational unfolding of an imbuing mind or spirit whose workings would make sense of past events, joining them into a comprehensible unity with the present, and projecting their culmination in the future. In contrast, Ranke and his students and followers, who occupied the new university chairs in history, declared the science of history to be founded in "particulars," philological and archival facts, "memoirs, diaries, letters, ambassadors' reports, and original accounts of eyewitnesses," all to be meticulously documented in footnotes. "A strict presentation of facts," observed Ranke, was the "highest law," and to these one must bring "impartial understanding, objective narration."[15]

To be sure, Ranke often sounded like Hegel and the earlier generation of idealist, progressivist historians too emphasizing, for example, that a historian's destination was an understanding of the "unity and the progress of events." To arrive at this destination, Ranke stressed the importance of form and beauty to a historian's craft.[16] Ultimately history, as this scion of a family of Lutheran pastors described it, was also a kind of calling. Like his colleagues on the science faculties, Ranke firmly believed that above and beyond the particulars, unattainable yet omnipresent, was "the hand of God": "Over everything," he wrote, "hovers the divine order of things."[17] Even despite this methodological embrace of factual particulars, and the familiar connection to a divine order arranged by the hand of God, academic historians' incursions into church history made them more of a threat to their colleagues on theology faculties than were the mechanist biologists, who carefully avoided any historical or humanist relevance whatsoever.[18]

The biologists' scientific methodology, embraced and institutionalized in the German universities, had emerged over the previous two centuries in natural theology through arguments from design, which rendered the natural world and all its inhabitants as passive machinery and gave God a monopoly on purpose and agency. Classical brute mechanism was thus a congenial, in fact, indigenous form, of "Protestant *Wissenschaft*." A dangerous mode of scientific explanation, in contrast, from the perspective of Protestant *Wissenschaft*, was one that would naturalize agency, denying God's monopoly, and insisting upon the mutual relevance of scientific and humanist, historical understanding. In short, Lamarckian-Darwinism.[19] The new biology faculties constituted themselves against this mode of science. Du Bois-Reymond coined a slogan for the new approach in his famous 1872 speech to the Leipzig Association of German Natural Scientists and Physicians, "On the Limits of Natural Knowledge," in which he renounced any claim on the part of science to the big questions, in particular questions of agency: how force moves matter, or how consciousness animates a brain. The slogan with which Du Bois-Reymond ended the speech, and which quickly became a rallying call, was "ignoramus ignoramibus": ignorant we are and shall ever remain.[20]

"Present-day science does not worry about the whole"; wrote another participant observer. "It thus no longer strives after a worldview." In fact this abandonment, or really rejection of a worldview itself constituted a worldview. Scientific acclaim now lay in applying refined methods to specific problems.[21] A successful investigator must be entirely devoted "to one science, nay, often to only a part of one science. He looks neither to the right nor to the left so that what is going on in his neighbor's field may not prevent him from burying himself in his specialty to his heart's content."[22] Again, Weber agreed: "Science has entered a phase of specialization that has never been known before, and this will not change for the indefinite future." Only a narrow specialist could hope to achieve something worthwhile, and "anyone who does not have the ability to put on blinkers . . . should stay well away from science."[23] Weber admonished that to be successful, one must accept that science was not a path to meaning, art, God, or happiness. Science dealt neither in the revelations of "seers and prophets," nor in the truths of "sages and philosophers." What science had to offer was technology, the means to control the conditions of daily life, and also clear thinking.[24] A key role of this German Protestant mode of positivist, mechanist science was to puncture the authority of generalized visions.

Accordingly, the new research universities, with unprecedented support from the German states and from Prussia, spawned scientific disciplines, subdisciplines and sub-subdisciplines.[25] These represented a new and formidable kind of power, at once intellectual, institutional, and political. In this climate of state-funded discipline formation, the natural sciences first caught up with the humanities in cultural prestige and social authority, and then began to surpass them.[26] The emerging authority of the natural sciences rested upon a particular representation of their nature and mission, one that set them firmly apart from the humanities.

The sciences must be specialized, strictly confined in methodology and purview, and segregated from humanistic concerns of every kind, philosophical, moral, historical, aesthetic. This quarantine would allow the sciences to uphold the core principle of Protestant *Wissenschaft*, presuppositionlessness. The quarantine would also act as a guarantee of the sciences' political neutrality: they would be untainted by the tendencies most threatening to the consolidating imperial regime, socialism and Catholicism (though by the turn of the century, as we will see, a socialist German-American Jew named Jacques Loeb had come to personify the new positivist-mechanist biology).

Thus, when Helmholtz exhorted his colleagues against the fragmentation of the sciences and their segregation from the humanities, he fought a losing battle.[27] The emerging sub-subsciences derived their considerable and growing influence from a specifically antihumanist mechanism: a mechanism that denied rather than incorporated philosophical, aesthetic, and especially historical modes of explanation.

The Mechanical Egg and the Historical Egg

Among the first crusaders for mechanism and against history was the son of a silk merchant and jurist from Basel, the anatomist and embryologist Wilhelm His.[28] One measure of the importance His attributed to this question is that when he became rector of the University of Basel in 1869, he chose for the topic of his inaugural address the importance of treating eggs and embryos as machines and, specifically, not as developing historical entities.[29]

Defending a patch of academic turf, in particular, defending the integrity of embryology as a discipline, was His's supervening concern. He thought that the institutional independence of embryology relied upon a particular sort of mechanical model of eggs and embryos, a model that appealed only to immediate physical causes—bits of matter pushing and pulling one another—and not to remote, phylogenetic, evolutionary ones. Otherwise, embryology would become a subfield of evolutionary theory.[30] To understand the urgency with which His pressed his case, and his choice of so general and public a forum in which to do so, consider the climate in which he had conducted his career thus far.

Twenty years earlier, one year into his medical training at Basel, His had made a pilgrimage to Berlin.[31] There he had attended lectures at the University of Berlin by the anatomist Johannes Müller. He had studied with the embryologist Robert Remak, who had recently received special entitlement from the king to lecture at the university despite being Jewish (although His's colloquium with Remak took place at Remak's house). In Berlin, His had also befriended students in theology and law. After these experiences, he had returned to Basel with a keen sense of how scientific and academic life should be constituted. Disciplines must have clear methodological and territorial boundaries. Anatomy, physiology, and embryology were exact sciences, he believed. They were mechanical, empirical, and in urgent need of defense against speculative philosophical incursions, especially those coming from evolutionists.

The burgeoning scientific departments of the research universities represented a new kind of power, commanding money, prestige, and administrative clout. One's ability to wield this new power rested upon one's ownership of a particular mode of scientific explanation. By imposing sharp limits on what counted as an explanatory cause in embryology, allowing only immediate physical causes, His identified a discrete mode of scientific explanation to which he and his discipline could lay claim, thereby securing him a seat in the halls of this new power. The tactic was successful: in 1872, he was invited to chair the anatomy department at the University of Leipzig. In his inaugural address there, he again argued for an exact mechanical approach to anatomy and embryology.[32] Having established his laboratory at Leipzig, His pressed his case aggressively for mechanism and against history as forms of biological explanation.

Today, His is mostly remembered among biologists for having built a better microtome, an instrument used for slicing tissue into thin sections, and having used it to advance the technique of tissue sectioning.[33] He used this technique to develop microscopic images of developing embryos in support of his view that the differential growth of the germ was the proximate, and therefore the only relevant, "mechanical" source of all embryonic development. That is, some parts of the embryo, growing faster than the adjacent parts, would push and pull these adjacent parts, inducing the various rollings and bendings that one saw as the embryo went through its successive stages. The neural tube, for example, emerged from this sort of folding over of a part of the germ tissue as a result of differential growth.[34]

Imagine holding the embryonic tissue, His instructed his readers, and then imagine pulling it this way or that. You will see in your mind's eye how these pulls will induce bends and folds. To make it more apparent, His modeled the process in leather, clay, paper, and rubber. Make a slit in the middle of a rubber tube, then push against its two ends and you will see the slit gap open, just like the neural tube at a certain stage of development. Attach a thread to the open end of a rubber tube and pull it: the pulled part will fold downward and the fold will cause the tube to jut out on either side, just like the optic vesicles jutting from a fold of the neural tube. His tugged and pushed at a leather embryo on a wire frame; he folded pieces of paper; he crinkled bits of cloth; he prodded lumps of clay and wax.[35] He also asked colleagues from the Basel mathematics faculty to help him derive equations for the differential growth of the embryo based upon the treatment of imperfectly elastic plates in analytical mechanics.[36]

An observer of one such collaboration in His's laboratory, the wax modeler Adolf Ziegler, at work with His on some wax models of embryos, described the scene mischievously in a letter to the leading German Darwinist, who was also a Lamarckian, Ernst Haeckel. "I sat beside him for a long time," wrote Ziegler, "on that occasion in Basel when he had the university mathematicus come to determine with him the formula of growth; the stocky gentleman comprehended only with difficulty, and they made drawings, folded letters and finally experimented with the calf's skin nailed on a board, until the two gentlemen at last agreed about things which—I no longer understood!"[37] All these procedures were meant to establish that pressures such as those exerted by differential growth could cause materials to roll and buckle in ways that resembled a developing embryo.

His aimed to show, in the words of the American embryologist and early geneticist Thomas Hunt Morgan, that cells did not "by their individual activity or movement change their shape. . . . For instance, a cell becomes conical not through its own initiative but because surrounding pressure forces it into a conical shape."[38] In His's scheme, cells were inactive, without "initiative."

Why, then, did different parts of the germ grow at different rates? His reckoned dismissively that it was because they had different dispositions (*Anlagen*) to grow.[39] This was not much of an explanation, but he didn't care much; the important thing was that only these growth rates and the forces they exerted counted as sufficiently proximate causes for embryology. In His's microsectioning hands, in other words, mechanist science was redefined negatively, in terms of what got sliced away, which was everything beyond the horizon of neighboring parts. Mechanist science of His's variety could include the rather mysterious dispositions-to-grow, provided these acted as proximate causes, pushing and pulling the immediately adjacent tissue, and providing also that one was careful never to attempt to explain their origins with reference to any more distant causes.

In defending this approach to embryology with its strict limitation to immediate causes only, His took aim at its opposite, the open, encompassing, active, historical-mechanist approach of the evolutionists. He did this in part by equating the historical approach with "vitalism," which made living phenomena into a mystical exception to natural law: "The mystics have not died out who imagine heredity causes the material of developing organisms to move according to other laws than those which are valid in nature as a whole."[40] Among the "mystics" who had not yet died out, His likely had in mind Haeckel, a vigorous sixty-year-old who was in fact three years junior to His, and His's lifelong target. (As part of his campaign against Haeckel, two decades earlier, His had famously accused the Lamarckian-Goethian-Darwinian of dishonestly manipulating his published images in order to make human embryos appear more similar to other animal embryos.)[41]

Haeckel himself described his own science as an anti-mystical, "mechanical" system of nature according to which the entire natural world, living and nonliving, constituted a single, rationally comprehensible organization of moving parts.[42] By casting Haeckel as a dinosaur, though he was younger than His, and a "mystic," though he offered mechanism specifically as an antidote to mysticism, and by characterizing Haeckel's philosophy as "vitalist" despite Haeckel's insistence upon the mechanist principle of continuity between living and nonliving nature, His developed an efficient rhetorical microtome. He sectioned the scientific world into just two possibilities: his own form of drastically limited, adjacent mechanical pushes and pulls; or outmoded magical nonsense.

For his part, Haeckel derided His's folded pieces of paper and tugged-upon rubber tubes. These he thought absurdly inadequate given the "infinitely subtle and complex nature of the mechanical problem." Nature's mechanisms were not patched and cobbled together as by an artisan. They emerged in time, "historically," through countless tiny developments driven from within.[43] In the active-mechanist tradition of Leibniz, La Mettrie, Buffon, Diderot, Goethe, and Lamarck, Haeckel assumed that the natural world was wholly made of the same stuff, at once matter and spirit, having the properties of each.[44] He therefore argued that evolutionary theory was something in between the natural sciences (*Naturwissenschaften*) and the humanities and social sciences (*Geisteswissenschaften*).[45]

When he insisted that phylogenesis (the evolutionary development of an organism) was an absolutely "mechanical process," Haeckel meant something different from what His meant by "mechanical." Haeckel meant that phylogeny resulted from the "movements of the atoms and molecules that compose organic matter," and also from an "infinite manifoldness" in the "active plasma" that made up all organisms. Phylogenesis, was accordingly "neither the premeditated purposeful result of a thinking creator, nor the product of some unknown mystical power of nature, but the simple and necessary effect of . . . known physical-chemical processes."[46] These processes acted from within the substance of natural entities to develop and proliferate them.

In Haeckel's view, natural machinery could be understood only historically, in terms of the activity over time of organic substance. His's approach with its rubber tubes and bits of leather "fancies it has found the real *mechanical* causes of the facts of embryology when it has traced them to simple physical processes, such as the bending and folding of elastic plates." But this approach was "pseudo-mechanical," in attempting "to reduce the most complex historical processes to simple physical phenomena." These physical phenomena were incomprehensible unless one placed them in the context of their evolutionary causes and realized "that each of these apparently simple processes is the recapitulation of a long series of *historical* changes."[47] Biology itself was a fundamentally historical discipline, the greater part of the phenomena with which it dealt being "complicated *historical processes*."[48]

To Haeckel, the historical and the mechanical were inseparable. "Phylogenesis is the mechanical cause of ontogenesis,"[49] he liked to maintain, the evolutionary development of an organism is the mechanical cause of its individual development, so that the true mechanical causes of embryonic development were historical causes. The machinery acted historically and the history took place as the growth and ramification of nature's mechanisms over time. "I am one of those scientists who believe in a real "natural history," Haeckel wrote, "and who think as much of an historical knowledge of the past as of an exact investigation of the present. The incalculable value of the historical consciousness cannot be sufficiently emphasised at a time when historical research is ignored and neglected, and when an 'exact' school, as dogmatic as it is narrow, would substitute for it physical experiments and mathematical formulae. Historical knowledge cannot be replaced by any other branch of science."[50]

With his insistence on the necessity of history to biology, Haeckel violated several tenets of the university departments' Protestant *Wissenschaft*: the strict separation between natural and human sciences; the cloistering of scientific subdisciplines; and the new model of historical scholarship itself as founded in the excavation of discrete, particular philological and archival facts whose proximate relations reflected a transcendent divine order. The form of history to which Haeckel appealed was empirical, and oriented toward combining empirical particulars to uncover patterned developments over time, but in these developments no external or transcendent God had a hand. Haeckel's historical mechanism invested nature with agency, and nature's agency with historical and moral meaning. He indeed went so far as to propose a new "religion of Nature" founded in "the monistic conviction of the unity . . . of mind and body, of force and matter, of God and Universe." This "Monistic religion" would require, among other things, a knowledge of the true origin of human beings as descended from apes by a process of "historical development."[51]

History was Haeckel's naturalist reply to the supernaturalism of certain of his opponents. One critic, for example, the Strasbourg zoology lecturer Alexander Goette, rehearsing the standard objection that neither Haeckel nor Darwin had offered a mechanism for variation or heredity, insisted that the emergence and preservation of form through these processes could only have a "supernatural" origin.[52] Haeckel responded to such critics that the agency responsible for bringing about nature's machinery acted over time from within the world's substance rather than without. "The explanatory path of His is fundamentally different from mine. I turn to phylogeny in order to explain the historical emergence of different forms of growth, and seek their completely sufficient explanatory basis in the interaction of heredity and adaptation."[53] Such interactive causation was a crucial feature of Haeckel's active mechanism and distinguished it from His's mechanism of passive adjacent parts.

In sum, to Haeckel, each apparently simple mechanical event in the development of a given individual organism was "a highly complicated historical result" of thousands of evolutionary modifications.[54] Moreover, the naturalism of evolutionary theory rested upon this combination of mechanism and history. Nature's machinery developed through the activity in time of its own agency distributed everywhere throughout the works.

Haeckel drew upon this vision of biological causation to answer the old question, which came first, the chicken or the egg? "We can now give a very plain answer to this riddle," Haeckel wrote. It was the egg. "The egg came a long time before the chick." It did not exist at first as a bird's egg, of course, but "as an indifferent amoeboid cell of the simplest character." The egg had spent thousands of years as an amoeba and had not become an egg in the modern physiological sense until the descendants of the unicellular protozoa had developed into multicellular animals, which had then undergone sexual differentiation. "The bird's egg we have experience of daily is a highly complicated historical product."[55]

This was a losing view, at least in the medium-term. Haeckel himself remained enormously prolific, widely read, and he retained his international place of honor in natural history. But as the century wore on, in part thanks to His's rhetoric, Haeckel's understanding of what constituted a "mechanical" process began to fall into the shadows.

It was one of Haeckel's own students, the Breslau zoologist Wilhelm Roux, who proved the next, most influential vanguard of the new mechanism.[56] Like His, Roux saw the egg as a "piece of machinery which, when once set in motion," would carry out its assigned task autonomously.[57] Roux was intrigued by His's rubber-tube method and came up with some inventive versions of his own. For instance, to test the notion that differential growth in different parts of the embryo exerted the pressures that shaped its development, Roux made balls of bread dough with various amounts of yeast in them. Sticking these together in different combinations in imitation of chick embryos, he then incubated them and confirmed that the resulting forms depended upon how the balls were arranged and how much yeast they contained.[58]

Roux used oil drops to simulate the cleavage of fertilized egg cells. In a wine glass half filled with diluted alcohol, he poured enough oil (he recommended olive or paraffin) to form a large drop, then poured a stronger alcohol on top, resulting in a spherical oil drop suspended between the two alcohols. Using a glass rod, he then sectioned the drop, forming smaller drops that were held in place against one another and compressed by the glass. He tried various equal and unequal sections and found that he could get the drops to assume forms reminiscent of the blastomeres (the cells at an early stage of division) of a frog embryo. Often the agreement appeared perfect, though not always, and Roux attributed the differences to various factors, such as the apparent stickiness of the blastomere walls.[59]

T. H. Morgan, describing Roux's simulation, mentioned in passing what one might have taken to be the most momentous of factors, namely that the blastomeres, unlike the oil drops, were alive: they were "living contractile bodies, and through their internal activity may interfere with the mechanical tendencies of the system."[60] Indeed, Roux himself hesitated between a passive-mechanist approach and one that ascribed agencies of various sorts to living mechanisms. He offered his own notion of mechanical causation as more sophisticated than His's push-me-pull-you approach. Roux explained that a "mechanist" science was simply one that offered causal explanations, but these were by no means restricted to the billiard-ball model of chunks of matter knocking or pushing against one another. At least not for the moment. The modern disciplines of physics and chemistry were working to reduce all sorts of phenomena to movements of parts: magnetic, electrical, optical, and chemical forces. Accordingly, a mechanist science of embryology could encompass "formative forces or energies" on the assumption that physics and chemistry would one day reduce these to movements of parts as well. But then again the mechanisms that Roux expected to feature in the reductive embryological science of the future retained an active sound in his description: "growth energies"; the "self-movement of the cell"; and the "creative modes of action" of the organism.[61]

Another example of Roux's ambivalence about the agency of embryonic cells came about when he identified a striking phenomenon: the separated blastomere cells of a frog embryo, when placed in a fluid medium of filtered egg whites and/or salt water, would spontaneously move toward one another. "They began to become pointed at the ends turned towards each other and to approach by jerks until they were in contact."[62] Despite this appearance of agency, Roux ultimately stuck with a passive-mechanist account of the phenomenon. He called it "cytotropism," by analogy with other tropisms such as heliotropism or geotropism, a name he found "purely objective" in its avoidance of all speculation about what the cause of the motion might be.[63] A "tropism" designated an apparently goal-oriented movement without ascribing any kind of agency to the moving entity.

Exercising his talent for neologisms, Roux came up with a German mouthful of a name for his new science, *Entwicklungsmechanik* (Development-mechanics). He published its manifesto in 1890, *Die Entwicklungsmechanik der Organismen, eine anatomische Wissenschaft der Zukunft* (The Developmental Mechanics of Organisms, an Anatomical Science of the Future); and in 1894 he founded a journal to serve as its official organ, the *Archiv für Entwicklungsmechanik* (Archive for Development-mechanics, now published by Springer as Development, Genes and Evolution).[64]

The embattled Haeckel understood Roux's "science of the future" as largely a continuation of His's antihistorical program, that "pseudo-mechanical school" guilty of bringing the "narrowest specialism" to the study of life.[65] Being lumped together with His frustrated Roux, who protested that he did not mean "mechanics" in the "school physics" sense as simply "the theory of matter in motion." Those who construed him that way were misunderstanding him altogether, mistakenly believing him to want, "like His, to trace back developmental events solely to the pressure and stress of growing parts on the neighboring areas; they did not recognize that it was a question of ascertaining *all* the mechanisms operating in development."[66] Indeed, Roux even defended a place, though subordinate, for phylogenetic, evolutionary explanations in studying these mechanisms.[67]

Yet Roux also explained that all of embryonic development would ultimately prove to reduce to a host of distributed, independent, and predetermined processes. In this sense, Roux's, like His's, was a homuncular theory, in the tradition of those who had believed embryonic development to be the growth of a tiny, preformed being in the egg or sperm. The homuncular theory had appealed to classical mechanists and authors of arguments from design during the seventeenth century, who saw the embryonic machine as further proof of a divine Engineer. Roux's version retained the essential homuncular axiom that the fertilized egg did not actively constitute an embryo, but merely housed a predetermined being. The constitution of the embryo fell outside the bounds of nature's process, before the beginning of the story and beyond science, whose only job was to describe a scripted sequence of passive-mechanical events.

Apart from his institution building (coining titles, founding journals), Roux's most important contributions to the emerging mechanist-antihistorical biology were what came to be known as the "half-embryo experiments," a series of experiments begun in 1888. In the course of these, Roux killed one of the two cells of a frog embryo at the two-cell stage by pricking it with a hot needle. He wanted to see whether the surviving cell would develop as it would have when partnered with another living cell, generating half an embryo, or whether it would instead shoulder the task of generating a whole embryo. A whole embryo would imply that development took place through a dynamic overall interaction of parts, such that one half could respond, for example, to the absence of the other. But Roux expected a half embryo, which would confirm his notion that development took place instead through a host of smaller and simpler, predetermined, independent, and differentiated processes in each part of the embryo.[68]

He did manage, with difficulty, partially to achieve his desired outcome. That is, in a small minority of cases, he persuaded the surviving cell to develop into a half embryo to the blastula or gastrula stage.[69] These embryos, he took to confirm his theory that each part of the egg was pre-constituted to produce a corresponding part of the embryo.[70] He surmised that a particular part of the nucleus of the egg contained the machinery responsible for producing the various pieces of the embryo. Over the previous decade or so, experimenters had begun to notice certain filaments in cell nuclei that, when exposed to a dye, readily took on its color, and had taken to referring to these as the "chromatic elements," or "chromatin." Around the time that Roux was performing his half-embryo experiments, "chromatin" gave rise to a further neologism, "chromosome," from the Greek roots for "color" and "body."[71] Observing that these colorable filaments in the cell nuclei separated longitudinally with each cell division, Roux concluded that these divisions parceled out the specialized machinery of the filaments, each daughter cell receiving the particular pieces of mechanism that would constitute an assigned part of the embryo.[72]

Roux updated the homuncular theory by locating the homunculus in the chromatin and by surmising that it was distributed among the dividing cells of the developing embryo. This modern passive-mechanist version of the ancient homunculus theory proved influential by informing what is generally considered to have been the first attempt to bring Darwinian natural selection together with a chromosomal theory of inheritance.

The Homunculus Machine and the Self-Made Embryo

An admirer of Roux, August Weismann was the first to reinterpret Darwinian natural selection in light of the idea he shared with Roux that chromosomes contained the machinery of inheritance.[73] The University of Freiburg, where Weismann taught from 1863 until his retirement in 1912, although it was one of the minority of traditionally Catholic universities in the German empire, nevertheless inhabited the larger culture of Protestant *Wissenschaft*. During the 1760s and '70s, the university had been opened to Protestants, both as students and as faculty, and had added a natural sciences faculty. Later the university became an organ of the state, in effect shrugging off the control of the Church. After a period of financial crisis, the university had been rescued in 1818 by the Lutheran Ludwig I, Grand Duke of Baden, who became its patron and for whom it renamed itself, becoming the Albert-Ludwigs-Universität Freiburg. By then, with its Protestant population, its history as a state institution, and its eponymous Lutheran patron, to call it a traditionally Catholic university no longer had much meaning. It was a key site for the development of the new institutional world of the scientific research university, shaped by the culture of Protestant *Wissenschaft*.[74]

Like Roux, Weismann believed that the development of the form of the embryo depended upon discrete hereditary units within the chromosomes of the egg's nucleus. Weisman's units were nested, with the "biophors" being the smallest "vital units" composing living matter. Biophors were grouped into "determinants," which constituted the individual cells of the body, and the determinants were in turn aggregated into higher-order units. It is important not to read either "biophor" or "determinant" as "gene," because these were importantly unlike the later concept of genes: they did not correspond with heritable traits. Instead, they corresponded with parts of the body, as the machinery responsible for constituting the cells of the various organs and body parts. With each cell division, these units separated into the daughter cells until each unit occupied its allotted cell. Although Roux's experiments involved only the early development of frog embryos, Weismann took them as confirmation of his conviction that such a "mosaiclike" process was responsible for the overall development of all organisms.[75]

At the heart of Weismann's theory of development was his insistence upon an absolute, radical separation between the bodily changes of individual organisms and hereditary, evolutionary changes. His first principle was that no kind of experience or behavior of individual organisms could alter the prearranged machinery of inheritance and development. Accordingly, Weismann identified two kinds of cells in multicellular organisms: "germ cells," responsible for reproduction and heredity, containing the hereditary units described above; and "somatic cells," making up the rest of the body and responsible for everything else. The "Weismann barrier," as it became known,[76] divided the reproductive germ cells from the somatic cells, shielding the germ cells from any possible repercussions of somatic transformations. Weismann believed that the reproductive cells constituted an immortal, continuously evolving "germ plasm," an eternal substrate of life that produced generation after generation of mortal individuals. Although the germ plasm transformed continually, it remained impervious to any change within individual organisms (with certain exceptions as we shall see), which principle Weismann referred to as the "*continuity of the germ-plasm*."[77]

In defining his notion of the germ plasm, Weismann worked partly from an earlier idea of Nägeli (chapter 7), who was about a generation older than Weismann and had also taught at the University of Freiburg and then at Munich. Nägeli had given the name "idioplasm" to a substance he thought responsible for inheritance and transmitted by the germ cells. Like Weismann's germ plasm, Nägeli's idioplasm was continuous and immortal, an evolving substrate of life. Still, Nägeli understood the idioplasm in essentially Lamarckian terms as transforming itself by two means: primarily, through an "automatic perfecting process" that made it continually more complex; and secondarily, through adaptation to the environment by the direct action of environmental "stimuli" over long periods of time.[78] Weismann retained this idea of an immortal, continually transforming and inexorable substrate of life. However, he eliminated the internal agency of Nägeli's perfecting process and the idea that animals might have the "power of being transformed" adaptively in response to external stimuli.[79]

Above all, Weismann rejected the "inheritance of acquired characteristics," a phrase that would ever afterward be associated indelibly with Lamarck, but which Lamarck himself had never used. In fact, the expression emerged only in the second half of the nineteenth century, several decades after Lamarck's death, and then initially in connection with Darwin, not with Lamarck.[80] Weismann, however, while he regretted Darwin's endorsement of the idea, attached it to the legacy of Lamarck. The neo-Lamarckian movement of the 1890s also crystallized around this representation of Lamarckian science. The Lamarckian-Darwinians, notably Herbert Spencer, who confronted Weismann, made the "inheritance of acquired characteristics" central to their position.[81]

Despite the efforts of these Lamarckian-Darwinians, Lamarck and the phrase "inheritance of acquired characteristics" took on an increasingly bad odor around the turn of the twentieth century. Morgan observed (approvingly) that "to Weismann more than to any other single individual should be ascribed the disfavor into which this view has fallen."[82] Weismann became ultimately responsible for casting Lamarck out of respectable science and recasting "Lamarckism" as mystical nonsense.

Weismann had some collateral victims too. On October 17, 1887, he chopped off the tails of twelve white mice, seven females and five males. They were still willing to breed and on November 16, the first two litters were born, eighteen babies in all. Since pregnancy in white mice lasts twenty-two to twenty-four days, Weismann pointed out that these babies were all conceived by two tailless parents. Each had a fine tail, eleven or twelve millimeters long. Over the course of fourteen months, the original twelve de-tailed yet productive parent mice bore 333 offspring, all with entirely normal tails. Meanwhile, to test the effects of continued mutilations over several generations, Weismann transferred fifteen of the babies to a second cage and lopped off their tails too. They produced 237 offspring over the course of the experiment, all with unremarkable tails. Fourteen of these had tail-ectomies of their own, went into a third cage, and produced 152 normal-tailed babies. Weismann did the same with a fourth generation in a fourth cage; these, like their forebears, produced normal-tailed offspring. By January 16, 1889, when Weismann concluded the experiment, he had de-tailed five generations of mice and produced 901 typically tailed young.[83]

Weismann considered other evidence, experimental, observational, and folkloric. These included various breeds of dogs, cats, and sheep whose tails were docked, but persisted in bearing offspring with tails; practices such as circumcision and foot binding, neither of which had apparent hereditary effects; and a "most respectable and thoroughly trustworthy family" with interesting deformations of the ear. The mother had a cleft earlobe on one side resulting from a childhood accident. Of her seven children, one had a cleft earlobe on the same side as his mother. Weismann, however, upon examining the two earlobes closely (see figure 8.1) concluded that not only were the scars in question entirely different from one another, but so too were the earlobes themselves "in their outlines as well as in every detail." The mother's ear was broad where the son's was pointed, "in short everything in the two ears, is as different as it can possibly be in the ears of two individuals." Not only had the son not inherited his mother's injury, Weismann concluded; he had not inherited his mother's ear: he had his dad's ears.[84]

The deformed earlobes and amputated mouse tails disproved a mere caricature of what Lamarck and Darwin had proposed.[85] In Lamarck's theory, for one thing, the principle that animals could inherit changes undergone by their parents was a secondary source of transformation of living forms, the primary source being the same inner force that drove all beings to compose and maintain themselves and to elaborate and complicate their organization over time.[86]

Lamarck did argue, and Darwin endorsed his view,[87] that the "habits," "ways of life," and circumstances of animals very gradually shaped their organs. These behaviors and conditions slowly enacted bodily changes in individual creatures that could then be reproduced in the next generation if they were common to both parents. In this way, Lamarck assigned an important role in his theory of the transformation of living forms to the creatures' own agency: "When the will determines an animal to perform a given action, the organs that must execute this action are immediately provoked by the affluence of subtle fluids" to carry it out. Many repetitions of these "acts of organization" would then "fortify, extend, develop and even create the necessary organs."[88]

Darwin's "pangenesis," the process he hypothesized to explain inheritance, also provided for the gradual inheritance of bodily changes. According to the theory of pangenesis, the "gemmules," tiny reproductive particles by means of

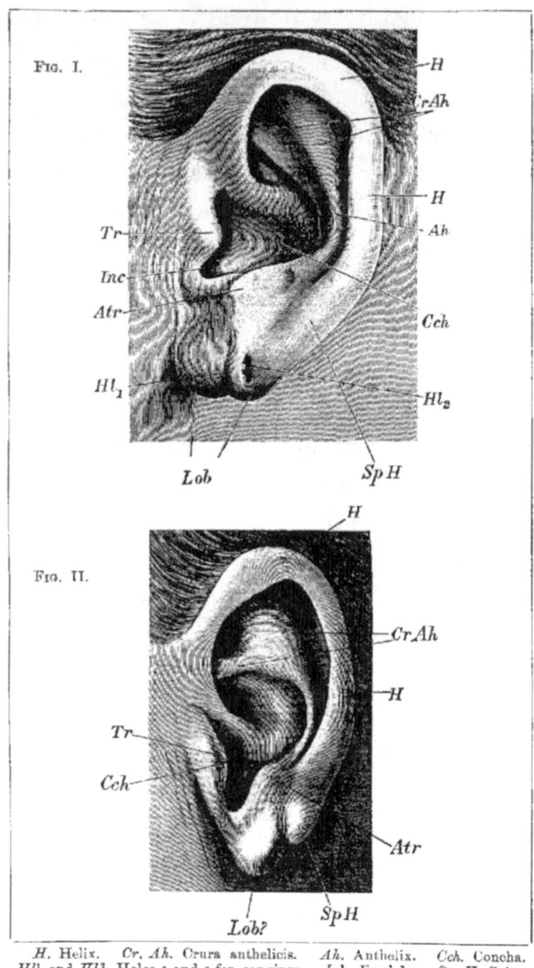

Figure 8.1 An earlobe mutilation that was not transmitted to the next generation, from August Weismann's "Supposed Transmission of Mutations" (1888).

which each part constituted its own counterpart in the offspring, registered changes in the adult organism and transmitted these to the progeny, but only very gradually over many generations, as the modified gemmules came to outnumber the older forms of gemmule. No single gemmule or generation of gemmules, but rather a convocation of gemmules of various forms and stages, collectively shaped the offspring. The persisting role of older generations of gemmules was how Darwin explained the non-inheritance of abrupt changes such as amputations. The multigenerational convocation of gemmules took place by means of an "elective affinity," which drew the gemmules to their appropriate dispositions in the reproductive elements, attracting them to the most similar and neighboring cells.[89] "An organic being," Darwin concluded, "is a microcosm—a little universe, formed of a host of self-propagating organisms, inconceivably minute and numerous as the stars in heaven."[90]

In Weismann's cases of amputated tails and torn earlobes, no repeated acts of "will" or "organization" such as Lamarck invoked could play a role; an external agent inflicted these transformations abruptly. As for Darwin's minute self-propagating entities and their elective affinities, bringing generations of varied gemmules together to determine collectively the form of the offspring, Weismann rejected any such possibility as "mysterious" and "unintelligible."[91]

He did acknowledge that amputations and other mutilations constituted only one kind of bodily transformation.[92] He also considered instances that more closely resembled Lamarck's core examples of habits, ways of life, and acts of will, like the recurrence of musical talent in the Bach family and mathematical genius among generations of Bernoullis. These cases might seem to suggest that the exercise of mathematical or musical ability produces a heritable increase in aptitude. By way of refutation, Weismann observed that Carl Friedrich Gauss's father was not a mathematician, and that George Frideric Handel's father was a surgeon.[93]

This was all neither here nor there. What the amputee mice lacked in relevance, they made up for in clarity, and they made a strong impression on Weismann's readers. The mouse-tail experiment is taught to this day as the decisive refutation of "Lamarckian" biology, even following the return of "Lamarckian" phenomena to the fold of mainstream biology in the late twentieth century.[94]

For Weismann, it became not only a matter of right science to reject stories of "Lamarckian" transformations, but an occasion for a kind of moral outrage and "contempt:" "No one can be prevented from believing such things, but they have no right to be looked upon as scientific facts or even scientific questions." The supposition that such effects existed was "entirely visionary."[95] Weismann rejected all of Lamarckian biology categorically: "The whole principle of evolution as proposed by Lamarck, and accepted in some cases by Darwin, entirely collapses."[96]

By identifying Lamarckism with the old wives' tales of inherited mutilations, and insisting that proper "mechanist" science must utterly reject Lamarckian biology, Weismann contributed crucially to setting the terms for a polarization of "Darwinist" and "Lamarckist" positions in the 1880s and '90s. In an 1882 essay entitled "On the Mechanical Conception of Nature," written several years before he had arrived at his ultimate view of Lamarck but while he was in the process of evolving it, Weismann declared: "For the naturalist the mechanical conception of Nature is the only one possible."[97]

Thus far, Lamarck himself might well have agreed that the only legitimate science was a "mechanist" one. But by "mechanist," Weismann meant a science devoid of internal directive forces such as Lamarck's "force of life." Weismann did not say how he would characterize the many directional forces of contemporary physics, including gravity and electrical and magnetic forces. He maintained that to assume internal directive forces was to renounce not only "mechanical explanation" but "all explanation conformable to law" and, indeed, "all further inquiry; for what is investigation in natural science but the attempt to indicate the mechanism through which the phenomena of the world are brought about? Where this mechanism ceases science is no longer possible."[98]

By the turn of the twentieth century, despite the pervasive presence of forces and tendencies throughout the natural world and the physical sciences, this view of mechanist science and the associated redefinition of "Darwinism" to exclude "Lamarckism" became axiomatic. "Variations cease to be mechanical as soon as they are assumed to be directed and limited," reads a representative Darwinist, anti-Lamarckian zoology text published in 1904. "Natural selection is then no longer a mechanical principle, and a unified conception of the world is impossible. As soon as a Mechanist recognizes variations of that kind he abandons the possibility of a harmonious system, and is no longer a Mechanist."[99] A history of animal morphology written shortly afterward, in 1916, included the following characterization of the moment: "We tend to lay the causes of form-change, of evolution, as far as possible outside the living organism. With Darwin we seek the transforming factors in the environment rather than within the organism itself. We fight shy of the Lamarckian conception that the living thing obscurely works out its own salvation by blind and instinctive effort. We like to think of organisms as machines, as passive inventions gradually perfected from generation to generation by some external agency, by environment or by natural selection, or what you will."[100]

There were exceptions to these generalizations, but they tended to be cautious and circumscribed. For example, the next year, when the Scottish biologist D'Arcy Thompson published his monumental study of organic forms and their transformations, entitled *On Growth and Form* (1917), he treated the mathematics of biological forms and demurred from an overt claim about causation. "As students of mathematical and experimental physics," Thompson wrote, "we are content to deal with those antecedents, or concomitants, of our phenomena, without which the phenomenon does not occur." Like His and Roux, Thompson believed that organic matter shared the same physical constraints as inorganic matter, and that these provided a key to understanding how organic forms arose and transformed. He studied Roux's frog-egg and oil-drop experiments, and drew many of his own analogies between living and inanimate forms.[101]

Unlike His and Roux, though, Thompson described these form-giving constraints on living matter not merely as external, imposed from without, but as a kind of internal dynamic, a "*conformation* of the organism itself . . . [an] interaction or balance of forces." Thompson also argued that the mathematical regularities of organic forms undermined the Weismannian-Darwinian principle that these originated exclusively in tiny, continuous, random variations acted upon by natural selection. Instead, the regularities, in Thompson's view, represented a finite set of alternatives of "physico-mathematical possibility," a kind of internal logic that would tend to repeat itself throughout nature. The formal patterns among living organisms reflected "deep-seated rhythms of growth which, as I venture to think, are the chief basis of morphological heredity." These, however, could also be acted upon by "altered environment and habits."[102]

This idea of internal rhythms of growth modified by environment and habits was clearly in keeping with Lamarck's view of transformation. But Thompson drew back from the brink: "To discuss these questions at length would be to enter on a discussion of Lamarck's philosophy of biology." Rather than asserting the causal role of these "deep-seated rhythms of growth," in combination with changing environments and habits, he stuck to describing them mathematically.[103]

Following the spectacular rise and fall of the Soviet "Lamarckian" biologist Trofim Lysenko between the 1920s and 1960s, Lamarck's name became firmly associated with Lysenko's, and consequently with fraud. However, it was not Lysenko who gave Lamarck his bad name. It was more like the reverse: Lamarck's bad name helped give Lysenko his program, to champion the villain of Western biology. A good Stalinist, he knew a dialectic when he saw one: Weismann had "declared war" on the Lamarckian notion of use and disuse, Lysenko observed, inaugurating a "strife which is going on to this day."[104]

By the time Lysenko began his career in the 1920s, Darwinism had already more or less become Weismannism.[105] The living organism had become a machine devoid of internal forces and powers, strictly passive. Possessing in itself "no principle of variability," it was "the statical element" in development, Weismann argued, and would always produce an exact replica of itself were it not for the external influences that interfered and complicated things. Environmental influences only, and not the static organism-machine, constituted the "dynamical elements" of evolution.[106] But how could environmental influences cause changes in the germ plasm? Didn't Weismann's principle shield the germ plasm from changes taking place in the somatic cells? And how could environmental influences act on the germ plasm other than via the bodies of individual organisms?

Weismann did acknowledge an influence of somatic upon germ cells, provided this influence was generalized and not directed. For example, the quality of nourishment of the adult organism could cause changes in the germ cells. Nutrition worked by transmitting particles that became assimilated, whereas "inheritance of acquired characters" would require that somatic cells transmit particles unchanged to the correct reproductive cells. This would be an unthinkably complex process requiring various agencies: an "affinity on the part of the molecules for the reproductive cells" and a further "unknown controlling force" to "marshal the molecules which enter the reproductive cell" to their places. "In short," Weismann concluded, "we become lost in unfounded hypotheses."[107]

Nutritional changes, being generalized and not directed, were consistent with his understanding of mechanism.[108] On the other hand, adaptive changes within an individual organism of the sort that Lamarck and Darwin had described as directed from within, by the agency of the organism itself or of its parts, were in clear violation of Weismann's ban on ascriptions of will or purpose to nature's machinery.

Weismann debunked what he saw as instances of apparent purposefulness at all levels of living nature. The ubiquitous appearance of purpose was everywhere deceptive. Consider, for example, those instincts that came into use only once or very few times in a lifetime, such as the egg-laying instincts of certain insects. The perfection of these instincts could not result from practice, since no individual had a chance to practice them. Therefore, they did not represent purposeful action, but the contrary, a total absence of discretion. The butterfly *Vanessa levana* laid her green eggs in a line projecting from the stem or leaf of the stinging nettle, perfectly camouflaged among the flower buds (which the baby caterpillars then ate), but got no credit for doing it from Weismann. "Of course," he wrote, "the butterfly is not aware of the advantage which follows from such a proceeding; intelligence has no part in the process. The entire operation depends upon certain inherent anatomical and physiological arrangements."[109]

A common objection to Weismann's form of Darwinism, especially lodged by Lamarckians, was that natural selection needed a source of advantageous variations to act upon. What influence produced these? Weismann found it as implausible as did his critics that a purely random process could work, but he was determined to avoid any kind of internal "directive power."[110] Instead, drawing upon Roux's idea that the competition for survival took place not just among individuals but among the parts of an individual organism,[111] Weismann proposed an extension of selection to the level of the biophors. The biophors themselves must be competing for nutrients in ever-fluctuating conditions. He first applied this idea to explain the disappearance of disused organs.[112]

Suppose a particular organ becomes useless to an organism, such as the eyes on a species of newt or crustacean that comes to live in darkness in a cave or at great depths. The determinants (groups of biophors) that build the eye are competing with the neighboring determinants, responsible for other organs that have remained useful. Perhaps eyes are not actually disadvantageous to the creatures, so although they will not be selected for, neither would they be selected against. Why, then, should the eyes gradually shrink and disappear, as disused organs do? Weismann explained this in terms of the competition among the determinants. Any strengthening of the determinants corresponding to the eyes will be disadvantageous to the crustacean, because it will concentrate nutrients away from those determinants building useful organs. Individuals with strong eye determinants will therefore be selected against, while individuals with weak eye determinants will be selected for. Meanwhile within the individuals, the weak eye determinants will be getting ever weaker as they lose the struggle with neighboring determinants for, say, claws or feet.

What about the positive case of an organ that becomes advantageous? Since the determinants for that organ are always varying and always struggling with the others, they will initially vary in a stronger direction for reasons independent of the advantageousness of the organ, simply because of fluctuations in their nutrients. However, once these determinants have grown stronger, since they are producing a useful organ, they will continue to do so at an accelerated pace, helped along by selection of the organism enjoying that organ.

Weismann thought that this process of "germinal selection" could explain how variations might be directed by utility in the absence of any internal directing force or agency. For his paradigm case, he chose the leaflike markings on a butterfly's wing, because this was a "passively functioning part," meaning that it worked simply by existing in a given environment. The markings were unrelated to the structure of the wing, so they could not be connected with a Lamarckian internal process of increasing structural complexity or perfection. The butterfly could not change the color of its wing by any sort of behavior, habit, or act of will, ruling out the other Lamarckian force of variation. "The Lamarckian principle," wrote Weismann, "is excluded here *ab initio*." The markings must depend "not on an *internal*, but on an *external* directive power."[113]

Germinal selection, Weismann proposed, could provide that external directive power. The biophors, despite their competition for nutrients, were essentially passive with respect to changes in their nutrient flow and the butterfly's external conditions. The determinants producing greenish-tinted areas of the wing would happen to grow stronger in some butterflies, purely as a result of changing nutrient flows within the germ plasm. The greenish-tinted butterflies would have a selective advantage, giving their strengthened determinants an extra boost. In turn, these determinants would begin to out-compete the neighboring ones for brownish-tinted areas on the wing, and thus the internal selection process among the determinants would begin to provide disproportionate numbers of greenish variations for natural selection to select. In other words, Weismann explained, the variations were not random; they were directed by utility. Germinal selection would produce a "directed progressive variation" in a beneficial direction.[114]

Because he thought the process of variation was directed by utility, certain historians, philosophers, and biologists have shown themselves more Weismannian than Weismann and suggested that Weismann succumbed to a form of Lamarckism.[115] But Weismann himself presented the theory of germinal selection as a way to avoid any Lamarckian internal directive force and to maintain passivity on the part of the organism. The selective process in the determinants originated in a variation unrelated to utility, "purely passive," and germinal selection allowed for an explanation of adaptations solely in terms of external directive forces.[116]

Directive forces were always external to the passive living organism in Weismann's biology. Ultimately, intelligence and agency were supernatural, operating outside of nature's machinery. Weismann did not reject teleology. Quite the contrary: he assured his readers that a mechanist biology did not just leave room for a "directive power" but, as participants in the emerging university world of Protestant *Wissenschaft* understood, it actually demanded one. Mechanism and teleology were mutually dependent: "Without teleology, there would be no mechanism, but only a confusion of crude forces; and without mechanism there would be no teleology, for how could the latter otherwise effect its purpose?" The teleological agency, however, acted only outside nature. Weismann returned to Paley's example of the watchmaker, who does not interfere once the watch is "combined harmoniously and the spring wound up." Weismann's directive force in nature acted before the beginning of every natural process, setting it in motion and then stepping aside.[117]

Weismann referred to a "Universal Cause"—sometimes the "Final Cause"—operating "behind" the mechanism of the universe. Regarding this Final Cause, "we can only say one thing with certainty, viz., that it must be teleological." Accordingly, Weismann emphasized that his mechanist conception of the universe was "absolutely opposed to that of the materialist."[118] The Lutheran theologian Rudolf Otto made a similar observation with regard to Weismannism. "Even an entirely naïve, anthropomorphic, 'supernatural' theology," Otto observed, would be compatible with "the whole system of causes and effects, which, according to the Darwin-Weismann doctrine, have gradually brought forth the whole diversity of the world of life." Although Otto did not endorse Weismann's view, he deemed Weismann worthy of serious engagement because he remained within the proper bounds of science, so that one did not have to "purge his theories from [*sic*] irrelevant, vitalistic, or pantheistic accessory theories, as we have, for instance, in the case of Haeckel. [Weismann's] book, too, is kept strictly within its own limits, and does not attempt to formulate a theory of the universe in general, or even a new religion on the basis of biological theories." Whereas Haeckel broke all the rules of Protestant *Wissenschaft*, in other words, and aroused the ire of the theologians, Weismann respected these rules, and Otto judged his to be "the clearest and best statement of the theory of selection."[119]

Weismann's mechanist science was the culmination of two centuries of intellectual and scientific development: a teleological, antimaterialist, in fact, dualist form of mechanism. It had first emerged in conjunction with the argument from design, and it now fit comfortably in the world of the new research universities where it supported the limited partnership established there between the science and theology faculties.

With his impermeable "barrier" between individual agency and evolutionary transformation, Weismann became the face of modern mechanist Darwinism. In Germany, his passive-mechanist, antihistorical rendition of evolution supplanted Haeckel's active-mechanist Lamarckian-Darwinian historicist one.[120] When Weismannism came to dominate biological thinking in the United States in the early-to-middle decades of the twentieth century, it was once again crucially transmitted by someone who had come up within the same world of the German research university, Ernst Mayr. Mayr studied biology at the University of Berlin in the mid-1920s, receiving his PhD in 1926. His adviser was the ornithologist Erwin Stresemann, who was an admirer of Weismann and a firm subscriber to the passive-mechanist view of living forms. Stresemann wrote, for example, "An animal does not act for itself, but under a higher commission: *animal non agit, sed agitur* [an animal does not act, but is acted upon]," and he endorsed the view that "the animal does not think, does not reflect, does not establish aims for itself, and if it nevertheless behaves purposively, then someone must have thought for it."[121] Note in both of these expressions how the passive mechanism of the animal implies a higher power, an external Someone.

Stresemann credited Weismann with having "vanquished the psycho-Lamarckism of the post-Darwinian period" with his barrier between somatic and germ cells. This barrier meant that "instincts are produced purely by selection; they are rooted not in the experiences provided by the individual life, but in the variations of the germ plasm."[122] When Mayr moved to New York in 1931, he brought his training and influences along. By then, at any rate, the United States had been importing the German model of the research university, its institutional culture and practices, particularly in biology, for more than half a century.[123] American biology was a well-prepared soil into which to plant Weismann's ideas and the model of science they represented.

Weismann's interpretation of Darwinian evolution was crucial to its establishment at the heart of twentieth-century biology. That Darwinian evolution should hold such a position was by no means a sure thing in the decades before it took place, in Germany or elsewhere. A few months after Darwin's death in 1882, Du Bois-Reymond delivered a heartfelt eulogy to the Prussian Academy of Sciences (of which Darwin had been a corresponding member) calling him the "Copernicus of the organic world." Whereas Copernicus had died before seeing his theory succeed, Du Bois-Reymond observed, and his followers had been persecuted by the Church, Darwin had lived to see his own theory triumph. Du Bois-Reymond's assurance was premature. His speech was followed by attacks in the press and in the Prussian Chamber of Deputies, on Du Bois-Reymond himself and on Darwinism, as irreligious and immoral.[124]

The period between about 1880 and 1930 saw an "eclipse of Darwinism," in the phrase of the English evolutionary biologist Julian Huxley, grandson of Darwin's friend and proselytizer T. H. Huxley. Writing retrospectively in 1942, Julian Huxley observed that Darwinians had been "looked down on by devotees of the newer disciplines, whether cytology or genetics, Entwicklungsmechanik or comparative physiology, as old-fashioned theorizers." Criticisms of Darwinism had centered upon the old familiar problem of agency. "It has been for some years the fashion among certain schools of biological thought," Huxley wrote, "to decry the study or even to deny the fact of adaptation. Its alleged teleological flavour is supposed to debar it from orthodox scientific consideration, and its study is assumed to prevent the biologist from paying attention to his proper business of mechanistic analysis." The charge was unjust, Huxley argued. The teleology involved in Darwinian adaptation was only "pseudo-teleology, capable of being accounted for on good mechanistic principles, without the intervention of purpose, conscious or subconscious, either on the part of the organism or of any outside power."[125]

Weismann had rescued Darwinism by ridding it of Lamarckian agencies, rendering it properly brute-mechanist. His anti-Lamarckian distinction between the somatic cells and germ plasm, individual action and hereditary machinery, according to Huxley, had constituted a "great clarification" of the Darwinian position. This "clarification" had been crucial to the establishment of what Huxley named the "Modern Synthesis": a new approach to biology formed by biologists bringing Darwin's theory of evolution together with the science of genetics.[126] Francis Crick later codified the Weismannian organizing axiom of the Modern Synthesis as the "Central Dogma" of molecular biology. Essentially a restatement of the Weismann barrier, Crick's Central Dogma holds that "information cannot be transferred back from protein to either protein or nucleic acid"; hence somatic changes cannot inscribe themselves in the DNA.[127]

Subsequent histories of Darwinism have maintained Huxley's view of Weismann's role in "sharpening" the definition of Darwinism by establishing the principle of "'hard' heredity: that is, the complete inability of the organism to influence the genetic information passed on to the next generation." With this principle of mechanist passivity, according to the traditional histories, Weismann effectively "purged" Darwinism of any taint of Lamarck.[128]

Thus this first reinterpretation of Darwin's theory of natural selection in terms of a *chromosomal* theory of inheritance, Weismann's hereditary mechanism of biophors and determinants acting within the chromosomes, was a brute-mechanist, homuncular theory in which development took place by means of an essentially passive, preset machinery. On the other hand, the first attempt to integrate a *genetic* theory of inheritance into a Darwinian framework looked very different from Weismann's passive-mechanist theory. The author of this early genetic theory, the Dutch botanist Hugo De Vries, both an admirer and a critic of Weismann, assumed an epigenetic process according to which the machinery of the developing embryo was active and self-constituting.

De Vries grew up in Haarlem, the son of a Baptist lawyer and minister of justice, and the grandson of a University of Leiden archaeologist. He spent most of his career at the University of Amsterdam[129] which, unlike the other new universities, was a municipal institution funded by the city. Its professors were appointed by the city council, which made for a less orthodox intellectual culture.[130]

Like Weismann, De Vries was very taken with Darwin's idea of natural selection. Unlike Weismann, De Vries also found much to admire in Darwin's proposed mechanism of inheritance, "pangenesis," in which "gemmules" from every part of the organism gathered by mutual affinity in the reproductive organs. De Vries endorsed the idea of minute particles responsible for inheritance and was unbothered by the clinging whiff of agency. However, instead of imagining that the particles of inheritance traveled throughout the body, De Vries proposed that copies of them existed in every cell, though each was only active in certain cells. To distinguish his own version of pangenesis, in which the particles of inheritance remained within a given cell rather than traveling from cell to cell, from Darwin's version, De Vries called his own theory "intracellular pangenesis" and his particles of inheritance "pangenes" rather than gemmules.[131]

Essential to De Vries's understanding of his "pangenes" was what he called their "historical" dimension, by which he meant their evolutionary action over time. "The chief point for the theory of heredity," De Vries wrote, was that "protoplasm always offers us certain historical characters besides physical and chemical properties." These "historical characters" were indispensable to understanding how the mechanism of inheritance worked. In support of the primary importance of the "historical" aspects of organic matter, De Vries explained that while it might be possible to produce the physical and chemical properties of organic matter synthetically in a laboratory—that is, to create an artificial composition of proteins—he thought it very doubtful "whether we shall ever succeed in obtaining living protoplasm in any other than the phylogenetic way."[132] He assumed that history constituted a crucial dimension of living nature: one could no more bypass a living organism's historical features than omit its physical or chemical properties and hope to understand how it worked.

Weismann disparaged De Vries's notion of "historical characters" of living matter. Protoplasms, he replied, possessed historical qualities "not *in addition to*, but *within* their physico-chemical ones," meaning simply that adaptation and selection had shaped their structure over many generations.[133] The mere fact that the physical and chemical properties of protoplasm had arisen gradually made no difference to understanding how they operated or to reproducing them artificially. But to De Vries, this fact made all the difference. That protoplasm was a product and an agent of history meant that its current physico-chemical properties constituted only a momentary, static, and insufficient glimpse of its functioning. Its current state was the result of its self-transforming action over time. One must supplement physical and chemical with historical explanation. De Vries proposed "pangenes" as the material medium for the historical dimension of living substance. He declined to specify the pangenes' structure or composition other than to say that they were extremely complex "material bearers for the individual hereditary characters."[134]

Otherwise, his pangenes were historical: "they are life-units, the characters of which can be explained in an historical way only. We must simply look for life-attributes in them, without being able to explain them." To Weismann's claim that a system of discrete bearers of individual hereditary characters would be unthinkably complex, De Vries responded that this was "no objection": "It is not to be thought that to-day we already stand at the end of our investigations concerning the nucleus." If one regarded the phenomena of heredity historically, one saw that hereditary characters varied independently of one another. The pangenes represented not "a morphological member of the organism, a cell or a part of a cell, but a special hereditary character."[135]

The essential principles of De Vries's historical-mechanist theory of heredity were first, that each hereditary character corresponded in an as-yet-undefined way with an individual material "bearer" of as-yet-unspecified composition, a "pangene"; and second, that every pangene existed in each cell of the body; but, third, that each was "active" only in certain cells. De Vries agreed with Weismann on some essential principles: a material mechanism of inheritance; the location of this mechanism in the chromatic filaments of the nucleus; and the primary importance of Darwinian natural selection in shaping the development of species. Nevertheless, they adopted sharply different operating principles. Weismann's basic assumption was the passivity of mechanism: the machinery of inheritance must be preexisting and predetermined. The chromatin must divide into discrete pieces of machinery responsible for fabricating corresponding parts of the embryo. To the contrary, De Vries assumed a selectively active "material bearer" for each hereditary character, present in every cell but active only in certain cells. He supposed further that one could only understand these agents of heredity in historical terms: in terms of their own actions over time.

In pursuit of a historical investigation of the discrete material agents of heredity, De Vries embarked upon an experimental study of plant hybrids, which led him to introduce the notion of "mutations" to evolutionary biology. According to De Vries's "mutation theory," organisms did not merely transform by tiny and random increments, as earlier evolutionists suggested, but also changed discontinuously, due to "latent" characters or "potentialities" in the organism that could become active abruptly in given environmental conditions.[136]

The distinction between "latent" and "active" characters was the informing principle of De Vries's theory of mutations. The capacity to mutate was itself a heritable character that existed in a latent or an active state. And this capacity, when it awakened and became active, was the source of all new characters. Selection could produce "*nothing new*" but could only augment or diminish those characters supplied by mutations. According to De Vries, Darwin himself had often reiterated that natural selection could only work upon a prior source of variation and had emphasized the importance of sporadic, discontinuous variations as well as of continuous, incremental ones, until "the pressure of criticism" had led him to renounce the view.[137]

De Vries considered that he was consistent with Darwin's original formulation of natural selection when he claimed that selection worked, not only upon infinitesimal, incremental changes, but upon real, qualitative transformations originating in the dynamic mechanism of the organism, a mechanism that was neither passive nor predetermined but the opposite: active, contingent, mutable, and "spontaneous."[138]

This earliest genetic theory of inheritance, then, was unlike Weismann's theory in being decisively active-mechanist and historical. However, soon after De Vries's publication of his ideas about "intracellular pangenesis" and mutation, when the Danish botanist Wilhelm Johannsen coined the term "gene," he did so to distance these "elements of inheritance," as he understood them, *both* from Weismann's homuncular, preformationist conception *and* from De Vries's active-mechanist, historical one. The Weismannian idea that the elements of inheritance corresponded with organs or parts of the organism had "no support in experience," Johannsen argued. Therefore, he proposed a new term, innocent of any such implications, "gene," which would "prejudice nothing," and was simply "a very applicable little word, easily combined with others."[139]

Do not be fooled: the term "gene," like all terms, certainly did carry implications. Along with the anti-homuncular implication, it carried an even more overt antihistorical one. Regarding a "genotype," the "sum total of all the 'genes' in a gamete," Johannsen wrote, "its history is without influence upon its reactions, which are determined by its actual nature. The genotype-conception is thus an 'ahistoric' view of the reactions of living beings." Johannsen recommended his genotype conception of heredity, a "'radical' ahistoric" conception, in its "strict antagonism" to the contrasting, historicist, "phenotype-view" of—guess whom!—yes, of course, Lamarck (and, to a lesser degree, Darwin).[140]

The genetic theory of inheritance may be said to have originated in De Vries's "active" and "historical" bearers of heredity, but the coining of the term "gene" expressed as brute-mechanist, antihistorical, and anti-Lamarckian an impulse as Weismann's mice.

The Intelligent Egg and the Dead Sparrow

The debate over whether the machinery of embryonic development was passive and predetermined or actively self-constituting gave rise, not only to competing theories, but also to real animals. They were the first clones, artificially generated from a manipulation of embryonic cells. These animals owed their lives to the intensifying struggle over the proper role of agency and mechanism in the scientific explanation of life. Their creator had set out to define a "machine-theory of life" but he ended up with a "vitalist" approach to living forms that specifically rejected machinery as passive, rote, inert, unresponsive, altogether antithetical to life.

In 1891, two years after completing his doctorate in natural science in Haeckel's laboratory at Jena, a twenty-four-year-old from Hamburg named

Hans Driesch traveled from the Stazione Zoologica in Naples, where he was working, to Trieste to gather sea urchins from the Adriatic.[141] He used the sea urchins to produce the first artificial living creatures and ultimately, with them, what their author saw as evidence against a passive, machinelike model of life and in support of an intrinsically active, nonmechanical model.

Initially, Driesch made the journey hoping to provide empirical support for Roux's mechanist manifesto, *The Developmental Mechanics of Organisms*. Performing a modified version of Roux's half-embryo experiments with sea urchin embryos, Driesch separated the cells at the two-cell stage simply by shaking them briskly in a bottle and then placed the separated cells in separate dishes. Crucially, this meant that the separated cells did not remain attached to dead partners, as in Roux's experiments. Driesch looked forward to an "extraordinary sight" of half embryos swimming about, although he also "thought the formations would probably die," as he later recounted. "Instead, the next morning I found in their respective dishes typical, actively swimming blastulae of half size."[142] He reared some of these all the way to the larval phase. Thus, rather than performing a predetermined sequence of actions, it seemed that the embryonic sea urchin cells had what Driesch later named "totipotency": the ability to become whatever the whole organism required.[143]

Driesch proceeded to mistreat his sea urchin embryos in many ways—by heating them, by diluting their sea water medium, by squashing them flat between glass plates—each time radically altering their sequences of cleavages. "No matter," he wrote, "the larva is bound to be typical. . . . All our subjects resulted in *absolutely normal* organisms." The cells rose to each occasion by recovering a normal course of development.[144] Driesch also tried using delicate scissors to snip embryos at the blastula stage into parts and found that, provided each part was no smaller than one quarter of the whole, it would form a complete and normal (though small) larva.[145]

Struggling to interpret his surprising results, Driesch first tried a form of active mechanism. Soon after his initial sea urchin experiments, while summering in the Swiss Alps, he wrote out a theory of embryonic development, the *Analytische Theorie der Organischen Entwicklung* (Analytical Theory of Organic Development). The first part of the text offered a mechanist explanation of ontogeny (embryonic development) in keeping with His's and Roux's understanding of mechanism, that is, explaining each stage of development in terms of proximate factors in the preceding stage only. The second part of the book, however, was called "Ontogeny as Development in Light of Teleological Considerations."

Driesch wrote that one could understand the processes of development only partially by identifying their proximate mechanical causes. Ontogeny was ultimately "a causal regularity striding along in a pronounced mystery." Proximate mechanical causes of the His-Roux-Weismann variety could offer "only fragmentary pieces" of an explanation.[146]

To make sense of these pieces, one needed an understanding of their purposes. Watching the construction of a ship, one would be baffled by all the bits and parts unless one knew what they were for. Similarly, one could never understand the early stages of embryonic development without reference to purpose. Driesch invoked Kant's third *Critique* in support of this claim, and used Blumenbach's term, *Bildungstrieb*, to describe the inner, natural intelligence responsible for arranging embryonic development. The neo-Darwinian idea that complex adapted forms could come about in the absence of any form of purposeful agency in nature was absurd. The egg was a complex machine and, like an artificial machine, it indicated the presence of a creative force or agency. A proper scientific account of the egg must therefore be at once causal, in the mechanist sense, and teleological. Driesch called this mechanist-teleological approach his "*machine-theory of life*."[147]

However, in the science of Driesch's day, the idea of a creative agency acting from within nature's machinery no longer seemed legitimate. Mechanist science as constituted in the new departments of the research universities construed natural mechanisms as did His, Roux, and Weismann, as passive and inert, confined to proximate pushes and pulls. The purposefulness made manifest in the machinery must be external, outside of nature: beyond the purview of mechanist science, but by the same token, its precondition. One should assume and never address the external agency that made this mode of mechanist science possible. These were the rules of Protestant *Wissenschaft* as practiced, by the end of the century, in universities of any religious affiliation or none, in Europe and in America.

Just a few years later, in 1899, Driesch declared his rejection of the machine theory and his conversion to "vitalism," meaning that he abandoned mechanist explanation for the argument that living things functioned according to their own principles. His contemporaries received this conversion as historians continue to describe it: as a departure from science.[148] A mechanism, in the sense of a rationally intelligible system of interacting parts, had become the antithesis of an entity with inner agency. Driesch accepted the choice between the two and chose natural inner agency over mechanism. He later recalled the choice in his autobiography in Romantic language, writing that he had had his first glimmerings of "vitalism" while strolling in the woods near Zurich.[149] But he also cited further division experiments carried out at Naples in the winter of 1894–95, in which he snipped sea urchin gastrula in half, showing that each half "sealed over the wound and developed into a healthy pluteus larva of reduced size," restoring the normal proportions of parts. This restoration of proportions, especially in the gut, "points far more to a type of phenomenon which is essentially not of a mechanical but of a specific vitalistic sort."[150]

Every element of the early embryo could become any element while maintaining certain proportions. "I call such a living system . . . as the cleaved sea urchin's egg: a harmonious equipotential system."[151] Some sort of regulatory agency governed the embryo, through which it responded to situations such as being snipped in half. This was in striking contrast with the regulatory mechanisms of artificial machines, which could make only certain prescribed, limited adjustments.[152] Driesch called this *Antwortsgeschehen* or "regulatory response" a kind of "agent" and later redesignated it using the word "entelechy."[153] This was an originally Aristotelian term meaning the fulfillment of a potential; for example, the soul was the body's entelechy, the fulfillment of its potential for life.[154] Leibniz had taken "entelechy" from Aristotle to use as another word for his monads, the tiny perceiving substances that made up the world, and he described monads as "entelechies" particularly in connection with living things: the monad or entelechy of an organism was its soul, the source of its action, order, and purpose.[155] For Driesch, the "entelechy" was the agency that directed the development of the "harmonious equipotential system" of the embryo, such that each element could become any element needed.

Driesch turned to the idea of a self-regulating "harmonious-equipotential system" precisely, he explained, in order to naturalize the purposeful agency he saw at work in embryonic development. "Only in this way can we free the 'causes,' in the strong sense that we give to the word, from teleological impurities." The causal agency became an "integrating ingredient" of the physical system, an "elementary or indivisible characteristic." Better that the teleological force be an "ingredient," an "element," or a "characteristic" of the natural system—as it had been for Blumenbach and Aristotle—than an external agent operating upon a passive machine.[156]

The triumphant sea urchin larvae, Driesch ultimately came to believe, conclusively disproved Roux and Weismann. Driesch characterized what came to be known as the "Roux-Weismann theory" as follows: "There is a very complicated machine in the egg and in particular in its nucleus—so Roux and Weismann said,—and the development of the embryo is carried out by the disintegration of this machine during the great number of cell cleavages which occur during the embryological process." No such disintegrating machine could account for the embryonic development of the sea urchin. Nor, indeed, could such a model explain the development of the frog, since Roux's half-embryo results had by then been toppled even for frogs. In frogs as in sea urchins, a single embryonic cell from the two-cell stage could give rise to a complete embryo at a later stage. The first cell division must not separate distinct parts of a preexisting machine in the nucleus, but rather two equivalent entities either of which could generate a whole frog.[157]

Weismann's insistence upon a homuncular, preexisting, and predetermined machinery in the egg now began to appear "*dogmatic*" to Driesch, since it rested upon a conviction about what science must be like, more than it did upon empirical evidence of what living phenomena were actually like. Driesch explained that Weismann believed "an epigenetic theory would lead right beyond natural science,"[158] and so rejected such theories on principle rather than on the evidence.

The conviction that embryos did not form by the division of preexisting machinery in the nucleus led Driesch, together with his friend and colleague, T. H. Morgan, to test the relative importance of the nucleus and the protoplasm in development. They experimented on the eggs of ctenophores, a kind of jellyfish-like creature, leaving all the nuclear material in the egg cell but cutting away some of the protoplasm. They got partial embryos similar to those that sometimes resulted from developing one of the first two blastomeres. The protoplasm, they concluded, played at least as crucial a role in the development of the embryo as the nucleus.[159] Meanwhile, Morgan, in a separate set of experiments, was able to get separated embryonic frog cells to form whole organisms simply by turning them over.[160] Driesch thought this might again indicate the "power of regulation" contained in the protoplasm, which was somehow sensitive to changes of position in the gravitational field.[161]

One could say the same of many an artificial machine: a pendulum clock does not work well upside down. But Driesch considered the regulatory capacity of the protoplasm, and whatever gravitational or other sensitivities it involved, to be non-machinelike. More generally, he found that the process of development as it appeared in his experiments on the sea urchin and other embryos belied the essence of machinery. These experiments indicated that each embryonic cell could play any role in generating a whole organism. Thus if the embryo were a machine, it must be a machine of which any given part, no matter how large or small, could do the job of the whole. "A very strange sort of machine, indeed, which is the same in all its parts!"[162] In the face of these results, Driesch wrote, "the machine theory as an embryological theory becomes an absurdity. These facts contradict the *concept* of a machine; for a machine is a specific arrangement of parts, and it does not remain what it was if you remove from it any portion you like."[163]

Yet, as we have seen in previous chapters, within the tradition of active mechanism from Harvey and Willis to Leibniz, La Mettrie, Lamarck, and Haeckel, the machinery of living creatures was constituted of active, changing, responsive parts. This model might have pleased Driesch, who not only saw living nature as intrinsically active and self-constituting, but was also drawn to mechanist analogies between living and inanimate phenomena. Indeed, even after his conversion to "vitalism," he continued to reach for explanations of the triumphant embryos that rested upon such analogies. He proposed a kind of force or "intimate structure" in the egg, belonging to its very smallest parts, which arranged these in response to one another and to contingencies such as being shaken, squashed, heated, or snipped. This force or structure might work analogously to magnets, attracting the parts together in constant relations and proportions no matter how they were reduced or deformed.[164] It is a measure of the absence of active mechanism from established science at the time Driesch was writing that, even while persisting in such analogies, he nonetheless represented his theory as "vitalist."

To Driesch, as to his mechanist opponents, describing living phenomena as intrinsically active meant making them exceptions to nature's machinery rather than integral to it, because he assumed that common matter was essentially passive and inert. Also like the mechanists, Driesch defined a "scientific" approach to living beings by its contrast with "historical" ones such as Lamarck's and Darwin's, which explained the development of living structures in terms of the contingent engagements of many natural agents over time. Of course, Driesch did not object to the agency that such accounts ascribed to natural mechanisms, but rather to their assumption that organic forms were essentially contingent. In this regard, he agreed with the brute-mechanist authors of the argument from design: "How could the origin of so complete an organ as the eye be due to contingent variations?" To Driesch, the organizing power at work in the development of living forms acted not contingently but inexorably.[165]

This organizing power demanded "scientific" rather than "historical" explanation. History, for Driesch, was an enumeration of successive events. It described but did not explain. Whenever history became actually explanatory, it borrowed its explanations from the sciences. For example, one might consider the account of an "evolution" such as the development from an egg to an adult to be a kind of history. But the explanations tying the succession of events together were scientific. History must "always lose its 'historical' aspect in order to become of importance to human knowledge." It must "become science." For science had no room for the contingencies or multiplicities of historical accounts. Here too Driesch was in agreement, ironic though it may seem, with the current generation of mechanists: "There is only one science, and only one kind of logic too." Biological knowledge was accordingly nonhistorical in essence, the knowledge of individual processes isolated in time: "It is from the study of the living *individual* only that we have so far gained elemental principles in biology."[166]

Even after he announced his embrace of vitalism, and while complaining of "mechanist dogmatism" among German orthodox Darwinists ("but not Darwin himself!"), Driesch had much in common with his mechanist counterparts. He agreed that natural knowledge was definitively ahistorical, its business being to explain natural processes that were essentially timeless. Driesch too rejected the contingency of historical explanations and insisted that living nature was the inexorable manifestation of a powerful agency.

Occupying the extreme mechanist position opposite Driesch was Driesch's friend and colleague, the German-American physiologist Jacques Loeb.[167] Perhaps no one so fully and successfully personified the new ideal of passive-mechanist biology for the dawning twentieth century as Loeb. Born the same year as Darwin's *Origin of Species* (1859), in the town of Mayen, in the Rhineland, Loeb was the son of a Jewish importer and wholesale merchant. He became a devoted follower of the Austrian physicist and positivist philosopher Ernst Mach, whose banishment of metaphysics from science gave Loeb his intellectual direction.[168] Loeb applied the positivist ideal of science in his research career in experimental physiology, which migrated from Berlin, Würzburg, and Strasbourg to the University of Chicago, the University of California, the Rockefeller Institute, and Woods Hole.

"If a sparrow flies down to a seed lying on the ground, we speak of an act of will," observed Loeb in 1912, "but if a dead sparrow falls upon the seed this does not appear to us as such." Yet the only difference between the two cases, according to Loeb, was the nature of the forces involved. In the case of the dead sparrow, "purely physical forces" caused it to drop onto the seed. In the case of the living, flying sparrow "chemical reactions are also taking place in the sense-organs, nerves, and muscles of the animal." These reactions delivered the living sparrow onto the seed as surely as gravity deposited the dead sparrow.[169] This reduction of an apparently self-directed, purposeful creature to an inanimate, passive entity epitomized Loeb's approach to physiology.

Moreover, Loeb also became, for a pivotal generation of American believers, the popular face of this ideal of brute-mechanist science, strongly associated with the German universities. In his novel *Arrowsmith*, Sinclair Lewis borrowed Loeb as a model for the testy, irreverent Max Gottlieb whose "mind burned through appearances to actuality." The novel's Pulitzer, as Lewis himself observed in rejecting the prize, was not so much for its literary merit as for the prize committee's admiration of the way of life it portrayed: a heroic moral and social vision of a modern, scientific way of life, imported from Germany and endangered in America (though in fact Loeb's model of experimental biology was then ascendant there). "I don't pretend to know anything," exclaims the novel's hero to a roomful of misguided fellow medical students,

> except I do know what a man like Max Gottlieb means. He's got the right method, and all these other hams of profs, they're simply witch doctors. You think Gottlieb isn't religious, Hinkley. Why, his just being in a lab is a prayer. Don't you idiots realize what it means to have a man like that here, making new concepts of life?[170]

Mark Twain called Loeb a champion of scientific truth against prejudice and "Consensus of Opinion."[171] And the French intuitionist and vitalist philosopher Henri Bergson called him one of the most dangerous men in America, or so Loeb boasted to a friend.[172] When Haeckel had a letter to Loeb in New York returned to him, he joked, "I am astonished that the New York post office does not know the international reputation of Jacques Loeb!"[173] While Loeb was representing German positivist-mechanist science in America, he was simultaneously personifying American positivist-mechanist science to German colleagues. Roux was a great admirer and appealed to the Prussian Ministry of Education for a professorial appointment for Loeb. His work was reviewed effusively in German journals and when he lectured in Europe in 1909, he collected honorary doctorates as he went.[174]

Adopting the proximate-mechanist mantle of Entwicklungsmechanik (Development-mechanics), Loeb praised His's work on differential growth in embryos. (His's approach is the one Loeb characterizes favorably in the epigraph to this chapter.) Loeb did regard His's and Weismann's mosaic theory as having been overturned by Driesch's experiments on sea urchin embryos, but nevertheless endorsed their approach to biological explanation. He himself performed a series of experiments following up on Driesch's, in which he burst fertilized sea urchin ova in distilled water and then returned them, with the escaped droplets of protoplasm, to seawater, where the drops produced twin, triplet, and quadruplet embryos. For Loeb, unlike for Driesch, these experiments did not imply that all the parts of the egg were "equipotential" and could develop into a whole organism. On the contrary, Loeb agreed with His and Weismann that embryonic development followed a predetermined path. In response to the multiple, artificially produced sea urchin embryos, Loeb imagined that the different parts of the egg responsible for producing corresponding parts of the embryo might be, in fact, not regions of chromatin, but liquid substances all mixed together in the ovum, which then separated according to their molecular properties under given osmotic conditions. This idea could explain the multiple cloned sea urchin embryos without recourse to Driesch's "totipotent," "equipotential," and active parts.[175]

Loeb's most widely read work was a manifesto, *The Mechanistic Conception of Life* (1912), which made the reductive elimination of agency from living nature into both the condition and the supreme goal of mechanist science. Loeb repudiated those physiologists who ascribed consciousness to anything that was apparently "purposeful": "If an animal or organ reacts as a rational human being would do under the same circumstances, these authors declare that we are dealing with a phenomenon of consciousness." This would mean attributing conscious agency to all sorts of things. By that logic, "plants must therefore have a psychic life, and, following the argument, we must ascribe it to machines also." His own approach was the opposite. "The contents of life from the cradle to the bier," he observed, "are wishes and hopes, efforts and struggles, and unfortunately also disappointments and suffering." This stuff of "inner life" might seem to suggest that life was all about agency, expressed or thwarted. But all that welter of felt, active, struggling selfhood was in a sense illusory, reducible to a complex of rote responses to external forces. The reduction must be extremely complex at the level of human hopes and desires, so Loeb began with an easier case, conducting extensive experimental studies of what he called "animal tropisms."[176]

Aphids, for example, were "heliotropic" animals that, in moving toward sources of light, had "no more will than [heliotropic] plants," or than planets obeying the force of gravity. (Note the contrast with earlier mechanists: La Mettrie, Diderot, Erasmus Darwin, and others followed the same analogy to the opposite conclusion: that plants must have a kind of basic will.) Heliotropic animals such as aphids were in fact nothing but "photometric machines." The intensity of light, Loeb explained, varied with the sine of the angle at which it struck the surface of the animal, who was "oriented by the light in such a way that symmetrical elements of its photosensitive surface are struck at about the same angle." When the creature was illuminated from one side, the light brought about a "compulsory" turning of the head, and the animal became "a slave of the light."[177]

As for the larvae of the brown-tail moth *Porthesia chrysorrhoea* awakening hungry from their winter sleep, Loeb wrote that a strong heliotropism left them "no freedom of movement, but force[d] them" upward toward the top of the tree to munch the first buds of spring. Once sated, the larvae lost their heliotropism and became "restless," but even this state, to Loeb, was non-volitional: "The restlessness which is characteristic of so many animals force[d] them to creep downward" until they came to another leaf, "the odor or tactile stimulus of which . . . [stopped] the progressive movement of the machine and . . . [set] their eating activity again in motion." Consider the difference between saying that the larvae were stopped by the odor or tactile stimulus of a leaf and saying that they stopped when they smelled or felt a leaf. In short, perception was to have no part of Loeb's science. He described vision too as "a kind of telephotography," entirely putting aside the question of visual perception.[178]

Humans were more complicated mechanisms than aphids, of course, but even their "wishes and hopes, disappointments and sufferings" all originated in "instincts" comparable to those of heliotropic animals: a need for food, an instinct for sex "with its poetry and chain of consequences," the maternal feelings. It was only a matter of time before all of these human impulses would be revealed to be tropisms. As with earlier human-machine manifestos, Loeb's presented an overtly ethical theory. The fact that human beings were "chemical mechanisms" entirely governed by "the play of blind forces," he thought, must inform a modern, scientific ethic at the root of which lay the instincts: "We eat, drink, reproduce not because mankind has reached an agreement that this is desirable, but because, machine-like, we are compelled to do so." The same must be true of the very highest and noblest of impulses, human beings' willingness "to sacrifice their lives for an idea." These ideas might give rise to "chemical changes" in the body that made people "slaves to certain stimuli just as the copepods [small crustaceans] become slaves to the light when carbon dioxide is added to water."[179]

"Slaves" with no "freedom of motion": this was Loeb's ethical-scientific vision of living creatures, including human beings. He compared his own age favorably with the "prescientific days" when people explained mysteries by invoking further mysteries, citing T. H. Huxley's joke that one no longer accounted for evaporation by the disappearance of "aquosity." To describe a natural world of mechanical compulsion and inescapable slavery, devoid of freedom or purposeful action, was to take the side of modern science against primitive mysticism and as such, it was an ethical duty. Loeb imagined that a "savage" might well attribute a will to the planets following their patterned course across the sky, whereas a modern mechanist would refrain from ascribing a will even to animals: "The word 'animal will' [is] only the expression of the forces which prescribe to animals the direction of their apparently spontaneous movements just as unequivocally as gravity prescribes the movements of the planets." Experimentally, the way to discover these forces was to control them, directing animals here and there by means of their complexes of "tropisms."[180]

For Loeb, recognizing the mechanical enslavement of all living beings was important not only scientifically but morally and politically, as an inoculation against the Romantic, metaphysical mythologies by which human beings enslaved one another. World War I persuaded him that a brute-mechanist theory of life was essential for a Germany that had fallen into the metaphysical and Romantic tendencies of its aristocratic Junker class. Junkerism had utterly contaminated German culture and science. Loeb saw Junker Romanticism at work even among colloid chemists who ascribed to proteins the ability to aggregate and disaggregate in response to environmental changes and to interact with ions and molecules in ways that eluded the rules of organic and inorganic chemistry. Bad science and brutal warfare alike followed from the failure to cleave to a mechanistic conception of life.[181]

In their contrast with one another, Driesch and Loeb personified the radicalization of their respective positions around the turn of the twentieth century:

dead-sparrow mechanism on one side, an inexorable vital force on the other. On neither side does one find the active mechanism of old: natural mechanism as a rational system of parts operating over time in a limited and contingent, yet intrinsically active way.

Between the two positions, dead-sparrow mechanism and inexorable-clone vitalism, we can locate T. H. Morgan, who was a friend, colleague, and collaborator of both Driesch and Loeb. On the one hand, Morgan took "mechanism" to be the only "proper" mode of modern science and stood firm against Driesch's dramatic turn to "vitalism." He rejected any sort of "unfolding principle"—"Bildungstrieb, nisus formativus, vital force," or "entelechy"—calling these so much "mystic sentiment." Even comparatively neutral-seeming words were suspect to Morgan: he avoided the term "individual" on the grounds that individuals were but sums of "chemico-physical reactions," so that the word served "to make a mystery out of a mechanism." "Creative" fell afoul of Morgan's eagerness to root out "supernatural and mystical implications," by implying that something could come from nothing. Even "force," he judged to be "loose" in physics and downright archaic in biology.[182]

On the other hand, despite these convictions, Morgan denied the Roux-Weismann theory of development through the partitioning of a preexisting machine in the chromatin, and he endorsed Driesch's notion of "totipotent" blastomeres.[183] He hesitated over whether it made sense to compare living phenomena to artificial machines. Although Morgan regularly referred to "the living machine" and to physiological systems as "machinery,"[184] he also questioned the value of such analogies. No artificial machine, he thought, had "the property of reproducing itself by means of parts thrown off from itself." Using a familiar old word to describe a material entity capable of reproduction and other feats of living machinery, Morgan called it an "organization." But whereas La Mettrie and the other early users of this term drew continual analogies to artificial mechanisms, Morgan worried that these analogies constituted a kind of irresistible stumbling block: "We almost unavoidably think [of an organization] as a structure having the properties of a machine, and working in the way in which we are accustomed to thinking of machines as working."[185]

The eighteenth-century mechanist theorists of living "organization" had assumed a kind of primitive agency spread throughout material nature connecting the animate with the inanimate, the machinery of humans with human-made machinery. But Morgan subscribed to his generation's definition of machinery as essentially passive and rote. It seemed apparent to him that organisms were not so. How then did they work? For example, if early blastomeres were totipotent, able to become any part of a developing embryo, what brought about the later differentiation of cells? "Driesch has pointed out that the egg seems to act like an intelligent being," he wrote. Did the causes of differentiation and regeneration then "belong to the category of intelligent acts, and can these latter be accounted for by the known principles of chemistry and physics? The plain answer is, we do not know."[186]

The idea that eggs acted like intelligent beings sat uneasily, in Morgan's thinking, with the idea that eggs were machines. He lived in a moment of absolutist possibilities: eggs must be either deterministic (passive) mechanisms or they must be deterministic (inexorable) agents. He lacked the older model of active mechanism, according to which nature was a machine made of constrained but active parts, whose contingent engagements continually transformed over time: a machine in perpetual self-constitution.

Morgan agreed with Driesch that history had no place in science. Indeed, he placed great emphasis on this point, with which he opened his *Critique of the Theory of Evolution* (1916). "When a biologist thinks of the evolution of animals and plants," Morgan warned, he is "thinking as the historian thinks, but sometimes he gets confused and thinks that he is explaining evolution when he is only describing it." Knowledge of a "consecutive series of events" did not constitute an explanation. For an explanation, one needed determinative causal connections tying the consecutive events together, and this was the business of science. Moreover, when evolutionary theorists had posited causal connections, these had tended to be based on analogies with the principal cause operating in the "inorganic world," namely "human purpose." Though earlier he had considered that an egg might appear to act intelligently, now he ruled humanlike purposes right out of science.[187]

In his travels to and fro between the conviction that living structures acted purposefully and the brute-mechanist banishment of purpose from nature, Morgan noted, as Claude Bernard had done several decades earlier on the same round-trip journey, a "power of self-adjustment," or regulation, by which living entities repaired their losses and preserved their integrity in the face of environmental changes. Most machines had no such power. But Morgan did consider a steam engine's flywheel governor and a thermostat regulator as possible exceptions.[188] These examples would soon take on a crucial importance in helping define cybernetics, an emerging field with two opposing pole stars that one might well characterize as the intelligent egg and the dead sparrow.

~ 9 ~

Outside In

Cyberneticists reenacted and so updated the old struggle between the intelligent-egg and the dead-sparrow models of the science of living beings. Cybernetics was a new field of research and engineering that emerged fully in the wake of World War II, with beginnings stretching back to the late 1910s and '20s. It was devoted to pursuing analogies between living creatures and machines, especially robots.

In pursuit of these analogies, the cyberneticists developed new mechanical systems, and also new ways of understanding mechanism, surrounding the key concepts of feedback and self-regulation. Using these concepts, cyberneticists claimed to be able to build the living agency of an organic being into artificial machinery. They represented their projects as constituting a new approach, not only to the design of machines, but also to the sciences of life, an approach that would include even living agency under the rubric of mechanism. In the end, however, cybernetics did not so much *explain* the agency of a living creature as explain it *away*.

The First Robots

The first robots were not actual machines, but fictional characters representing an indictment of industrialization (see figure 9.1). On the evening of January 25, 1921, three decades after the first clones swam around Driesch's dishes and eleven years after De Vries wondered whether he might bypass history to create living protoplasm artificially, an audience at the National Theatre in Prague applauded a play whose main characters included clone-like beings made of artificial protoplasm. The play introduced into Czech, and soon after into every other European language, the word "Robot," the name of these fictional synthetic people. The play was *R.U.R.*, for "Rossum's Universal Robots," and the playwright, the Czech science fiction author Karel Čapek, coined the neologism, or rather his brother and close collaborator Josef coined it, from the Czech word *robota*, which means "drudgery," or "servitude." *Robotnik* is Czech for "serf," and the word "robot" in both German and English, before it came to mean robot, referred to a kind of serf specific to central Europe.

These original robots are strikingly unfamiliar today. For one thing, the word was initially pronounced in Czech. To an anglophone ear, it sounded a lot like what a frog says in a children's story: "Robot (pronounced Rubbitt)," specified a reviewer for the *New York Times*; though his counterpart at the *Chicago Tribune* thought it was "robbut."[1] For another thing, current readers and spectators, when they encounter Čapek's Robots, are immediately struck by the lack of clunky metallic parts: these beings have neither wheels nor gears, nor do they clank or whir or grind. The word "mechanical" and its variants appear only twice in the play: the robots are described as "slightly mechanical in their speech and movements," and later as "mechanically much better than we are." But they are made of a soft stuff: a synthetic "dough" mixed in giant mixers.[2]

The story takes place several decades in the future on an island occupied by a corporation that produces "Robots." Rossum, the founder of the corporation, who is deceased when the play opens, has a life story reminiscent of Hans Driesch (chapter 8): a marine biologist with a hobby of trying to synthesize living tissue. Rossum had successfully produced a substance that was chemically distinct from protoplasm but functionally identical. This synthetic living matter is self-constituting: the factory manager explains that the protoplasm itself creates the Robots' organs, bones, and nerves.[3] Čapek's Robots were more like what would come several decades later to be called clones:[4] they were made of organized matter, soft and alive although synthetic, actively self-constituting.

Insofar as Čapek took an interest in the nature of life and the possibility of synthesizing it artificially, it makes sense that he described the Robots in these active-mechanist terms. His main philosophical influences were the American pragmatist William James, who argued that animal-machinery required intrinsic

Figure 9.1 Sylvia Field as Helena and Albert Van Dekker as Radius, a robot, in a scene from the stage production of *R.U.R.*, and a robot uprising from the stage production of *R.U.R.*, photos by Vandamm Studio © Billy Rose Theatre Division, the New York Public Library for the Performing Arts.

consciousness as a "steering" mechanism,[5] and the French vitalist Henri Bergson, who invoked a Lamarckian "*élan vital*" (vital impulse) as the motor driving the evolution of organic machinery.[6] However, although *R.U.R.* borrowed some ideas from contemporary biology and philosophy, it was not, as its author himself maintained,[7] really about the Robots. Instead, these artificial beings illustrated Čapek's real point, which was an indictment of the burgeoning system of industrial capitalism. In addition to the dough mixers, Rossum's plant included vats of liver and brain, a bone factory, a spinning mill to make nerves and veins, an intestine mill, and an assembly room. Writing six years after the first assembly line began operating at the Ford Model-T plant in Detroit, Čapek had his Robot factory manager describe Robot manufacturing as "just like making a car really. Each worker contributes just his own part of the production which automatically goes on to the next worker."[8]

Despite their flesh and bone, intestines, liver, nerves, and brains, the Robots immediately became in the public imagination "mechanical" things: boxy, metallic, grinding, and clanking. Reviews of *R.U.R.* described the Robots as "mechanical men," "a sort of clockwork," "the mechanical workers of our mechanized civilization—human cogs and levers."[9] The first cartoon images of robots looked like Cubist renditions of a cross between a Model-T and a telephone switchboard (see figure 9.2).[10] Or, they resembled the Tin Woodman of Oz, who was a generation their senior and also presented the argument that industrialization was dehumanizing.[11]

R.U.R. was an international sensation and immediately translated into French, English, and German. It opened in New York on Broadway at the Garrick Theater in October 1922, where it was a huge hit, then the following year in London, Chicago, and Los Angeles. The Robots were trendsetters: their "stiff, swollen, sculpture-like garments" intrigued clothing designers. A reviewer reported that the fashionable coats of the season were inspired by these, "made of a material which looks like embossed leather. They are long close-fitting affairs with huge bolster collars. To complete the sculptured and mechanical illusion they are usually to be seen in deep bronze or stone colours."[12] *R.U.R.* would be among the most widely performed works of the century and was adapted several times for television and radio during the 1930s and '40s.[13]

The word "robot" was even more sensationally successful and enduring than the play that introduced it. The neologism appeared in English-language newspapers worldwide during the New York opening and lost its capital letter as it traveled beyond reviews into articles on other topics entirely. In 1926,

Section 9 SPECIAL FEATURES AUTOMOBILES—AVIATION—RADIO

The New York Times.

SPECIAL FEATURES AUTOMOBILES—AVIATION—RADIO Section 9

XX SUNDAY, OCTOBER 23, 1927. XX

SCIENCE PRODUCES THE "ELECTRICAL MAN"

The "Televox," Nearest to a Robot, Obeying the Human Voice, Can Control Power Stations, Motors and Switches From a Central Point Hundreds of Miles Away—How the Device Does Its Work

By WALDEMAR KAEMPFFERT.

THE MECHANICAL MAN AND HIS CREATOR.

Figure 9.2 Early robot cartoon from "If There Really Were Robots!" *Life* (February 8, 1923), and "Science Produces the 'Electrical Man,'" *New York Times* (October 23, 1927): Westinghouse's Mr. Televox in the news.

the *Manchester Guardian* described the machines at a local Business Efficiency Exhibition —an electric typewriter, an automatic telephone, a letter-stamping device, a coin-sorter, some tabulation machines—as constituting a "'Robot' Office."[14] By the following year, the term "robot" had supplanted both of the older terms "automaton" and "android."[15]

Many early robots, like their ancestors, the automata and androids, were responsive and even sentient. "Robot Dream Realised," trumpeted the headlines in reference to the "marvelous electrical 'mechanical man'" exhibited by the Westinghouse Electric Company in New York in October 1927. The electrical mechanical man, one of several Westinghouse robots on display, could hear the cry of "Open Sesame" and open a door. Another robot turned on lights, an electric fan, and a searchlight, operated an automatic sweeper and lit a signal lamp. These sentient robots publicized the work of their synthetic colleagues, three "electrical men" who checked the water levels in Washington, DC, reservoirs."[16]

These earliest robots were the work of Roy James [R. J.] Wensley, a Westinghouse engineer. His idea had been to build an automatic machine, "Mr. Televox," controllable by sound, voice, or pitch pipe over a telephone (see figure 9.3). The sound activation meant that Westinghouse employees could telephone their reservoir-control robots to find out the water levels. Wensley also envisioned a modern housewife calling home from her bridge party to have the robot start the roast and open the furnace draft.[17] "Mechanical men are springing up all around," observed a writer for the *New York Times*, "and some people are promptly becoming terrified," but the fear was misplaced. Robots were neither new nor scary: "The sewing machine is a robot, the automobile is a robot . . . the phonograph and the telephone are all robots."[18] Mr. Televox's author concurred. "Really it is largely the idea that is new," Wensley insisted. "The principles and devices were in the main previously known."[19]

In September 1928, a robot named Eric (see figure 9.4) personified the emerging idea of a "robot" as he opened the Model Engineer Exhibition at the Royal Agricultural Hall in London. The machine was a native of Surrey and the progeny of a Captain W. H. Richards, the exhibition's secretary. Richards had designed Eric in a fit of pique after the public figure who had agreed to open the festivities canceled. "I said, in exasperation, I could make a man of tin," Richards later explained, "and that is how Eric came to be made." In a demonstration before the opening, the "all-steel man who look[ed] like nothing so much as a suit of armor" stood up, sat down, turned his head, and gave a Fascist salute.[20] On opening day, his performance was more impressive yet. "Eric, the Robot, lurched a little unsteadily to his metallic feet today," reported the *New York Times*,

> stretched out one metallic arm for silence and then made a speech opening an exhibition of engineering models. Before him stood an incredulous crowd,

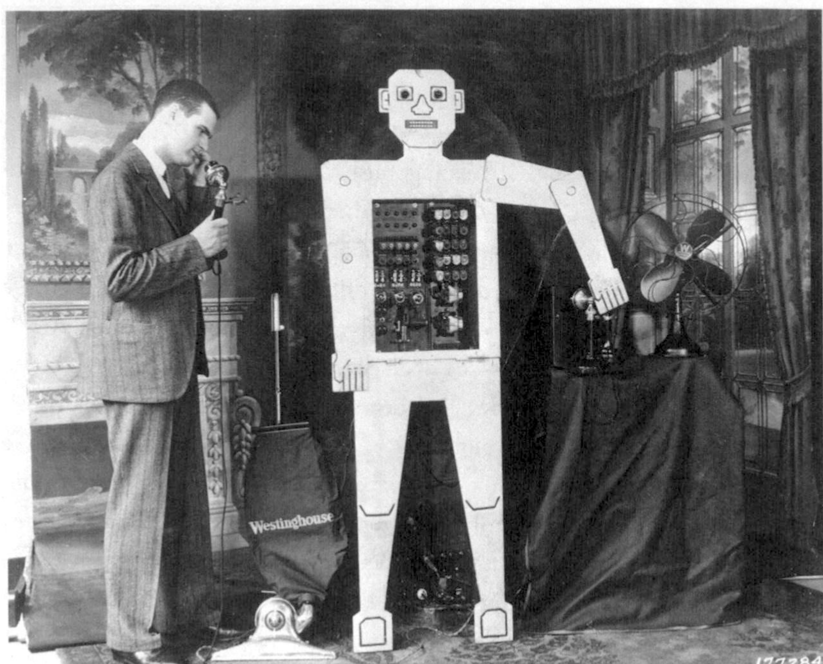

Figure 9.3 Westinghouse's Mr. Televox, Detre Library and Archives, Sen. John Heinz History Center.

> which seemed to include most of the small boys of London, and all grew wide eyed with wonder at what they saw. . . . They had not realized he would be so terrifying. His entire anatomy was sheathed in gleaming tin—the program said steel, but it did not matter which. His face had the horrible immobility of Frankenstein's monsters they had seen in films. It had slanting eyes with electric eyeballs, a toothless mouth without lips, armor-plated chest and arms and sharp metal joints at the knees such as armored knights wear at the Metropolitan Museum.

When Eric spoke, yellow lights flashed in his eyes and when he had finished, he "bowed stiffly once, turning his leering head to right and left. Then with clanking of his armorplates he sat down." Altogether it was awe-inspiring, although Eric's speech, consisting of the usual mumbled platitudes delivered by radio via a diaphragm in his head, had been unremarkable and somewhat undermined the flashing yellow eyes and tin physique.[21]

A ROBOT TO OPEN AN EXHIBITION: THE NEW MECHANICAL MAN.

Drawn by our Special Artist, G. H. Davis, from Sketches Made at Gomshall with the Assistance of the Inventors, Captain Richards and Mr. A. H. Reffell.

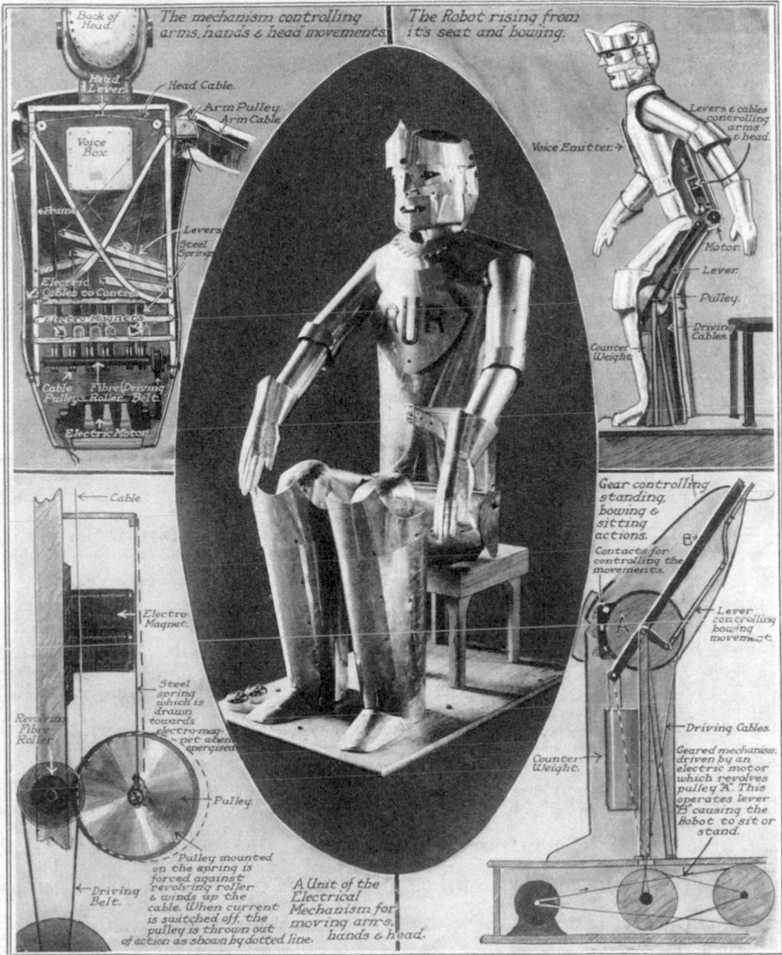

AN ALUMINIUM "MAN" THAT RISES, BOWS, AND MAKES A "SPEECH": A KNIGHT-LIKE ROBOT.

There has just been completed at Gomshall, near Dorking, the first British Robot, a gleaming thing of aluminium, not unlike a mediæval knight in armour, whose first duty will be to open a Model Engineering Exhibition to-day at the Royal Horticultural Hall. Concealed in the body is an electric motor which drives a fibre roller. Just above are several electro-magnets, with steel springs. To the base of these springs are fixed pulleys carrying cables that operate levers which move the Robot's arms and head. When the electro-magnets are energised the springs are drawn to the magnets pulling the edges of the pulleys against the revolving fibre roller. The pulleys revolve, winding in the cable and moving the head or limbs as desired. By cutting off current the wheel-face is detached from the roller, and the arm falls back to its normal position. For raising the Robot from its seat, causing it to bow to the audience and resume its seat, another motor is concealed in the platform below the figure's feet. This operates large pulley wheels concealed in the knees. When these wheels are slowly turned, a lever attached to each raises or lowers the man as required. Three contacts on the pulley give the desired positions to the operator. A second lever tilts the body and gives the "bowing" movement. To ease the work of the motor, counter-weights in the legs balance the weight of the body and interior mechanism. An ingenious electrical gear (which is the jealously guarded secret of the inventors) enables the Robot to hear questions and answer in a human voice. The Robot has been designed and made in under six months, so that it is but an infant and not yet able to walk, but the inventors state that in time it will be able to use its legs. At present, however, its chief work will be in the realms of publicity.

B

Figure 9.4 Eric the Robot, from the *Illustrated London News* for Saturday, September 15, 1928.

Eric embodied the idea of "robot" that had arisen around the image of Čapek's fictional mass-manufactured artificial people, but he did not reflect the contemporary generation of actual robots, most of which were capable of responsive action. Just over a year later, indeed, Eric the strapping English robot was outdone by a Parisian poodle: a *chien électronique* that walked, yapped, and followed its owner's flashlight. The electronic poodle was the creature of Henri Piraux, an engineer at the French subsidiary of the Dutch electronics firm Philips, which specialized in electrical, X-ray, and radio technologies. Piraux would later become the "head of technical propaganda" at Philips, and he designed the creature, which he named "Philidog" (in honor of both Philips and the eighteenth-century composer and chess master François-André Danican Philidor), to publicize the possibilities of photoelectric cells.[22] Philidog performed in Paris in the autumn of 1929, first in September at the Exposition internationale de TSF (International Wireless Exhibition) at Magic-City,[23] a dance hall otherwise known for its drag balls, and then in late October at the Salon national de la TSF (National Wireless Salon) in the Grand Palais.

Philidog ran on wheels, was powered by batteries, and had a photoelectric cell in each eye, connected to two motors, one powering the right wheels, the other the left. When light shone on both photoelectric cells at once, both motors turned and the dog moved forward. If only one eye received light, only the corresponding motor ran, and the dog turned in the opposite direction. When the cells were exposed to a very bright light, as when the flashlight was held right up close to them, both motors stopped and the mechanical poodle "barked," first through a horn, and later by means of an internal phonograph that also allowed it to growl. This enhanced version of Philidog also had a set of long, steel, electrically charged touch-sensitive whiskers.[24]

Philidog's purpose was to be a watchdog, but he was too undiscriminating: he could not distinguish between an intruder and the man of the house coming home late and "a trifle unsteady." Also, "nowadays, everybody, including the flapper daughter, has a latch key and sometimes 'checks in' a little later than she wants to advertise." So Piraux proposed other ideas, such as a bison's head on a museum wall "that will bellow and keep it up till the watchman comes and opens a switch, or a suit of armor that can be made to swing a sword or battle axe."[25] However, Piraux and his generation had been scooped in their plans for electronic butlers and watchdogs by over half a century.

Flashback: A Nineteenth-Century Smart House

Responsive electromechanical machines originated in the mid-nineteenth century. During the 1850s and '60s, the French magician Jean-Eugène Robert-Houdin,[26] debunker of von Kempelen's chess-playing Turk (chapter 4), filled his house in Saint-Gervais, near Blois in the Loire valley, with contraptions that made Piraux's electronic watch-poodle look like a toy. From his workshop, the wizard governed every detail of his house and grounds by means of electromechanical "servants":[27] the positions of doors, windows, and gates; the temperature in the hothouse and woodbin; and, through the settings of clocks and alarms, the activities of his human servants.

An inventor who won awards for innovations in clockwork and automaton design,[28] Robert-Houdin designed some of the earliest electromechanical devices and patented crucial parts for electrically driven technology: an electric regulator, an interrupter, a distributor, a vibration motor, an electromagnetic plunge battery. He gave electrically lit soirées and perhaps anticipated Thomas Edison by devising the first incandescent electric lightbulb, used just once, in June 1863, on the occasion of his daughter's first communion when he gave a celebratory dinner under arbors illuminated by electricity.[29] In the 1840s, Robert-Houdin also sketched an idea for a "musical telegraph" or "daguerreotype of sounds"; and some thirty years later, on April 2, 1878, his theater in Paris, the Soirées Fantastiques, then operated by his son, Emile, would host Paris's first public demonstration of the telephone. Still later, the French filmmaker Georges Meliès, who had been captivated by Robert-Houdin's theater from the age of ten, would buy it and transform it into the world's first public cinema.[30]

Robert-Houdin was also among the first to create electric clocks, for which he also held patents, and coordinated systems for electrical timekeeping. He dreamt "of the day when the electric wires, issuing from a single regulator, will radiate through the whole of France, and will bear the precise time to the largest towns and the most modest villages."[31] He designed electric alarms and warning devices to protect ships from leaks, houses from fire and theft, his hothouse from temperature fluctuations, and his filly from an empty feed bin.[32]

One of Robert-Houdin's first patents was for an alarm clock with a candle concealed in a little compartment. When the carillon sounded, the door to the compartment opened and the candle emerged already lit. The Reveil-Briquet sold briskly and helped to launch Robert-Houdin's business in clock making.[33]

By the end of his career, he had traveled much of the distance from the Reveil-Briquet to Bill Gates's "smart house." The Priory, the house in Saint-Gervais to which Robert-Houdin retired, was nicknamed L'abbaye de l'Attrape (Trap-ist Abbey) by his friend, the artist Jean-Pierre Dantan.[34] The sorcerer's mansion was tricked out with electromechanical contrivances in every corner. Buried in trenches, electric wires connected every part of the property.[35]

The front gate, a quarter of a mile from the house, included a side door with a gilt knocker. Over this, a small gargoyle held a sign that said KNOCK. The knocker triggered an electric bell that rang inside the house until a servant opened the lock (which could be done remotely by pressing a stud in the hall). As the bolt was withdrawn, a nameplate on the door, inscribed ROBERT-HOUDIN, was replaced by a sign saying WALK IN. Then, as the door opened, it set off a further sequence of electric bells. Robert-Houdin claimed he could tell from the speed and rhythm of these bells how many guests he had and what sort: friends (rapid), strangers (a bit slower), beggars (timid, hesitant). The door then sprang shut, replacing the walk-in sign with the nameplate. The door could not be opened without the preliminary of knocker and electric bell, so that no servant, "either by mistake, in fun, or through carelessness" could press the stud and open the gate in the absence of a visitor. Meanwhile a coachman, upon entering the side door, would find a key to the carriage gate identified by another sign. His opening and closing of the carriage gate would be announced in the front hall of the house on a plate reading "*THE CARRIAGE GATES ARE*—. At the end of this incomplete inscription follow alternatively the words *OPEN* and *SHUT*, according to whether the gates are in one or the other of these conditions."[36]

The gate also contained a letter box whose flap rang a second electric bell inside the house. The postman was instructed first to put all the newspapers and circulars through the flap at once, followed by the letters, one at a time. "We are therefore notified in the house of the delivery of each article, so that if we are not inclined for early rising, we may, even in bed, reckon up the different items of the morning post-bag." The mail signal could also be reversed. When Robert-Houdin had letters to send, he would set the bell to ring, not in the house, but at the gate, alerting the postman to come to the house and fetch the outgoing mail. "My electric porter . . . ," Robert-Houdin concluded, "leaves me nothing to desire. His duties are discharged with the utmost precision, his fidelity is beyond proof, and he tells no secrets. As to his salary, I doubt whether it would be possible to pay less for so perfect a servant."[37]

A single electric regulator directed all the clocks in the house and two large clock faces outside, one mounted on the main house for "inhabitants of the valley generally," and one mounted opposite, on the gardener's cottage, for residents of the house. The clocks shared a common bell housed in a miniature bell tower on the roof, rung by "a clockwork arrangement." An electric current, distributed by the regulator, withdrew the catch of the striking movement and caused the hammer to strike the bell the right number of times. To wind the ringing mechanism, Robert-Houdin connected it with a winding-up apparatus over the swing door of the kitchen such that the servants would be "unconsciously winding up the striking movement" every time they passed through the door.

He also regulated his household staff with a coordinated timekeeping system. Automatic alarms rang to wake "three different persons at three different hours, beginning with the gardener." To silence the alarm, the sleeper was obliged to press a stud at the far end of his room. Moreover, when the master wanted to "advance or retard the hour of a meal," he did so by pressing an electric button in his study that would put all the clocks forward or back. "The cook often fancies that the time has passed somehow very quickly, and I myself have gained a quarter of an hour."[38]

Upon finding that a stable boy was stealing his horse's fodder, Robert-Houdin took to feeding Fanny automatically. Now, he wrote, she "has an affectionate person to serve her meals; he is a most honest boy who, precisely because of his probity, takes no offense at my . . . electrical procedures."[39] The new stable boy was in fact a mechanism on the feed bin, connected with the Priory's central regulator, and set to deliver oats three times a day. The trigger that released the oats from their box could operate only when the stable door was locked, which could only be done from the outside, and an alarm rang in the house whenever the door was opened while there were oats remaining in the bin.[40] "Do I not have in electricity and mechanics intelligent auxiliaries on whose service I can count?"[41]

Similarly, a "thermo-electrical arrangement" in the hothouse alerted the magician when his gardener allowed the temperature to vary too widely, and another in the woodshed warned him of fires. A burglar alarm system rang a large bell whenever any of the doors or windows were opened. In the shooting gallery, the marksman who hit a bull's-eye was rewarded by a leafy crown delivered from above with the help of "the electric fluid." In the park, a mechanical system of "flying seats" delivered passengers from one side to the other of a small ravine.[42]

These first responsive, sentient electromechanical machines were born just a few decades after Babbage's projects in mechanical computing, and a good half century before the first "robots." The idea that human functions, especially the functions of socially inferior humans, might be carried out by artificial machinery, originated as we have seen in the early stages of the Industrial Revolution, and therefore long predated cybernetics. So too did the idea that artificial mechanical servants could be sentient and responsive: could be endowed with a kind of mechanical agency. Indeed, if Robert-Houdin's career is any indication, artificial, mechanical agency was a guiding preoccupation of nineteenth-century technical innovation. Our sorcerer worked right at the cutting edge of his century's new technologies, from electric timekeeping to lightbulbs to the telephone to cinema. The abiding project of this father of modern scientific magic was artificial machinery that could perceive and reply.[43]

Neither the idea nor the reality of sensitive and reactive electromechanical automata was new with cybernetics; both predated cybernetics by a full century. What was new, as we shall see, was the set of implications the cyberneticists drew from such devices, and upon which they founded their movement.

Agency in the Passive Voice

In the same years that Mr. Televox, Eric the Robot, and Philidog were making their debuts as "robots," the Harvard physiologist Walter Cannon coined another new word. This one, "homeostasis," named what Claude Bernard had identified a half century earlier as the distinguishing feature of living machinery: its ability to respond to the outside world, interact with it, and through this interaction, to maintain its own stability and order.[44] Cannon had studied physiology as a Harvard undergraduate in the laboratory of Henry Pickering Bowditch, dean of the Harvard Medical School, who had studied with Bernard in Paris, and both Americans were very much influenced by Bernard's ideas.

A host of "agencies," Cannon wrote, were always "acting, or ready to act," to maintain "constancy" in a living organism. However, he described these agencies almost exclusively in the passive voice, having apparently absorbed the sciences' ban on agency despite himself. Referring to the disturbing effects of environmental changes on an organism, Cannon wrote, "Such disturbances are normally kept within narrow limits, because automatic adjustments within the system are brought into action, and thereby wide oscillations are prevented and the internal conditions are held fairly constant." He summed up: "The previous pages have illustrated various agencies employed in the organism."[45]

Characterizing agencies like these in the passive voice made for a curiously contorted kind of prose. Who or what accomplished all of this: "limited," "adjusted," "prevented," "held," "employed"? Rather than declaring the living organism to be the subject of his many verbs, Cannon hid it behind a wall of passive-voice constructions. Here is the same passage edited, as professors in the humanities wearily continue to edit their students' papers, to avoid the passive voice so as not to obscure agency: "The living being normally keeps such disturbances within narrow limits, automatically adjusting its system so as to prevent wide oscillations and keep internal conditions fairly constant." The first formulation implicitly invoked an unknown agency outside the living being; this one explicitly identifies one internal to it: the being itself!

The advocates of this notion of "homeostasis," from Bernard through Cannon, had long struggled over the question of whether this ability to interact and compensate was what distinguished living creatures from artificial machines. Bernard sometimes maintained that self-regulation was a distinctive characteristic of organisms, but also pointed out that artificial machines such as steam engines exhibited the same capacity. During World War II, a number of people in Britain and the United States, many of whom were working as researchers or consultants on experimental technologies for military uses, adopted the latter view that artificial machines could exhibit the very same homeostatic capacities, and indeed that homeostasis constituted a link between animal and artificial machinery.[46]

These were the first cyberneticists. In Britain, this group included such people as the experimental psychologist Kenneth Craik,[47] who worked with the Royal Air Force in connection with the Flying Personnel Research Committee, designing an artificial cockpit that was one of the first flight simulators; the neurophysiologist William Grey Walter,[48] pioneer of the electroencephalogram, who worked on scanning radar and guided missile technology; and the psychiatrist Walter Ross Ashby,[49] who built his "Homeostat," a simulation of a living creature generally identified as the first cybernetic machine, out of former RAF bomb control units.

In the United States, the central figures included the mathematician Norbert Wiener,[50] who conducted wartime research on systems for aiming and firing antiaircraft guns; and the mathematician and engineer Claude Shannon, then working at Bell Laboratories on fire-control systems. Among the most important cyberneticists was also the Hungarian-born Princeton mathematician John von Neumann, who sought a theory of automata that would embrace both "artificial automata, specifically computing machines" and "natural automata," by which he meant primarily the human nervous system, but also including biological mechanisms of reproduction, self-repair, evolution, and adaptation.

All these phenomena, artificial and natural, von Neumann observed, crucially involved a single, lifelike feature: a system that shaped and maintained itself in response to its changing environment. He initially wanted to test these analogies by building a robot able to build another robot, that is, reproduce itself, but he soon turned to creating an abstract version of the experiment instead. The result was a species of mathematical devices known as "cellular automata." These consisted of grids of geometrical cells that evolved in complex patterns by means of a simple set of rules governing the relations between each cell and its neighbors. These became, and remain, central experimental models throughout the sciences.[51]

Each of these cyberneticists worked to build machines capable not only of carrying out a discrete purpose, but also of adapting their function purposefully in the manner of a living organism adjusting to environmental changes. It was Wiener who coined the term "cybernetics" to describe the unified science of homeostatic artificial and organic machinery. Automata, "whether in the metal or in the flesh," must enact a continual exchange with the outside world, and "cybernetics" was to denote the study and manipulation of this engagement between machine or organism and environment.[52] The word "cybernetics" came from the Greek for "steersman" (κυβερνήτης), which is also the etymological source of the word "governor." Wiener had in mind James Watt's steam engine governor (1788), which maintained the speed of a steam engine between certain limits, and which Wiener identified as the earliest artificial homeostatic device. He also referred to James Clerk Maxwell's essay "On Governors" (1868), which Wiener took as the first theoretical treatment of cybernetic (i.e., artificial, homeostatic) mechanisms.[53]

But Wiener didn't know much about history. In his manifesto, *Cybernetics: or Control and Communication in the Animal and the Machine* (1948), he offered a sketch of what he took to have been the prehistory of cybernetics and declared that ancient Greek automata lay "outside the main lines of the direction of the modern machine." In fact, ancient texts in Renaissance translation had provided the principal model for the early modern explosion of automata across the European landscape. The ancient automata, he wrote, did not "seem to have had much influence on serious philosophical thought."[54] But ancient automata had been an important reference not only for Aristotle but also for Aquinas and the entire tradition of Scholastic philosophy to which Descartes and other seventeenth-century mechanists responded, a tradition that assumed mechanical things worked by means of innate agencies and responsive capacities.

Skipping over this ancient and medieval mechanist tradition, Wiener considered Descartes the first "serious" philosopher to have taken an interest in early modern clockwork automata, with his analogy to living beings. But "just how these living automata function," Wiener wrote, "is something that, as far as I know, Descartes never discusses."[55] But as we know, Descartes not only did discuss the functioning of living automata in exhaustive detail, but was a primary founder of the new mechanist physiology.

On the other hand, Wiener was a great admirer of Leibniz, anointing him the "patron saint of cybernetics" for having pioneered the notions of a universal symbolism and a calculus of reasoning, ideas that had provided the conceptual foundation for the "*machina ratiocinatrix*," or reasoning machine, of the cyberneticists. Still, he was not a fan of Leibniz's monads, which he misconstrued as having each existed in splendid isolation, "in its own closed universe": "They have no real influence on the outside world, nor are they effectively influenced by it. As he says, they have no windows." Wiener compared the monads to the "passively dancing figures on top of a music box" and to "a Newtonian solar system writ small," a passive clockwork cosmos.[56]

But monads were the opposite of isolated, passive, Newtonian atoms: they were universally connected, pure agency, and explicitly *anti*-Newtonian. Far from existing in its own closed universe, each monad reflected or "expressed" every other monad. Leibniz's remark that the monads had "no windows"—no way for anything to come in or out, and therefore no way to alter one another—was part of his argument that they were not material objects but perceiving souls. As a perceiving soul, a monad could neither act materially upon others nor be transformed materially by others. It engaged with other monads, not by pushing or pulling, but by perceiving and knowing. Leibniz deplored the passivity of Newton's clockwork cosmos and its dependence upon a supernatural source of action. When he called monads "incorporeal automata," he did not mean that they were like the passively pirouetting figures on top of a music box, but just the opposite: that they were self-moving, the sole sources of their own actions.[57]

A last mistake: Wiener concluded that according to Leibniz, the "apparent organization of the world we see is something between a figment and a miracle.[58] But the apparent mechanical organization of the world was, to Leibniz, precisely not a miracle. The virtue Leibniz saw in his doctrine of "preestablished harmony"—holding that all things in nature acted according to a single, coherent set of laws—was that it "casts off every notion of miracle from purely natural actions and makes things go their course ruled in an intelligible manner."[59] The rational, mechanical system of the universe was intelligible to Leibniz only if it could be self-sufficient, containing its own agency.

Why belabor Wiener's upside-down readings of history? He was a mathematician and philosopher, after all, not a historian. Yet he founded his field of cybernetics upon a vision of history. An axiom of cybernetics was that until then, no one from Hero of Alexandria onward—neither Aristotle nor Descartes nor even Leibniz—had considered the possibility that responsiveness might be mechanical: that animals and artificial machines might share mechanisms of sentience and response. Cyberneticists maintained their new machines were the first to transcend the limits of the passive, scripted, "blind, deaf and dumb" automata of old. Although they did make an exception for Watt's self-regulating steam engine and the other governors in Maxwell's 1868 paper, these, they thought, had gone unrecognized as sentient and responsive machines and had thus failed to spark a revolution in philosophy and engineering.

Cybernetics was this deferred revolution coming at last. Accordingly, Wiener emphasized that the new generation of "sensitive automata" had sense organs: photoelectric cells able to tell light from dark, and instruments for measuring tension in a wire or changes in temperature. Any scientific instrument for measuring any quantity whatsoever was potentially a sense organ. "Sensitive automata" included thermostats, automatic door openers, controlled missiles, proximity fuses, gyrocompass ship-steering systems, and antiaircraft fire-control systems.[60]

But the cyberneticists were mistaken in thinking that theirs were the first sentient, responsive automata, not only because of Philidog (1929), Robert-Houdin's electric porter and stable boy (ca. 1860), or even Watt's steam engine governor (1788). Responsive automata extended at least back to the eye-rolling, grimacing, water-spraying mechanical tricksters of the thirteenth century and, since these involved devices that had been in existence since antiquity, far earlier yet. Even mechanisms employing what the cyberneticists named "feedback" existed in antiquity.[61] A philosophical interest in mechanical sentience and responsiveness, and technological attempts to achieve these, were ancient traditions.

Not knowing history, the cyberneticists were condemned to repeat it. As a result, their mechanist models of sentience, responsiveness, and living agency became just the opposite: so many confirmations of the absence of agency in biological creatures. They offered an updated version of the classical mechanist picture in which any appearance of purposeful action in living beings was simply an appearance.

The founding manifesto of cybernetics, written by Wiener, his colleague Julian Bigelow, and the Mexican physiologist Arturo Rosenblueth,[62] announced that they had made purpose and teleology safe for modern science. They reckoned that teleology in the form of Aristotelian final causes had fallen out of favor "chiefly because it was defined to imply a cause subsequent in time to a given effect."[63] They meant that the accomplishment of the final purpose or end result of an event must follow the event in time.

But an Aristotelian final cause was not subsequent to its effect, which would indeed be strange. Final causes were purposes that preceded their effects the way my purpose to drink precedes my filling a glass with water. The Aristotelian doctrine of final causes supposed that will, purposefulness, acted throughout nature. It was this permeation of nature by agency, rather than any problem of causal chronology, that had led the classical mechanists to reject Aristotelian final causes.

To counter this worrisome Aristotelian teleology, Rosenblueth, Wiener, and Bigelow introduced "negative feedback," borrowing a term from electrical engineering and lending it a more broadly philosophical purpose. "Negative feedback," they explained, occurred when an object's "behavior" was "controlled by the margin of error at which the object stands at a given time with reference to a relatively specific goal."[64] They had in mind devices such as antiaircraft or missile-guidance systems, in which an object is guided to a target. This negative-feedback kind of teleology, the authors proposed, involved no time-reversal problem in which causes acted subsequent to their effects, since the goal (the target) acted upon the behaving object in the present. The concept of negative feedback thus avoided the problem cyberneticists ascribed to Aristotelian physics of reversing cause and effect in time.

Negative feedback also eliminated the actual problem that Aristotelian physics had posed for classical mechanists, in that it imputed no agency to the behaving object. The "teleology" of Rosenblueth, Wiener, and Bigelow was only apparent and not real. The object did not have an inner purpose, but only external influences. Note the passive voice: the object's behavior "is controlled by" its distance from a goal. The "behaving" object was passive. It did not seek its goal, but was drawn by it. Signals from the goal, for example, a light-source, restricted outputs from the behaving object so that these would converge on the goal. A tropism such as those Jacques Loeb studied was, according to Rosenblueth, Wiener, and Bigelow, a simple kind of negative-feedback-governed behavior. So was a bloodhound following a scent.

A more complicated negative-feedback-governed behavior required "predictions" of the future, the way a cat, when chasing a running mouse, ran to where the mouse was heading rather than where it currently was. According to Rosenblueth and his colleagues, even so active-seeming a process was directed from outside the cat. As in a simple tropism, signals from the mouse governed the cat's behaviors. "All purposeful behavior," the authors maintained, "may be considered to require negative feed-back. If a goal is to be attained, some signals from the goal are necessary at some time to direct the behavior."[65]

From an engineering perspective, this was an extremely fruitful conception of how a machine might function autonomously to accomplish a given purpose even in changing conditions, an effective approach to designing a guided missile or an antiaircraft gun. Not only engineers but also philosophers and biologists seized hold of the idea. Ernst Mayr hailed Rosenblueth, Wiener, and Bigelow's paper as a "breakthrough in our thinking about teleology."[66]

Yet this first cybernetic construal of purposeful action was philosophically very limited. For one thing, as early critics of cybernetics pointed out, a purposeful action on this model required a "goal" that existed in the form of a signal-emitting object.[67] Approaching or otherwise achieving certain relations to an object in the landscape is only one form of purposeful action, plenty of which (writing an opera, solving a math problem) involves no signal-emitting goal object. Cyberneticists expanded their understanding of the "goal" of negative-feedback-driven behavior to include the homeostatic purpose of maintaining stable internal conditions for parameters such as temperature and osmotic pressure in response to varying external conditions.[68] However, in both the object-seeking and the homeostatic versions of this cybernetic view of purposeful action, the source of action was external to the acting object, in its environment.

In order to locate the source of action externally to the acting object, Rosenblueth, Wiener, and Bigelow adopted what they named a "behavioristic" analysis of willful action. Applicable to both artificial machines and living organisms, a behaviorist approach considered the internal nature and organization of an object to be irrelevant. In order to evaluate whether an object "behaved" purposefully, one had merely to consider whether it changed in response to its surroundings in such a way as to achieve and/or maintain a given set of relations to those surroundings. This was an external evaluation. If the object did transform in such a way, it would be behaving purposefully.[69] By this definition, for example, a thermostat behaves purposefully.

Cyberneticists, with their behaviorist approach, described a world of apparent agency, but no actual agency. Their fellow traveler, the French philosopher of biology Georges Canguilhem, gave a behaviorist interpretation of the central cybernetic principle that living organisms and machines were essentially the same, similarly reducing agency to behavior. In a 1952 essay entitled "Machine et Organisme," Canguilhem called for an end to considering mechanism "only as an intellectual operation of man," an application of knowledge. One must understand mechanism more generally as "a behavior of the living." When people built machines, they merely extended the mechanical functions of their bodies outward, behaving as all living entities did, that is, mechanically. Artificial machines, seen this way, were natural ramifications of the mechanical nature of organisms.[70]

Removing agency, along with all subjective experience, from psychology was the core principle of the behaviorist school then prevalent among psychologists. The Harvard behaviorist B. F. Skinner wrote that the study of reflex action had undermined the "prevailing theories of the inner agents responsible for behavior": "An external event had been identified which could be substituted, as in Descartes's daring hypothesis [he meant Descartes's construal of animals as machines], for the inner explanation. The external agent came to be called a *stimulus*."[71] Skinner and other behaviorist psychologists in the United States and elsewhere described an animal's ability to respond to its environment as "machine-like" and presented their behaviorist approach as scientific in its construal of creatures as machines.[72]

By describing animals as "machine-like," Skinner and other behaviorists meant that they imputed neither inner agency nor even experience to their subjects. If psychology were to become properly scientific, psychologists must stop trying to study ineffable pseudo-entities, notably consciousness, an invisible and nonmechanical phenomenon. Instead they must set their sights on behavior, which, the behaviorists argued, was both observable and fully mechanistic, and was therefore the only fit object of scientific study. The behaviorists prepared the ground for cyberneticists to import a behaviorist externalization of agency, a rendition of agency in the passive voice, as a founding principle of their own new discipline. Wiener adopted this principle when he wrote that the environment of an organism could "modify the behavior into one which in some sense or other will deal more effectively with the future environment."[73]

The cyberneticists' central project was the mechanization of perceptive agency. They described their approach as behaviorist in its evaluation of agency by outward behaviors only, and Darwinist in that over time their mechanisms acquired complexity from the environment. They offered an account of perceptive agency as only an external appearance: the impression an observer gathered from watching as the environment shaped and directed the actions of an essentially passive, mechanical creature.

The Passive-Active Cybernetic Menagerie

Two decades after the coining of the terms "robot" and "homeostasis," and coeval with the founding of the discipline of cybernetics to unite the two concepts, the British psychiatrist Walter Ross Ashby built a robot whose entire purpose was to be homeostatic, that is, to maintain a steady state in response to certain environmental disturbances (see figure 9.5).[74] Like many of the automaton and robot makers from the fourteenth century onward, Ashby was connected with clock- and watchmaking. On his mother's side, he was the grandson, great-grandson, and great-great-grandson of London watchmakers, all named Henry Lemmon, and was himself an amateur watchmaker.[75] At the time he built the Homeostat, as he called his machine, he was also director of research at Barnwood House Private Mental Hospital in Gloucester, where he conducted research on electroconvulsive therapy, a new practice the hospital had recently introduced to England along with lobotomy.

The Homeostat was made from four control units taken from RAF bombs.[76] Each unit included a coil and pivoting magnet on top, and each emitted a DC output proportional to the deviation of its magnet from a central position. Ashby wired the units together so that each sent outputs to, and received inputs from, the other three. Each of the pivoting magnets moved in response to the inputs it received, changing its outputs, which in turn acted upon the other magnets, causing them to move and change their outputs, and so on.[77]

In each unit, a potentiometer (a variable resistor that can control how much

Fig. 1—The homeostat, with its four units, each one of which reacts on all the others.

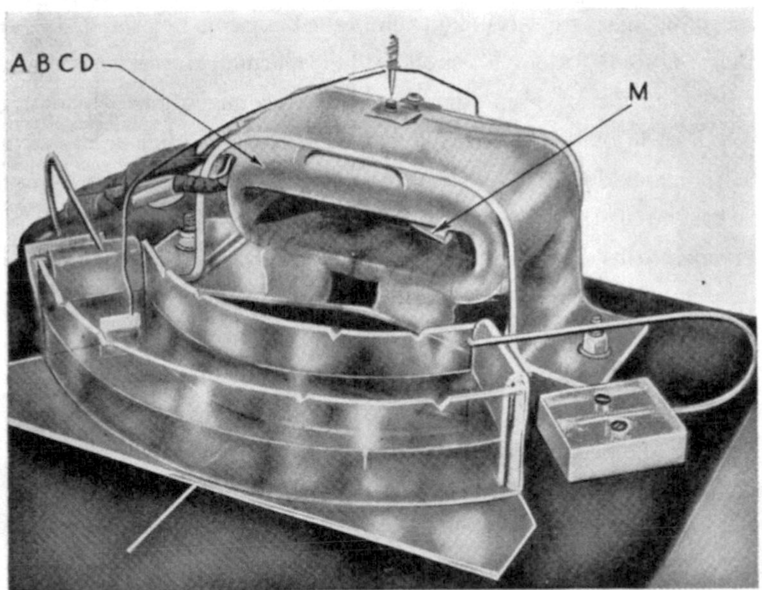

Fig 2—Quadruple coil ABCD encircles magnet M which is suspended by the needle pivot. The suspending wire extends forward on its end into the water in the semicircular plastic trough which has electrodes at each end. Potential for the grid is taken from the pivot socket.

Figure 9.5 Ross Ashby's Homeostat, from "The Electronic Brain," *Radio-Electronics* (March 1949).

current flows through a circuit) and a commutator (a switch that can change the direction of current) determined, respectively, what polarity and fraction of the inputs arriving at each unit would reach the coil. How one set these parameters, the potentiometer and the commutator, would determine the configuration of the machine. In some configurations, the four magnets would eventually come to rest at the centers of their pivots where, Ashby said, "they actively resist any attempt to displace them." When they were displaced, "a *co-ordinated* activity brings them back to the centre."[78]

In other configurations, however, the magnets might go wild, swinging faster and faster on their pivots until they hit the ends of their arcs. The machine included an additional, crucial feature: each unit had a switch that could be set so that its incoming current, rather than passing through its manually operated potentiometer and commutator, instead passed through a uniselector stepping switch, a device that takes a single input and connects it to one of multiple outputs. Each uniselector was activated only above certain output values, that is, when its corresponding magnet exceeded a certain range in its oscillations. When activated, the uniselector would change the values of the parameters: the direction of flow and volume of current. Each uniselector had twenty-five output positions, taken from a random numbers table, generating 390,625 possible randomized combinations. These allowed the machine to reset itself randomly to a new configuration, mimicking the random reconfiguration of living organisms in Darwin's nature. If a magnet tripped its corresponding uniselector, the uniselector would reset the parameters until the output and magnet position came back within the prescribed range.[79]

Thanks to its uniselectors, the Homeostat, according to Ashby, shared a "remarkable property of the nervous system," namely its "power of self-reorganization": the ability to adapt to environmental changes, even those inflicting "surgical alterations" upon its own structure. For example, one could manually reverse a commutator and the Homeostat would soon recover equilibrium. This new equilibrium would be qualitatively different from the one the machine had held prior to being accosted, with different relations among the units and different parameter settings. The machine, according to Ashby, would thus have "adapted" itself to the change.[80]

In other words, the Homeostat had what Ashby called "ultrastability" due to its "double feedback," a feedback system within a feedback system. The first connected the magnets to one another, allowing them to move in response to one another's motions. The second feedback system connected each unit's uniselector switch with the position of its magnet, tripping the uniselector when its magnet attained a position outside a given range. This allowed the machine not only to tend toward equilibrium in given conditions, but to reconfigure itself in response to changed conditions.[81]

In describing the Homeostat and the nervous system as "self-organizing," Ashby borrowed from Kant's *Critique of Judgement* the property of self-organization, which Kant had believed was specific to living beings. Ashby now extended "self-organization" to artificial machinery as well, preserving Kant's original distinction between appearance and reality.[82] According to Ashby, self-organizing machines, whether living or artificial, were only apparently self-organizing. To understand self-organization as an external appearance would "resolve the conflict about the nervous system" between the "requirement that it should be a strictly determinate mechanistic system" and the fact that it "seems to change its organisation spontaneously." These seemingly spontaneous changes in organization "referr[ed] to the externally observable behaviour," and not to what went on inside the nervous system.[83]

Like a dog, the Homeostat could also be "trained" by means of "punishments" for bad behavior and "rewards" when it corrected itself. Suppose that you, as the Homeostat's trainer, want to train it to move its magnet number 2 to the right whenever you force magnet 1 to the left. You begin by manually turning magnet 1 to the left. Let us say that the Homeostat reacts in the wrong way, that is, magnet 2 moves also to the left. You would then "punish" the Homeostat by yanking magnet 3 all the way into the forbidden zone, thereby tripping the corresponding uniselector, which would reconfigure the parameters of the machine.

You try again: you move magnet 1 once again to the left. Perhaps, in the new configuration of the machine, a forced leftward movement of magnet 1 results in a rightward movement of magnet 2, in which case the "training" session has already reached a successful conclusion, and the machine has shown itself to be a quick study. You reward it by leaving it in peace. If it has not yet learned its lesson, you repeat the punishment and test it again, and continue until the Homeostat arrives at a configuration in which the forced leftward movement of magnet 1 results in a rightward movement of magnet 2.[84] The Homeostat has "learned" something in that it has achieved a new configuration in response to an environmental pressure, your thumb on magnet 3.

One could describe the same process differently by saying that you have forced the machine into successive new configurations until it arrived at one you like. In other words, in Ashby's contraption, learning took the behaviorist form of a reconfiguration imposed from outside. Was the Homeostat a model of mechanized agency, or of mechanized passivity? Ashby's friend and colleague, the neurologist Grey Walter, called the Homeostat "*Machina sopora*" because it reminded him of "a fireside dog or cat which only stirs when disturbed, and then methodically finds a comfortable position and goes to sleep again."[85]

Ashby often attributed agency to his creature, as when he wrote that the machine resisted attempts to deflect it from equilibrium and, once disrupted, restored itself to equilibrium by means of coordinated actions.[86] But might one not just as well say that a brick actively resists attempts to lift it? The machine was active, in Ashby's view, not only because it *did* something, but more importantly because it did so autonomously: the Homeostat did not follow a script, but rather acted as it was in its nature to act. No program determined the magnets' positions at any given moment; they moved themselves and one another as the general constitution of the machine led them to do.[87] But overall, the Homeostat did one thing and one thing only, as Ashby himself observed, although he maintained that it did so "in many intricate and interesting ways": it moved "to a state of equilibrium."[88] Its sole purpose and only function was to lapse into inactivity, a curious model for the essence of life and thought.

To its creator and his cybernetic colleagues, however, the thing looked purposeful, and they placed an emphasis on "looked." Ashby, we have seen, defined self-organization as an external appearance only. Wiener agreed. In Ashby's machine, he said, "as in Darwin's nature, we have the appearance of purposefulness in a system which is not purposefully constructed."[89] Ashby himself, with Wiener's encouragement, represented this appearance of purposefulness as a mechanical model of Darwin's answer to Descartes. Descartes had argued that there must be at least as much perfection in the cause as in the effect (he had meant that the idea of a perfect and infinite God must have a source outside the imperfect and limited human mind). But Darwin had shown, Ashby wrote, that a "quite simple rule, acting over a great length of time, could produce design and adaptation far more complex than the rule that had generated it." The Homeostat was a piece of "Darwinian machinery" (Ashby thanked Wiener for suggesting the term), using "an evolution-like process in its working," because it "selected" information from the positions of its magnets.[90]

The thing was very heavy and Ashby needed the help of his laboratory assistant to lug it around to conferences. But he brought it with him everywhere, including to California in 1955, where he participated in the inaugural year at the

Center for Advanced Study in the Behavioral Sciences at Stanford University. Did the Homeostat make the trip alongside Ashby and his wife, Rosebud (Elsie Maud Thorne) in their new two-seater Triumph TR2? (Their three daughters, at least on the return trip, traveled by Greyhound, via the Rockies, the Grand Canyon, and their great aunt Clara in Toledo.) Ashby later brought the machine to the United States once again, where it spent the 1960s, while he held a chair in electrical engineering in the Biological Computing Laboratory at the University of Illinois.[91]

Ashby's machine received tremendous publicity. In late December 1948, when he first announced it to the world, it generated more headlines in Britain than the christening of the one-month-old Prince Charles. A journalist for *Time* magazine described it as a "modest black contraption" that looked just like "four storage batteries in a square," but its creator called it "the closest thing to a synthetic brain so far designed by man." This was because unlike a mechanical calculator, which simply acts upon instructions in a "predetermined" manner, the Homeostat "really thinks . . . at least in the sense that it takes action on its own." The journalist pointed out that so did a seesaw, a compass needle, or a sunflower. But Ashby replied that his machine acted in a more complicated way than any of these. Its action, wrote the journalist, "he holds, constitutes thinking."[92]

While he was at the mental hospital, Barnwood House, Ashby spent Thursdays with Grey Walter, then director of physiology at the Burden Neurological Institute, an hour's drive away in Bristol. Ashby was developing his Homeostat and Walter was working on machines with which he intended to simulate, in a rudimentary way, the purposeful, sensitive, responsive, active, and interactive aspects of living creatures. The resulting "electro-mechanical creature" behaved "so much like an animal," Walter wrote, "that it has been known to drive a not usually timid lady" (he meant his wife, Vivian Dovey) "upstairs to lock herself in her bedroom."[93]

Walter called his creatures "Tortoises," and named them "Elmer" and "Elsie" (for **el**ectro**me**chanical **r**obots, **l**ight **s**ensitive) (see figure 9.6). The Tortoises had two senses each: touch sensors as well as photocells. These enabled them to navigate a simple environment consisting of lightbulbs and obstacles. They were attracted to moderate light sources but repelled by very bright ones, and when they encountered an obstacle, they backed up and shifted around it. When they were hungry, that is, in need of recharging, they developed an appetite, which is to say that they were no longer repelled by very bright lights

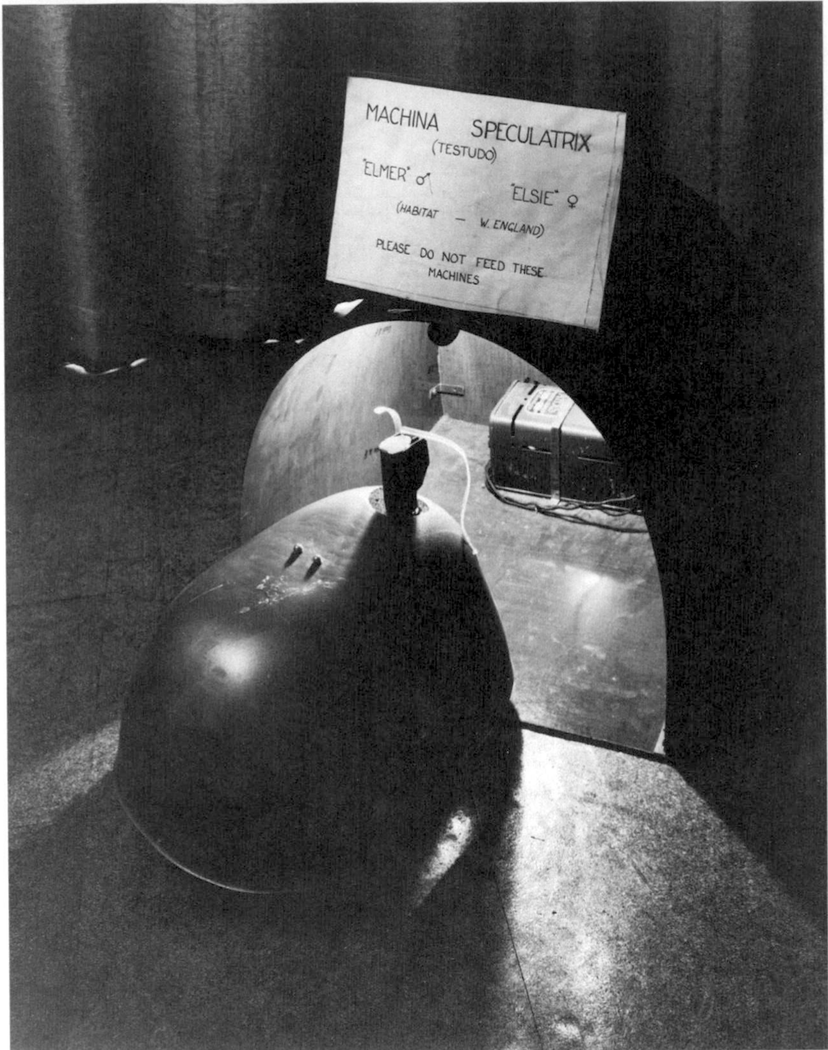

Figure 9.6 One of Grey Walter's electromechanical tortoises, Elmer and Elsie.

and would tend toward the bright light in their hutch, going home to "eat" and "rest."[94]

A kind of restless agency was the characteristic feature of the Tortoises. They exhibited, Walter said, a "typical animal propensity to explore the environment rather than to wait passively for something to happen." They were in constant motion, never still except when feeding, that is, having their batteries recharged. Like the "restless creatures in a drop of pond water," Elmer and Elsie bustled about "in a series of swooping curves, so that in an hour [they] . . . will investigate several hundred square feet of ground."[95]

The Tortoises surprised their creator and his colleagues. For one thing, they appeared to recognize their own images in a mirror. This was thanks to the small flashlamp that each wore on its head, which turned off automatically whenever the photocell was exposed to a strong enough light. When Elmer or Elsie encountered its image in a mirror, it was drawn, first of all, to the reflected light of its own headlamp. When it got close enough to the mirror, the reflected light became bright enough to cause the headlamp to switch off and the Tortoise to move away. As soon as the headlamp switched off, however, it was no longer exposed to its own reflection, and it therefore would switch back on, causing the Tortoise to be attracted again toward the mirror. "The creature therefore lingers before a mirror, flickering, twittering and jigging like a clumsy Narcissus. The behaviour of a creature thus engaged with its own reflection is quite specific, and on a purely empirical basis, if it were observed in an animal, might be accepted as evidence of some degree of self-awareness."[96] Another surprise was that the Tortoises seemed to recognize one another in much the same way as they recognized their own mirror images. They were attracted to one another's headlamps, but each of these would cause the other to switch off when they got too close together. Switching off would cause them to switch back on so that the creatures appeared to dance little jigs of mutual greeting. "Therefore, when no other attraction is presented, a number of the machines cannot escape one another; but nor can they ever consummate their 'desire.'" A population of such machines, Walter mused, would form "a sort of community, with a special code of behaviour." If an independent source of light appeared in such a situation, "the community [would] . . . break up" as the machines wandered off toward it, but would form once again as they converged upon the stimulus and blocked one another's paths.[97]

As a behaviorist, Walter interpreted the Tortoises' behaviors the way he would the behaviors of a natural animal. He and his fellow cyberneticists defined responsiveness, interaction, and sentience in terms of their outward manifestations, and set aside questions of internal mechanism or subjective experience. Thus Elmer's and Elsie's apparent recognition of themselves and of one another constituted recognition itself. Some of their behaviors had been predictable, Walter wrote, but others had been "quite unforeseen." The faculties of self-recognition and mutual recognition had been accidental, since the headlamps had been intended only to indicate when the steering mechanism was in operation. "It may be that they are only 'tricks,' but the behaviour in these modes is such that, were the models real animals, a biologist could quite legitimately claim it as evidence of true recognition of self and of others as a class."[98]

The Tortoises' various features and tendencies constituted, in Walter's eyes, "an exploratory and ethical attitude toward the universe as well as a purely tropistic one."[99] Wiener concurred enthusiastically, writing that the Tortoises exhibited "mutual reactions which would be interpreted by most animal psychologists as social behavior if they were found encased in flesh and blood instead of brass and steel." Wiener predicted that Walter's unexpected observation of his mechanical creatures' apparent social life was "the beginning of a new science of mechanical behavior."[100]

All of the Tortoises' defining features—restlessness, exploratory curiosity, self-recognition, sociability—were believable only when one disregarded the internal mechanisms that caused these appearances. One had to equate appearance with reality, to decide that something seemingly conscious, curious, restless, or sociable, was indeed conscious, curious, restless, or sociable. Walter wrote that the Tortoises' patterns of behavior illustrated a key principle in the study of animal behavior, namely that "any psychological or ecological situation" in which a feedback loop with the environment existed could "result in behaviour which will seem, at least, to suggest self-consciousness or social consciousness."[101]

Wiener too devised sentient, purposeful electromechanical creatures. He worked in collaboration with his colleague at MIT, the electrical engineer J. B. Wiesner, one of those whom Wiener had drawn into a cybernetics supper club, and who would later become the president of MIT,[102] and with Wiesner's student, Henry E. Singleton, the future founder of the electronics and aeronautics company Teledyne. Their idea was to build a "tropism machine" demonstrating what they named "voluntary feedback," that is, a machine that could choose among feedback systems constituting different tropisms. In addition to its primary tropic feedback mechanisms, such a machine would need a secondary feedback mechanism that Wiener and his collaborators called "postural feedback": one that kept the various joints and parts not directly associated with the machine's "purposive movement" at a given moment in an appropriate state of tension and readiness. The three were interested in modeling, not only the successful functioning of postural feedback in conjunction with voluntary

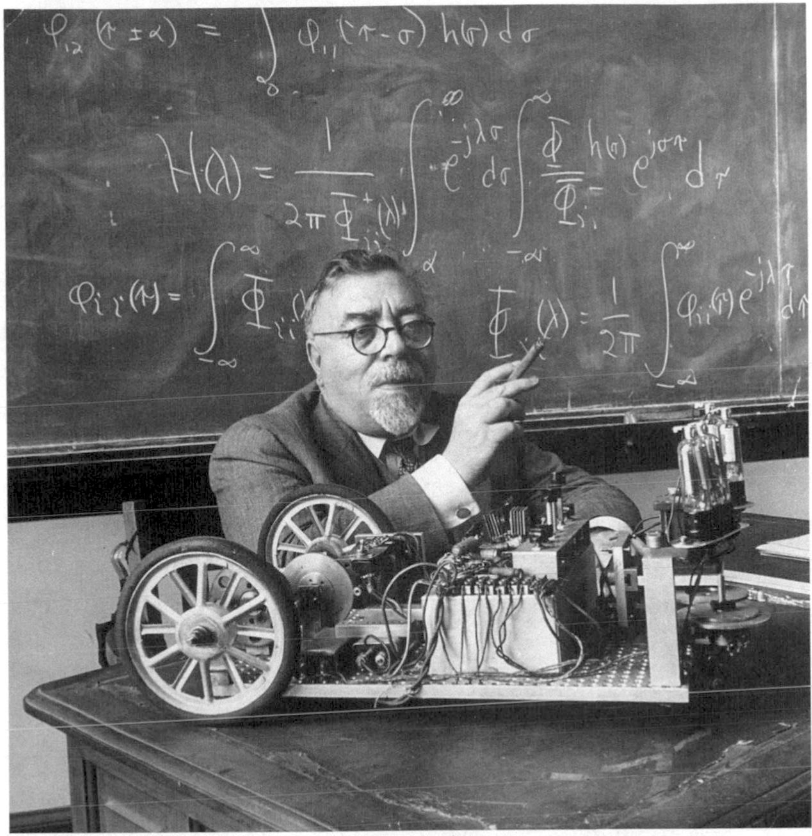

Figure 9.7 Norbert Wiener with his tropism machine, the Moth/Bedbug, courtesy MIT Museum.

feedback, but its breakdown, which they thought would mimic the tremors typical in a Parkinson's patient.[103]

The result of their labors, which Wiener pronounced "brilliant and successful," was a little three-wheeled cart equipped with a propelling motor, a rudder, two positioning motors, and two photocells (see figure 9.7). Depending on how the output from the photocells was connected with the positioning motors, and in turn to the tiller, the machine could be set up to be either positively phototropic, light-seeking, in which mode its creators named it the Moth, or negatively phototropic, light-avoiding, whereupon it became the Bedbug. In either mode, the machine tended to balance and regulate its motions so as to keep them on the correctly tropic path. Wiener called this tropic mechanism "voluntary feedback" and judged it analogous to a voluntary action in a human being, which he took to be "essentially a choice among feedbacks."[104]

The Moth/Bedbug machine could be arranged in such a way, with its adjustable amplifier set to too great an amplification, that it would oscillate more and more wildly in trying to correct itself, exhibiting what Wiener judged to be analogous to an "intention tremor" in patients with injuries to the cerebellum. Meanwhile, a second feedback system on the positioning mechanism for the rudder, which Wiener described as analogous to a "postural feedback," by which animals maintain a given posture or postural range, had a "zero point regulated by the output of the first feedback." This second feedback became overloaded and provoked a second sort of tremor "in the absence of light: that is, when the machine is not given a purpose." Wiener compared this second sort of tremor to a Parkinsonian tremor, which tended to be worse when the patient was at rest and better when the patient was engaged in a deliberate act.[105] Intentional behavior and autonomic regulation alike, in Wiener's view, originated in a hierarchy of feedback loops by which the environment acted upon the mechanical creature.

Shortly after Wiener built the Moth and Bedbug, his friend and colleague Claude Shannon contributed another species to the genus of sentient electromechanical creatures. Theseus was a life-size, wooden Mouse, with a magnet inside and copper-wire whiskers, who solved a maze (see figure 9.8). However, he had some help through the labyrinth, not from Ariadne's ball of string, but from the labyrinth itself. Theseus sought "cheese," a switch that shut off his motor, by making his way through a five-by-five labyrinth of twenty-five squares that could be reconfigured at will by moving the partitions to form different labyrinths. Once the Mouse had solved a given configuration of the maze, he had learned it, meaning that he was able to proceed directly to the bait from any spot he had previously occupied. Placed upon a spot he had never before occupied, Theseus would search for a previously occupied square and thence proceed directly to the bait, adding the new origin point to his repertoire of learned routes.[106]

However, it was not really Theseus who solved the maze, but the maze that employed Theseus as a pawn to solve itself. The "brain" was not in the Mouse's head, nor indeed in any other part of his body, but rather beneath his feet, in a relay circuit running below the floor of the labyrinth. Each square had a "memory" consisting of two relays recording one of four directions, north, south, east, or west. Each time the magnetic Mouse left a square, the square's

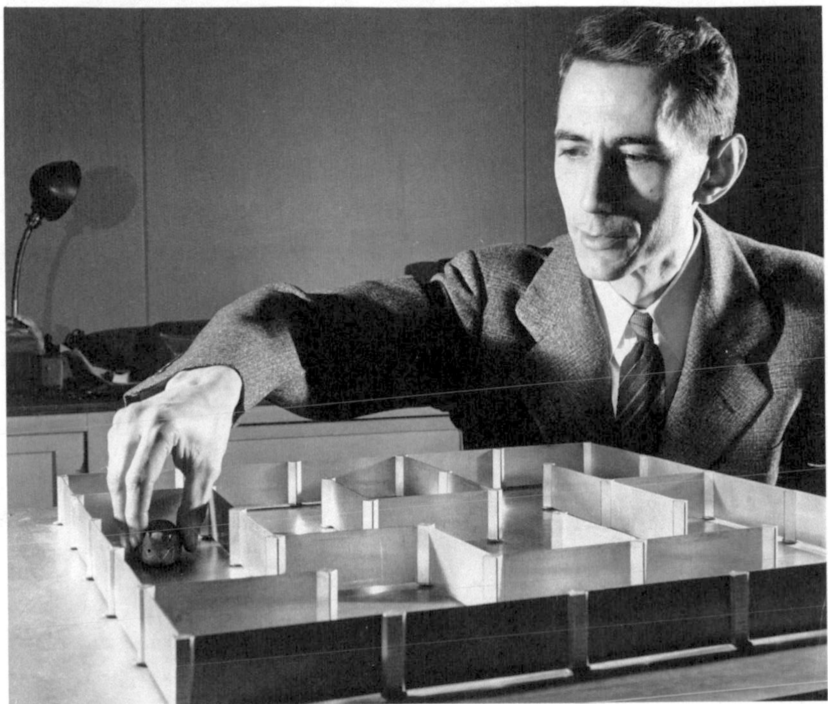

Figure 9.8 Claude Shannon with his mouse, Theseus, courtesy MIT Museum.

associated memory recorded the direction it had taken. The maze employed its memory in one of two modes: in an "exploration strategy" or a "goal strategy." Before the Mouse had found the "cheese," the machine was in "exploration" mode, which meant that each time the Mouse entered a given square, an electromagnet beneath the floor propelled his magnetic body out at a ninety-degree angle counterclockwise to the direction he had previously taken when exiting the same square. If his copper-wire whiskers brushed a barrier, the electromagnet would send him back to the center of the square and rotate him again ninety degrees.[107]

Once the Mouse had arrived at the bait a first time, the maze entered its "goal strategy," which meant that whenever Theseus returned to a square he had already occupied, he would leave in the direction he had last taken. The machine also had a "forgetting" feature: as long as it was in "goal strategy" mode, it kept track of how many moves it made, and if these exceeded a certain maximum number, the maze would "decide" that the bait must have been moved or the maze otherwise reconfigured, and would revert to its "exploration strategy" mode.[108]

Yet when Shannon presented his maze-solving machine, he tended to locate the seat of agency and motion in the Mouse itself. Even at a professional conference, a sufficiently serious setting to warrant replacing the cute Mouse with what he called a "sensing finger," he explained the machine by attributing its actions to the finger. This was in March 1952, at the eighth of ten Macy Conferences, a series of annual meetings on cybernetics in New York City funded by the Josiah Macy Jr. Foundation. As Shannon switched on his machine, he told his audience, "You see now the finger exploring the maze, hunting for the goal." After the finger had arrived at the goal, Shannon said, "if I now move the finger to a part of the maze that it has not explored, it will fumble around until it reaches a known region."[109]

Next, he demonstrated what happened when he changed the partitions to reconfigure the maze: the finger entered a "vicious circle" because its remembered solutions were no longer appropriate. "A neurosis," commented an audience member, the neurophysiologist Ralph Gerard. "Yes," Shannon agreed. "It can't do this when its mind is blank," intervened the mathematician Leonard Savage, "but it can do it after it has been conditioned?" Shannon replied, "Yes, only after it has been conditioned. However, the machine has an antineurotic circuit built in to prevent just this sort of situation" (he meant the reversion to "exploration strategy" mode after a certain number of moves).[110]

Had Shannon described the whole machine, including the maze, as the simulated intelligence, it would no longer have seemed to model a sentient creature engaging with an environment. It would simply have been a machine going through a sequence of configurations. By focusing attention on the Mouse (or finger), Shannon shaped how people understood the simulation, including himself. He and his viewers could regard the moving entity as having a "mind" and even possibly suffering from a "neurosis." Yet the whole "mind"—decisions, strategies, neurosis, anti-neurotic correction, all of it—was in the Mouse's environment, its maze, and not in its quite empty self.

A final example of the essential passivity of cybernetic creatures is Job, the electronic Fox presented by the French science journalist Albert Ducrocq the year after Shannon's introduction of Theseus. Job, now at the Centre national des arts et métiers in Paris, is a device covered in fox hide and has five senses: two photoelectric cells for eyes, a microphone connected with an amplifier for an ear, touch-sensitive electrical contacts, a potentiometer in the neck that registers the position of the head, and even a nose, an electrical capacitor at the front of the head that detects metal and other conducting objects at a distance. Ducrocq styled the machine a "fox" because he judged it cleverer than earlier electronic creatures, especially Walter's Tortoises: unlike these, it had a memory allowing it to construct an evolving graph of its surroundings.[111]

Nevertheless, apart from some basic, regular motions, such as a slow pivoting of the head through 180 degrees, all the action originated in the Fox's environment: "We have constituted, to confront this electronic fox, a micro-universe whose fundamental actions produce the excitations [of the sense organs]," Ducrocq explained. These excitations connected in turn to a servo-network (a network that transforms in response to feedback), and thereby constituted the evolving graph of the Fox's limited world.[112] Even the wily Fox was an instrument upon which its surroundings inscribed their own image.

Outside In

In November 1946, Ross Ashby received a cordial letter from the mathematician Alan Turing,[113] then working in the Mathematics Division of the National Physical Laboratory in London, and soon to be a member, along with Ashby and Walter, of the Ratio Club, a cybernetic supper club that met between 1949 and 1958 in London. Turing's friend the cryptographer Donald Michie later described him as a kind of "Robinson Crusoe" who tackled every problem, from his pollen allergies to theorems in probability theory, as though he were alone with it on a desert island.[114] Yet the Crusoe of early electronic computing did share certain core assumptions with Ashby and the other cyberneticists. Turing had spent the war years at the Government Code and Cypher School at Bletchley Park in Buckinghamshire, developing an electromechanical code-breaking machine, the Bombe. Since the end of the war, he had been designing the ACE, or Automatic Computing Engine, whose name was a nod to Charles Babbage's Analytical and Difference Engines of the 1830s. The ACE was to be a "universal machine" of the sort that Turing had described in his landmark paper "On Computable Numbers," in which he set out the theoretical basis for computer science.[115]

Turing had seen a letter from Ashby to Sir Charles Darwin, grandson of the naturalist and director of the National Physical Laboratory, describing his plans for the Homeostat and inquiring about the ACE. In the letter, Ashby speculated that the ACE would be incapable of regulating and correcting its own functioning the way Ashby intended the Homeostat to do.[116] But Turing assured Ashby that the ACE could in fact be configured so as to "try out variations of behaviour and accept or reject them in the manner you describe." Moreover, Turing himself had been "hoping to make the machine do this," since he was "more interested in the possibility of producing models of the action of the brain than in the practical applications to computing." He thought that it was possible to configure the ACE to regulate its functioning without altering its design or construction because the ACE was a "universal machine" capable of modeling any other machine, provided it had the right instructions. Thus, just as an actual brain could change physically, remaking its "neuron circuits by the growth of axons and dendrites," the ACE could transform itself equivalently by remembering different data. Turing ended his letter by urging Ashby to consider using the ACE to model the Homeostat "instead of building a special machine. I should be very glad to help you over this."[117]

Ashby did not accept Turing's offer, being keen precisely to build his "special machine." Historians of computing have recently emphasized the contrast between Ashby's machine as a quivering, physical entity and Turing's as a primarily, though not uniquely, mathematical "universal machine."[118] Turing himself occasionally referred to his universal computer as a "paper machine," since it could be modeled by a person following a set of rules,[119] while Ashby presented his Homeostat as a partially sentient, material being whose movements of bodily parts constituted the very focus of his experimental program.

Nevertheless, Ashby's and Turing's approaches to building a thinking machine shared several defining features. First, Turing, like Ashby, maintained that a thinking machine must acquire its ability to think through a more basic capacity to learn. A machine, like a human being, could not exhibit intelligence without being educated; Turing repeated in every one of his writings on machine intelligence this crucial point,[120] whose importance was such that, in the summer of 1947, he applied to Darwin for a one-year sabbatical from the National Physical Laboratory to go to Cambridge and work on the question, as Darwin described it, of whether "a machine [could] be made that could learn by experience," a kind of "theoretical work . . . better done away from here."[121]

Instead of trying to build an adult intelligence directly, Turing urged, it made better sense to try to simulate a child's mind.[122] He imagined building, to begin, what he called an "unorganised machine," that is, a bunch of neuron-like components connected "in a relatively unsystematic way" through "connection modifiers," their configuration subject to some random variation. He thus laid out a blueprint for an approach to artificial intelligence that a decade and a half later would be called "connectionism."[123] The machine would become organized by means of an education, consisting of systematic "interference" that altered, fixed, or disrupted the connections among the components.[124]

Turing imagined two sorts of "interfering" input, "one for 'pleasure' or 'reward' (R) and the other for 'pain' or 'punishment' (P)." Pleasure interference would work to fix in place the current configuration of the machine while pain interference would disrupt it, causing previously fixed features to change or to vary randomly. You might then simply allow the machine "to wander at random through a sequence of situations, applying pain stimuli when the wrong choice is made, pleasure stimuli when the right one is made," acting as the machine's trainer in just the way that Ashby described training the Homeostat. One could even program the "teaching policies" right into the machine, Turing suggested, so that it would learn all on its own and direct its own process of organization.[125]

Thus, the first principle Turing shared with Ashby and other cyberneticists, notably Wiener, was the conviction that life and mind were crucially diachronic mechanical phenomena, developing over time. As Wiener expressed it, the time of cybernetics was not that of classical physics, in which events were causally linked and "reversible," but instead that of thermodynamics, in which events were statistically correlated and irreversible.[126] Second, however, Turing, again like Ashby, reached for a form of development in time, of learning, that required no actual self, no inner agent. He too drew upon behaviorist principles to describe how learning might be directed from the outside in.

In order to learn, a machine must be fallible: embracing fallibility would crack the problem of machine intelligence. "If a machine is expected to be infallible," Turing wrote in 1947, early in his work on the ACE, "it cannot also be intelligent." As Turing saw it, the Austrian logician, mathematician, and philosopher Kurt Gödel's incompleteness theorem entailed this point.[127] The theorem states that any axiomatic proof system sufficient to generate elementary arithmetic will also be capable of producing sentences that are neither provable nor disprovable within the system, self-referential statements such as "this sentence cannot be proven."[128] Gödel's theorem, Turing reasoned, meant that if one wanted to build a machine to determine the truth or falsity of mathematical theorems, any machine would in some cases be unable to give an answer, if one refused to tolerate an occasional mistake.[129]

But this intolerance, Turing argued, was arbitrary and even misguided. Infallibility was no requirement for human intelligence. On the contrary, one could make very intelligent and interesting mistakes; indeed, human intelligence developed precisely through such mistakes. Gödel's theorem, along with related theorems, had nothing to say about how intelligent a machine might be if it made no pretense to being infallible.[130]

How, in time, might a fallible machine achieve intelligence? According to Turing, in addition to discipline, in the form of an ability to carry out instructions, an intelligent entity must have "initiative." As a model of initiative in intellectual activity, Turing proposed the ability to conduct "various kinds of search." He identified three kinds. First came "intellectual" searches, defined as "searches carried out by brains for combinations [of variables] with particular properties," in which the seeker systematically tries out possible combinations in a given order. A second kind of initiative was the "genetical or evolutionary search by which a combination of genes is looked for, the criterion being survival value" (others, notably the Michigan psychologist and computer scientist John Holland, would develop this idea into the programming technique of genetic algorithms).[131] Finally, there was the "cultural" search, which was "carried out by the human community as a whole, rather than by individuals."[132]

"Initiative" suggests agency, the ability to act spontaneously, independently, and purposefully. Turing's interpretation of initiative as the ability to conduct searches seems, at first glance, to require just these abilities. Yet each of Turing's three definitions uses the passive voice: searches were "carried out by" brains or by the human community, and a combination "[was] looked for." Here once again is an account of agency in the passive voice, and again, the grammar is symptomatic of a deeper philosophical stance.

Like the cyberneticists' tropic and homeostatic creatures, Turing's thinking computer, even in its hypothetical form, was meant to appear intelligent from the outside, but its apparent intelligence would not be founded on any intrinsic quality of will or innate agency. The machine would search simply by trying out possibilities, partly systematically, according to a rule or set of rules, and partly at random. Turing compared a learning process having a random element to the evolutionary process that created human intelligence.[133] Searching at least partly at random, the machine would be corrected by "interference" from outside or by a systematic, internalized version of such interference. A machine with the ability to conduct searches, whether systematic or random, and capable of changing as a result of "interference," might, Turing thought, arrive at something that looked like human intelligence.

Turing's machine's "initiative," then, did not ultimately involve actual initiative, in the form of autonomous purposeful behavior, but rather the appearance of initiative. Turing addressed the question himself in January 1952 in a panel discussion on the BBC on the subject of machine intelligence. Turing's interlocutor was his friend and former teacher the mathematician and cryptologist Max Newman. The moderator was Richard Braithwaite, lecturer in moral science at Cambridge and, like Turing, a fellow of King's College.

At points in the conversation, Braithwaite returned to his own view that, in order to learn from experience, a machine would need "springs of action," something like "appetites" or "interests," to enable it to pay attention to the relevant factors in its environment. Newman too described the essence of human thinking in active terms as the "sudden pounce on an idea." But Turing responded to Braithwaite that he thought a machine, even without appetites or interests, could try out combinations "more or less at random" and then "receive marks for various merits." It would thus evolve intelligence the way neo-Darwinian evolutionists assumed nature had done, essentially at random, corrected from without.[134]

Of course, the machine would have to "appear to behave as if it had free will" in order to imitate a human brain convincingly, and Turing proposed two ways one might accomplish this. The first was reminiscent of Babbage's parlor trick in which he set his calculating engine to print a patterned sequence of numbers with occasional discontinuous jumps. Turing's idea was that a random element in the machine's behavior —"something like a roulette wheel or a supply of radium"—might give it the appearance of acting willfully. Here he seemed to conflate the appearance of acting willfully with that of acting arbitrarily.[135]

Anyway, Turing did not show much interest in the possibility of including a random element. He preferred to base the appearance of autonomy and free will in the observer's ignorance, both of what was going on in the machine, and of the set of consequences of any set of facts or principles. Turing invoked this ignorance in response to something that the mathematician Ada Lovelace had written with reference to Babbage's Analytical Engine. The engine, she had observed, had "no pretensions whatever to originate anything. It can do whatever we know how to order it to perform." Turing pointed out this assumed that "when we give it its orders, we know what we are doing, what the consequences of those orders are going to be,"[136] an example of "a fallacy to which philosophers and mathematicians are particularly subject."

> This is the assumption that as soon as a fact is presented to a mind all consequences of that fact spring into the mind simultaneously with it. It is a very

useful assumption under many circumstances, but one too easily forgets that it is false. A natural consequence of doing so is that one assumes that there is no virtue in the mere working out of consequences from data and general principles.[137]

Turing pointed out that one can plant a seed effectively without understanding the mechanism of germination. Similarly, one could program a machine to do interesting, unanticipated things, in which event "I should be inclined to say that the machine *had* originated something."[138]

In other words, intelligence was in the eye of the beholder, unless it was in the private experience of the intelligent being, where it was inaccessible to science. Pressed to define thinking itself, as opposed to its outward appearance, Turing reckoned he could not say much more than that it was "a sort of buzzing that went on inside my head."[139] Ultimately, the only way to be sure that a machine could think was "to *be* the machine and to feel oneself thinking."[140] But that way lay solipsism, not science. From the outside, Turing argued, a thing could look intelligent as long as one had not yet found out all its rules of behavior.[141]

Accordingly, for a machine to seem intelligent, at least some details of its internal workings must remain unknown. Turing thus carried behaviorism a step beyond the behaviorists themselves.[142] The behaviorists had argued that it was a mistake to try to study the inner workings of consciousness and intelligence directly, rather than through their behavioral effects, because these inner workings were not directly observable. Turing argued that a science of the inner workings of intelligence was not only methodologically problematic but also essentially paradoxical, since any appearance of intelligence would evaporate in the face of such an account.[143]

Newman concurred, drawing an analogy to the beautiful ancient mosaics of Ravenna. If you scrutinized these closely, you might be inclined to say, "Why, they aren't really pictures at all, but just a lot of little coloured stones with cement in between." Intelligent thought could similarly be a mosaic of simple operations that, when studied up close, disappeared into its mechanical parts.[144] Ashby would later write, similarly, "What I am saying is that . . . 'real' intelligence does not exist. It is a myth. It has come into existence in the same way that the idea of 'real' magic comes to a child who sees conjuring tricks."[145]

The necessity of measuring intelligence from outside and not inside was the crucial point of Turing's landmark essay "Computing Machinery and Intelligence" (1950). The essay famously proposed a method for deciding whether a machine was intelligent or not, since known as the Turing Test. If the machine in question could beat a human being in the "Imitation Game" by persuading a human interviewer that it, the machine, was the true human being and its human opponent the computer, then one must grant the machine the courtesy of considering it intelligent. Of course, the interviewer can have no direct contact with either player, each of which (whom?) ought to remain in a separate room from the interrogator, with all communication taking place by means of typewritten messages. The conversation should be entirely unrestricted, encompassing anything the interlocutor cares to discuss: arithmetic, chess, poetry, Dickens, Christmas.[146]

Turing specified that a machine must be admissible for the test even if its engineers could not fully describe how it worked "because they have applied a method which is largely experimental." He imagined the successful machine as a "child" machine that would become intelligent by being taught, and he emphasized that even the "teacher" of the machine would "often be very largely ignorant of quite what is going on inside." The triumphant machine's designers themselves would remain at least partially in the dark regarding its inner workings.[147] This was necessarily the case, since to be able to describe the machine's inner workings completely would be to lose the sense that it was intelligent. But then, the same would be true in the case of a man: if one could attribute his behavior entirely to a set of explicit rules, Turing argued, one would come to see him as a machine and, by implication, without intelligence.[148]

Turing believed that such a set of rules existed. In response to the view that humans could not be machines because there existed no "complete set of rules of conduct by which a man regulates his life," Turing removed the assumption of individual agency and restated the problem. There was no complete list of "precepts such as 'Stop if you see red lights,' on which one can act, and of which one can be conscious." On the other hand, Turing thought it likely that there existed a complete set of "laws of behaviour," by which he meant "laws of nature as applied to a man's body such as 'if you pinch him he will squeak.'" Humans did not regulate their lives by rules of conduct. Instead they were regulated by laws of behavior: they were the objects, not the agents, of regulation.[149]

Several current fields of research emerged from these mid-twentieth-century projects: cognitive science, artificial intelligence, robotics, artificial life. Historians and practitioners of these fields generally describe their history as shaped by a struggle between two competing approaches. The first, “cognitivism,” or “computationalism,” construed thought as information processing: the rule-governed manipulation of symbolic, representational structures. Turing’s work provides the defining paradigm for this first approach.[150] The second approach, still according to the consensus view, is “embodied.” In contrast with the cognitive-computationalist tradition, it does not construe thought as abstract reasoning, but as a form of physical engagement with the world. Early cyberneticists such as Ashby and Walter typify this second, embodied, non-cognitivist tradition.[151]

But this dichotomy hides deeper commonalities. Turing’s disembodied, cognitivist program included his “P-machine,” trainable through “pleasure and pain,” on the principle that an intelligent machine must learn from experience.[152] Ashby’s active, “embodied” program employed the same abstract notion of “pleasure” and “pain” as Turing: the positions of certain levers. The “cognitivist” and “embodied” programs shared the axiom that intelligence must be an epiphenomenon of pleasure and pain, which were in turn mechanisms by which the environment directed the mechanical creature. Fundamentally, they agreed upon the essential passivity of the living, intelligent being.

Both Turing and the cyberneticists described intelligence as a diachronic phenomenon, occurring in time through learning. Both appealed to Darwinian evolution as the guiding model for how intelligence can emerge. Both described their Darwin-inspired learning processes, in keeping with neo-Darwinist principles, as fundamentally passive: drawn from outside rather than driven from within. The founders of cybernetics and artificial intelligence proposed ways to use artificial mechanisms to study what makes an intelligent being intelligent and what makes a living being lifelike. They identified agency (purposeful behavior, initiative) as the defining feature of both life and intelligence.

In their physical experiments, thought experiments, and interpretations, they understood agency in behaviorist terms, as a secondary appearance rather than a primary reality. They flirted with the intelligent egg but they married the dead sparrow.

~ 10 ~

History Matters

> If we allow ourselves the license of talking about genes as if they had conscious aims, always reassuring ourselves that we could translate our sloppy language back into respectable terms if we wanted to, we can ask the question, what is a single selfish gene trying to do? It is trying to get more numerous in the gene pool.
>
> —Richard Dawkins, *The Selfish Gene* (1976)

This book has traced the history of a paradox at the heart of modern science, a paradox of particular significance for scientific accounts of life and mind. The paradox originated in the seventeenth century, with the emergence of modern science, in its mechanical clockwork model of nature. This model banished from nature all purpose, sentience, and agency, leaving behind a brute mechanical world that was fully intelligible without reference to mysterious forces or agencies. But, just as a watch implies a watchmaker, this passive mechanical artifact world relied upon a supernatural, divine intelligence. It was inseparably and in equal parts a scientific and a theological model.

These chapters have also traced the parallel development of a competing form of science that naturalized rather than exported purpose and agency. This alternative, active-mechanist tradition, though overshadowed by the brute-mechanist one, developed in ongoing dialectic with it. This final chapter shows that the history of these competing forms of science is profoundly relevant to current debates in artificial intelligence, cognitive science, and evolutionary biology.[1] To read current scientific debates by the light of history is to recognize a deep-seated principle, so elemental as to be essentially invisible, which holds considerable sway even among those who work to question it: that agency cannot be a primitive, elemental feature of the natural world; that the idea of natural forms of agency has no place in legitimate science; and that scientific knowledge is therefore definitively distinct from humanistic, particularly historical, understanding, whose objects include conscious agents.

This ban on agency in scientific explanation precludes rudimentary material tendencies, such as the Lamarckian idea of a tendency in living matter toward greater complexity. Current evolutionary biology prohibits such primitive tendencies. The only spontaneous action on the part of living forms to which it assigns an explicitly causal role is a capacity for random changes. The ban on agency in science also applies to higher, more complex forms of agency, such as the behaviors of conscious animals. These behaviors are often doubly inert in current biology: reducible to brute-mechanical causes and, since their effects can never be inherited according to neo-Darwinist theory, having no evolutionary consequences of their own.

The same antihumanist, antihistorical turn at the end of the nineteenth century that banned historical forms of explanation from natural science, eradicating not only Lamarckism but even key aspects of Darwin's own theory from neo-Darwinist science, also produced a culture of indifference among scientists to the histories of their own disciplines. But history matters to science; historical knowledge is integral and essential to scientific knowledge. Most evolutionary biologists, cognitive scientists, and artificial intelligence researchers exhibit only a cursory interest in the long histories of their fields. There is in the scientific community a widespread assumption that historical knowledge is not only fundamentally distinct from scientific knowledge, but also less sure, less real, less solid. This assumption constricts the domain of scientific explanation. In evolutionary biology, the antihistorical bias has narrowed the science on two levels: it has contributed to the ruling out of a mode of explanation, the active-mechanist historical mode of the Lamarckian-Darwinians, and it has also largely blinded scientists to the complexity and the stakes of their own intellectual heritage.

By recognizing the historical roots of the almost unanimous conviction (in principle if not in practice) that science must not attribute any kind of agency to natural phenomena, and recuperating the historical presence of a competing tradition, we can measure the limits of current scientific discussion, and possibly, think beyond them.

Armies of Idiots

For the last thirty years or so, the interdisciplinary field of cognitive science, which brings together people from various fields interested in studying how brains work, has been divided into two camps: the "embodiment" camp and the "representation" camp. Both sides build, use, and otherwise invoke machines in support of their arguments. Those in the embodiment camp have assumed the rebellious position. Beginning around 1980, they rebelled against the mix of neuroscientists, linguists, philosophers, and psychologists who tried to understand the way brains worked in terms of linguistic and perceptual representations of the world.

"Elephants Don't Play Chess" is the title of a manifesto published in 1990 by the MIT roboticist Rodney Brooks.[2] Brooks warned that the field of artificial intelligence was foundering, overtaken by Cartesian rationalism in the form of the "*symbol system hypothesis*," namely the assumption that intelligence consisted of the rule-governed manipulation of symbols constituting representations of the outside world. Artificial intelligence had assumed, in other words, that intelligence resided in "*thought* and *reason*."[3] Chess-playing provided the paradigmatic intelligent act for the symbol-system hypothesis. The rationalist, symbolic approach to artificial intelligence had reached the limit of what it could achieve because it overlooked much if not most of what constitutes intelligence. It would be "unfair," for example, "to claim that an elephant has no intelligence worth studying just because it does not play chess."[4]

In evolutionary terms, chess-playing came comparatively quickly on the heels of what Brooks takes to have been the hard part, the core of intelligence, simply "the essence of being and reacting." He therefore inaugurated a "nouvelle AI" that, he says, works by embodiment rather than abstraction, forgoing the rational manipulation of symbolic representations of the world in favor of direct, physical engagements with the world itself. The intelligence of the resulting artificial creatures is more like that of an elephant than a chess master. It is the opposite of abstract: it is "grounded" and "embedded" in the creature's body and environment. All of the creature's "goals and desires" must take the form of "physical action," and it must "extract all its knowledge from physical sensors."[5]

Instead of decomposing intelligence into "information processing modules," Brooks and his fellow researchers in nouvelle AI work to evolve intelligence as an emergent result of assemblages of "behavior generating modules."[6] For this reason, Brooks's nouvelle AI also goes by the name of "behavioral" or "evolutionary" robotics. He and his colleagues call the design scheme by which they assemble their behavior-generating modules "subsumption architecture," because discrete behaviors are subsumed together, with no highest-level behavior to oversee and govern all the others. The behaviors, Brooks says, are "abstraction barriers": each consists of a direct physical engagement with the outside environment. By assembling such behaviors, the subsumption architecture "tightly connect[s] perception to action, embedding robots concretely in the world."[7]

A crucial feature of Brooks's robots is that they have no robot equivalent of a Cartesian, unified, knowing self, no central repository of information, no overall governing module.[8] Their intelligence is without a discrete locus: "There is no homunculus."[9] This use of "homunculus" might be confusing: Brooks does not mean the tiny, preformed, mechanical embryo of seventeenth-century embryology, but rather something like its intellectual counterpart, the rational soul. Brooks's creatures have no rational soul. Their behaviors do not "know about" one another. Brooks's account of "Herbert" exemplifies this lack of a knowing subject. Herbert was a robot that wandered around the lab, going into people's offices and stealing empty soda cans. It avoided obstacles, followed the walls of corridors and rooms, recognized soda cans and picked them up. "The remarkable thing about Herbert," Brooks writes, "is that there was absolutely no internal communication between any of its behavior generating modules." Each behavior module connected with sensors, and through these, with the physical world of offices and soda cans. The soda-can-finding behavior drove the robot toward soda cans but "did not tell the arm controller that there was now a soda can ready to be picked up." Instead, the arm behavior responded to the wheels: when these stopped, it moved the hand in search of a soda can in front of it.[10]

The hand in turn had a "grasp reflex" causing it to close in response to something breaking the infrared beam between its fingers. No single module contained instructions for the overall project. This arrangement allowed for flexibility and opportunism. Brooks explains that one could hand Herbert a soda can that it had not been seeking and it would grasp it. This makes Herbert a successful design for a soda-can-stealing robot. Is it also a persuasive model of natural intelligence? In Brooks's view, Herbert implied that natural intelligence too could exist without a unified, knowing subject. "It is very easy," Brooks writes, "for an observer of a system to attribute more complex internal structure than really exists. Herbert appeared to be doing things like path planning and map building, even though it was not."[11]

But did Herbert truly lack a unified, knowing subject? The answer must depend upon what one means by a unified, knowing subject. The robot's autonomous behaviors all connected with an "arbitration network" on the output side, which drove the actuators that allowed Herbert to act on the information its sensors brought it.[12] Moreover, a unified program in the minds of the authors of Herbert clearly contained the project of stealing soda cans from start to finish. To argue that apparently purposeful behavior emerges from a welter of purposeless movements, would one not need to see apparent purposefulness emerge from actual purposelessness rather than, as in the case of Herbert, be assembled from apparently disconnected parts? In other words, is it not the purposelessness that is apparent here, rather than the purposefulness?

Likewise Genghis, a walking robot able to navigate rough terrain, had no single control system, no "central representation" or "repository which modeled the robot's configuration" in space. Brooks attributed its "robustness" in the face of bumps and potholes to the "distributed" nature of the control.[13] The same was true for Toto, a robot that traveled to a designated spot on command without having a traditional "map," a static representation of a given domain. Instead, Toto used a map composed of "nodes" each of which was a "behavior." These behaviors activated themselves as Toto moved about a landscape and encountered landmarks.

"Nodes become more active if they believe they correspond to the place at which the robot is currently located. Thus the robot has both a map, and a sense of where it is on the map, but a totally distributed computational model,"[14] that is, no unified self. Brooks does not dwell on what it might mean for a behavioral "node" to have a "belief" about its location in the world. But he argues that Toto did not represent anything "circumscribed in its head" and yet acted as one imagines it would if it did, thereby constituting "the nail in the coffin of traditional representationalism."[15]

Each of his robots, according to Brooks, looks as though it were acting upon a centralized set of purposes, but this is only an illusion. In the creatures, as in nature, "intelligence is in the eye of the observer."[16] The robots are manifestations of this idea, the guiding principle of Brooks's research program, which "turns the old approach to intelligence upside down. . . . It is only an external observer that has anything to do with cognition, by way of attributing cognitive abilities to a system that works well in the world but has no explicit place where cognition is done. . . . Cognition is only in the eye of the observer."[17] This is a frequent refrain in Brooks's writing: "It is only the observer of the Creature who imputes a central representation or central control. The creature itself has none; it is a collection of competing behaviors. Out of the local chaos of interactions there emerges, in the eye of the observer, a coherent pattern of behavior."[18] In his conviction that to be a purely material creature is to have no unified, rational self, Brooks is an orthodox Cartesian despite himself. Where he departs from Descartes is in his belief that there is no such thing as a unified, rational self.

Not only are the robots chaotic, according to their author, they are also essentially passive. The source of their actions is external: they do not move themselves through the environment, but are drawn through it: "To a large extent the state of the world determines the action of the Creature. . . . There is no central locus of control. Rather, the finite state machines are data-driven by the messages they receive."[19] The robots' passive, decentralized mode of action, Brooks acknowledges, seems like "the antithesis of thought," but he argues that it is in fact the very essence of thought. "Real biological systems are not rational agents that take inputs, compute logically and produce outputs. They are a mess of many mechanisms working in various ways, out of which emerges the behavior that we observe and rationalize."

At the heart of Brooks's philosophy is the notion of emergence: an agencyless mode of coming-to-be (about which more presently). "My feeling is that thought and consciousness are epiphenomena of the process of being in the world." The more complicated the world Brooks's creatures inhabit, the more complicated their actions as they are drawn through it, so that their actions will appear increasingly intelligent, thoughtful, and conscious. "Thought and consciousness will not need to be programmed in. They will emerge."[20]

Thought and consciousness are to emerge especially in interactions with an environment that includes other thoughtful and conscious beings. "Cog" was a humanoid robot consisting of a torso, head and neck, arms and hands (see plate 9). It had sensory systems that included vision (video cameras), hearing (microphones), touch (resistive force sensors), and proprioceptive systems to orient the head and maintain balance. Cog, when fitted out with a particular head system called "Kismet," also had a cartoonish face for projecting emotions including anger, fatigue, fear, disgust, excitement, happiness, interest, sadness, and surprise. These expressions corresponded to systems in the robot that Brooks designated as "drives," "behaviors," "emotions," and "states."

For example, as long as the robot's "drives" remained within a given "homeostatic" range, the robot projected "interest" and "happiness." If any one of its drives left the range—for example, if it was under-stimulated for a time—it would begin to express a distressed emotion such as "sadness." Brooks and his collaborators played the role of "caretakers" to Cog in a process modeled on the interaction of adult humans with babies. As "caretakers," they taught Cog to do rudimentary things such as pointing to a visual target and imitating head nods.[21] However, despite these successes, Brooks reported that, as one might expect in a creature designed to avoid a unified self, Cog lacked "coherence": he had no way to choose among his various behaviors such as, for example, turning to look at a face or reaching to grasp an object. "This problem is multiplied with the square of the number of basic behaviors available to Cog, so the problem grows rapidly."[22]

Problems of selfhood and agency haunt Brooks's descriptions of his machines, which begin in the passive voice but lapse regularly into the active, particularly when he is describing the "higher" layers of the subsumption architecture. In Allen, for example, an exploring and obstacle-avoiding robot, "every ten seconds or so, a desire to head in a random direction *would be generated.*" However, a higher layer of Allen's subsumption architecture "made the robot *look* (with its sonars) for distant places and *try to head* towards them." The desires of this higher layer "*suppressed* the direction *desired by* the wander layer." Because of Allen's obstacle-avoidance behavior, the "upper layer had *to watch* what happened in the world, through odometry, in order *to understand* what was really happening in the control layers, and *send down* correction signals."[23] This account of Allen's functioning implies that his "higher" behaviors constitute a relatively centralized, coordinated, and active form of intelligence.

But Brooks and his fellow proponents of a behavioral, evolutionary, "embodied" approach to artificial life and intelligence have more in common with their computationalist, representationalist counterparts in classical AI than either side seems to recognize. They share the Cartesian conviction that to be a material being is to have no inner rational agent, no unified self, which informs their anti-Cartesian conviction that there is therefore no such thing as a unified self. "There exists, inside your brain, a society of different minds," wrote Marvin Minsky, author of several landmark works in artificial intelligence and of the first neural network simulator. These can work together like "members of a family," but they can also be as mutually ignorant as "people whose apartments share opposite sides of the same walls. Like tenants in a rooming house, the processes that share your brain need not share one another's mental lives."[24]

Elsewhere Minsky used a different metaphor, describing the mind as "a sort of tangled up bureaucracy" in which all the agencies remain essentially mutually mysterious.[25] He described a person drinking a cup of tea precisely the way Brooks describes Herbert stealing a soda can: "Your GRASPING agents want to keep hold of the cup. Your BALANCING agents want to keep the tea from spilling out. Your THIRST agents want you to drink the tea. Your MOVING agents want to get the cup to your lips."[26] You are brimming with "agents" and for that very reason, utterly lacking in agency.

There really is "no proper self"[27] at all, concurs Daniel Dennett, a philosopher of mind and biology and a cognitive scientist at Tufts University who has a foot in both communities, classical artificial intelligence and behavioral robotics. Dennett was an early defender of the idea that one could understand human cognition by modeling it using digital computers.[28] Later, he reconsidered and stipulated that any artificially intelligent creature would need "eyes, hands, ears and a history,"[29] coining the phrase "High Church Computationalism"[30] to criticize classical artificial intelligence in its orthodox, disembodied guise. But he retained the view that computational models cast light on "the deep theoretical question of how the mind is organized."[31]

Regarding the nonexistence of the self, Dennett agrees with colleagues on both sides of the divide, from Minsky to Brooks. He endorses Minsky's view that a person is "nothing" beyond the "various subagencies and processes in [the] nervous system."[32] Those who doubt it are just "refus[ing] to suspend their intuitive judgements." If you are one of those who find it "hard to imagine how it can be right to talk of choices made without a chooser, disapproval without a disapprover, even thoughts occurring without a thinker," well then you should just "pause to consider the possibility that this barely conceivable step might be a breakthrough, not a mistake."[33]

Descartes's rational self is thus as roundly banished from current artificial intelligence and cognitive science as from behavioral robotics. Classical artificial intelligence defenders, computationalists such as the Harvard cognitive scientist Steven Pinker, dismiss what they take to have been the traditional view of the mind as a "little man" inside the brain who reads and makes sense of its representations of the world. This "homunculus," as Pinker calls it, meaning a rational agent rather than a miniature material embryo, has given way to legions of homunculi. In computers they are known as "agents," "demons," "supervisors," "monitors," "interpreters," and "executives," and Pinker assumes that these all have counterparts in the human brain. The armies of sub-homunculi count as legitimate elements of a scientific explanation rather than as magical nonsense, according to Pinker and likeminded cognitive scientists, because unlike the old homunculi, these new sub-homunculi are limited and obtuse.[34] They are "ignorant, narrow-minded, blind." The smallest, lowest-level homunculi are "so stupid that they can be, as one says, 'replaced by a machine.' One *discharges* fancy homunculi from one's scheme by organizing armies of idiots to do the work."[35]

According to the main advocates of classical artificial intelligence, intelligence itself is a myth, a mistaken impression. Minsky endorses Turing's view that intelligence is a figment of the observer's ignorance, as does Ray Kurzweil, who has served in the last couple decades as the discipline's chief propagandist.[36] "Intelligence," Minsky observes, denotes "the momentary horizon of our ignorance about how minds work."[37] This is because the very smallest, lowest-level agents of thought are unintelligent and indeed "cannot think at all." Arriving, in one's reduction of mind, at the level of these dumb, foundational agents, one will have lost all sense of intelligence as a feature of mind.[38]

Classical AI researchers, like the old cyberneticists and their present-day heirs, the behavioral roboticists, also argue the sources of living and intelligent action lie outside the sentient creature in its environment. To explain how one can speak scientifically about purposes and other mental entities, Dennett devised what he calls the "intentional stance."[39] To adopt the intentional stance is to explain the behavior of an entity, whether living or artificial, "by relying on ascriptions to the system of beliefs and desires (and hopes, fears, intentions, hunches, . . .)." In so doing, one must keep firmly in mind that the entity in question cannot be intentional in itself, but only "in relation to the strategies of someone who is trying to explain and predict its behavior." In other words, if one wants to understand the "purely physical system" that is a living creature, one might well "find it convenient, explanatory, pragmatically necessary for prediction, to treat it as if it had beliefs and desires and was rational."[40]

In justifying the ascription of desires and purposes to a purely physical system, Dennett reasoned that we regularly talk about computers as though they have understandings and goals of various kinds. So "at least one sort of purely physical object" has intentions, though only in the sense that people regularly ascribe intentions to it. A computer "can only be said to be believing, remembering, pursuing goals, etc., relative to the particular interpretation put on its motions by people." If computers are intentional, they are so only in virtue of their creators' intentions. Dennett acknowledges that people, unlike computers, have no such intentional creator, at least if one is to avoid "the God hypothesis." He finds the solution in natural selection, which he judges capable of "endow[ing]" the physical states of living systems with "content," that is, purpose.[41] In his later work on evolutionary theory, though, Dennett has insisted upon the "mindlessness" of Darwinian natural selection too: "the crudest, most rudimentary, stupidest imaginable lifting process. . . . Each step has been accomplished by brute, mechanical, algorithmic climbing."[42]

Whatever the external source of ascription, Dennett affirms that agency cannot inhabit the acting creature, but can only be ascribed to it from the outside. He shares in a widespread consensus in classical AI as in behavioral robotics. According to this consensus, agency can only be apparent, and as such, a result of the creature's environment rather than its insides. Herbert Simon, who was an economist and cognitive and computer scientist at Carnegie Mellon University, compared the complexity of human behavior to the path of an ant across a sandy beach. The path, which Simon copied onto a piece of paper and showed to an uninitiated friend, looked to the friend like the trail of an expert skier navigating a slalom course or a sailor tacking in the wind. This was because the path had "an underlying sense of direction, of aiming toward a goal." But in fact, Simon argued, the goal-directedness originated outside the creature. "[The] complexity is really a complexity in the surface of the beach, not a complexity in the ant."[43]

Simon was the coauthor with Allen Newell of two landmark programs in artificial intelligence, the Logic Theory Machine and the General Problem Solver, and he was one of the most influential proponents of the computational model of human thought and behavior.[44] But on the matter of agency, he was in full agreement with the cyberneticists, old and new. As with the ant so too with human beings, Simon thought. The "apparent" goal-directed complexity of their behavior was "largely a reflection of the environment."[45]

The computationalists also share with the behavioral roboticists the axiom that cognition does not reside in any particular kind of stuff. In the case of behavioral roboticists, it is surprising, given their emphasis on physical engagement and embodiment, that the physical aspects of the materials they use do not figure in their theoretical discussions of what constitutes responsive, engaging, lifelike behavior. Instead, in their theoretical discussions, they dwell on the structural features of the machinery. Computationalists, similarly, maintain that no particular substance is responsible for cognition. "Intelligence," writes Pinker, "has often been attributed to some kind of energy flow or force field," a view he associates with "spiritualism, pseudoscience, and science-fiction kitsch." What makes a device or creature smart is not the stuff from which it is made, but its "truth-preserving" arrangements of parts, which in turn stand for elements of the world outside.[46]

The principle that the stuff of a sentient creature is irrelevant to its sentience appears to be axiomatic in both of these two research programs. Yet other natural forces and phenomena—electrical, magnetic, chemical, gravitational—are associated with particular substances. Why should the idea that intelligence is associated with a particular kind of substance be tantamount to, as Pinker puts it, "spiritualism, pseudoscience, and science-fiction kitsch"? The reason I propose is that such an idea violates the classical mechanist ban on agency in nature. Those who continue to endorse this ban on natural agency do not seem to recall that its original justification was as much theological as scientific.

Heated Agreement

> Teleology is like a mistress to a biologist: he cannot live without her but he's unwilling to be seen with her in public.—Ernst Mayr, *Toward a New Philosophy of Biology* (1988)

At the beginning of this book, I described a conversation with a biologist friend in which she agreed with me that biologists continually attribute agency—intentions, desires, will—to the objects they study (for example cells, molecules).[47] But, she said, this attribution was only a manner of speaking, a kind of placeholder that biologists use to stand in for explanations they can't yet give. Once they fully understand the processes they study, they will have no more need to ascribe agency of any sort to their objects, even as a manner of speaking. Historians, as I have said, unlike scientists, generally believe in agency. But they do not generally believe in "manners of speaking." Speaking is never just speaking, historians think, at least, I do, because it is inseparable from thinking. It seems implausible that scientists' thinking could be so fully independent of their speaking.

Consider, then, the manners of speaking that biologists and philosophers of biology employ when they speak about agency. According to the philosopher of biology Peter Godfrey-Smith, the question of how to treat agency, or apparent purposefulness in living systems, remains an organizing problem of the discipline. He gives the name "agentialism" to a way of speaking about evolution "in terms of a contest between entities with agendas, goals and strategies." Even though this language is ostensibly metaphorical, Smith warns, it is dangerous because once one starts "thinking in terms of little agents with agendas—even in an avowedly metaphorical spirit—it can be hard to stop."[48]

Yet even those whom Godfrey-Smith identifies as "agentialists" never ascribe agency to natural entities in an overt, literal way, but only in a "metaphorical" one. Indeed, they derive a certain rhetorical freedom from offering their metaphors as purely metaphorical, and by no means literal. The metaphors, being metaphors, assert their great distance from literalness. An example is how the Harvard biologist David Haig justifies using game theory to model the action of genes in evolutionary processes. Game theory became a standard model in evolutionary biology in the early 1970s with the work, notably, of the British theoretical evolutionary biologist John Maynard Smith, who applied a game-theoretic model to the behaviors of organisms in the struggle for survival.[49] Haig and other evolutionary theorists including Richard Dawkins now argue that the gene, rather than the organism, is the fundamental unit of natural selection and so they apply their metaphorically agential, game-theoretic language to genes rather than to whole organisms.[50] "Why use strategic thinking, which anthropomorphizes genes," asks Haig rhetorically,[51] before explaining that game theory is a "pragmatic" solution. It allows for the modeling of complicated situations with many variables in which some arrangements of variables lead to more successful outcomes than others. One can call these arrangements "strategies," Haig implies, without positing a strategist.

Haig borrows from evolutionary psychology the principle that human beings are particularly smart at thinking about agents and strategies, and he applies this principle to conclude that biologists will be smarter about genes if they treat them as strategic actors, without of course implying that genes *are* strategic actors.[52] Accordingly, Haig customarily describes genes as having "interests" and as acting upon "information" to exert "power" and "influence."[53] But he does not mean any of this in a literal sense. Indeed, the metaphors themselves act as constant reminders that to take them literally would be a mistake: they ascribe and revoke agency in the same figure of speech.

The application of game theory to evolutionary biology has followed the same old oscillation between construing life as agency, on the one hand, and evacuating agency from life, on the other. Maynard Smith, in his initial applications of a game-theoretic model to natural selection, attached the word "strategy" to organisms in the struggle for survival and made a living organism, at least for the purpose of his computer simulations, into a strategic agent. At the same time, Maynard Smith emphasized that his technical definition of "strategy" was strictly behaviorist: "A 'strategy' is a behavioural phenotype," he wrote; "i.e. it is a specification of what an individual will do [in a given situation]."[54] These "strategies," therefore, ostensibly involved no ascription of internal strategic thinking or agency, but merely outward observations of behavior. Dawkins, in adapting the game-theoretic approach to his theory of gene functioning, has also emphasized that the "strategies" in question are behaviorally defined and do not require the ascription even of consciousness, let alone agency, to the strategic agent. His is a theory of "unconscious strategists." The deliberate oxymoron encourages the reader to accept these apparent ascriptions of agency to genes as radical denials of any such agency.[55]

When Godfrey-Smith objects to what he calls "agentialism," then, in a sense it is a case of heated agreement with those he criticizes. He warns that metaphors tend to take over their users' thinking, resulting in "an uneasy mix of the metaphorical and the literal."[56] No doubt he is right about this mix, but it is not the whole story: at the very same time, these "agentialists," gene-centered theorists such as Dawkins and Haig, enlist their metaphors in a radically anti-agentialist program. They carry Weismann's construal of Darwinism, with its radical ban on agency in living organisms, to a logical extreme. Not only do these neo-"extreme"- Weismannians, as Dawkins himself characterizes his own views,[57] insist on a metaphorical and/or behavioral construal of agency, but precisely by applying this metaphorical/behavioral agential language to genes, they withhold it from organisms, which are denied even a metaphorical form of agency.

From what Dawkins calls the "gene's eye view," organisms, including people, are simply the "vehicles" that they "construct" in order to protect and replicate themselves. Having "discovered" how to do this, the genes "swarm inside huge colonies, safe inside gigantic lumbering robots," namely organisms. From within us, their robots, they are always "communicating with [the outside world] by tortuous, indirect routes, manipulating it by remote control. They are in you and in me; they created us body and mind; and their preservation is the ultimate rationale for our existence." Accordingly, Dawkins endorses the conviction that we have already seen embraced by cyberneticists, computer and cognitive scientists. and artificial intelligence researchers, that the unified self is a figment of the human condition. The feeling of being "a unit, not a colony," he argues, is a "subjective" artifact of the fact that natural selection "has favoured genes that cooperate with others."[58]

In passages such as these, genes certainly do seem to have not only agency, but an agency that is sinister, hidden, all-embracing.[59] They create, communicate, cooperate, and use people and other organisms as tools for their own ends. Their agency, moreover, clearly derives its salience from its contrast with the organisms' "robot"-like lack of agency.[60] Elsewhere Dawkins emphasizes that whenever he seems to be ascribing motives to genes, he is doing so in purely "*behavioural*" terms, and because it is "convenient, as an approximation." Genes appear from the outside to perform actions, but they have no will or purpose in doing so; their purposefulness is only "apparent." Although surviving genes have struggled to survive, they "did not know they were struggling, or worry about it; the struggle was conducted without any hard feelings, indeed without feelings of any kind." Successful genes have not striven for survival. They have struggled in the sense that any advantageous mutation (here Dawkins moves into the passive voice) "was automatically preserved and multiplied." Overall, "nothing actually 'wants' to evolve," he explains. "Evolution is something that happens, willy-nilly."[61] Dawkins's selfish genes with their emphatically metaphorical agency thus serve essentially to remove agency from the evolving organism by making these the tools of their blind, rote parts.

Another gene-centered evolutionary theorist whose work Godfrey-Smith characterizes as "agentialist" for his metaphors is Dennett who, we have seen, is also a radical anti-agentialist. Dennett's major philosophical purpose in writing about evolutionary theory has been to eradicate every instance of what he takes to be a "skyhook," that is, "a 'mind-first' force or power or process, an exception to the principle that all design, and apparent design, is ultimately the result of mindless, motiveless mechanicity."[62]

For example, Dennett has redescribed the so-called Baldwin Effect so as to purge it of appeals to agency.[63] The Baldwin Effect is named for the American psychologist James Mark Baldwin, though several contemporaries described it around the same time. In 1896 Baldwin wrote a paper proposing that, in addition to the physical action of the environment by means of "physico-genic" forces such as temperature or chemical action, each organism was subject during its lifetime to two other sorts of modifying forces. The first of these, which Baldwin called "neuro-genic," consisted of "the spontaneous activities of the organism itself in the carrying out of its normal congenital functions." These activities, according to Baldwin, displayed "a readiness and capacity on the part of the organism to 'rise to the occasion,' as it were, to make gain out of the circumstances of life." The final category of modifying forces, "psycho-genic" ones, included "the great series of adaptations secured by conscious agency," all of which involved "intelligent" processes such as imitation, instruction, learning from experience, and reasoning.

Neuro-genic and psycho-genic forces, according to Baldwin, constituted a capacity on the part of a living organism to "accommodate itself" to favorable conditions, to "repeat movements which are adaptive, and so to grow by the principle of use." He gave the name "organic selection" to this form of active, vital selection. He specified that organic selection was a distinct phenomenon from natural selection, and changed nothing about how natural selection operated. Natural selection was an "entirely negative agency," a term for what occurred when an organism did not have what it took to survive in a given environment. Variations constituted the positive side of Darwinian evolution, and here was where organic selection entered the picture, as a source of the variations that were then subject, in turn, to the law of natural selection.[64]

The overt appeal to a form of conscious, intelligent agency made the Baldwin Effect controversial from the start and the debate has now continued for over a century.[65] Dennett has proposed what he sees as a way past the controversy by purging the Baldwin Effect of its appeals to mind, consciousness, and vital agency. He translates Baldwin's "neuro-genic" and "psycho-genic" forces of modification into a capacity on the part of organisms "to adjust or reverse their wiring, depending on what they encounter." They might do this adjustment, Dennett imagines, by cycling through different "wirings" randomly, as long as they also have "an innate capacity to recognize (and stay with)" an advantageous modification "when they stumble upon it." Dennett does not specify how this innate capacity operates other than to say that it is mindless and that he understands it in entirely "*behavioristic* terms:" "All it requires is some brute, mechanical capacity to stop a random walk when a Good Thing comes along, a minimal capacity to 'recognize' a tiny bit of progress, to 'learn' something by blind trial and error."[66] How does a brute mechanical entity stop at Good Things but not bad ones? What is the difference between "recognizing" and recognizing, "learning" and learning? These are not questions Dennett takes up.

The accusation of smuggling agency back into biological explanations is a charge that all have leveled against all. We have seen that gene-centered evolutionary theorists such as Dawkins and Dennett imply that their gene-centered explanations offer an escape from the fallacy of attributing evolutionary agency to organisms.[67] At the same time, their critic, Godfrey-Smith, charges these same evolutionary theorists with "agentialism" in their accounts of the behavior of genes.

It is often more telling to examine what the two sides of a debate tacitly agree upon than what separates them. In this case, those who use "agential" language and those who reject it share a conviction that such language can only be behavioristic and metaphorical at best, and that words such as "recognize" and "learn" must have implied or actual scare quotes. This shared conviction is in fact what allows the metaphors to run amok, when allowed to run at all, since it precludes any serious consideration of what form a naturalized agency might take.

For another example of a debate in which each side accused the other of smuggling agency into biology, consider the controversy of the late 1980s to mid-'90s between Dawkins and Dennett, on one side, and on the other, the Harvard evolutionary biologists Stephen Jay Gould and Richard Lewontin, along with Niles Eldredge, a paleontologist and curator of invertebrates at the American Museum of Natural History in New York City. Eldredge and Gould assigned the epithets "ultra-Darwinism"[68] and "Darwinian fundamentalism"[69] to Dawkins and Dennett's approach: their explanation of every structure of an organism as an adaptation produced by natural selection acting upon its genes in their struggle for reproductive success.

Against this mode of explanation, Gould argued for what he called "Darwinian pluralism": the view that although natural selection (acting at the level of the organism, not the gene) offered the central explanation of living structures, it was not the only explanatory factor. Other "neutral" sorts of causes acted at levels both below and above the organism. These included contingent events such as a mass extinction caused by an asteroid hitting the earth; "neutral, nonadaptive changes" including some that population geneticists had recently found in the evolution of nucleotides, the building blocks of DNA; and the forces of stability that Gould believed held species constant for extended periods, as well as the generally nonadaptive causes leading new species suddenly to branch off "in a geological moment." Because it describes biological change as taking place through long periods of stability punctuated by abrupt transformations, Gould and Eldredge's theory of evolution is known as "punctuated equilibrium."[70]

In Gould's view, in contrast with Dawkins and Dennett's adaptationist determinism, evolutionary theory must describe the interplay of these various causes and contingencies, constituting an essentially historical account of the development of life. "You cannot argue that all rises from objective realities of adaptivity: you need also the connectivity of history."[71] A historian of science as well as a paleontologist and evolutionary biologist, Gould integrated history and science in his understanding of living forms.

Nevertheless, despite his historical view of scientific explanation, Gould was no less adamantly opposed to the idea of natural agencies than were his opponents. In this controversy once again, each side's principal accusation against the other concerned agency. Dennett characterized Gould and Eldridge's theory of speciation as appealing to a forbidden "creative force." Of Gould himself, Dennett wrote, "my diagnosis . . . is that he has all along been hoping for skyhooks."[72] Gould answered that the neutral, nonselective causes he cited, including causes of speciation, were nothing like Dennett's "skyhooks": "These other causes are not, as the ultras often claim, the product of thinly veiled attempts to smuggle purpose back into biology." On the contrary, they were "as directionless, nonteleological, and materialist as natural selection itself."[73]

Meanwhile, Gould, for his part, rejected Dawkins and Dennett's adaptationist approach according to which "the entire history of life becomes one grand solution to problems of design" and biology is essentially reduced to engineering.[74] Extreme adaptationism of this kind, transforming natural selection into an "optimizing agent," Gould and Lewontin argued, amounted essentially to the approach of the seventeenth- and eighteenth-century natural theologians.[75] To be sure, "natural selection" now stood in for "divine intelligence," but this made little difference. In particular, strict adaptationists, like the natural theologians of old, assumed that all structures in nature existed for reasons of optimal design, and that the orderings and arrangements of the natural world were therefore normatively correct and good.

The then-nascent field of evolutionary psychology, in Gould's view, was carrying the fallacy of adaptationism—assuming a ubiquitous and omnipotent designing agency in nature—to a new extreme by extending selectionism beyond biological structures to human behaviors and cultural forms. This meant overlooking not only nonselective causes such as asteroids, but also the basic mechanisms of human cultural change and inheritance, which Gould described as "fundamentally Lamarckian" rather than Darwinian (for he too assumed a stark opposition between Darwin and Lamarck): "Whatever we invent in our lifetimes, we can pass on to our children by our writing and teaching," whereas a correct view of Darwinian inheritance, according to the prevailing view that

Gould shared, precluded the passing on of traits acquired during an individual organism's lifetime.[76]

Against evolutionary psychology, Gould together with the MIT linguist Noam Chomsky and cognitive scientist Massimo Piattelli-Palmarini, argued that human language was a by-product rather than a product of natural selection, what Gould called a "non-adaptive side consequence." Their argument rested partly on Chomsky's theory of language, which posits an innate structure in the brain corresponding to a universal grammar underlying all human languages. In reference to this assumption of a rational center in the human mind, Chomsky presented his approach as "Cartesian Linguistics."[77] One implication of the theory, as Piattelli-Palmarini expressed it, was that there was really "no such thing as learning" (notably language learning). That is, the learner did not really acquire anything in the process of learning: there was no "transfer of structure *from* the environment *to* the organism."[78]

Instead, the environment configured a preexisting complex system in the human brain: all the structure was already in place, but merely needed some "parameter setting," a flipping of "internal 'switches.'" This complex, internal structure was what Piattelli-Palmarini, citing Gould and Chomsky, called a "spandrel."[79] This architectural term, denoting the space between two arches, became synonymous with "non-adaptive side-consequence" or "exaptation" following Gould and Lewontin's landmark 1979 article, "The Spandrels of San-Marco and the Panglossian Paradigm." Here they compared nonadaptive biological structures to the spandrels of the Basilica di San Marco in Venice, whose beautifully harmonious paintings provoke the impression that they were the purpose of the whole structure—the way Dr. Pangloss, in Voltaire's parody of Leibnizian philosophy, maintains that noses exist to support spectacles and legs to carry shoes—when in fact they were just a necessary "by-product of mounting a dome on rounded arches."[80] To say that the complex, innate structure in the brain responsible for language was a "spandrel" was to say that, although it appears to us to be the selective purpose of the whole architecture, it is just a side consequence, or a compilation of side consequences, of other, selected features.

Piattelli-Palmarini emphasized that he and his colleagues were no Lamarckians. "For reasons that are mysterious to me," he wrote (but I hope they will not be mysterious to the reader, who has been following their history through the previous chapters) "*every* critique of the canonical neo-Darwinian theory is invariably and obdurately received as an incitation to espouse some brand of neo-Lamarckism. This is always a very bad move and it would be particularly ill-advised in our case. The picture I am presenting here is far removed from *any* Lamarckian assumption or implication." The organism would seem to be as passive in Gould and Chomsky's theory, as rendered by Piattelli-Palmarini, as their opponents could possibly desire: "language 'learning' is something that 'happens' to the child and not something that the child 'does.'"[81] Yet Piattelli-Palmarini's protestations of anti-Lamarckism, and of the passivity of the learning child, and even his quotes around "learning" were all unavailing. In response to Piattelli-Palmarini's presentation of his own, Gould's, and Chomsky's view of language as a nonadaptive side consequence, and only indirectly a product of natural selection, Dennett charged them with "skyhooks": Gould and Chomsky were "supporting each other over an abyss" of anti-Darwinism and miracle-mongering. In Gould, Dennett diagnosed a hidden desire to restore a "Mind-first, top-down vision" of evolution.[82]

In evolutionary biology and cognitive science as in artificial intelligence, whatever the specifics of the disagreement, when you want to get out the big guns, you accuse the opponent of ascribing agency to nature's machinery.

Lamarckophobia and Its Treatment: A Dose of History

Weismannism remains strong among currently active philosophers (Dennett, Godfrey-Smith), cognitive scientists (Dennett), practicing biologists (Haig), biologists who have turned mostly to popularization (Dawkins), and as a result, also in popular understanding of how evolution works. However, in the last two decades or so, a subfield of biology has arisen at the opposite end of the ideological spectrum from Weismannism and from the gene-centric theorists. This subfield is "epigenetics."[83] The following discussion of epigenetics is drawn largely from the work of two of its leading proponents, Eva Jablonka, a geneticist, theoretical biologist, and historian and philosopher of biology at Tel Aviv University, and Marion Lamb, a biologist emerita at the University of London.[84] Rejecting the "gene-centered view of evolution," or what we have seen Dawkins describe as the "gene's-eye view," epigeneticists including Jablonka and Lamb emphasize the importance of the whole context in which genetic material functions, beginning with the cell outside the nucleus and extending through the organism and its environment.

Epigeneticists study the ways in which environmental factors can produce heritable changes in an organism by changing the epigenetic factors that influence the way its genome operates. One group of researchers defines "epigenetic" as referring to "all molecular signals that are literally on top of DNA."[85] Because it studies heritable somatic (bodily) changes undergone during the lifetime of an individual organism, the field of epigenetics has often been called "Lamarckian" by both its critics and its proponents. (The proponents are likelier than the critics to point out that Darwin too believed in heritable changes undergone during the lifetime of an individual organism.)[86]

A paradigmatic example of epigenetic effects is the phenomenon of "methylation," in which methyl groups (CH_3, a carbon atom bonded with three hydrogen atoms) attach to a segment of DNA, often to its cytosine, thereby changing the degree of its effects in the organism. In general, the greater the degree of methylation, the lesser the effects of the methylated DNA. Even small changes in DNA methylation can have important effects on the relevant cells. For example, methylation makes the difference between genetically identical but phenotypically dramatically different variants of flowers, such as the difference between the ordinary toadflax flower, *Linaria vulgaris*, and the variant *Linaria peloria* dubbed "monstrous." Methylation, influenced by the diet of pregnant mice, also makes the difference between offspring all sharing the same coat-color genes, some of which turn out to be brown, others yellow, and still others mottled yellow and brown.[87]

The regulation and maintenance of methylation patterns are crucial to normal development: embryonic mice that lack this capacity die before birth, and tumor cells often have different patterns of methylation from the corresponding normal cells. Environmental factors such as diet or temperature can influence the degree of methylation of a segment of DNA, and methylation patterns can be inherited from generation to generation. More than two centuries after the Swedish botanist Carl Linnaeus identified monstrous toadflax, it is still growing in the area where he first spied it, suggesting that methylation patterns are not only inherited but persist from generation to generation.[88]

Another example of a heritable epigenetic factor is known as "histone modification." Histones are small proteins instrumental in the packaging of DNA into units called "nucleosomes." Certain enzymes can modify the amino-acid tails of the histone molecules, which protrude from the nucleosome, by adding or removing groups such as acetyls or methyls. This affects how tightly the DNA is packed, which in turn affects how likely it is to be transcribed and therefore the degree of gene expression.[89]

The field of epigenetics has arisen partly in the context of work in genetics itself that has also tended to undermine the Weismann barrier isolating germ cells from changes in somatic or bodily cells, and therefore from bodily changes in individual organisms. Biologists now talk about many examples of mutations occurring in nonrandom ways, in response to environmental factors. They describe, for example, bacteria responding to stresses such as starvation by rearranging their DNA using "transposons," mobile elements in a cell's genome that can alter their relative positions with heritable phenotypic effects.[90] Transposons were the discovery of Barbara McClintock, a biologist at Cold Spring Harbor and the 1983 Nobel Laureate in physiology and medicine. McClintock introduced the concept of "genome shock" to describe the state of genomic response to adverse experiences, such as starvation or population collapse.[91] According to current science, bacteria also develop resistance to antibiotics, not through a purely random process of mutation followed by natural selection, but in important part by moving their DNA around.[92]

Biologists explain the functioning of the mammalian immune system by its ability to rearrange DNA sequences, for example, in the antibody molecules produced by B lymphocytes, a type of white blood cell. Proteins modify DNA sequences through processes such as one called "SOS mutagenesis," a response to damage in which the cell cycle halts while it repairs and mutates its DNA. Summarizing the situation, James Shapiro, a bacterial geneticist in the Department of Biochemistry and Molecular Biology at the University of Chicago, writes that virtually all cells "possess the basic biochemical tools for modifying DNA: proteins that cut, unwind, polymerize, anneal, and splice DNA strands." Moreover, cells use these techniques responsively, not purely randomly. "It is difficult (if not impossible) to find a genome change operator that is truly random in its action within the DNA of the cell where it works. All careful studies of mutagenesis find statistically significant nonrandom patterns of change."[93]

McClintock first observed the phenomenon of transposition in maize, as a response to stressful conditions, for example, following sudden changes in temperature.[94] This finding would have undermined the Weismann barrier between germ plasm and soma as he originally intended it, since he imagined it to be universal, applying to plants as well as to animals.[95] The applicability of the principle to plants was uncertain during the 1920s when McClintock was in graduate school and beginning her research career, but by the early 1930s, no one believed plants to have a segregated germ line, so McClintock's results in maize did not have direct bearing on the then-current understanding of the Weismann barrier, as applying only to animals with segregated germ lines.[96]

But McClintock's transposons turned out to be important in animals too, from fruit flies to human beings. Transposition has also proven to be importantly related to epigenetic factors such as methylation, which implies that epigenetic changes induced by environmental factors seem to be instrumental in bringing about heritable changes in the genome.[97]

In her Nobel lecture, McClintock described the genome as "a highly sensitive organ of the cell" that monitored its own activities and corrected itself, "sensing the unusual and unexpected events, and responding to them."[98] Shapiro, a radical who predicts a coming "upheaval in biological thinking,"[99] was a close friend and colleague of McClintock and describes her as having had a great influence on his own research. He has embraced and extended her way of describing cells as active agents of their own evolution. "Living cells and organisms," he writes, "are cognitive (sentient) entities that act and interact purposefully to ensure survival, growth and proliferation." Shapiro has coined the term "natural genetic engineering" to denote the processes by which a cell does this, and emphasizes the importance of what he calls "cell cognition": "life," he writes, "requires cognition at all levels."[100]

An implication of the current work in epigenetics and on genetic phenomena such as transposition is that variations do not occur only randomly, but can happen in response to conditions and events in the lives of individual organisms. This idea, sometimes called "adaptive" or "directed mutation," has been extremely controversial (though it may be becoming less so) since it violates the core principle of the "Modern Synthesis," the strict separation of somatic from genetic changes. Debate surrounding the idea of "directed mutation" has been about whether and how such a notion attributes agency to organisms. Gene-centered theorists such as Dawkins, who characterizes his own position as "extreme Weismannism," have rejected the possibility absolutely, at least until recently.[101]

Mark Ridley, for example, a zoologist, writer, and former student of Dawkins, has been categorical on the subject of theories of "directed variation." Writing in 1985, he insisted, "We must rule them out. There is no evidence for directed variation. . . . Whatever the internal plausibility of these theories, they are in fact wrong."[102] Dawkins himself wrote around the same time, in reference to the possibility of directed mutation, "To be painfully honest, I can think of few things that would more devastate my world view." This confession appears in a book section entitled "A Lamarckian Scare," dedicated to an apparent instance of directed mutation, the work of an Australian immunologist named Edward J. Steele during the 1970s.[103]

Interpreting four experimental studies of rabbits conducted by other researchers, Steele concluded that baby rabbits appeared to inherit immunities from their parents: they inherited, that is, the ability to form antibodies to certain antigens to which their parents had been exposed, and to which they had consequently formed antibodies, before mating to produce the offspring.[104] In further experiments, working with Reginald Gorczynski, an immunologist at the University of Toronto, Steele found evidence that baby mice seemed to inherit immune tolerances, acquired by their parents before they were conceived, in a similar way. Steele hypothesized that such inherited immunities and tolerances might involve a process of reverse transcription from the messenger RNA in the relevant antibodies, which might be picked up and carried to the germ line by viruses, back into DNA.[105] His own original experiments were not successfully replicated, and the idea of an inheritance of immunological changes has been controversial, but it has nevertheless continued to be taken seriously in mainstream scientific discussion.[106] The authors of a recent work in evolutionary biology have written, concerning an array of findings involving processes of reverse transcription, that these "clearly demolish the alleged central dogma" (referring to Francis Crick's Central Dogma of molecular biology, that there can be no reverse transcription from protein to RNA to DNA).[107]

Dawkins was able to avoid devastation by reexplaining Steele's data in a way he judged consistent with Weismann's ban on somatic changes effecting germ-line changes. First, Dawkins argued that although it looked as though the parent rabbit acquired a capacity that it then passed on to its offspring, this was an illusion caused by looking at the process at the organism level rather than the "replicator" level, meaning the gene level. The adult rabbits came to form successful antibodies not by "instruction" —in which their immune systems learned to make a new kind of antibody—but rather by a process of "selection" taking place within their bodies, in which the struggle against the antigen selected for successful antibodies generated at random.[108] Thus nothing in the system was actually acquiring an ability or characteristic; the organism view just made it appear that way, but it was certain replicators being selected in the struggle of antibodies against antigen.

Next, Dawkins argued that the selected antibodies were in fact "vehicles" for "*germ-line* replicators" and not somatic replicators. True, he acknowledged, the antibodies would not traditionally have been considered as parts of the germ line, "but it is a logical implication of the theory that we have simply been mistaken as to what the germ-line truly is." Any gene in a somatic cell that could be conveyed into a germ cell was, "by definition, a *germ-line* replicator." Dawkins offered this definitional principle, which guarantees Weismann's barrier a priori, in support of an argument that Steele's result had in fact carried Weismannism to a whole new level: the apparent inheritance of an apparently acquired characteristic became instead natural selection taking place at the "replicator" level. "Far from being uncomfortable to neo-Weismannists," Dawkins concluded, "it turns out to be deeply congenial to us."[109]

Dennett has agreed with Dawkins and Ridley regarding directed mutation, placing the idea (along with the long-suffering Lamarck) in a chapter on "losers" and writing that both would be "fatal to Darwinism," but were "safely discredited." To be sure, Dennett qualifies the ban on directed mutation by granting that it is permissible to ascribe accelerated mutation rates to certain "nonmiraculous mechanisms" (implying that other sorts of nonrandom mutations would be miracles). As examples of such "nonmiraculous mechanisms," Dennett proposes sexual reproduction, which continually crosses and combines genes in new ways; and the transfer of DNA between species, such as when a mite that feeds on fruit flies happens to puncture the egg of one species of fly and transfer some of its DNA into the egg of another species.[110]

These are acceptable to Dennett because they are after all essentially random: the mutations do not follow any particular pattern of advantage to the organisms. In fact, it is hard to see in what sense they might be directed. But as I noted earlier, biologists currently cite many examples of mutations occurring in nonrandom ways. Jablonka and Lamb have proposed the term "interpretive mutations" to designate those that are neither fully random nor entirely directed by environmental circumstances. Examples are some of the ways in which bacteria adapt quickly to hostile environments. In some instances, they increase their rates of mutation throughout their genome when they encounter such environments. In other instances, they have permanently higher rates of mutation in the regions of their genomes where changes are likeliest to be useful in stressful conditions. Richard Moxon, a researcher in pediatrics and infectious diseases at Oxford University, for example, showed that the meningitis bacterium, *Haemophilus influenzae,* changes its surface structure according to where it is in the host's body. It does so by means of what Moxon calls a process of "discriminate" mutation involving "contingency genes": highly mutable genes that influence surface structure, which generate many possibilities to be selected among.[111]

There are also instances of bacteria mutating both at increased rates when they encounter hostile conditions and in the regions likeliest to be useful. Barbara Wright, a developmental biologist at the University of Montana, has identified such a phenomenon in *E. coli* bacteria. She found that when these bacteria had a defect in a gene needed to synthesize amino acids, they would respond to a shortage of amino acids in their environment by increasing the rate of mutation of exactly that gene.[112]

When evaluating the scientific significance of this research, it is important to understand the origins and history of the prohibition on patterned (nonrandom) mutations and heritable changes to the soma of individual organisms. It is important to understand where and why these prohibitions originated. Scientists and philosophers have shown little patience for historical discussion. Dawkins, for instance, in his analysis of Steele's claims regarding directed mutation, wrote, "There are some historical points to get out of the way first." First, he continued, Lamarck placed no special emphasis upon what later came to be called "the inheritance of acquired characteristics"; and second, such inheritance was in fact "conventional wisdom" at the time Lamarck wrote and extended to Darwin's own thinking. "But," wrote Dawkins, "I mention history just to get it out of the way." A bit later he again expressed reluctance to get "caught up in the historical details of what Lamarck and Darwin said."[113]

But a lot of the science is in the historical details. Historical understanding is integral to scientific understanding. As an example, let us see how a more deliberate historical analysis could enlighten this discussion of directed mutation.

We know, for one thing, that in Lamarck's time, the idea that animals could undergo changes that would be inherited by their offspring was *not* conventional wisdom, but quite a radical notion, since it violated the long-standing conviction among naturalists that species were fixed and unchanging. The first theories of species-change began to emerge in the second half of the eighteenth century, and the idea was new and controversial when Lamarck was writing. Georges Cuvier, for example, rejected it vehemently.[114] Lamarck's version of the idea of species-change, residing primarily in an intrinsic force of increasing complexity and, secondarily, in the heritable effects of animals' will and "habits" on their bodily forms, was, in the tradition of theories of species-change from La Mettrie and Diderot onward, rigorously naturalist.[115]

Dawkins casts Lamarckians as miracle-mongers, but far from appealing to a supernatural, divine power, Lamarck himself did just the opposite: he proposed a way to avoid such an appeal, upon which the argument from design, with its passive-mechanist natural world, rested. Instead he proposed to naturalize a kind of living agency, to consider it a force like gravity, electricity, or magnetism.

From Weismann onward, biologists, philosophers, and even historians have assumed that Darwin's Lamarckian beliefs were oversights, failures to carry his own theory to its logical, anti-teleological conclusion. We might call this a neo-Darwinist or a Weismannian approach to history. Currently, Weismannism seems to be spreading in the discipline of history even beyond histories of evolutionary theory. Consider a recent book, *On Deep History and the Brain*, by Daniel Smail, a history professor at Harvard, a medievalist, and now also practitioner of what he calls "deep history," an approach founded in evolutionary neurobiology. Smail argues for the historical application of "Darwinian" methods to history, by which he means neo-Darwinist explanations of the brain in terms of "blind variation and selective retention." He deploys these methods against a "Lamarckian" approach that he characterizes using Dennett's phrase as "Mind-first," that is, an approach that identifies intentional actions as historical causes. In contrast, a model compatible with Darwinism is one that "explicitly discount[s] the goals and intentions of rational actors."[116]

Smail does not argue that we should eliminate "Lamarckian processes" from historical understanding altogether, since one can hardly deny that individuals pass on acquired social and cultural capacities to future generations (by teaching, writing, drawing, composing, building, and most other human activities). But, Smail says, Darwin wins anyway: "Lamarck will always be dogged by Darwin, for it matters not one whit how traits are acquired and transmitted if natural selection continues to sift out the adaptive traits." He does not make explicit the logic behind his assertion that the source of variations is irrelevant, a view not shared, as far as I know, by biologists on any side of the question about how variations originate, nor behind his conclusion that a Darwinist "model of historical change" can therefore allow a circumscribed "place for Lamarckian acquisition" and still "banish intention." Smail's ultimate purpose is to "minimize the role of genius and forethought" so beloved by an "older generation of Lamarckian-style historians" and "abandon . . . [the] Mind-first model."[117]

In short, Smail's approach to history is perfectly Weismannian. He makes the Weismannian assumption that Lamarck's and Darwin's theories were essentially opposites and that true Darwinism is pure of Lamarckian-style agencies (whereas we have seen that Darwin believed unwaveringly in the evolutionary importance of use and disuse, which mattered more than a whit to his theory). Smail is also Weismannian in his call to eradicate agency from historical explanation. Just as Weismann and his contemporaries argued for the elimination of history from biology, Smail seems to advocate the elimination of history from history, and its replacement by Weismannian biology. I propose the opposite, a de-Weismannization in the form of a restoration of history to evolutionary biology.

Let us return to the "historical details" that Dawkins declined to get caught up in and that Smail too leaves aside. These indicate that the neo-Darwinist history is wrong. Insofar as Darwin adopted Lamarck's forces of change, he did so not out of a failure of nerve or an inability to carry his own revolution all the way, but on the contrary because he too sought a rigorously naturalist theory and was determined to avoid the mechanist solution of externalizing purpose and agency to a supernatural god.

In short, Lamarckism represented its era's most naturalist, nontheological account of species-change. Those who rejected Lamarckism in the first instance—and through the turn of the twentieth century—were arguing in a mechanist tradition that, as this book has shown by tracing it back to the seventeenth century, was conjoined with a theological tradition founded in the principle that design implied a Designer. The biologists and philosophers of biology of the twentieth to twenty-first centuries who have categorically excluded Lamarckian explanations and all nonrandom variation as beyond the bounds of legitimate science have believed they were expressing a commitment to an ongoing naturalism. But history places them in the opposite camp. In fact, they are the heirs to the tradition with which they meant to do battle: the argument from design.

To get around the historical messiness of the fact that Darwin was a Lamarckian who believed in the heritability of somatic changes, Dawkins left aside the terms "Lamarckism" and "Darwinism," replacing them respectively with "the instruction theory," in which animals heritably adapt themselves to their environments by some other means than randomly generating forms to be selected among, and "the selection theory," in which the struggle for survival selects the best adapted among randomly generated possibilities both among and within individual organisms. "My world-view will be overturned," Dawkins specified, "if someone demonstrates the genetic inheritance, not just of an 'acquired characteristic' but of an instructively acquired adaptation. The reason is that . . . [this] would violate the 'central dogma' of embryology."[118] As Dawkins saw it, a true violation of Weismann's barrier, as restated in Crick's Central Dogma, would thus be an individual organism not only heritably adapting itself to its environment, but doing so in some other way than by randomly generating structures for selection.

Why should such a finding devastate Dawkins and overturn his worldview? I think it must be because the Weismann barrier, to him, does not just separate germ plasm from soma, or genotype from phenotype in Johannsen's reformulation: it is the all-important firewall between science and religion, keeping scientific explanation free of teleological appeals to purpose or agency. But a wall has two sides and a historical understanding of this wall reveals that it was first built to serve the reciprocal function as well: to be a firewall protecting theology from science, ensuring a monopoly on agency to a supernatural God and, to His theologian-interpreters, ownership of the ultimate questions of the meaning and essence of life.

Dawkins repudiates any attempt to breach the Weismann barrier as an assault by the forces of religious "fanaticism" and "zealotry."[119] Dennett, we have seen, shares the conviction that Lamarckism is tantamount to miracle-mongering. Here, again, is why history matters to current science, for in historical terms, Dawkins and Dennett are mistaken. Lamarck's was the naturalist theory, while the passive-mechanists they champion, notably Weismann, justified their brute-mechanist account of nature with the assumption of a supernatural, guiding intelligence. From a religious perspective (and specifically, from a turn-of-the-century German, Protestant perspective), Lamarck's was the more dangerous idea.

Yet many scientists and philosophers among the present generation of neo-(neo)-Lamarckians seem ultimately to accept that natural selection is the only defense against an essentially religious account of life's development. Jablonka and Lamb, for example, assure their readers that they have not "slyly introduced some mysterious intelligence into evolution," and in doing so, they appear to travel to a kind of extreme of selectionism (attributing all living forms ultimately to natural selection acting on random variations). They argue that all those systems they call "instructive elements" of evolution, including systems responsible for "interpretive mutations" and for epigenetic inheritance, are likely products of natural selection alone. Indeed, Jablonka and Lamb extend this claim to what they call the "symbolic inheritance system" of human culture. Cultural developments too, they propose, are the results of natural selection, thereby endorsing the central (selectionist-reductionist) axiom of evolutionary psychology. Do Jablonka and Lamb then share with their gene-centered counterparts the conviction that natural selection acting upon randomly generated forms is ultimately the only legitimate scientific explanation for a living form?[120]

When I asked Eva Jablonka this question, she emphasized the importance of distinguishing among different forms and levels of agency. A very primitive kind of material agency, she explained, is not only admissible in current biological explanations, but indeed many biologists assume it to exist. Jablonka proposed, referring to my "restless clock" image, that one might call this primitive, material agency the "restlessness of matter." As I understood Jablonka's explanation, the "restlessness" of matter is in part simply a tendency in matter toward random transformation, a tendency of the sort that Darwin assumed when he first postulated his theory of natural selection. So far, this is safely within the bounds of the brute-mechanist ban on agency, which allows (indeed, requires) accidental variation. But is any additional, nonrandom force of change admissible?

Jablonka directed me to the work of another pair of epigenetics researchers, Gerd B. Müller, a theoretical biologist at the University of Vienna, and Stuart A. Newman, a cell biologist at New York Medical College. Müller and Newman argue that random variation and natural selection alone do not account for the presence of organic forms in nature. Evolutionary biology, they write, has neglected the origin of organic form: natural selection can only work on what is already there.[121] But pioneering work on the origins of organic forms such as that of D'Arcy Thompson (chapter 8) has been "kept on the sidelines, muscled out by the notion that genes could do everything by themselves."[122]

To explain how organismal forms originate, Müller and Newman invoke an "inherent plasticity" in living matter, comparable to that of various forms of nonliving matter such as clay or lava. Additionally, living tissues are "excitable media." The term originated in physics, to describe "materials that respond in active and predictable ways to their physical environments."[123] Examples of inanimate excitable media include oscillating chemical reactions that produce patterns in response to external factors such as light.[124]

Plasticity and responsiveness together, according to Müller and Newman, constitute an intrinsic, "primitive, physically based" property, a capacity for "morphological variation in response to the environment." This primitive tendency, during the very earliest stages of the development of life, was the dominant force in generating new forms. Müller and Newman emphasize that the mechanisms that give rise to new organic forms have themselves been evolving, a possibility they say neither Darwin nor successive generations of Darwinians considered. As more complex organisms emerged, Müller and Newman suggest, the "inherent material properties" of plasticity and responsive excitability ceded importance to genetic factors, obscuring the early significance of these epigenetic mechanisms.[125]

For this reason, according to Müller and Newman, biological explanation must include a historical dimension, whereas the "reverse-engineering" approach of neo-Darwinism, in its assumption of reversibility, is essentially ahistorical. "Instead," they write, "we need an 'archaeological' perspective that recognizes that modern activities of form generation are based on principles whose origins have often become obscured by history."[126] Like Kant's "archaeology of nature" (chapter 6), Müller's and Newman's archaeological, historical approach to the history of life assumes inherent natural agencies whose action over time has produced a history that is neither designed nor random, but contingent.

Yet the intrinsic forms of natural agency described by Müller and Newman also retain at their core a version of the same dialectic whose history this book has been tracing. They rely upon the notion of "emergence," an organizing idea that has permeated the natural and social sciences during the last three decades (we have seen it at work in Rodney Brooks's work). The term's users trace its lineage back to an array of origins including Aristotle, nineteenth-century British positivism,[127] and the cyberneticists of chapter 9. "Emergence" denotes any complex phenomenon that seems irreducible to the sum of its parts: phase transitions such as from solid to liquid or liquid to gas, hurricanes, the stock market, living organisms.[128] The key to such phenomena, according to theorists of emergence, is complexity itself; so that no reductive account can ever capture their essence.

John von Neumann (chapter 9) wrote, regarding the difference between a brain and a computer, that "the human nervous system is roughly a million times more complicated. . . . Below a certain minimum level of complexity, you cannot do a certain thing, but above this minimum level of complexity, you can do it."[129] Among the paradigmatic emergent phenomena are the inherent physical properties that Müller and Newman refer to, such as excitability,[130] and especially "self-organization."[131] This is a concept intimately related to emergence. Self-organization now appears in scientific discussion variously as the primary emergent property in nature, as another phrase to express the same idea as emergence, or as a separate but largely overlapping phenomenon.[132]

We saw the origins of the notion of self-organization with Kant's struggles over the role of purpose in nature and natural science (chapter 6). In Kant's formulation, self-organization meant the apparent purposiveness of each part of a living being with regard to every other part. Kant wrote that this apparent purposiveness did not mean that organisms were in fact intrinsically purposive, but rather that human beings could only apprehend them as such. Cyberneticists, notably Ross Ashby, established the idea of self-organization at the heart of cybernetics (chapter 9), where it continued to imply a distinction between appearance and reality. The cyberneticists' self-organizing machines appeared to observers to behave in spontaneously purposive ways, as long as the observers did not dwell on the internal mechanisms producing these behaviors.

In recent and current biology, self-organization is an "emergent" property of complex biological systems.[133] Kant's gulf between epistemology (how we explain the world) and ontology (how the world actually is in itself), which the cyberneticists made over into a behaviorist disjuncture between objective appearance and subjective experience, has now become a discontinuity between the micro- and the macro-level of living nature.

Consider how the theoretical biologist Stuart A. Kauffman describes self-organization in work that Müller and Newman cite to argue for self-organization as an inherent source of organismal form. Kauffmann has proposed that self-organization is a primary source of order in living nature[134] and constitutes a form of "minimal biological agency," attributable, for example, to cells. These minimal agents, Kauffman specifies, are at the bottom of a spectrum of agency at the top of which is the "robust, conscious agency" of human beings. A minimal agent is one able simply to "act on its own behalf," such as a bacterium swimming toward higher concentrations of glucose. Because this agency does not reside in any of the parts of the agent, but only in their organization starting from a certain level of complexity, Kauffman calls it "emergent."[135]

To say that agency is an emergent feature of organizational complexity, as Kauffman does, is thus also to affirm that it is not itself a basic element of the natural world, but rather a product of basic elements. What sort of a product? Kauffman writes that to answer this question, biologists will need what they do not yet have, a "theory of organization." At the same time, however, he also returns to Kant's original dichotomy between how one describes living nature and how living nature is in itself. Kauffman calls ascriptions of agency a "language game," borrowing the term from the Austrian philosopher of language Ludwig Wittgenstein, and he emphasizes that this game has no necessary, provable connection to reality. "Minimal molecular agents" look like they are referring, choosing, and acting, so biologists are justified in attributing these actions to them, writes Kauffman. However, "we are hesitant to make this claim as a metaphysical assertion, since it's not clear how one would 'prove' the appropriateness of playing the agency language game."[136] According to Kauffman's argument here, then, it is an emergent feature of organizational complexity, not to be an agent, but to be an entity with respect to which biologists can legitimately play the "agency language game."

In sum, even the revisionists among the current generation of biologists seem mostly, ultimately, to agree that naturalism precludes treating agency as an elemental feature of the natural world, or indeed as anything beyond an irresistibly compelling appearance. To violate this ban has seemed tantamount to lapsing right out of scientific explanation into a religious or mystical one. Yet we have seen this core principle of modern science—banning agency from natural processes—emerge historically from a tradition that denied agency to nature in order to ascribe it instead to a designer God.

Schrödinger's Clock

What, then, about the unabashedly active-mechanist tradition, whose practitioners have sought to naturalize agency rather than banishing it from nature and science, and whose shadowed history we have followed throughout this book? Although the situation may be changing, this tradition appears still to be mostly absent from mainstream evolutionary biology. Yet during at least one pivotal moment, it played a crucial role. It did so, moreover, in important part thanks to a brief but influential essay written, not by a biologist, but by a physicist.

That the author is a physicist makes sense. In physics, the problem of natural versus supernatural forms of agency came to a head in the late seventeenth and early eighteenth centuries. During that period, Newtonians largely succeeded in establishing a Newtonian understanding of force at the heart of mechanist science, beginning with the force of gravity and continuing through the eighteenth century with electrical and magnetic forces. Against the Cartesian reduction of force to moving streams of matter, and the Leibnizian notion of a "living force" or *vis viva* permeating the natural world, Newtonians instituted their understanding of forces at once as mathematical generalizations, rather than causal hypotheses, and as the manifestations of an external, divine presence.[137]

Unlike in biology, therefore, where, the ontological question of natural forces and agencies became ever more heated during the eighteenth and nineteenth centuries, in physics this question had lost most of its heat by the early nineteenth century. Moreover, the twentieth century opened in physics with a dramatic departure from the realm of brute-mechanist explanation, namely the emergence of quantum mechanics, forgoing classical mechanical causes altogether.

In 1944, one of the principal authors of quantum mechanics, the Viennese physicist Erwin Schrödinger, published a short and momentous treatise entitled *What Is Life?* in which he proposed a way of bringing quantum mechanics to bear upon the defining question of biology. Molecular geneticists generally credit this treatise with having inaugurated their discipline. Its influence has been controversial from the first, to be sure, but the very controversy is a testament to its importance.[138] In the treatise, Schrödinger proposed that the essence of life was a form of agency. "What is the characteristic feature of life?" he asked. "When is a piece of matter said to be alive? When it goes on 'doing something,' moving, exchanging material with its environment and so forth."[139]

Specifically, the most salient "something" that organisms characteristically did, Schrödinger proposed, was to resist entropy, to avoid decay into equilibrium. They achieved this feat by means of their various metabolic functions: eating, drinking, breathing, photosynthesizing. In a world that tended toward disorder and decay, these functions enabled a living thing to maintain its order and energy, "continually sucking orderliness from its environment," "drinking orderliness," and "concentrating a 'stream of order'" upon itself.[140]

Because of this very particular form of agency, the ability to produce and maintain order, Schrödinger considered living creatures to be essentially machines. Producing and maintaining order, he proposed, was the defining feature of machines, whether natural or artificial. Machines represented one of the two ways, Schrödinger wrote, in which order occurs in nature: either statistically, as in thermodynamic events such as the magnetization of a gas or the diffusion of a liquid; or else mechanically, as in the motion of the planets or the working of a clock. Mechanisms, whether natural or artificial, seemed not to follow the statistical "order from disorder" principle of thermodynamics but rather to enact an "order from order" principle. Living organisms, with their active ability to produce order from order, were therefore like "clockwork" mechanisms. This might seem like a "ridiculous conclusion," Schrödinger wrote, but it was possibly the "clue to the understanding of life."[141]

To see how this might be, one had to consider that in reality neither organisms nor clocks did in fact produce order from order strictly according to a set of deterministic laws. Rather, both were real, physical entities, made of matter, and therefore subject to entropic forces. On the most fundamental level, that is, organisms and clocks too, like the rest of the natural world, were statistical regularities rather than law-determined systems. An ideal clock would never need to be wound: "Once set in motion, it would go on forever." One could see that a real clock was subject to entropy in the simple fact that it ran down. And the opposite was also theoretically possible, Schrödinger emphasized: an unwound clock might experience "an exceptionally intense fit of Brownian movement" and begin to move spontaneously. It might indeed "all at once invert its motion and, working backward, rewind its own spring."[142]

A clock was accordingly to Schrödinger, as it had been to Leibnitz, a beleaguered, disquiet, and restless thing, perpetually striving to maintain an orderly and balanced state of affairs. In this sense, it resembled a living organism. Both artificial clockwork and the clockwork of living organisms maintained their order, for a time, against the forces of disorder that continually beset them, and here was the true "Relation Between Clockwork and Organism."[143] How did they do it?

According to Schrödinger, both kinds of mechanism, artificial and organic, relied upon the stability and order-producing features of molecules, features that the new field of quantum mechanics, of which he himself was a principal author, had recently begun to identify. According to quantum theory, Schrödinger explained, atoms were systems of subatomic particles that could exist only in certain discrete states, corresponding with certain energy levels. In passing from one to another of these states, they did not transform in a continuous way, since they could not occupy the intermediate states even momentarily, but rather jumped discontinuously: a "rather mysterious event" called a quantum jump. During a quantum jump, the atoms also absorbed or emitted energy (depending upon whether they were moving to a higher or lower energy level), not continuously, but in discrete chunks, or quanta.[144]

This accounted for the order- and stability-producing feature of molecules. Schrödinger explained that molecules were configurations of atoms occupying their lowest energy level. In order to change configurations, they needed to receive at least a minimum quantum of energy. The stability of a molecule could perhaps account for the order-producing capacity of genes, responsible for maintaining the structure of living organisms both within individuals and across generations. "We shall assume the structure of a gene," wrote Schrödinger, "to be that of a huge molecule, capable only of discontinuous change." In this case, for a gene-molecule to change would require a "comparatively enormous" explosion of energy.[145]

Schrödinger further surmised that a gene-molecule was what he called an "aperiodic crystal," one in which "every atom, or group of atoms, plays an individual role," as opposed to a periodic crystal whose parts formed a repeating pattern. How this molecule worked remained a mystery to him, but he was certain that it did so in a way that allowed living organisms temporarily and actively to elude the general tendency of nature toward disorder. This ability to imbibe order from their surroundings, while continually giving off disorder in return, was their defining feature. Artificial clocks produced order from order thanks to the quantum stability of the molecules that composed them; similarly, organisms produced order from order by means of the stable, "aperiodic" molecule that organized them.[146]

It is a striking feature of Schrödinger's picture that in trying to explain what he saw as the defining activity of living organisms, he looked in the first instance to a kind of stasis, the genetic molecule's resistance to change. Certain aspects of his picture were dramatically confirmed by subsequent research, notably by James Watson and Francis Crick's identification of the structure of the DNA molecule in 1953, which is indeed well described as an "aperiodic crystal." But by the end of the century, with the emergence of the field of epigenetics, biologists had for the most part come to understand the stability of inheritance, not as the result of the static nature of the DNA molecule, but rather in terms of a dynamic system of maintenance, regulation, and repair involving not only the DNA, but the entire cell and its environment in and outside the body.[147]

Still, in Schrödinger's picture, even with the tendency toward stasis on the part of the gene-molecule—or rather, importantly, thanks to this tendency—living entities had an essential, irreducible, and defining activeness. Moreover, the gene-molecule's resistance to change constituted only part of its role in explaining the "order-drinking" activity of organic beings. Living entities worked not only by remaining stable over long periods, but also by transforming, and indeed, by transforming in a particular way. Schrödinger embraced De Vries's view that organisms underwent rare "jump-like" changes (he called this a "quantum theory" of biology) and that these quantum shifts supplied the material for natural selection, in place of the continuous, tiny accidents emphasized by neo-Darwinians, "a long pearlstring of chance events, independent of each other." The quantum explanation of the hereditary substance had the further advantage, in Schrödinger's view, not only of explaining its stability, but also of explaining its changes, that is, mutations. He speculated that these might be due to "quantum jumps in the gene molecule."[148]

These quantum jumps provided the material, in Schrödinger's view, for a process of evolution in which natural selection functioned crucially but not exclusively: rather, natural selection worked in constant conjunction with the behaviors of living organisms. Schrödinger regretted the "gloomy and discouraging" nature of contemporary representations of Darwinism that insisted upon the "passivity of the organism." Weismannian neo-Darwinists, as we have seen, made this passivity a primary condition of their science: the individual living entity was to have no role to play in evolution, no influence on the hereditary material it received from its parents or left to its offspring, and thus, wrote Schrödinger, "its activity during its lifetime seems to be biologically irrelevant": "Any skill or training is lost, it leaves no trace, it dies with the individual, it is not transmitted. An intelligent being in this situation would find that nature, as it were, refuses his collaboration—she does all herself, dooms the individual to inactivity, indeed to nihilism."[149] In contrast, Lamarck's theory, which, as Schrödinger saw it, ascribed evolutionary progress to individual agency, was "beautiful, elating, encouraging and invigorating." What a shame that it was wrong.

For of course Lamarck's theory was "untenable" . . . or was it? Schrödinger considered that it did contain a "very true kernel," namely the causal connection between the use of an organ or characteristic and its improvement over generations. In fact, he mused, it was almost "as if Lamarckism were right": even neo-Darwinist theory associated the use of an organ with its improvement. The difference, as Schrödinger saw it, was that whereas Lamarck had surmised that the behavior came first and caused physical change, modern Darwinists put physical change at the beginning of the process: a useful mutation would rouse an organism to new behaviors.

These behaviors, however, would in turn reinforce and enhance the usefulness of the mutation, and therefore its selective value and improvement, so that behavior would indeed lead to further physical change. Parents would transmit the changed behaviors to their offspring not directly, through their genetic material—Schrödinger too assumed the integrity of the Weismann barrier—but "by example, by learning." Thus behavior and learning, Schrödinger concluded, had "an important say" in the process of evolution by natural selection.[150]

In fact, Schrödinger went so far as to argue that the very distinction between structure and use, which the neo-Darwinists insisted upon, had no real meaning. "Behavior and physique merge into one." It made no sense to distinguish between an organ and "the urge to use it and to increase its skill by practice." With respect to natural selection, these were not two separate characteristics of an organism, but one. Selection could never transform a mutation into a new organ if it were not "aided all along by the organism's making appropriate use of it":

> And this is very essential. For thus, the two things go quite parallel and are ultimately, or indeed at every stage, fixed genetically as one thing: *a used organ*—as if Lamarck were right.[151]

Moreover, the development of used organs by means of selection gave the irresistible impression of a tendency toward ever greater specialization, as though each species had "found out in which direction its chance in life lies and pursues this path."[152] While refraining from calling it a *vis viva* or *Bildungstrieb*, Schrödinger identified in the essential agency of living organisms their tendency to use what they had, a complicating force at the heart of evolutionary development.

The essential activeness of living mechanisms, the indistinguishability of part from use, offered Schrödinger a source of scientific hope. He returned to it in addressing what he took to be the worst "impasse" of contemporary science, namely, how to bring "mind"—sensation, experience—back into the world picture from which it had systematically excluded itself. Somehow, he wrote, science must overcome this "exclusion principle" by which natural philosophers and scientists over the preceding three centuries had systematically exempted themselves from the natural world, producing a "horrible antinomy. I maintain that it cannot be solved on the level of present-day science which is still entirely engulfed in the 'exclusion principle.'" To get beyond the confines of brute mechanism, Schrödinger urged, "scientific attitude would have to be rebuilt, science must be made anew."[153]

Science has not been made anew since Schrödinger wrote these lines, nor can it be: history is as ineradicable from science as agency appears to be from nature. But history also indicates that Schrödinger himself represents a limit to his "exclusion principle." He exemplifies the regular recurrence of a dialectical tradition at the heart of scientific explanations of life and mind: one that has sought to naturalize agency rather than eradicating it from scientific explanation. Our current state of understanding is a product of the struggle between these two traditions, one explicit, the other implicit and often hidden from view, but still active.

Three and a half centuries after the natural sciences assumed their modern forms—intellectual, institutional, instrumental, and social—their relations to religious beliefs and institutions remain fiercely embroiled. This book has tried to show that their embroilment was built into modern science from its inception, built into a definition of science that made a supernatural God necessary to naturalist explanation by banishing agency from nature's mechanism. Practitioners of the sciences have regularly, from the beginning, challenged this brute-mechanist definition of science and violated its ban on agency. Yet they have not succeeded in dethroning the brute-mechanist ideal of science, whose supernaturalism is deeply implicit despite being long hidden. To say that a human being works like a machine, whether one accepts or rejects the idea, sounds like science. But it sounds less like science when one describes the machinery as restless, moved by its own inner agency. Historical analysis, by explaining why the first sounds more like science and the second less, can help to reopen scientific possibilities that the classical clockwork model of nature foreclosed, but that tick on within the model of a living being as a *restless* clock.

ACKNOWLEDGMENTS

The help of colleagues and friends, students and family, has sustained the *nisus formativus* of this book through its long period of generation. *The Restless Clock* began as a graduate essay on Jacques Vaucanson's Defecating Duck written for Professor Randy Starn's seminar on historical methods at the University of California, Berkeley, in 1992. Though the project then took a backseat to several others for more than a decade, it has been with me in one way or another for a very long time: as I completed graduate school and embarked on a teaching career, through several moves, the publication of my first book, and the tenure process, through the births and early growings-up of two children. Where to begin?

For their time and attention, those most precious currencies, I am grateful to Ken Alder, Keith Baker, David Bates, Mario Biagioli, Marianne Constable, Lorraine Daston, Daniel Edelstein, Paula Findlen, Daniel Garber, Stephen Gaukroger, Denise Gigante, Deborah Gordon, Nir Hacohen, the late Roger Hahn, John Heilbron, Eva Jablonka, Alain Jehlen, Myra Jehlen, Evelyn Keller, Daniel Kevles, Christopher Kutz, Thomas Laqueur, George Levine, Andrea Page-McCaw, Robert Richards, Carl Riskin, Richard White, Lora Wildenthal, and Caroline Winterer, and three anonymous readers for the University of Chicago Press, whose responses to the work under way proved essential to its completion. David Haig generously elucidated his views about biological applications of game theory for me. Daniel Dennett conferred with me about our mutual interest in old French robotic dogs. Simon Schaffer kindly gave me guidance during my first research forays. His own work on the history of automata has been a reference throughout this project. Simon also introduced me to David Secrett, whom I thank for showing me his reproduction of Vaucanson's Duck, then still in progress. Seeing this new Duck, majestic even in pieces, helped me to imagine the momentousness of its historical model.

Ken Alder's friendship and intellectual generosity have been an inestimable gift. Through his readings and responses to my work, he has helped me always to think better. His erudition has also regularly enriched my reasoning by reattaching it to the messy world. Myra Jehlen's presence as asker of questions that challenge, complicate, and deepen the analysis has attended this book at every stage. Only she can know the full importance of her help, but I might convey a sense of it by saying that she has likely committed substantial portions of the manuscript to memory. Chris Kutz has accompanied this book and its author through thick and thin. In addition to serving as photographer and technical crew on research trips, he has had the patience to read drafts, offer suggestions, be told these are all wrong, see them shape the subsequent draft without demur, and then tirelessly repeat the process, giving new meaning to the phrase "for better, for worse."

Colleagues and students at the Massachusetts Institute of Technology and at Stanford University provided the ongoing conversations that are so necessary to writing a book. As I was just beginning to think my way into this one, at MIT, Justine Cassell, Evelyn Keller, and David Mindell helped me to consider the history and current state of attempts to measure humans against machines and vice versa. Rodney Brooks talked with me about his research program and granted me a private tour of his laboratory, which was then the AI Lab at MIT. Since my arrival at Stanford, Paula Findlen has offered countless kinds of help including conceptual responses to drafts, editorial and bibliographic suggestions, logistical aid and advice, and great administrative support. Keith Baker and Richard White have provided important readings and responses, encouragement and support. Denise Gigante has sustained my writing with her conversation and company; on several occasions, she and Julian Rovee provided a luxurious haven in which to work. My students at Stanford have been a constant source of inspiration and new ideas, and I thank especially Gautam Dey, Jenny Pegg, Greg Priest, Stephan Risi, and Rebecca Wilbanks, who read portions of the manuscript in draft. I thank Charlotte Weissman for letting me read her dissertation on August Weismann and for helping me think through his theory of heredity.

To colleagues in Paris at the Laboratoire SPHERE, Université de Paris 7-Diderot, and at Sciences Po-Paris, I am abidingly grateful for their warm welcome during an extended visit and for allowing me to be an ongoing participant in their conversation. Thanks especially to Marie-Noëlle Bourguet, Karine Chemla, Olivier Darrigol, Nadine de Courtenay, Liliane Hilaire-Perez, Bruno Latour, David Rabouin, Paul-André Rosental, Stephane Van Damme, and Koen Vermeir. Also in Paris, heartfelt thanks for the same to Daniel Ander and Bernadette Bensaude-Vincent (Université de Paris 1), and Pierre Jacob and François Recanati (Institut Jean Nicod, CNRS, EHESS, and ENS).

For their adept and unstinting research assistance, I thank Ian Beacock, Laura Monkman, and Andrew Schupanitz. Laura initiated the final stage of manuscript preparation before leaving to conduct her own research in Paris; Ian resolved bibliographic snafus, held proficient control over an often unruly manuscript, and made regular suggestions that are incorporated throughout; and Andy achieved masterly feats of image capture, securing reproductions and permissions, while his own photography graces many of the illustrations. Maria Kager and Julia Roever graciously helped me interpret the subtleties of the German language in some of its more opaque applications to turn-of-the-century embryology, theology, and academic culture. I am equally thankful for the expert help of librarians, archivists, and curators in the Department of Special Collections at Stanford University Libraries, in particular John Mustain, Tim Noakes, and Mattie Taormina. I have relied especially on John's great generosity with both his knowledge and his time.

Many other librarians, archivists, and curators in several countries have helped me in the research for this book. They include: Chantal Delepouve, Archives départementales de Pas-de-Calais, France; Nadia Bouzid and Bruno Galland, Archives nationales, Paris, France; Jochen Kirchhoff, Eva Mayring, and Harmut Petzold, Deutsches Museum, Munich, Germany; Hervé Lefebvre and Jean Safarian, Musée des arts et métiers, Paris, France; Pierre Buser, Musée d'horlogerie, Locle, Switzerland; Jean-Michel Piguet, Musée international d'horlogerie, La Chaux-de-Fonds, Switzerland; Peter Widmer, Musik Automaten Museum, Seewen; Margaret Kiechefer, Library of Congress; and Margaret Hughes, Archives and Special Collections, University of California at San Francisco. Oliver Riskin-Kutz provided excellent companionship and research assistance during a trip to the hydraulic automata at the Schloss Hellbrunn in Salzburg.

For expertly shepherding this book from its earliest days, I warmly thank Melissa Chinchillo and Christy Fletcher at Fletcher and Company, as well as Emma Parry, who oversaw the first stages. Edwin Barber gave me adroit editorial help during the last revisions. Thank you to the University of Chicago Press for continuing courageously to represent what book publishing is and should ever remain. In particular, having Karen Darling's editorial collaboration in my work seems an extraordinary kind of luck. I thank also Jennifer Rappaport for her thorough and graceful copyediting, and Tadd Adcox, Jenni Fry, Levi Stahl, Evan White, and the rest of the Press production team. A concluding stroke of luck has provided this book's readers with a Virgil to guide them: Derek Gottlieb has crafted the index with great artistry, insight, and philosophical rigor.

The research for this book has been supported by grants from Iowa State University, the Massachusetts Institute of Technology, Stanford University, the American Council of Learned Societies Charles A. Ryskamp Research Fellowship Program, and the National Science Foundation Scholar's Award program. Essays and articles relating to this book project have appeared in *Critical Inquiry, Representations, Studies on Voltaire and the Eighteenth Century, Republics of Letters, The American Historical Review, The Intellectual History Review,* and in three edited volumes: *A Cultural History of the Human Body,* edited by Carole Reeves; *Early Modern Things,* edited by Paula Findlen; and *Nature Engaged,* edited by Mario Biagioli and me.

The final thanks go to my family for their love and their sense of humor: Myra Jehlen, Carl Riskin, Ruth Kutz, and Chris Kutz. Oliver and Madeleine Riskin-Kutz have enthusiastically cheered me on in writing this book for as long as they can remember; it is dedicated to them with all my love.

ABBREVIATIONS

ADA	Ross Ashby Digital Archive
AdN	Archives départementales du Nord, France
AdPC	Archives départementales du Pas-de-Calais, France
AE	Academy Edition of Kant's Works
ANF	Archives nationales de France
APS	American Philosophical Society
AS	Archives de l'Académie des sciences, Paris, France
AT	Adam and Tannery, eds., *Œuvres de Descartes*
BL	British Library, London, England
BMHN	Bibliothèque centrale du Muséum national d'Histoire naturelle, Paris, France
BNF	Bibliothèque nationale de France, Paris, France
CNAM	Centre national des arts et métiers, Paris, France
CW	Aristotle, *Complete Works*
DCP	Darwin Correspondence Project
JLP	Jacques Loeb Papers, Library of Congress, Washington D.C.
MAH	Musée d'art et d'histoire, Neuchâtel, Switzerland
MRH	Fechner, *Magie de Robert-Houdin*
NA	National Archives, Kew, England
OCM	*Œuvres complètes de Montesquieu*

OCV	*Œuvres complètes de Voltaire*
PE	Leibniz, *Philosophical Essays*
PPL	Leibniz, *Philosophical Papers and Letters*
PS	Leibniz, *Philosophischen Schriften*
SSB	Leibniz, *Sämtliche Schriften und Briefe*
SWH	Ranke, *Secret of World History*
TA	Turing Archive for the History of Computing

~ NOTES ~

Introduction

1. Huxley, "Physical Basis," 129. On James Cranbrook and his radical lecture series, see Stratham, "Real Robert Elsmere."

2. Huxley, "Physical Basis," 139–40.

3. Morley, *Recollections,* 1:90.

4. Some recent examples: Hunter, *Vital Forces,* 70; Wayne, *Plant Cell Biology,* 5; Berkowitz, *Stardust Revolution,* 120.

5. Andrea Page-McCaw, personal communication, March 2015.

6. Campbell et al., *Biology,* chaps. 5.4, 7.6, and 11.4.

7. Leibniz, *Nouveaux essais* (ca. 1690–1705), in his *Philosophischen Schriften* (hereafter cited as *PS*), vol. 5, book 2, chap. 20, at 153.

8. See Helmholtz, "Recent Progress." (The original German version is *Fortschritte.*)

9. Dawkins, *Selfish Gene.*

10. Dennett, *Darwin's Dangerous Idea,* 76.

11. Brooks, "Elephants."

Chapter 1

1. Corley, *Lancelot,* 104–5, 118. This edition is based on Elspeth Kennedy's translation of the prose Lancelot: see Kennedy, *Lancelot.* The original texts on which Kennedy's translation is based are of unknown authorship and written in the early to mid-thirteenth century. They were also the basis of Thomas Malory's *Le Morte d'Arthur* (1485).

2. Corley, *Lancelot,* 192–94.

3. Cooper and Koschwitz, *Pèlerinage,* line 352ff.; Grave, *Eneas,* line 7692ff.; Constans,

Roman, 2:374–94; Britain, *Roman*, 1:309–13; Tors and Bernay, *Romans*, 343–44, 445; Troyes, *Perceval*, 1:64, 201–3, 2:13, 353ff.; Sommer, *Lestoire*, 83; Löseth, *Roman*, 223; Guessard and Grandmaison, *Huon*, 136; Gennrich, *Romans*, 239. See Bruce, "Human Automata," 515; and Truitt, "Knowledge and Automata." I thank Elly Truitt for helpful references on this subject.

4. Wriothesley, *Chronicle*, 1:74. On the Boxley Abbey Rood of Grace, see also Chapuis and Gélis, *Monde*, 1:95; and Jones, "Theatrical History," 243–44.

5. Lambarde, *Perambulation*, 205–6. See also Wriothesley, *Chronicle*, 1:75; and Herbert of Cherbury, *Life*, 494.

6. Lambarde, *Perambulation*, 209–10.

7. Later examples of automaton Christs include an eighteenth-century one in Dachau, Germany, that has human hair. Hidden in its beard are strands that control the movements of the eyes, mouth, and head. Another, in Limpias, Spain, moves its lips, rolls its eyes, blinks, and grimaces. See Chapuis and Gélis, *Monde*, 2:95–96, 1:104. See also Tripps, *Handelnde Bildwerk*, 159–73, 292–93 (plates 10e and f), 325 (plates 42a and b), 326 (plate 43a).

8. Chapuis and Droz, *Automata*, 119–20.

9. Du Camp, *Paris*, 1:375–76. This devil was at the Musée de Cluny in Paris during the nineteenth century. Others are depicted in drawings by the fifteenth-century engineer Giovanni Fontana: Bayerische Staatsbibliothek Cod.icon. 242 (MSS. mixt. 90), 59v–60v, 63v; printed in Battisti and Battisti, *Macchine*, 134–35. On Fontana's automaton devils, see especially Grafton, "The Devil as Automaton." See also Chapuis and Gélis, *Monde*, 2:97–101.

10. Thanks to Paula Findlen for bringing this devil to my attention.

11. D'Ancona, *Origini*, 1:526; Monnier, *Quattrocento*, 2:204.

12. Vasari, *Lives*, 2:229–32. On Brunelleschi's and other *ingegni*, see Prager and Scaglia, *Brunelleschi*; Galluzzi, *Ingegneri*; Grafton, "New Technologies"; and Buccheri, *Spectacle*, chap. 2. Jacob Burckhardt was scathing on the subject of the *ingegni*: the drama, he said, "suffered from this passion for display." See Burckhardt, *Civilization*, 260. The moving figures in Heaven that Vasari mentions were in fact little boys; nevertheless, he considers them as components in an overall work of "machinery." The Paradise apparatus was so heavy that it pulled down the roof of the Carmine monastery where it was installed and necessitated the monks' departure. See Vasari, *Lives*, 2:229. On moving Annunciations, see also Tripps, *Handelnde Bildwerk*, 84–88.

13. See Vasari, *Lives*, 3:194–96.

14. "La Passion et Résurrection de nostre saulveur et redempteur Jhesucrist, ainsi qu'elle fut juée en Valenchiennes, en le an 1547, par grace demaistre Nicaise Chamart, seigneur de Alsembergue, alors prevost de la ville (sixteenth century): BNF, MS. Fr. 12536; Cohen, *Histoire*, 97; Chapuis and Droz, *Automata*, 356–57.

15. Monnier, *Quattrocento*, 2:204; D'Ancona, *Origini*, 1:514–15. These mechanized performances were elaborations of an older tradition of religious puppet shows, or "motions." An individual puppet might also be called a "motion" and the puppeteer was known as the "motion master." William Lambarde described a motion that took place in Witney, in

Oxfordshire, in which the priests used articulated figures to enact the Resurrection. These puppets might have included some mechanized features. The figure of the waking watchman who saw Christ rise made a continuous clacking noise that earned him the nickname "Jack Snacker of Wytney." See Lambarde, *Dictionarium*, 459; and Hone, *Every-Day Book*, entry for September 5, 1825, entitled "Visit to Bartholomew Fair."

16. Rothschild, *Mistère*, 4:112; Cohen, *Histoire*, 147. The *Mistère du Viel Testament* was composed around 1450 by the brothers Arnoul et Simon Gréban and first performed in 1470.

17. Méril, *Origines*, 253; Cohen, *Histoire*, 31.

18. Gréban and Gréban, *Mystère*, 27; Cohen, *Histoire*, 147.

19. Cohen, *Histoire*, 31; Gasté, *Drames liturgiques*, 10.

20. Van Emden, *Jeu d'Adam*, 23 (line 292); the original is in the Bibliothèque municipale de Tours, MS 927. Cohen, *Histoire*, 54.

21. Gréban and Gréban, *Mystère*, 6–7, 20–21, 23; Cohen, *Histoire*, 147. The *Mystère des Actes des Apôtres* was an assemblage of dramatizations of stories from the Old Testament collected around 1450 and staged in Paris at the beginning of the sixteenth century.

22. Cohen, *Histoire*, 143–44; Chapuis and Droz, *Automata*, 356–57.

23. Vasari described the high point of Florentine festivals as having coincided with the career of the architect Francesco d'Angelo (known as Cecca) in the second half of the fifteenth century: Cecca "was much employed in such matters at that time, when the city was greatly given to holding festivals." These took place, according to Vasari, not only in churches "but also in the private houses of gentlemen." There were also four public spectacles, one for each quarter of the city. For example, the Carmine kept the feast of the Ascension of Our Lord and the Assumption of Our Lady. See Vasari, *Lives*, 3:194.

24. See Vidal, *Notre-Dame*; and Chapuis and Gélis, *Monde*, 1:102.

25. See Auguste, "Gabriel de Ciron," 26–29; Chapuis and Gélis, *Monde*, 1:103.

26. This practice was thriving in the mid-seventeenth century, when Madame de Mondonville (Anne-Jeanne Cassanea de Mondonville) recalls in her memoirs having built a Virgin ascender with her brothers. She used her precious glittering bits of quartz crystal to carve the Virgin. See Auguste, "Gabriel de Ciron," 29; and Chapuis and Gélis, *Monde*, 1:103. On mechanical Assumptions, see also Tripps, *Handelnde Bildwerk*, 174–90.

27. The Dieppe mechanical figures were built for a local festival of the Virgin, celebrated on the day of the Assumption and the following day, called the "mitouries de la mi-août." See Desmarquets, *Mémoires*, 1:36; Maindron, *Marionnettes*, 99–102; Vitet, *Histoire*, 45; and Chapuis and Gélis, *Monde*, 1:103. This performance, like the others discussed here, combined living and mechanical actors.

28. Lambarde, *Dictionarium*, 459; Hone, *Every-Day Book*, entry for September 5, 1825; Chapuis and Gélis, *Monde*, 1:104.

29. See Heilbron, *Sun in the Church*; and Landes, *Revolution in Time*, esp. chap. 3.

30. Chapuis and Gélis, *Monde*, 1:114; Chapuis and Droz, *Automata*, 53. The name may have come from Saint John the Baptist. See Wins, *Horloge*.

31. The current clock in the Piazza San Marco is the result of many modifications that have taken place over the centuries, but these central elements have remained. On the his-

tory of the clock, see Erizzo, *Relazione*; Chapuis and Gélis, *Monde*, 1:118; and Zamberlan and Zamberlan, "The St Mark's Clock, Venice."

32. For other examples, see Chapuis and Gélis, *Monde*, vol. 1, chap. 7.

33. "Le coq, qui aux temps les plus reculés était déjà symbole de fierté et vigilance, est sans contredit l'animal que les constructeurs de machines horaires à sujets animés, ont mis le plus souvent à contribution." In Chapuis and Gélis, *Monde*, 1:172–73.

34. Chapuis and Gélis, *Monde*, 1:120. Cluny Abbey was the biggest church in Europe before Saint Peter's Basilica in Rome; it was almost entirely destroyed in 1790 during the French Revolution.

35. "Description d'une horloge merveilleuse," par "Jean BOUHIN," BNF, fonds français, MS no. 1744, published as Duret, "L'horloge historique."

36. Chapuis and Gélis, *Monde*, 1:120–27; Price, "Automata," 18, 22; Bedini, "Role of Automata," 29, figs. 2 and 3.

37. On the connection between organs and automata in the medieval period, see Sherwood, "Magic," 588–89.

38. Hamel, "Notice historique," l-li; Chapuis and Gélis, *Monde*, 1:106. A 1541 bill records: "A Nicolas Quesnel, ymaginier, pour faire deux ymages des anges mouvantz, pour mettre sur l'amortissement des orgues." In Laborde, *Notice*, 2:342; Chapuis and Gélis, *Monde*, 1:105.

39. Chapuis and Gélis, *Monde*, 1:106. Joseph Gallmayr built a Saint Cecilia in Munich powered by organ pedals in the late eighteenth century. See ibid.

40. "Rohraffen" meant "roaring apes" and referred to the grotesque, bellowing figure of the Bretzelmann. He had help, for the vocal part of his act, from a Münsterknecht, a servant of the cathedral hidden in the pendentive that held the organ. But the physical motions were the Bretzelmann's own and were controlled by the organ strings. See Gass, *Orgues*; Conseil Régional d'Alsace, *Orgues*; Chapuis and Gélis, *Monde*, 2:108–9; and Sherwood, "Magic," 585–86.

41. Item rusticanam quondam imaginem: in sublimi sub organis: in ecclesia maiori collocarunt. Qua sic abutuntur. In ipsis sacris diebus Penthecostes: quibus ex tota dyocesi populus processionaliter cum sanctorum reliquijs: deuocionis et laudandi dei gratia canens et iubilans: matricem ecclesiam subintrare consueuit. Nebulo quispiam se post iiam imaginem occultans: incomptis motibus: voce sonora: prophana et indecora cantica eructans: veniencium hymnis obstrepit: eosque subsannando irridet: ita ut non solum illorum deuocionem in distractionem: gemitus in cachinnos vertat: sed et ipsis clericis diuina psallentibus: sit impedimento: immo diuinis missarum solemnijs (quas non longe inde celebrare contingit) ecclesiastici immo diuini cultus gelatori longe abominandam et execrandam afferat perturbacionem [cited passage underlined]. Peter Schott and Emmerick Kemmel (ca. 1482–85), in Wimpheling, *Patricii*, fols. 116a–117b. Published in both the original Latin and in English translation in Cowie and Cowie, "Geiler von Kayserberg," here at 489–90, 494.

42. Bourdette, Monastère; Chapuis and Gélis, Monde, 1:106–7.

43. Salies, "Lettre sur une tête automatique," 114. Three similar heads continue to grace the organ, built in 1557, of the Church of Saint-Savin-en-Lavedan in the Hautes Pyrénées.

44. Ibid., 99.

45. Ibid., 98. On the basis of this head's markedly Lancastrian features and its nickname Gallima (which could mean "he who brings death to the French"), Salies suggests that the head may have been an effigy of Henry V, Henry VI, or Humphrey, Duke of Gloucester, who acted as Protector during the regency of Henry VI.

46. The text was edited by Giovanni Sulpizio da Veroli, a close collaborator of the young Cardinal Raffaele Riairo, to whom the edition was dedicated, and who was a central actor in the Renaissance renovation of Rome. Most of Vitruvius's illustrations had been lost; the Veronese friar Fra Giovanni Giocondo reconstituted these in a 1511 edition printed under the auspices of Pope Julius II. Vitruvius's treatise continued to inform the great building projects of the Renaissance popes through the reigns of Leo X and Paul III, not least Saint Peter's Cathedral. See Ingrid D. Rowland's introduction to *Vitruvius,* 1–31.

47. The astronomical tower clock in Bern, the Zytglogge, built in 1530, continues to play a four-minute mechanical pageant every hour: a rooster crows; a dancing jester rings the bells; Chronos turns his hourglass; bears (the city mascot) parade around.

48. See Cipolla, *Clocks and Culture*; Le Goff, *Medieval Imagination*, esp. chap. 1; Landes, *Revolution in Time*; and, for a slightly later period, Hanafi, *Monster in the Machine.*

49. Chapuis and Gélis, *Monde,* 1:165–66. The sovereign and his seven electors reappeared on the clock of the Marienkirche in Lübeck where they paraded before Christ. See ibid., 1:170.

50. Vasari, *Lives,* 4:99. The other known primary mentions of this event are Giovanni Paolo Lomazzo, *Libro dei Sogni* (1564) and *Trattato dell'arte della pittura, scoltura et architettura* (1584), both in Lomazzo, *Scritti,* 1:53 and 2:96. See also Lomazzo, *Idea,* 17, for a mention of "andari Leoni per forza di ruote"; and Buonarroti, *Descrizione,* 10. See Burke, "Meaning." Accounts vary as to the French king in question, the occasion, the date, and the place. The currently established view, which I have adopted here, is that of Carlo Pedretti, who discovered the Buonarroti the Younger account in 1973. See Pedretti, *Leonardo architetto,* 322. See also his "Leonardo at Lyon" and his *Study in Chronology and Style,* 172. On the mechanical lion and other creations of Leonardo that might be seen as automata—flying machines, the figure of a knight with an internal cable-and-pulley system enabling it to wave its arms and move its head and mouth, and a wheeled cart that might have been a programmable robot platform— see also Rosheim, *Leonardo's Lost Robots.* On Leonardo's uses of cords and wires to model muscles, see Galluzzi, "Leonardo da Vinci." An even earlier instance of an automaton designed for purely secular political purposes is the iron fly purportedly presented by Johannes Müller (Regiomontanus) (1436–76), to Maximilian I, and an eagle that escorted the emperor to the city gates of Nuremberg. The first account of these machines is by Peter Ramus in *Scholarum,* 62. John Dee then mentions Regiomontanus's automata in his "Mathematicall Praeface" to Henry Billingsley's 1570 English translation of Euclid's *Elements.* See Dee, *Mathematical Praeface.*

51. Doppelmayr, *Historische Nachricht,* 285. See Chapuis and Gélis, *Monde,* 2:181; and Bedini, "Role of Automata," 31.

52. See Chapuis and Gélis, *Monde,* 1:179ff.

53. Bedini, "Role of Automata," 34; Chapuis and Gélis, *Monde,* 1:192–97, 2:152–53;

Chapuis and Droz, *Automata*, 76–77, 242. Aristocratic dining tables were also set with automaton-embellished fountains that poured out wine and perfumed liquor. See Bedini, "Role of Automata," 33. Bedini mentions a fourteenth-century example at the Cleveland Museum of Art. See Penzer, "Fourteenth-Century Table Fountain."

54. Chapuis and Droz, *Automata*, 67. With the development of springs rather than weights as the motive force, clocks got ever smaller and more elaborate. See ibid., 77. Clocks and ultimately, with the addition of pinned barrels to organize the motions, watches too, often included carillons. See ibid., 265–66.

55. The Italian architect and stage designer Nicola Sabbattini regularly included mechanized elements in his sets: crank-operated ocean waves; clouds on pulleys. See Sabbattini, *Manual*, 132–33, 149–50. Other principal builders of mechanical sets for secular theater include Giacomo Torelli, the seventeenth-century Italian stage designer and engineer who invented the "chariot-and-pole" system of changing sets and who directed the fêtes of Louis XIV. See Guarino, "Torelli a Venezia"; Gamba and Montebelli, *Macchine*. Another important figure in this area was Ferdinando Galli Bibiena (1657–1743), the most illustrious of an Italian dynasty of architects and set designers. See Bibiena, *L'architettura civile* for a modern edition of his work with an introduction by Diane M. Kelder.

56. Dee, *Autobiographical Tracts*, 5–6. As mentioned above, Dee discusses several legendary automata—Archytas's wooden dove, Albertus Magnus's brazen head, Regiomontanus's iron fly and artificial eagle—as examples of the mathematical art of "thaumaturgike" in his *Mathematical Praeface*. On Dee's automaton, see Woolley, *Queen's Conjurer*, 12–15; and Harkness, *Conversations with Angels*, 121.

57. Honnecourt, *Sketchbook*, 58c, 58e. Also in BNF, MS. Fr. 19093. Nothing is known about the artist apart from what can be inferred from his portfolio itself.

58. My discussion of the monk is derived from the work of Elizabeth King. See her "Perpetual Devotion" as well as her "Clockwork Prayer." The monk now resides in storage at the Smithsonian Institution's National Museum of American History; it has a twin at the Deutsches Museum in Munich.

59. On Turriano, see García-Diego, *Relojes y autómatas*; García-Diego, *Juanelo Turriano*; King, "Clockwork Prayer"; Bedini and Maddison, *Mechanical Universe*, 56–58; Bedini, "Role of Automata," 32; and Chapuis and Gélis, *Monde*, 1:90–91. Originally from Cremona, Turriano was also known as Giovanni Torriani and as Gianello della Torre.

60. Strada, *De bello belgico*, 7. See also Sterling-Maxwell, *Cloister Life*, 116, 178–80, 499; Montañes Fonteñla, "Relojes del Emperador"; and Bedini, "Role of Automata," 32. One of Turriano's automata, a woman playing a lute, remains in existence. It is in the collection of the Kunsthistorisches Museum in Vienna.

61. King, "Perpetual Devotion," 264–66, 274–75.

62. On the complex relations of matter to spirit in late medieval theology, philosophy, and religious practice, see Bynum, *Christian Materiality*.

63. At the beginning of the twentieth century, the German sociologist Max Weber described the Protestant Reformation as having accomplished a "disenchantment of the world." In Weber, "Psychology of World Religions," 290. See also his *Protestant Ethic* and *Sociology of Religion*. This disenchantment or desacralization, Weber argued, had brought a newly intellectualized, transcendental form of religious belief together with a rationalist,

ethical approach to everyday life that had been conducive to the development of capitalist enterprise. For over a century now, historians have been discussing Weber's characterization. First, they sought to broaden it. In 1938, for example, the American sociologist Robert Merton extended Weber's claim by applying it to the Scientific Revolution in England; see Merton, *Science*. In 1971, Weber's thesis informed the British historian Keith Thomas's anthropological study of the history of religion: *Religion and the Decline of Magic*. Most recently, historians have been amending, revising, and denying Weber's claim, in part by showing evidence of the persistence of magical beliefs and practices through the Reformation. For a cogent overview of this long conversation and an assessment of its current state, see Walsham, "Reformation." See also Latour, *We Have Never Been Modern*.

No one contests, however, that the core Protestant doctrines asserted a new set of distinctions between the material, earthly, natural world and the divine, spiritual realm. In this book I am interested in how these distinctions informed the new scientific model of nature and of science that emerged around the middle of the seventeenth century. These Reformation-era distinctions had the effect of building two assumptions into the foundations of modern science: the assumption of a transcendent designer God and of his artifact, a passive world-machine. This book follows the ramifications of these assumptions, in particular for the sciences of life.

The Reformation is a vast topic with a commensurately vast literature. For overviews, see Hillerbrand, *Protestant Reformation*; McGrath, *Intellectual Origins*; Muir, *Ritual*; and MacCulloch, *Reformation*. For a recent representative of the revisionist perspective on Weber's disenchantment thesis, see Soergel, *Miracles*.

64. The art historian David Freedberg has described a certain kind of viewing that comes in response to powerful religious representations. The devout beholder "reconstitutes" the thing being represented, turning its representation into a presence: "The slip from representation to presentation is crucial . . . from seeing a token of the Virgin to seeing her there." See Freedberg, *Power of Images*, 28.

65. Leaders of the Reformed movement, notably Ulrich Zwingli, denounced the use of organs and other musical instruments in church. See Garside, *Zwingli*; Faulkner, *Wiser than Despair*, chap. 9; and MacCulloch, *Reformation*, 146, 590.

66. Chapuis and Gélis, *Monde*, 1:104–5.

67. Chamber to Cromwell, February 7, 1538, in Brewer, *Letters and Papers*, 13:79, no. 231.

68. Lambarde, *Perambulation*, 210.

69. For synoptic discussions of this vast and complex subject, see Besançon, *Forbidden Image*, chaps. 5 and 6; Eire, *War against the Idols*, chaps. 3, 4, 6, and 8; Freedberg, *Power of Images*, chap. 8; Michalski, *Reformation and the Visual Arts*; Dyrness, *Reformed Theology*; Muir, *Ritual*, chaps. 5 and 6; and Latour and Weibel, *Iconoclash*. For a study of the relationship between Reformation art and iconoclasm, see Koerner, *Reformation of the Image*.

70. Page, *County of Kent*, 2:154.

71. On the popularity of the Rood of Grace as a tourist attraction, see the letter of William Warham, Archbishop of Canterbury, to Thomas Wolsey on May 3, 1524 [R.O.] in Brewer, *Letters and Papers*, 4:127, no. 299.

72. Page, *County of Kent*, 2:154.

73. Wriothesley, *Chronicle*, 1:74.

74. Page, *County of Kent*, 2:153–55.

75. Wriothesley, *Chronicle*, 1:75; see also Stow and Howes, *Annales*, 575: "The 24, of February being Sonday, the Rood of Boxeley in Kent, called the Rood of Grace, made with divers vices, to move the eyes and lips, was shewed at Pauls Crosse by the Preacher, which was the Bishop of Rochester, and there it was broken, and plucked in pieces."

76. Lambarde, *Perambulation*, 207, 206, 210, 208.

77. Chapuis and Gélis, *Monde*, 1:104.

78. Auguste, "Gabriel de Ciron"; Chapuis and Gélis, *Monde*, 1:103.

79. Waterworth, *Canons and Decrees*, 235–36.

80. On the history of the *presepio*, see Gargano, *Presepio*; and De Robeck, *Christmas Crib*, chap. 10 and fig. 39. On mechanical nativity scenes, see also Chapuis and Gélis, *Monde*, 2:200–202.

81. From Kircher's *Ars Magnesia*, quoted in translation from Gorman, "Between the Demonic and the Miraculous," 68. On Kircher's automata, see Bedini, "Role of Automata," 35; Haspels, *Automatic Musical Instruments*; Findlen, "Scientific Spectacle"; Findlen, "Introduction," 34–35; Hankins and Silverman, *Instruments and the Imagination*, chaps. 2–4; Gorman, "Between the Demonic and the Miraculous"; and Gouk, "Making Music."

82. The mechanical nativity scene had been donated by Ferdinand of Bavaria, the Elector of Cologne. It and the other automata that Trigault shipped in 1618 did not arrive directly at their destination since the Emperor Wan-Li was even then expelling the Jesuits from China. But a couple of decades later, after the Ming dynasty gave way to the Manchurian T'sing dynasty, many of the automata arrived belatedly as the Jesuits returned under Emperor Shun-che and set up clockmaking workshops. See Lamalle, "Propagande"; Chapuis and Droz, *Automata* (1958), 77–84; and Spence, *Memory Palace*, 180–84. On the Jesuits' clockwork gifts in China, see also Ricci, *China in the Sixteenth Century*, book 1, chap. 4 and book 4, chap. 12.

83. Pfister, *Notices*, notice 88; Chapuis and Droz, *Automata*, 315.

84. Chapuis and Gélis, *Monde*, 1:192; 2:141–43, 152–53; Chapuis and Droz, *Automata*, 242.

85. Principal builders of animated paintings during the last decades of the seventeenth century included the clockmakers Abraham and Christian-Théodore Danbeck and Christophe Leo of Augsburg; during the early eighteenth century, Jean Truchet (le père Sébastien). See Chapuis and Gélis, *Monde*, 1:319.

86. Late medieval and early modern palace waterworks were informed by translations of ancient texts, notably the works of Hero of Alexandria, and made virtually no changes to the ancient mechanisms. See Bedini, "Role of Automata," 26.

87. Richard, *Petite-nièce*, 333; Sherwood, "Magic," 589–90. On Hesdin, see also Tronzo, *Petrarch's Garden*, 101–10.

88. ANF, KK 393; Richard, *Petite-nièce*, 308.

89. AdPC, A 297; Richard, *Petite-nièce*, 336.

90. AdPC, A 548; Richard, *Petite-nièce*, 341.

91. AdPC, A 648; Richard, *Petite-nièce*, 342.

92. *Ve compte de Jehan Abonnel dit Legros, conseilleur et receveur général de toutes les finances de monseigneur le duc de Bourgoingne . . .* , in the *Recette générale des finances, Chambre des comptes de Lille*, AdN, série B no. 1948 (Registre); Laborde, *Ducs de Bourgogne*,

1:268–71. Quoted passage taken from translated excerpt in Sherwood, "Magic," 587–90; Price, "Automata," 20–21; Chapuis and Gélis, *Monde*, 1:72.

93. On the fame and influence of the Hesdin engines, see Sherwood, "Magic," 590; and Price, "Automata," 21.

94. Montaigne, *Journal de voyage*, 125. See also Häberlein, *Fuggers of Augsburg*, chap. 6.

95. Chapuis and Gélis, *Monde*, 1:74.

96. Montaigne, *Journal de voyage*, 187. See also De Caus, *Raisons des forces mouvantes*, 1:29, 2:13.

97. Hülsen, "Ein deutscher Architekt," 164–65; Castellan, *Letters on Italy*, 92; Chapuis and Gélis, *Monde*, 1:75. On hydraulic entertainments in Italy, see also Morel, *Grottes maniéristes*.

98. Montaigne, *Journal de voyage*, 388.

99. Ibid., 270.

100. Ibid. The owl-and-birds arrangement was taken from a much-imitated design by Hero of Alexandria. See Woodcroft, *Pneumatics of Hero of Alexandria*, no. 15. On Hero's automaton, see Berryman, "Imitation of Life."

101. The primary source of information about the Francini fountains at Saint-Germain-en-Laye is a collection of engravings by Abraham Bosse, done from the original designs. Alessandro Francini is listed as the author of the collection: Alessandro Francini, *Recueil. Modèles de grottes et de fontaines. Dessins lavés*, in BNF Estampes et photographie, Réserve Hd-100(A)-Pet Fol; also in ANF, O[1] 1598. There is also John Evelyn's description in his diary: *Diary*, entry for February 27, 1644. See also Evelyn, *Elysium Britannicum*, book 2, chaps. 9 and 12, book 3, chap. 9. For other primary accounts of the hydraulic automata at Saint-Germain-en-Laye, see Houdard, *Châteaux Royaux*, vol. 2, book 3, part 3, chap. 2. For secondary descriptions, see Mousset, *Francine*, chap.1 and plates 2, 3, and 4; La Tourrasse, *Château-neuf*; Marie, *Jardins*; and Chapuis and Droz, *Automata*, 43–47. The fountains were abandoned after Louis XIV moved his court to Versailles in 1682 and virtually no trace remains. On the fountains at Versailles, however, as emblems of material power, see Mukerji, *Territorial Ambitions*, esp. 181–97.

102. Evelyn, *Diary*, entry for February 27, 1644.

103. Du Chesne, *Antiquités*, 221–24.

104. Evelyn, *Diary*, entry for February 27, 1644; Mousset, *Francine*, 35–41.

105. Du Chesne, *Antiquités*, 222; Evelyn, *Diary*, entry for February 27, 1644.

106. Evelyn, *Diary*, entry for February 27, 1644; Mousset, *Francine*, 38.

107. Hérouard, *Journal*, 1:370, entry for September 27, 1601: "A dix heures et demie et demy quart selon ma monstre faicte a Abbeville par M. Plantard."

108. See, for instance, Hérouard, *Journal*, 1:676, entry for June 6, 1605.

109. In April and May 1605 alone, visits to the grottoes are recorded on April 11, 13, 14, 17, and 29; May 2, 8, 9, 15, 27, and 29. Hérouard, *Journal*, 1:638, 639, 643, 653, 655, 657, 658, 660, 666, 668.

110. Ibid., 1:633, entry for March 20, 1605: "Faite semblan que je sui Ophée (Orphée) e vous le fontenié (-er), dite chante le (les) canarie." The text is my translation of the Dauphin's baby talk as recorded by Hérouard.

111. See entries for April 13, 14, and 17, 1605: Ibid., 1:638, 639, 643.

112. See entries for May 25, June 2, July 3 and 30, August 20 and 28, and October 25,

1605; May 9, June 15, and July 8, 1606, in ibid., 1:664–65, 672, 703, 722, 741, 759, 809, 943, 987, 1000.

113. See entries for April 16 and 18, June 1, 2, 6, 10, 13, 15, 22, 26, and 30, and August 3, 1605, in ibid., 1:640, 643, 671, 672, 673, 676, 681, 684, 686, 692, 696, 698, 725.

114. See entries for February 12, March 13, and September 27, 1605, in ibid., 1:596, 614, 767.

115. "*Maman ga, allon voir ma fontaine ché Francino*". M. "Mr., je n'ay point de carrosse". D. "*Nous iron bien a pié*". Madame luy dict: "Mousieu i a bien loin". D. "*Madame, nou passerons pa le jadin nou i seron incontinent.*" Entry for May 25, 1605, in Hérouard, *Journal,* 1:664–65. "Maman ga" was Madame de Montglat (Françoise de Longuejoue), the Dauphin's governess. See also the entries for May 30 and June 7, 1605, in ibid., 1:669, 678.

116. See entries for April 29 and June 2, 1605, in Hérouard, *Journal,* 1:653, 672.

117. "Dict qu'il y a ung robinet a son cul et ung autre sa guillery: '*fs fs.*'" Entry for April 18, 1605, in Hérouard, *Journal,* 1:643. For some repetitions of the joke, see entries for April 2, June 1 and 10, 1605, in ibid., 1:638, 671, 681. The eroticism that pervades Hérouard's journal has occasioned a good deal of analysis. See, for example, Ariès, *L'enfant.*

118. Hérouard, *Journal,* 1:1502, entry for September 12, 1608.

119. See entries for July 31, 1611, and October 27, 1612, in ibid., 2:1939, 2066.

120. Ibid., 1:699, entry for June 30, 1605: "*Maman ga, je sone les heure, dan, dan, i (il) sone come le jaquemar qui frape su l'enclume.*" The prince referred to the clock at Fontainebleau.

121. Ibid., 1:1396, entry for March 5, 1608.

122. Ibid., 1:1434, entry for May 16, 1608.

123. Antoine Morand's clock with automata, made in 1706 to honor Louis XIV, is currently in the Musée de Versailles. Chapuis and Gélis, *Monde,* 1:233–37. François-Joseph de Camus built a mechanical carriage for the future Louis XIV (then still the dauphin) with moving figures of soldiers and a lady. The carriage may have been melted down along with all other objects of precious metal, by order of the king, in 1709, when war had depleted the royal treasury. See Camus, *Traité des forces mouvantes,* 521–33; and Chapuis and Gélis, *Monde,* 2:13–18. Jean Truchet (le père Sébastien) built the mechanical opera. It is described in Fontenelle, *Suite des éloges,* 170; and in Fontenay, "Notice." See Chapuis and Gélis, *Monde,* 1:337.

124. The automaton army was built by Gottfried Hautsch in Nuremberg for Louis, le Dauphin du Viennois (le Grand Dauphin), born in 1661. See Doppelmayr, *Historische Nachricht,* 304; and Chapuis and Gélis, *Monde,* 2:12–13.

125. See Wharton, *Italian Villas,* 154.

126. On the Villa Aldobrandini (originally named the Villa Belvedere), see Benocci, *Villa.* See Giovanni Battista Falda's engraving of the Stanza dei Venti in *Le fontane di Roma,* plate 7.

127. On Sittikus and the Schloss Hellbrunn, see Chapuis and Gélis, *Monde,* 1:76–77; Chapuis, "Amazing Automata"; Bedini, "Role of Automata," 27; and Schaber, *Hellbrunn.*

128. See Falda's engraving of the Semicircular Water Theater of the Villa Mondragone in *Le fontane di Roma,* plate 18.

129. The story of the wedding and De Caus's work in Heidelberg is told in Yates, *Rosi-*

crucian Enlightenment, chap. 1. Yates suggests that De Caus invented the mechanical use of steam power at ibid., 19.

130. De Caus, *Raisons des forces mouvantes* and *Hortus Palatinus*.

131. De Caus, *Raisons des forces mouvantes*, 2: np, 1r, 2:16v, 17r; 1:34v, 35r; see also Chapuis and Gélis, *Monde*, 1:78–82.

132. Valla, *Georgii Vallae*; Commandino, *Aristarchi*; Aleotti, *Gli artificiosi*. Subsequent editions included Giorgi da Urbino, *Spirituali*. See also Bedini, "Role of Automata," 25.

133. Ramelli, *Various and Ingenious Machines*, chap. 186. See also chap. 187, which describes an essentially similar arrangement in which the effects are produced by a concealed person blowing through a pipe rather than by a flow of water. Another example is Della Porta, *Pneumaticorum*.

134. De Caus, *Raisons des forces mouvantes*, 1:30: "Pour faire representer plusieurs oiseaux lesquels chanteront diversement quand une choüette se tournera vers iceux, & quand ladite choüette se retournera, ils cesseront de chanter." See also De Caus, *Nouvelle invention*, 125, plate 13.

135. De Caus, *Raisons des forces mouvantes*, 1:34: "Machine par laquelle l'on representera une Galatee qui sera trainee sur l'eau par deux daufins, allant en ligne droite, & se retournant d'elle mesme, cependant qu'un ciclope Joüe dessus un flajollet."

136. Evelyn, *Elysium Britannicum*, 231, 242, 191. The markings indicate Evelyn's amendments in the manuscript.

137. Kircher, *Musurgia Universalis*, 2:347; Bedini, "Role of Automata," 35.

138. Price writes that in automaton clocks, "for the first time [around 1550], wheelwork is used instead of levers, gears instead of strings, organ-barrel programming instead of sequential delay devised hydraulically." Price, "Automata," 22.

139. Al-Jazari, *Book of Knowledge*. On al-Jazari and the medieval Islamic tradition of automaton-making, see Tabbaa, "Medieval Islamic Garden," 322–39; Dolezal and Mavroudi, "Theodore Hyrtakenos," 125–31; and al-Hassan, *Science and Technology in Islam*, 181–83.

140. This was a collaborative work. The goldsmith Randolph Bull, clockmaker to the queen, built the clock part of the device while the organ maker Thomas Dallam built the organ part and supervised installation in the sultan's palace. See Drover, "Thomas Dallam's Organ Clock"; and Bedini, "Role of Automata," 35.

141. De Caus, *Raisons des forces mouvantes*, 1:29.

142. Price, "Automata," 20. Langenbucher's Pomeranian Cabinet, presented to Duke Philip of Pomerania in 1617, held an automatic organ of twenty-one pipes that played four pieces of music. On the cabinet, see Haspels, *Automatic Musical Instruments*, 55–57.

143. Evelyn, *Elysium Britannicum*, 232–42.

144. Ibid., 241.

145. Ibid., 232, 244, 249–50. Evelyn cites Pliny, Philostratus, Pausanius, Lucian, Tacitus, Strabo, and Kircher as authorities on the legendary statue. See also Kircher, *Oedipus Aegyptiacus*, vol. 2, part 2, 326–27; and otherwise on speaking statues, vol. 3, 488.

146. On Martin Löhner and his automata, see Doppelmayr, *Historische Nachricht*, 306; and Chapuis and Gélis, *Monde*, 1:76.

147. Evelyn, *Diary*, entry for February 27, 1644.

148. Evelyn, *Elysium Britannicum,* 184, 439.

149. Montpensier, *Mémoires,* chap. 23 (July–September 1656). The owner of the estate, whom the writer identifies as "Esselin," was Louis Cauchon d'Hesselin, who served as Maître de la chambre aux deniers de la maison du roi. Madame de Lixein was Henriette de Lorraine, daughter of François de Lorraine, who had married the prince of Lixen.

150. Darnton, *Great Cat Massacre,* 78.

151. Bergson, *Le rire, "du mécanique plaqué sur la vivante"*: 39, 50, 58; laughter as a *"correction"*: 89, 174, 197, 200. For Freud's use of Bergson's theory of humor in his own very different account, see Freud, *Jokes,* chap. 7, esp. 259–60.

152. For an anthology of recent treatments of this subject, see Bremmer and Roodenburg, *Cultural History of Humour.* Peter Burke mentions Renaissance palace waterworks in passing in "Frontiers of the Comic," 84. The history of humor is an integral part of the discussion in Norbert Elias's classic work, *The Civilizing Process.*

Chapter 2

1. This according to Descartes's first biographer, Adrien Baillet; see Gaukroger, *Descartes,* 64. The earliest evidence of Descartes's interest in automata is a note in the *Cogitationes privatae* in which he describes designs for an automaton like the dove of Archytas or a tightrope walker. See Descartes, *Œuvres de Descartes* (hereafter AT), 10:231–32.

2. On the role of mechanism in seventeenth-century physiology, and on the development and influence of René Descartes's physiology in particular, see Brown, "Physiology"; Dear, "Mechanical Microcosm"; Duchesneau, *Modèles du vivant,* chaps. 2–3; Jaynes, "Problem of Animate Motion"; and Sloan, "Descartes."

3. On the misconstruals, see Gaukroger, *Descartes' System,* 181. On modern, Anglophone misconstruals, see Morris, "Bêtes-machines."

4. From Descartes's *Traité de l'homme,* in AT, 11:120. The *Traité* was written between 1630 and 1633 but was first published in 1662 in Latin. See AT, 6:54–56 for the argument in *Discours de la méthode* (1637) that God could build even better automata; and "Méditation sixième" (1641), in AT, vol. 9, part 1, 67, 69 for the claim of "the body of man as being a machine constructed in a certain way and composed of bones, nerves, muscles, veins, blood and flesh."

5. *Traité de l'homme,* in AT, 11:130–31. On Descartes's reference to palace waterworks such as Saint-Germain-en-Laye and the Palatine Gardens, see also Jaynes, "Problem of Animate Motion"; Des Chene, *Spirits and Clocks,* 13n1; and Descartes, *Monde,* 179n45.

6. *Traité de l'homme,* in AT, 11:130–31.

7. *Discours de la méthode,* in AT, 6:46, 48, 50. On the heart as the "the great spring, and principle of all the motions" of the body, see *La description du corps humain,* in AT, 11:226, 227. Descartes began writing this text in the winter of 1647–48 but died before completing it. Gaukroger sees it as a partial draft of the projected fifth book of the *Principia philosophiae,* which was to treat "living beings." See the introduction to Gaukroger, *World,* xxix.

8. Harvey, *Motion of the Heart,* 31. Harvey does not here use the word "pump," but he does describe the auricle as the "store-house and cistern" of the blood. He makes the water-pump analogy explicit in his *Second Disquisition,* 135. In general, however, despite this divergence from Aristotle's account of the heart, Harvey considered himself an Aristotelian.

For a recent treatment of the role of hydraulic pumps in Harvey's thinking, see Wright, *William Harvey*, essay 7.

9. *Description du corps humain*, in AT, 11:228–45; see also *Traité de l'homme*, in AT, 11:123. On Descartes's rejection of Harvey's view of the heart as a pump, see Bitbol-Hespériès, *Principe de vie*, 185–91 as well as her "Cartesian Physiology"; Gaukroger, *Descartes*, 271–72; Des Chene, *Spirits and Clocks*, 28–29; and Gilson, *Études*, 80–100.

10. *Description du corps humain*, in AT, 11:244–45.

11. *Traité de l'homme*, in AT, 11:130–31.

12. "Méditation sixième," in AT, vol. 9, part 1, 69.

13. *Les principes de la philosophie* (1644), part 4, sections 189–95, in AT, vol. 9, part 2, 310–14.

14. For the list of passions, see *Les passions de l'âme* (1647), part 2, articles 53–67, in AT, 11:373–78 and *Traité de l'homme*, in AT, 11:165–66. For first and last causes, see *Passions de l'âme*, part 2, article 51, in AT, 11: 371–72.

15. *Traité de l'homme*, in AT, 11:165–66. The physical model that Descartes offered for memory was a piece of cloth punctured by a needle: "If one were to pass several needles, or engraver's points, through a cloth, as you see (*Fig. 30*) in the cloth marked A, the little holes that one would make there would stay open, as around *a* and around *b*, after the needles were withdrawn; or if they did close again, they would leave traces in this cloth, as around *c* and around *d*, which would make them very easy to open again." *Traité de l'homme*, in AT, 11:178–79. On the sentience of the animal-machine, see also Gaukroger, *Descartes*, 287–88.

16. *Description du corps humain*, part 4, in AT, 11:252–72. On the importance of the material on embryology, see Gaukroger, *Descartes*, 405–6, as well as his "Resources," and his introduction to *World*, xxviii–xxix.

17. *Traité de l'homme*, in AT, 11:131–32.

18. Ibid., 11:131.

19. Ibid., 11:202. See also *Description du corps humain*, in AT, 11:226: "There is no more reason to postulate a soul to explain those movements we experience as being un-willed than 'to judge that there is a soul in a clock that makes it show the time'"; and *Passions de l'âme*, part 1, article 16, in AT, 11: 341–42: "All the movements that we make without our will contributing to them (as happens often when we breathe, walk, eat, and do all the actions we have in common with animals) depend only on the formation of our parts and on the course that the spirits, excited by the heat of the heart, naturally follow in the brain, in the nerves and in the muscles, in the same way that the movement of a watch is produced by the sole force of its spring and the configuration of its wheels." On Descartes's elimination of the vegetative and sensitive souls, and its implications, see Des Chene, *Spirits and Clocks*.

20. "It is Nature that acts in them, according to the disposition of their organs: in the same way a clock, composed only of wheels and springs, can count the hours, and measure the time, more justly than we with all our wisdom." *Discours de la méthode*, in AT, 6:59.

21. Gaukroger has emphasized this point in *Descartes*, see esp. 287–89.

22. *Passions de l'âme*, in AT, 11:330–31.

23. Descartes to Marin Mersenne, November or December 1632, in AT, 1:263.

24. Descartes, *Traité de l'homme*, in AT, 11:165.

25. Describing Descartes's hydraulic understanding of the nervous system, Jaynes writes: "The Francini brothers from Florence had built more than they knew; they had created the essential image behind modern psychology." See Jaynes, "Problem of Animate Motion," 225.

26. On the importance of hydraulic as opposed to clockwork mechanism in Descartes's physiology, see the introduction to Gaukroger, *World*, xxiv; and Des Chene, *Spirits and Clocks*, 39: "To think only of clocks and wind-up toys in imagining a Cartesian automaton is to be quite misled."

27. Perrault and Perrault, *Œuvres diverses*, 1:359–60, cited in Fearing, *Reflex Action*, 33; and Jaynes, "Problem of Animate Motion," 233.

28. *Discours de la méthode*, in AT, 6:54–56.

29. See Gaukroger, *World*, xxiii; and Gaukroger, *Descartes*, 270. For in-depth studies of the relations of Cartesian to ancient and Scholastic physiology, see Bitbol-Hespériès, *Principe de vie*, esp. part 1, chap. 2; part 2, chap. 2; and part 3, chap. 1; see also Sanhueza, *Pensée biologique*; and Gilson, *Études*, chap. 2. Certain areas of Descartes's physiology were newer than others, notably his account of embryology and the circulation of the blood. See Gaukroger, *Descartes' System*, 182–83.

30. Aristotle, *Movement of Animals*, 701b2–33, in *Complete Works* (hereafter *CW*), vol. 1. In the continuation of the passage, Aristotle distinguishes animals from automata by their capacity to change qualitatively: their parts could expand and contract, grow hotter and colder. Although he asserted a fundamental distinction between nature and art, machines and living things, Aristotle also regularly crossed this divide. On the assumption that art imitated nature, he found instruments and machines to be convenient models for understanding natural phenomena. Aristotle, *Parts of Animals*, book 1, 639b30, in *CW*. On the complicated relations between art and nature in Aristotelian and Scholastic philosophy, see Des Chene, *Physiologia*, chap. 7, esp. 243; Close, "Commonplace Theories"; Fiedler, *Analogiemodelle*; and Newman, *Promethean Ambitions*.

31. Aristotle, *On Youth, Old Age, Life and Death, and Respiration*, 480a16–22, 474a7–17, in *CW*.

32. Aristotle, *Generation of Animals*, book 2, 734b3–734b18, in *CW*; see also book 2, 741a32–b24: "Now the parts of the embryo already exist potentially in the material, and so when once the principle of movement has been imparted to them they develop in a chain one after another, as in the case of the automatic puppets."

33. Hippocrates, *De corde*, in *Hippocratic Writings*, 349–50. Galen's writings are filled with examples: see, for instance, his *Usefulness*, books 1 and 10.

34. Galen, *Usefulness*, book 7, chap. 408.

35. Galen, *Natural Faculties*, book 3, chap. 15.

36. Galen, *Usefulness*, book 17, chap. 418.

37. See Bitbol-Hespériès, *Principe de vie*, 169–70.

38. Plato, *Timaeus*, 77d, in *Dialogues*.

39. Aristotle, *Parts of Animals*, 647b2–7, 668a9–29, in *CW*.

40. The Scholastics followed the example set by ancient authors: while they insisted upon the essential difference between art and nature, they also imported examples from the arts and their devices into natural philosophy. Des Chene, *Physiologia*, chap. 7, section 3.

A prominent and much discussed example, because of the later importance of the clockwork analogy in astronomy, is Nicole Oresme's comparison of the cosmos to a mechanical clock in *Livre du ciel*, book 2, chap. 2. For other examples, see Jean Fernel's *De abditis rerum causis* (1567), translated into English as *On the Hidden Causes of Things*, 435–37.

In the Aristotelian scheme of things, nature was primary and art secondary, divine art the ideal and human art the mimic. For this reason, such seventeenth-century philosophical revolutionaries as Descartes and Francis Bacon represented their insistence upon the essential sameness of art and nature as a radical departure from tradition, and historians have traditionally seen it in the same light. See, for example, Rossi, *Philosophy*, appendix I. Yet parallels between natural phenomena and the instruments and machines of the arts were present throughout Scholastic as well as ancient writing.

41. Aquinas, *Summa Theologica*, 2-1, q. 13, a. 2. The clock and engine analogy is in ibid., ad. 3.

42. Gilson, *Études*, 52–64.

43. Fernel, *Physiologia*, 95, 99, 19, 21, 97. Galen develops the comparison at length in *Usefulness*, book 7, chap. 14. See Fernel, *Physiologia*, 618n100.

44. Descartes to Mersenne, November 11, 1640, in AT, 3:232. On Eustachius a Sancto Paulo and his relevance to Descartes, see the introduction to Ariew et al., *Descartes' Meditations*, 68; Ariew, *Descartes and the Last Scholastics*, 27.

45. Eustachius, *Summa*, quoted in translation in Ariew et al., *Descartes' Meditations*, 88.

46. Des Chene writes that the Cartesian Revolution was "*conservative* over the phenomena, but *radical* in its interpretation of them"; that is, Descartes maintained that life could be explained in fully material terms. See Des Chene's *Spirits and Clocks*, 9. He also writes: "The artifice of the body is the work of nature: in that sense the bête-machine exists long before Descartes." Ibid., 97.

47. A tradition of scholarship has examined Descartes's complex relations to the Scholastic philosophy. Important works in this tradition are Gilson, *Index*; Gilson, *Études*; Grene, *Descartes*; Sanhueza, *Pensée biologique*; and Ariew, *Descartes and the Last Scholastics*. Jean-Luc Marion has argued that one can only understand Descartes's *Regulae ad directionem ingenii* (1626–28) in the context of the fundamentally Aristotelian philosophical framework in which it was written. See Marion, *Grey Ontology*. On Descartes's theology in relation to Scholastics such as Franciscus Suárez, see Marion, *Théologie blanche*, book 1, section 7. Finally, urging historians to read Descartes in the late Aristotelian framework in which he developed his views, Des Chene has written that "the *novatores*, however strenuously they denied their debts to the past, were cognizant of, and made use of, the Aristotelian tradition, and for that reason the historian must understand it." See Des Chene, *Life's Form*, 15. I would add that the revolutionaries' relations to the Aristotelian tradition consisted not just of cognizant uses but also of the fact that their intuitions were formed in it.

48. Descartes to Mersenne, June 1637, in AT, 1:378.

49. Des Chene and Peter Dear have each called automata "models of intelligibility" in Descartes's philosophy. See Dear, "Mechanical Microcosm," 59; and Des Chene, "Abstracting from the Soul." See also Garber, *Descartes Embodied*, 2: "According to the mechanical philosophy of which Descartes was the founder, I would argue, everything in the physical world must be explained in the way in which we explain machines"; and Ariew et al.,

Historical Dictionary, 29: "The general point about these machines was methodological. According to Descartes, a phenomenon is explained if we indicate the way in which it *can* be produced."

50. For example, "The example of various bodies, composed by the artifice of men, much helped me: for I recognize no difference between machines made by artisans and the various bodies that nature alone composes, except that the effects of machines depend only on the arrangement of tubes, or springs, or other instruments, which, being in proportion with the hands of those who made them, are always big enough that their figures and movements can be seen, whereas the tubes and springs that cause the effects of natural bodies are ordinarily too small to be perceived by our senses." Descartes, *Principia philosophiae* (1644), in AT, 8:326.

51. Locke, *Human Understanding*, book 3, chap. 6, section 3 (see also section 9). Leibniz repeated the image in his *Nouveaux essais sur l'entendement humaine* (ca. 1690–1705); see *PS*, vol. 5, book 3, chap. 6, section 1, 283.

52. Locke, *Human Understanding*, book 4, chap. 3, section 25.

53. Georges Canguilhem has observed that the "Cartesian animal-machine remained as a manifesto, a philosophical war-machine, so to speak" in contrast with eighteenth-century physiologists' and mechanicians' "elaboration of detailed plans with a view to the construction of simulators." See Canguilhem, "Role of Analogies," 510–11.

54. Galileo was tried by the Roman Inquisition in 1633 for espousing Copernicanism, found "vehemently suspected of heresy," and, following his formal recantation, sentenced to house arrest. Quotation from Santillana, *Crime*, 306–10. On the Galileo affair, see also Finocchiaro, *Galileo Affair*; Westfall, *Essays*; Blackwell, *Galileo*; Langford, *Galileo*; and Shea and Artigas, *Galileo in Rome*. On Descartes's response to Galileo's trial and conviction, see the introduction to Gaukroger, *World*, xxvi–xxviii; and Gaukroger, *Descartes*, 290–92.

55. *The Treatise on Light* (*Traité de la lumière*) laid out a mechanist physics and cosmology and was published posthumously as *The World* (*Le monde*) in 1664.

56. Descartes to Mersenne, November 1633, in AT, 1:270–71.

57. Descartes to Mersenne, [April 1634], in AT, 1:286. I have taken all dates for Descartes's writings and correspondence from the Adam and Tannery edition. Brackets indicate some uncertainty in the dating. In each case, a full explanation can be found in the Adam and Tannery edition.

58. See Vermij, *Calvinist Copernicans*, esp. part 4; Van Nouhuys, "Copernicanism."

59. Elsevier published Galileo's *Discorsi e Dimostrazioni Matematiche, intorno á due nuoue scienze attenenti alla meccanica* in 1638.

60. The fact of Descartes's eagerness to please the Jesuits is well documented. See Gaukroger, *Descartes*, 355, 357, 364–80; Garber, *Metaphysical Physics*, 9; and Rochemonteix, *Collège*, 4:59–60.

61. See Gaukroger, *Descartes*, 292, 375, 380.

62. See ibid., 354–55.

63. Gassendi, *Syntagma philosophicum*, in *Opera*, 2:170–71. Gassendi was an anti-Aristotelian and an Epicurean: a subscriber to the philosophy of Epicurus (ca. 300 BCE), who was an atomist and materialist. Gassendi sought a union of Christian doctrine with atomist mechanism. See Giglioni, "Automata Compared"; and Fisher, "Gassendi's Atomist

Account." On the compatibility of mechanism and theology for Gassendi (and Descartes), see Osler, Divine Will.

64. On Caterus and his objections, see Verbeek, "First Objections"; and Armogathe, "Caterus."

65. On the authorship of these sets of objections, see Garber, "J.-B. Morin and the Second Objections."

66. The question of the relations between Descartes's view of the animal soul (or rather the lack thereof) and that of the Scholastics has been a major subject of philosophical discussion for three and a half centuries. Certainly Descartes's view that animals lacked any soul diverged from the Scholastic view that they had sensible but not rational souls. But commentators have valued this divergence varyingly. Pierre Bayle, in his dictionary article "Rorarius," represented Descartes's idea of the animal-machine as theologically advantageous, though untenable, and also as close in some respects to the Scholastic view, which Bayle found equally untenable. See Bayle, "Rorarius," in *Dictionnaire*, 4:76–77; Rosenfield, *Beast-Machine*, 213n65; and Des Chene, "Animal as Concept." Bayle also likened Descartes's view of animals to that of the Jesuit Théophile Raynaud, who argued in his *Calvinismus Bestiarum Religio* (1630) that animals must have only corporeal souls since they lacked free will. Bayle, "Rorarius," in *Dictionnaire*, 80. The Protestant theologian David-Renaud Boullier, in his *Essai philosophiques*, associated Descartes's *bête-machine* with Catholic theology. See Rosenfield, *Beast-Machine*, 45–49, 213n65, and appendix D; and Cohen, "Chardin's Fur," 50. For my purposes here, the significant point is that initial theological respondents to Descartes's work did not take the idea of animal automatism to be theologically dangerous.

67. Descartes to Huygens, March 9, 1638, in AT, 2:660. On Descartes and Plempius (Vopiscus Fortunatus Plemp), see Grene, "Heart and the Blood." On the world in which Descartes was educated, see the four-volume history of La Flèche: Rochemonteix, *Collège*, esp. vol. 4, which treats his "enseignement philosophique, scientifique et historique." Descartes's experiences at La Flèche and his relations with the school afterward are discussed in 4:51–113. See also Gilson, *Index*, esp. iv–v.

68. Arnauld, "Quatrième objections," in AT, vol. 9, part 1, 159–60.

69. Ibid., 167, 169.

70. Ariew, *Descartes and the Last Scholastics*, chap. 8; Gaukroger, *Descartes*, 356–57; Menn, "Greatest Stumbling Block." For the further development of the controversy over Cartesianism and the doctrine of the Eucharist, see Armogathe, "Cartesian Physics."

71. Ariew, *Descartes and the Last Scholastics*, 158–60. I am grateful to Dennis Des Chene for calling my attention to the absence of any mention of animal automatism in the Louvain condemnations.

72. Digby, *Two Treatises*, chap. 23, sections 4, 5, and 6. On Digby, see Dobbs, "Studies"; Janacek, "Catholic Natural Philosophy"; and Keller, "Embryonic Individuals."

73. Turriano, *Twenty-One Books*.

74. Digby, *Two Treatises*, chap. 23, sections 4 and 6.

75. On the origins and construction of the Segovia mint, see Murray, "Génesis."

76. Digby, *Two Treatises*, chap. 23, sections 5 and 6.

77. Descartes, "Méditation seconde" (1641), in AT, vol. 9, part 1, 21–22.

78. *Traité de l'homme*, in AT, 11:172ff.

79. The literature on this subject is vast. For a sample of perspectives, see Taylor, *Sources of the Self*, esp. chap. 8; Toulmin, *Cosmopolis*, 318; and Israel, *Radical Enlightenment*, esp. chap. 2. For recent scholarly treatments of Enlightenment musings on the "self," see Wahrman, *Making of the Modern Self*; and Goldstein, *Post-Revolutionary Self*.

80. See Gaukroger, *Descartes*, 318–19; and Taylor, *Sources of the Self*, chap. 10.

81. See Taylor, *Sources of the Self*, 146; and Baker and Morris, *Descartes' Dualism* on the cogito and Descartes's notion of consciousness. See also Maisano, "Infinite Gesture," for an argument that Shakespeare invented transcendent inner selves for his characters by exploiting the contrast between inner self and bodily machine.

82. Generations of scholars have discussed the interdependence of these two initially Cartesian notions—the incorporeal rational soul and the body-machine—and have located the origins of modernity, the disenchantment of the world, and unbelief (among many other things) in their pairing. An example is Taylor, *Sources of the Self*, 148.

83. See Locke, *Travels in France*, 6–7, 151–52, 154, 164–68, 173, 175, 178. Locke's passion for the state-of-the-art machinery he saw in France tried the patience of his twentieth-century editor John Lough, who lamented: "How gladly [an editor] would exchange the tedious pages in which Locke minutely describes the fountains of Versailles for a detailed account of his impressions of the Court of the *Roi Soleil*!" See ibid., xxiv. But surely Locke's impressions of the French court were right there in his minute descriptions of the Versailles waterworks.

84. Locke, *Human Understanding*, book 2, chap. 27, sections 5, 6, 9, 11. Locke's passing observation in book 4, chap. 3, section 6 that God "can, if he pleases, superadd to Matter a Faculty of Thinking" gave eighteenth-century readers an opportunity to misconstrue him as a materialist. See Vartanian, *Homme Machine*, 204.

85. On the much-discussed question of the relation between the *Essay concerning Human Understanding* and the *Two Treatises of Government* with regard to Locke's understanding of personhood and personal identity, see, for example, Jolley, Locke, chap. 10; and Simmons, Lockean Theory of Rights, 82–84.

86. Locke, *Two Treatises*, book 2, chap. 6, section 54. Some physical inequalities could, however, result in an inequality in rights. See Simmons, *Lockean Theory of Rights*, 81–82.

87. Locke, *Two Treatises*, book 2, chap. 5, section 26; Jolley, *Locke*, 205–11.

88. *Discours de la méthode*, in AT, 6:57. This line of argument continues to have influential advocates. In his *Cartesian Linguistics*, Noam Chomsky used it to argue not for a disembodied rational soul but for a single "universal grammar" common to all languages and reflecting fundamental properties of the human mind.

89. On the radicalism of Descartes's conception of soul, see Garber, "Soul and Mind," 764–69.

90. Arnauld, "Quatrième objections," in AT, vol. 9, part 1, 154; Augustine, e.g., "Of True Religion," paragraph 110. For treatments of the relationship between Cartesianism and Augustinianism, see Gouhier, *Cartésianisme et Augustinisme*; Menn, *Descartes and Augustine*; and Gilson, *Études*, 289–94. Specifically with regard to the theology of the Augustinian Pierre de Bérulle, founder of the Oratory, see Marion, *Théologie blanche*, book 1, section 8. On the "internalization" of religion on both sides of the confessional divide, see Gaukroger, *Descartes*, 24–25.

91. See Gaukroger, *Descartes,* 318; and Taylor, *Sources of the Self,* 148.

92. By means of the mechanist philosophy, we can make ourselves "maîtres et possesseurs de la nature [lords and masters of nature]." See *Discours de la méthode,* in AT, 6:62.

93. Descartes to Christina, November 20, 1647, in AT, 5:85. See also Taylor, *Sources of the Self,* 146–49.

94. Aquinas, *Summa contra gentiles,* book 2, chap. 68, paragraph 6: "The intellectual soul is said to be on the horizon and confines of things corporeal and incorporeal, in that it is an incorporeal substance and yet the form of a body"; chap. 86, paragraph 2: "Now, the nutritive and sensitive soul cannot operate independently of the body"; chap. 89, paragraph 12: "Therefore, in the soul of the brute there is nothing supra-sensitive, and, consequently, it transcends the body neither in being nor in operation; and that is why the brute soul must be generated together with the body and perish with the body." On the position of the sensitive and vegetative souls in Scholastic discussion, see Rosenfield, *Beast-Machine,* xxiii, 80, 84; and Des Chene, *Life's Form,* chap. 9.

95. See Vidal, "Brains," esp. 944–46 and 936; and Bynum, *Resurrection.*

96. See Aquinas, *Summa Theologica,* 1, q. 76, a. 1, ad. 6.

97. See, for example, Aquinas, *Expositiones in Job,* chap. 19, lectio 2, in Aquinas, *Opera omnia,* 18:119–20; Bynum, *Resurrection,* 256–58. On hylomorphism, resurrection, and medieval Christian conceptions of matter, see also Bynum, *Christian Materiality,* especially chap. 4.

98. The following discussion is derived from Des Chene, *Life's Form,* chap. 2.

99. Gnosticism was an ancient form of religious belief, founded on the principle that one could achieve salvation through esoteric knowledge of the mysteries of the cosmos, which informed the ideas of some early Christians and gave rise to various heresies. On the relations between Gnosticism and early Christianity, see Pagels, *Gnostic Gospels.*

100. Des Chene, *Life's Form,* 51, 69, 48–50.

101. Des Chene, *Life's Form,* 51. See also Des Chene, "Descartes"; and Garber, "Soul and Mind," 761.

102. Aquinas, *Summa Theologica,* 1, q. 75, a. 7, ad. 3.

103. Des Chene, *Life's Form,* 117.

104. Ibid., 181.

105. Caterus, "Premières objections," in AT, vol. 9, part 1, 80.

106. Arnauld, "Quatrième objections," in AT, vol. 9, part 1, 157.

107. Ibid. I thank Steven Nadler for pressing me to refine my thinking about Arnaud's position. On Arnauld's dualism and his form of Cartesianism, see Nadler, *Arnauld.*

108. Bourdin, "Objectiones septimae," in AT, 7:490. Translation quoted from Descartes, *Philosophical Writings,* 2:331.

109. "Sixième objections," in AT, vol. 9, part 1, 218–19. On Mersenne's and others' failure to understand why Descartes excluded the possibility of a material entity being capable of thought, see Garber, *Descartes Embodied,* 257. On the corporeality of thought in the medieval Christian tradition, see Pagel, "Medieval"; and Gaukroger, *Descartes,* 278–79.

110. Daniel, *Voyage,* 35–43. I thank Daniel Garber for calling my attention to this text.

111. More to Descartes, in AT, 5:238–39. See also Koyré, *Closed World,* chap. 5; Gaukroger, *Descartes,* 411; Coudert, "Henry More"; and Hutton, "Edward Stillingfleet."

112. On the relations between iconoclasm, Reformation visual culture, and Cartesianism, see Besançon, *Forbidden Image*, 171; and Dyrness, *Reformed Theology*, 5, 151, 305. On the Jansenist and Protestant resonances of Descartes's philosophy, see Nadler, *Arnauld*; and Schmaltz, "Cartesianism."

113. "But things which are not from nature, such as a bed and clothing and like things, which are spoken of in this way because they are from art, have in themselves no principle of mutation, except per accidens, insofar as the matter and substance of artificial bodies are natural things. Thus insofar as artificial things happen to be iron or stone, they have a principle of motion in them, but not insofar as they are artifacts. For a knife has in itself a principle of downward motion, not insofar as it is a knife, but insofar as it is iron." From Aquinas, *Commentary*, book 2, paragraph 142. For an account of the Scholastic understanding of artificial forms, see Des Chene, *Physiologia*, chap. 7, section 3. For a discussion of "what it meant to be a machine" for Descartes and his contemporaries, see Garber, "Descartes and the Scientific Revolution," 409.

114. Aquinas, *Summa contra gentiles*, book 3, chap. 104, paragraphs 8–9.

115. Descartes, *La dioptrique* (1637), Discours quatrième, in AT, 6:109. See also Descartes, *Sixième méditation* (1641), in AT, vol. 9, part 1, 62: Descartes argues that although he could clearly and distinctly conceive of himself without the faculties of imagining or sensing, he could not conceive of these faculties without himself, that is, without being attached to a thinking thing, since they involved a kind of thought. Therefore they must be incidental features of his intellectual self, not aspects of his body.

116. *Description du corps humain*, part I, in AT, 11:224. See also *Principes de la philosophie*, part 4, section 189 in AT, vol. 9, part 2, 310: "It is these various thoughts of our soul, that come immediate from the movements that are excited by the agency of the nerves in the brain, that we properly call our sentiments, or the perceptions of our senses."

117. Pollot to Reneri [Reniersz Régnier] for Descartes, February 1638, in AT, 1:512.

118. Descartes to Reneri for Pollot, March 1638, in AT, 2:40–41.

119. Descartes, "Réponses aux sixièmes objections," in AT, vol. 9, part 1, 229.

120. Descartes to Reneri for Pollot, April or May 1638, in AT, 2:40–41.

121. Descartes, *Passions de l'âme*, part 1, articles 1, 2, and 50, in AT, 11:328–29, 369–70.

122. Descartes to Gibieuf, January 19, 1642, in AT, 3:479. On the apparent reversal in this letter (up until this point, Descartes had consistently treated the imagination as corporeal), see Gaukroger, *Descartes*, 393. See also Bates, "Cartesian Robotics."

123. "Sixième objections," in AT, vol. 9, part 1, 219.

124. Descartes, "Réponses aux sixième objections," in AT, vol. 9, part 1, 228. See also Ariew et al., *Historical Dictionary*, 21: "If by soul one means the 'principle of life,' then, says Descartes, animals can be said to have a soul: namely, the blood or the heat of the heart."

125. Descartes to Regius, May 1641, in AT, 3:369–72. See also Wilson, "Descartes and the Corporeal Mind."

126. Descartes to William Cavendish, Marquis of Newcastle, November 23, 1646, in AT, 4:574–75. On Descartes's ascription of emotions to animals in this letter, see Gaukroger, *Descartes*, 392; and Gaukroger, *Descartes' System*, 213.

127. Descartes to More, February 5, 1649, in AT, 5:278. See also Canguilhem, *Connaissance*, 111; and Rosenfield, *Beast-Machine*, 16.

128. Locke, *Elements of Natural Philosophy*, in *Works*, 2:329. Locke found it "evident" that animals "do some of them in certain instances reason." See Locke, *Human Understanding*, 160. Of those who claimed the contrary, Locke wrote that they "seem to me to decree, rather than to argue." See Locke to Anthony Collins, March 21, 1703–4 (?), in *Works*, 10:283. See also Locke, "Mr. Locke's Reply to the Right Reverend the Lord Bishop of Worcester's Answer to His Second Letter," in *Works*, 4:463.

129. On Malebranche's version of the animal-machine, see Alquié, *Cartésianisme*, 49–55.

130. The abbé Trublet reported the story in his *Mémoires*, 115–16. See also Rosenfield, *Beast-Machine*, 70; and Woloch, "Huygens's Attitude," 416.

131. Rosenfield, *Beast-Machine*, 70.

132. Leibniz to Ehrenfried Walter von Tschirnhaus, [October 17, 1684], in Leibniz, *Sämtliche Schriften und Briefe* (hereafter *SSB*), series 2, 1:860–61.

133. Leibniz to Antoine Arnauld, October 9, 1687, in *SSB*, series 2, 2:248–61. For Leibniz's views on the beast-machine hypothesis, see also the *Nouveaux essais* in *PS*, 5:60 (Cartesians are "needlessly puzzled about the souls of beasts"); the *Système nouveau*, 66 (the Cartesian view of animals is possible but improbable); his letter to Walter von Tschirnhaus, [October 17, 1684], in *SSB*, series 2, 1:860–61 (animals have souls or substantial forms); "Reflections on the Souls of Beasts" (primitive entelechies are distributed in matter, becoming souls when they are joined to organic bodies, hence animals have perception); the *Essais de théodicée*, in *PS*, 6:3–436, at paragraph 250 (animals have less keen pleasures and pains than humans since they do not reflect); and the appendix to the *Essais* in *PS*, 6:400–36, at paragraph 20.

134. Huygens, *Celestial Worlds*, 59–60. See also Woloch, "Huygens's Attitude," 419.

135. Arnauld, "Quatrième objections," in AT, vol. 9, part 1, 141.

136. Régis, *Cours entier*, 2:504, 506. It was a vexed question in the seventeenth century whether Christian doctrine allowed the possibility of immortal souls in animals. For discussions of this debate, see Harrison, "Virtues of Animals"; and Fowler, *Descartes on the Human Soul*, esp. chap. 4.

137. Borelli, *On the Movement of Animals*, 8, as well as chap. 1 in general. On Borelli's use of an immaterial animal soul to enable his mechanist physiology, see Des Chene, "Mechanisms of Life."

138. Borelli, *On the Movement of Animals*, 319–20.

139. Perrault, *Mécanique des animaux*, in his *Essais de physiques*, 3:1. On Perrault's invocation of an immaterial animal soul, see Moravia, "Homme Machine," 48–49; and Des Chene, "Mechanisms of Life."

140. Des Chene has described this exclusion as in keeping with the older, traditional division between mechanics, which dealt with the mathematics of forces, and physics, which explained their origins. See Des Chene, "Abstracting from the Soul." I would add to this that by 1680, mechanists had contributed to a redefinition of physics such that the traditional exclusion had taken on a new meaning: to exclude something from mechanics (and in Perrault's writing this is explicit) was often to set it outside the domain of physics as well.

141. Perrault, *Mécanique des animaux*, in *Essais de physiques*, 3:2.

142. Ibid., 3:24.

143. Hobbes, "Troisièmes Objections," in AT, vol. 9, part 1, 134–35.

144. Hobbes, *Leviathan*, part 1, chap. 1, at 4.

145. Ibid., chap. 3, at 9, 11.

146. Ibid., chap. 5, at 20; chap. 6, at 26, 28; chap. 2, at 8; chap. 3, at 11.

147. Ibid., chap. 16, at 80–83.

148. Hobbes, *Leviathan*, part 1, introduction, 1.

149. See Skinner, "Hobbes," 24–27. A revised version of this piece is included in Skinner, *Visions of Politics*, 3:177–208. For a response to Skinner's emphasis on the "artificial" nature of the person of the Commonwealth, see Runciman, "Debate." Thanks to Kinch Hoekstra for guidance in this area.

150. Hobbes, *Leviathan*, part 2, chap. 17, at 87.

151. On this antecedent, see Schofield, *Plato*. On the main antecedent model of artificial persons, the Roman legal idea of the corporation, which during the Middle Ages became known as a *persona ficta*, see Buckland, *Text-Book*, 307; Buckland, *Main Institutions*, 88; and Runciman, *Pluralism*, 14, 16, 18, 49, 91, 93, 107–8, 243–44.

152. "Belluae sunt mera automata omni cognitione ac sensu carentia." Quoted in Sanhueza, *Pensée biologique*, 236n66. Even here, animal automatism appeared twenty-first on a list of thirty. The first was "that the human mind could and should doubt everything except its own thought and therefore its existence." See Sortais, "Cartésianisme"; and Sanhueza, *Pensée biologique*, 196–98n61. See also Rochemonteix, *Collège*, 4:89–94.

153. "Mettez une machine de chien et une machine de chienne l'une près de l'autre et il en pourra résulter une troisième petite machine, au lieu que deux montres seront l'une auprès de l'autre toute leur vie, sans jamais faire une troisième montre. Or, nous trouvons par notre philosophie, Mme. B. et moi, que toutes les choses qui, étant deux ont la vertu de se faire trois, sont d'une noblesse bien élevée au-dessus de la machine." See Fontenelle, *Lettres galantes de M. Le Chevalier d'Her**** (1683), in his *Œuvres*, 3:20.

154. Catherine Descartes to Madeleine de Scudéry (1689), quoted in Scudéry, *Madeleine de Scudéry*, 313; see also Van Roosbroeck, "'Unpublished' Poems."

155. Quoted in Buffon, *Histoire*, 20:129, note h.

156. Voltaire, "Bêtes," in his *Dictionnaire philosophique* (1764), in *Œuvres complètes de Voltaire* (hereafter *OCV*), 35:411–15, at 412–13.

157. Priestley, *Disquisitions*, 1:283.

Chapter 3

1. Galileo, *Sidereus Nuncius*; Van Helden, "Invention"; Van Helden, "Development"; Van Helden et al., *Origins*.

2. Freedberg, *Eye*, 169–170.

3. Swammerdam, *Book*; Hooke, *Micrographia*; Wilson, *Invisible World*.

4. Landes, *Revolution*, 136–37, appendix A; de Weck et al., *Engineering*, 24.

5. Coward, *Second Thoughts*, 105–6.

6. On natural theology in antiquity and the Middle Ages, see Vanderjagt and Berkel, *Book*.

7. On the persistence of this tradition into the nineteenth century, see Gillespie, "Divine Design."

8. More, *Antidote*, chap. 12, at 142–46.

9. Humours were the fluids and tunicles the membranes of the eye.

10. More, *Antidote*, 144–45.

11. Ibid., 145.

12. On More's and Boyle's relations, in particular, their dispute over More's interpretation of Boyle's air-pump experiments, see Greene, "Henry More"; and Shapin and Schaffer, *Leviathan*, 207–24.

13. Boyle, *Disquisition*, 148.

14. The quotation is from a 1691 codicil to Boyle's last will, in which he left funds to endow the Boyle lectureship. See Jacob, *Newtonians*, 144.

15. Boyle, *Disquisition*, 66, 218, 48.

16. Boyle, *Free Enquiry*, 125, 39. See also ibid., 40, 60, 102, 104, 134, 135, 146, and 160, as well as Westfall, *Science and Religion*.

17. Ray, *Wisdom of God*, 248, 254, 257. (Ray lifted some passages from More virtually verbatim.)

18. Scheiner, *Oculus*. See also Roman, "Discovery." It was Hermann von Helmholtz who demonstrated that accommodation to distance takes place through changes in the shape of the lens.

19. Ray, *Wisdom of God*, 252. More had also listed the ability of the pupil to dilate and contract among the eye's perfections, in his *Antidote*, 144–45.

20. Cheyne, *Philosophical Principles*.

21. Derham, *Physico-Theology*. See Davidson, "'Identities Ascertained.'"

22. Réaumur, *Mémoires*; Pluche, *Spectacle*. See also chapter 5.

23. On the history of the *camera obscura* and the analogy to an eye, see Wade and Finger (2001).

24. Cheyne, *Philosophical Principles*, paragraph 45.

25. Kepler, *Ad Vitellionem*, chap. 5.

26. Ray, *Wisdom of God*, 255–56.

27. Ibid., 260.

28. More, *Antidote*, 146.

29. See Brooke, *Science and Religion*, chap. 6. See also Brooke, "Fortunes of Natural Theology"; Brooke, "Scientific Thought": and Brooke, "'Wise Men.'" In Brooke et al., *Science in Theistic Contexts*, see esp. Cook, "Divine Artifice," and Osler, "Whose Ends." See also Jenkins, "Arguing about Nothing"; Osler, "Mixing Metaphors"; Roger, "Mechanistic Conception of Life"; Olson, "On the Nature of God's Existence"; and Schaffer, "Political Theology."

30. See Ospovat, *Development of Darwin's Theory*; Gould, "Darwin and Paley"; Phipps, "Darwin"; Richards, "Instinct and Intelligence"; Richards, *Darwin and the Emergence*, 128–34, 342–45. For a contrasting view, see Bowler, "Darwinism."

31. Boyle, *Disquisition*, 146–47.

32. Ray, *Wisdom of God*, 249. Emphasis added.

33. Boyle, *Disquisition*, 52–55 (see also 193), 59–64 (see also 201–2). Ray repeated the argument about the nictitating membrane in his *Wisdom of God*, 262. Cheyne repeated all of the arguments mentioned here; see Cheyne, *Philosophical Principles*, paragraph 45. And

for Paley's version of the same arguments, see Paley, *Natural Theology*, chaps. 3 and 9; the nictitating membrane, for example, is discussed on pp. 35 and 129.

34. Boyle, *Disquisition*, 66, and see also 218.

35. Ibid., 58.

36. Ibid., 56–59, 201.

37. The current field of aberration studies is devoted to studying this problem. See Shapiro, "Newton and Huygens," 273ff; Dijksterhuis, *Lenses and Waves*, 83–92; and Wilson, *Reflecting Telescope Optics*, chap. 1.

38. Mariotte, "Nouvelle." Cited passages in the English translation of the same year, "A New Discovery," 668–69. The anatomist Jean Pecquet, to whom Mariotte reported his discovery, responded: "Every one wonders, that no person before you hath been aware of this Privation of Sight, which every one now finds, after you have given notice of it." On Mariotte, see Mahoney, "Mariotte."

39. Ray, *Wisdom of God*, 254–55.

40. Cheyne, *Philosophical Principles*, paragraph 45.

41. Priestley, *Disquisitions*, 1:283. Here Priestley was not stating his own opinion on the matter, but characterizing the Cartesians' view, especially that of Nicolas Malebranche.

42. Harvey, *Anatomical Exercises*, 417; see Aristotle, *Generation of Animals*, 741a32–b24, in *CW*. On Harvey's theory of generation in relation to Aristotle, see Lennox, "Comparative Study of Animal Development."

43. Harvey, *De motu*, 99, 153: "Or as muscles in automata: in action movement is effected while the parts become greater and less by turns."

44. Harvey, *Motion of the Heart*, 31. In the original Latin: "una rota aliam movente, omnes simul movere videantur."

45. In the same way, Harvey also found the heart to be like the mechanism in firearms, in which the sequence of trigger, flint, steel, spark, powder, flame, explosion, and shot all seemed to take place "in the twinkling of an eye." See ibid., 31–32.

46. Ibid., 30–31.

47. Malpighi, *Dissertatio epistolica de formatione pulli in ovo* (1673), in *Opera omnia*, 1:1. (Each item paginated separately.) See Duchesneau, "Leibniz's Model," 399. On Malpighi, see also Adelmann, *Marcello Malpighi*; and Bernardi, *Metafisiche dell'embrione*.

48. Malpighi, *Anatome plantarum* (1675–76), in *Opera omnia*, 1:1. See also Malpighi, *Opera posthuma*, 19.

49. Harvey, *Exercitationes*, 148; for the English translation, see Harvey, *Anatomical Exercises*, 334ff. Descartes, *La Description du corps humain* (1747), in AT, 11:253. The debate between "preformationists" and "epigenicists" has been much studied. For some landmark and some recent treatments, see Bowler, "Preformation"; Bowler, "Changing Meaning"; Gould, "On Heroes and Fools in Science"; Roe, *Matter, Life, and Generation*; Maienschein, "Competing Epistemologies"; and Roger, *Life Sciences* (first published as *Sciences de la vie*), chap. 6.

50. Malebranche, *De la recherche de la vérité* (1674–75), in his *Œuvres*, vol. 3, book 1, chap. 6, section 1, 43–52. The watchmaker passage is on 168.

51. *Description du corps humain*, in AT, 11:253–54.

52. Borelli, *On the Movement of Animals*, part 2, ch. 14, proposition 186, 398–99.

53. On the widespread triumph of preformation in the later seventeenth century, see Roger, *Life Sciences*, 285–87.

54. Harvey, *De motu*, 145, 147.

55. For example, Fuchs, *Mechanization of the Heart*; Wright, *Circulation*.

56. Harvey, *Lectures*, 22.

57. Harvey, *Anatomical Exercises*, 508–9.

58. Harvey, *De motu*, 127.

59. Ibid., 43.

60. Ibid., 143.

61. The passage in Aristotle to which Harvey refers, describing male fish sprinkling their semen over eggs laid by female fish, is in *History of Animals*, 567a29–b6, in *CW*; Harvey, *Anatomical Exercises*, 359.

62. Harvey, *Anatomical Exercises*, 359–60.

63. Aristotle, *Generation of Animals*, book 2, 734b3–b18 and 741a32–b24; see also chap. 2. All found in *CW*.

64. Harvey, *Anatomical Exercises*, 345–46, 350, 359–60. See also *De motu*, 147 and 151, where Harvey canvasses the various models of order: a well-governed state; the work of masons, bricklayers and carpenters; the working of a ship; an army and a choir.

65. Harvey, *Anatomical Exercises*, 579.

66. Ibid., 372, 575–79, 585.

67. Harvey, *Lectures*, 219. In this context, Harvey meant that brains produced animal spirits, the medium of sensation and movement.

68. Harvey, *Anatomical Exercises*, 575–79, 585.

69. Boyle, *Disquisition*, 157–58. See also Hunter and Macalpine, "William Harvey and Robert Boyle."

70. Harvey, *Anatomical Exercises*, 369.

71. Willis, *Two Discourses*, 3. Willis attributed the same view to the Spanish doctor and philosopher Gómez Pereira before Descartes and to Kenelm Digby afterward.

72. Ibid., preface [n.p.], 56.

73. Ibid., 34, 24.

74. Ibid., 6.

75. Ibid., 24, 56.

76. Ibid., 34. Willis adds that "more perfect Brutes" are better able to learn and vary their actions.

77. On Lamy and his relations to Willis, see Thomson, *Bodies of Thought*, 86–88.

78. Lamy, *Explication méchanique et physique des fonctions de l'âme sensitive* (1677), in Lamy, *Discours anatomiques*, 167.

79. In the course of writing this book, I have often presented parts of it to audiences that include philosophers, and have come to realize that my Leibniz is different from most philosophers' Leibniz. Moreover, I have become interested in the reason for this difference. I think it has everything to do with the different purposes of history and philosophy, and accordingly, the different ways in which historians and philosophers read. Philosophers' purpose, as I understand it, is to arrive at a philosophically correct view of the problems that interest them. When they read Leibniz (or any other historical figure), they therefore

seek coherence: a view that is both internally consistent and also concurs with their own intuitions and manners of thinking. Accordingly, philosophers tend to look for ways to eliminate (resolve, correct, filter out) the ambiguities, contradictions and inconsistencies in the historical texts they read. Historians, in contrast, want to understand the ideas in their original context: the concerns that motivated them, the forces that shaped them, the implications that flowed from them. In their efforts to understand ideas in their original context, historians catch at the very things that philosophers try to eliminate: anything that seems unfamiliar, contradictory or inconsistent. These things, which get in the way of a philosophical reading, are essential to a historical reading: they are the jagged edges and fault lines that reveal the contours of the original context, and the forces at work in it. If my Leibniz seems less familiar than most philosophers' Leibniz, from a modern scientific perspective, it is because I have been interested in the unfamiliar aspects of his understanding of science, and the possibilities these represented.

80. Leibniz, *Principes de la philosophie [Monadologie]* (1714), in Leibniz, *Principes*, paragraph 64. See also Leibniz, *Nouveaux essais* (ca. 1690–1705), in *PS*, vol. 5, book 3, chap. 6, at 309: "organic bodies are actually machines."

81. Leibniz, "Reflections on the Souls of Beasts."

82. Leibniz, *Système nouveau de la nature et de la communication des substances* (1695), in Leibniz, *Système nouveau*, 66.

83. On the *vis viva* controversy, see Iltis, "Leibniz"; and Garber, "Leibniz: Physics and Philosophy," 309–14.

84. Leibniz, "A Discourse on Metaphysics" (1686), translated in Leibniz, *Philosophical Essays* (hereafter *PE*), 51. In the original French: "Car le mouvement . . . n'est pas une chose entièrement réelle. . . . Mais la force ou cause prochaine de ces changemens est quelque chose de plus réel." See Leibniz, *Discours de métaphysique*, section 18, 57–58. Leibniz first presented his refutation of Descartes's principle of conservation of motion in "Brevis demonstratio erroris memorabilis Cartesii . . ." in the Leipzig journal *Acta Eruditorum* (March 1686): 161–63; see the translation in Leibniz, *Philosophical Papers and Letters* (hereafter *PPL*), 1:455–63. See also Leibniz, *Specimen Dynamicum* (1695), translated in *PE*, 130.

85. Leibniz, *Specimen Dynamicum*, in *PE*, 119, 123.

86. Ibid., 125, 130.

87. Ibid., 126.

88. Leibniz, "Essay de dynamique sur les loix du mouvement, où il est monstré, qu'il ne se conserve pas la même quantité de mouvement, mais la même force absolue, ou bien la même quantité de l'action motrice," in his *Mathematische Schriften*, 6:215–31.

89. Châtelet, *Institutions de physique*, chap. 21, at 446, 449–50.

90. "Materialism" during the later seventeenth and eighteenth centuries could mean different things depending upon whether one ascribed intrinsic forces and tendencies to matter itself. For a recent treatment of this issue, see Thomson, *Bodies of Thought*. On active matter, see also Yolton, *Thinking Matter*; and Roe, *Matter, Life, and Generation*. "Mechanism" too had different meanings depending upon one's assumptions about the essential nature of matter and motion. Historians have tended to describe a shift over the course of the late seventeenth and eighteenth centuries from a (brute) mechanist approach to life to a (non-mechanist) vitalist one in which physiologists rejected mechanical models in favor of vital

spirits and forces. See, for example, Brown, "From Mechanism to Vitalism." Such accounts leave aside a mechanist tradition that is no less important for being harder to hear through the distorting acoustics of the eighteenth and nineteenth centuries: a mechanist tradition in which spirit and agency were constitutive aspects of the machinery itself.

91. On the role of *vis viva* in eighteenth- and nineteenth-century physics and engineering, see Hiebert, *Historical Roots*; Cardwell, "Some Factors"; Costabel, *Signification*; Grattan-Guinness, "Work"; Séris, *Machine et communication*; and Darrigol, "God, Waterwheels, and Molecules." I am grateful to Robert Brain for urging me to consider the physical and practical importance of *vis viva* and the Leibnizian tradition that formed around it.

92. Leibniz, *On Nature Itself* (1698), in *PE*, 156.

93. Leibniz, "Réponse aux réflexions continues dans la seconde édition du Dictionnaire critique de M. Bayle, article Rorarius, sur le système de l'harmonie préétablie" (1702), in Leibniz, *Système nouveau*, 194, 197–98.

94. Leibniz had read Spinoza's *Tractatus Theologico-Politicus* (1670). On Leibniz's relations and encounter with Spinoza, see Leibniz to Jean Gallois, September(?) 1677, in Leibniz, *SSB*, series 2, 1:566–71, at 568; Leibniz to Ernst von Hessen-Rheinfels, August 4/14, 1683, in *SSB*, series 2, 1:844; *Essais de théodicée* (1710), in *PS*, 6:3–436, at 339, paragraph 376; Bouveresse, *Spinoza et Leibniz*; Wilson, *Leibniz's Metaphysics*, 84–86; Woolhouse, *Descartes, Spinoza, Leibniz*; Ariew, "G.W. Leibniz," 27; Brown, "Seventeenth-Century Intellectual Background," 54; and Antognazza, *Leibniz*, chap. 3.

95. Leibniz to Gallois, September (?), 1677, in *SSB*, series 2, 1:568.

96. Spinoza, *The Ethics* (1677), in *Life, Correspondence, and Ethics*, part 1, 446–52; part 3, 500; part 1, 561–63. A tradition of historical writing has credited Spinoza with having originated modern philosophy, science, materialism, naturalism, and mechanicism. For a classic example, see Wolfson, *Philo*; for a more recent one, see Israel, *Radical Enlightenment*, 159, 243.

97. Spinoza to Henry Oldenburg, November 20, 1665, in *Correspondence*, 212; Spinoza, *Ethics*, in *Life, Correspondence, and Ethics*, part 3, prop. 9, 509; Spinoza to G. H. Schuller, October 1674, in *Correspondence*, 295.

98. On Spinoza's use of the term and concept of *conatus*, see LeBuffe, "Spinoza's Psychological Theory." On Leibniz and *conatus*, see Carlin, "Leibniz on Conatus."

99. Spinoza, *Ethics*, in *Life, Correspondence, and Ethics*, part 1, 439.

100. For "fatal" necessity and also the "excellent" aspects of Spinoza's system, see Leibniz to Henry Justel, February 4/14, 1678, cited in *PPL*, 195n6. On Spinoza's fatalism, see also Leibniz, *Essais de théodicée* (1710), in *PS*, 6:217–18, 336–37; Leibniz, *Principes de la nature*, in Leibniz, *Principes*, chap. 13, at 48, 58. For Leibniz's views about Spinoza's philosophy more generally, see Leibniz, "On the Ethics of Benedict de Spinoza" (1678), in *PPL*, 196–205; and Parkinson, "Philosophy and Logic," 211.

101. See Leibniz, " Discourse on Metaphysics," in *PE*, section 13.

102. I thank Tim Brook for drawing my attention to Leibniz's interest in Chinese natural philosophy in relation to his search for an alternative, active form of mechanism.

103. See the introduction by Cook and Rosemont in Leibniz, *Writings on China*, 10–18; and Perkins, *Leibniz and China*, chap. 3.

104. On Matteo Ricci, accommodationism, and Leibniz, see Cook and Rosemont in Leibniz, *Writings on China*, 13–14.

105. See Leibniz, "Remarks on Chinese Rites and Religion" (1708), in Leibniz, *Writings on China*, 68; and Leibniz, "Discourse on the Natural Theology of the Chinese" (1716), in ibid., 77–80.

106. Leibniz, "Discourse on the Natural Theology of the Chinese" (1716), in Leibniz, *Writings on China*, 85.

107. Ibid., 80.

108. Leibniz, "Réponse aux réflexions continues dans la seconde édition du Dictionnaire critique de M. Bayle, article Rorarius, sur le système de l'harmonie préétablie," in Leibniz, *Système nouveau*, 197.

109. Leibniz, *Nouveaux essais*, in *PS*, 5:59. See also Leibniz, "Lettre [à la Reine Sophie Charlotte de Prusse] touchant ce qui est indépendant des Sens et de la Matière" (1702), in *PS*, 6:507: "Perception cannot be explained by any machine that might exist." For secondary treatments of Leibniz's view of selfhood in relation to bodily mechanism, see Mates, *Philosophy of Leibniz*, 206–208; Mercer, *Leibniz's Metaphysics*, esp. chaps. 2 and 4; and Wilson, *Leibniz's Metaphysics*, esp. 238–43. For a comparison of Leibniz's to Descartes's views in this matter, see Duchesneau, *Modèles du vivant*.

110. Leibniz, *Monadologie*, in Leibniz, *Principes*, paragraph 17.

111. Leibniz, *Système nouveau*, 71.

112. On this trajectory in Leibniz's thinking, see esp. Duchesneau, *Modèles du vivant*, 315–72.

113. Smith emphasizes that Leibniz did not intend "these basic entities underlying body as a means of *explaining away* body. Instead they offer a means of *accounting for* body." See Smith, *Divine Machines*, 7. Just as Descartes undertook to account for life in terms of mechanism, Leibniz proposed to understand matter in terms of spirit. Neither intended to eliminate the very thing they sought to comprehend.

114. Leibniz, *Monadologie*, in Leibniz, *Principes*, paragraphs 19, 89; Leibniz to Antoine Arnauld, October 9, 1687, in *SSB*, series 2, 2:257.

115. Leibniz, *Monadologie*, in Leibniz, *Principes*, paragraphs 1–15.

116. Leibniz to Arnauld, April 30, 1687, in *SSB*, series 2, 2:182.

117. Voltaire, *Candide*, 3–6.

118. Voltaire, *Cabales*, 9. Relatedly, in his article on atheism in the *Dictionnaire philosophique*, Voltaire wrote that while a catechist announced God to children, Newton "demonstrates him to the learned," 43. On the other hand, in his *Traité de métaphysique*, chap. 2, Voltaire observed that the watch-implies-watchmaker argument could indicate only the probability of an intelligent and powerful being, but could not prove the existence of an omniscient and omnipotent God.

119. Leibniz, *Essais de théodicée* (1710), in *PS*, 6:229; Leibniz, "Streitschriften zwischen Leibniz und Clarke," in *PS*, 7:418. See also Leibniz, *Nouveaux essais*, in *PS*, vol. 5, book 3, chap. 6, at 309: "Organic bodies are really machines"; and Leibniz, *Monadologie*, in Leibniz, *Principes*, paragraph 64: "Thus each organic body of a living being is a sort of divine machine, or a natural Automaton, that infinitely surpasses all artificial automata."

120. Leibniz, *Nouveaux essais*, in *PS*, vol. 5, book 2, chap. 20, at 153. Leibniz was responding in these passages to John Locke's discussion of pleasure and pain in the *Essay concerning Human Understanding*; see Locke, *Human Understanding*, book 2, chaps. 20–21.

Leibniz began with Locke's notion of "uneasiness" (which Leibniz translated as "inquietude" or "unruhe") and built it into a physical yet non-reductive theory of the basis of human behavior. The key element in this transformation was Leibniz's idea of unconscious perceptions, which created a continuity between bodily and conscious responses.

121. Leibniz, *On Body and Force, against the Cartesians* (May 1702), in *PE*, 253; Leibniz, *Système nouveau*, 72; Leibniz, *Essais de théodicée* (1710), in *PS*, 6:110.

122. Bayle, "Rorarius," in *Dictionnaire*, 4:82nH.

123. See, for example, Leibniz, *Système nouveau*, 72; and "Postscript of a Letter to Basnage de Beauval" (1696), in *PE*, 147.

124. Leibniz, "Streitschriften zwischen Leibniz und Clarke," in *PS*, 7:418.

125. See Antognazza, *Leibniz*, chap. 9. The year after Leibniz's correspondence with Clarke, in the third edition to the *Opticks* (1717), Newton himself reached for an indigenous solution to the problem of the universe winding down, by appealing to "active principles" such as gravity and fermentation. See Dear, *Intelligibility of Nature*, 33–35.

126. For Hume's case against the argument from design, see Hume, *Human Understanding*, section 11; Hume, *Dialogues concerning Natural Religion*; and Gaskin, "Hume on Religion." Despite the current consensus among philosophers that Hume's argument was devastating, it was not received as such at the time; witness the success of Paley's book a half century later. On the fortunes of Hume's argument against the argument from design, see Pyle, *Hume's Dialogues*, 136: "The lack of impact of Hume's Dialogues in early nineteenth-century Britain is very clear from the immense popularity of Paley's *Natural Theology* (1802)." See also Gaskin's introduction to Hume, *Principal Writings on Religion*, ix–x: "William Paley's *Evidences of Christianity* (1794) and *Natural Theology* (1802) had in effect been refuted by Hume in the *Dialogues* (1779) and elsewhere before they were even written; but Paley, not Hume, was the standard reading on religion for students throughout the nineteenth century and into the twentieth. Even when the balance changed in Hume's favour, under the influence of Logical Positivism in the 1930's and philosophical analysis in the post-war decades, the fashion was still to discuss a single argument or section of Hume's work in isolation from the rest."

127. Boyle, *Disquisition*, 55.

128. Leibniz, "Against Barbaric Physics" (1710–16?), in *PE*, 318–19. On the stakes of the Leibniz-Clarke dispute, see also Baker, *Condorcet*, 95–109.

129. Leibniz, *Plus ultra* (1679), in *PPL*, 227.

130. Leibniz, "Doutes concernant la Vraie Théorie Médicale de Stahl" (1709), in Carvallo, *La controverse*, 81.

131. Leibniz argued that mechanist and teleological approaches to nature were therefore each (partially) correct, and so the sectarians ought to be able to resolve their differences. There should be no quarrel between a purely mechanist physiology and one relying upon final causes: "Both are good, both can be useful," and they converged in the larger truth of nature: mechanism was purpose, and purpose, mechanism. Leibniz to Nicolas Remond, January 10, 1714, in *PS*, 3:607; Leibniz, *Discours de métaphysique*, 65; and Leibniz, "Streitschriften zwischen Leibniz und Clarke," in *PS*, 7:416–18. See also the introduction by Ariew and Garber in *PE*, ix.

132. On Leibniz's distinctive form of mechanism as applied to the phenomena of life,

for which Smith coins the term "organics," see Smith, *Divine Machines* ("organics" appears at 19). See also Garber, "Leibniz and the Foundations of Physics" (for an argument that to Leibniz, physics must follow the study of living things, which was the foundational science); Garber, *Leibniz*, chapters 2–4 (on "Reforming Mechanism"); Duchesneau, *Modèles du vivant*, chap. 10 and Duchesneau, "Leibniz's model," 398. Duchesneau argues that Leibniz's theory of organisms grew out of his understanding of dynamics, a "science of power and action," which he developed in the last decade of the seventeenth century.

133. Leibniz, "Principles of Nature and Grace, Based on Reason" (1714), in *PE*, 207. See also Leibniz, *Système nouveau*, 70–71: "A natural machine still remains a machine in its least parts"; and Leibniz, *Nouveaux essais*, in *PS*, vol. 5, book 3, chap. 6, section 39.

134. Leibniz to Lady Damaris Masham, June 30, 1704, in *PS*, 3:356. Smith notes that this is Leibniz's sole use of "organism" as a count noun, to denote *an* organism. Otherwise he used "organism" as an abstract noun equivalent to "mechanism" to describe a state of bodies. See Smith, *Divine Machines*, 105, 119.

135. Leibniz, "On Nature Itself, Or, on the Inherent Force and Actions of Created Things, Toward Confirming and Illustrating Their Dynamics" (1698), in *PE*, 156. See also Leibniz, "Against Barbaric Physics: Toward a Philosophy of What There Actually Is and against the Revival of the Qualities of the Scholastics and Chimerical Intelligences" (1710–16?), in *PE*, 319.

136. Leibniz to Lady Damaris Masham, June 30, 1704, in *PS*, 3:356.

137. Leibniz, *Monadologie*, in Leibniz, *Principes*, paragraph 64.

138. In *De ipsa natura* (*On nature herself*) (1698), Leibniz defended his active-mechanist views against John Christopher Sturm, who argued on behalf of Robert Boyle's mechanist cosmology in his *Free Inquiry into the Vulgarly Received Notion of Nature* (1682). See Leibniz, *De ipsa natura* (1698) in English translation as *On Nature Itself, Or on the Inherent Force and Actions of Created Things*, in *PPL*, 498–508.

139. Leibniz, *Nouveaux essais*, in *PS* vol. 5, book 4, chap. 16, at 455; see also preface, 51. Leibniz's law of continuity, in which he adopted and transformed an ancient idea and transmitted it to contemporaries and followers, has been the subject of a long tradition of scholarship. A landmark contribution to this tradition, extremely influential and much-criticized, was Lovejoy, *Great Chain of Being*. For a recent take on the subject, see Smith, *Divine Machines*, 241–43; and Smith, " Series of Generations."

140. Leibniz, *Monadologie*, in Leibniz, *Principes*, paragraphs 52–55.

141. See Duchesneau, "Leibniz's Model."

142. On Leibniz's encounter with Jan Swammerdam, see Brown, "Seventeenth-Century Intellectual Background," 63n41; and Leibniz, *Monadologie*, in Leibniz, *Principes*, paragraph 74. On his meeting with Antoni van Leeuwenhoek, see Leibniz, *New System*, 13n35. On both, see also Ariew, "G. W. Leibniz," 27.

143. Leibniz, *Monadologie*, in Leibniz, *Principes*, paragraph 74.

144. Ibid., paragraphs 65–70.

145. Leibniz, "Principles of Nature and Grace," in *PE*, 207.

146. Leibniz, *Monadologie*, in Leibniz, *Principes*, paragraph 14.

147. Swammerdam, *Book of Nature*, part 2, 147n.

148. I discuss Stahl, phlogiston, and how phlogiston theory constituted a particular understanding of the nature of life in Riskin, *Science in the Age of Sensibility*, chap. 7.

149. Stahl, *Recherches sur la différence qui existe entre le mécanisme et l'organisme* (1706), in his *Œuvres*, 2:259–348; see esp. section 75.

150. Leibniz, "Doutes," in Carvallo, *La controverse*, 89. On Leibniz's and Stahl's dispute over the nature of the animal-machine, see also Smets, "Controversy"; and Carvallo, *La controverse*.

151. Leibniz to Arnauld, November 1671, in *SSB*, series 2, vol. 2, part 1, 284; translation from Duchesneau, "Leibniz's Model," 382.

152. Leibniz, "Doutes," in Carvallo, *La controverse*, 91.

153. Leibniz, "Repliques aux observations de Stahl" (1709?), in Carvallo, *La controverse*, 105.

154. Leibniz, "Doutes," in Carvallo, *La controverse*, 97.

155. Aristotle, *Physics*, book 2, 198b17–b33 and 199b10–b13; Aristotle, *On the Soul*, book 2, 412a27–b9; both in *CW*.

156. Locke, *Human Understanding*, book 2, chap. 27, sections 6, 8.

157. Ibid., sections 5, 6.

158. Leibniz, *Essais de théodicée* (1710) in *PS*, 6:40–42, 228–29.

159. Ibid., 6:42; see also 152, 229, 356. On Leibniz's particular version of preformation in embryology, see Smith, *Divine Machines*, chap. 5; Roger, "Leibniz"; and Bowler, "Preformation and Pre-existence." Smith underlines the difference, originally described by Roger and Bowler, between theories of preformation and preexistence: "Only the latter view for these authors has it that all creatures must exist since the creation, *and* that generation is only the unfolding of what has always already existed. Preformation, in contrast, for them holds that the fetus is formed before conception at some particular time, but not necessarily by God, and perhaps even by some natural means. Leibniz's theory involves not just some preexisting entity or other, but an entity with a particular organic *form* . . . and to speak of preformation rather than preexistence brings out more accurately this feature of Leibniz's theory" (170).

160. Leibniz, *Essais de théodicée* (1710) in *PS*, 6:152, 356; Leibniz to Remond, February 11, 1715, in *PS*, 3:635.

161. Leibniz, *Système nouveau*, 70–71.

162. Ibid., 68.

163. Leibniz, *Monadologie*, in Leibniz, *Principes*, paragraphs 71, 73, 76, 82.

Chapter 4

1. "Extrait du journal d'Allemagne: Machine surprenante de l'Homme artificiel du sieur Reyselius," in *Journal des sçavans* (1677), 252. For the swan, tree, cat-vomiting woman, and pig respectively, see *Journal des sçavans* (1677), 2, 154, 156, 204. The reference regarding the twelfth issue of the same year is to a description of a pump used to model the heart, in *Journal des sçavans* (1677), 127–29. On Reisel, see Bröer, *Salomon Reisel*; Hankins and Silverman, *Instruments and the Imagination*, 192; and Doyon and Liaigre, *Vaucanson*, 117–18, 162–63. Reisel also wrote the first systematic description of the Württemberg Siphon, invented by Jean Jordan of Stuttgart: a siphon with two equal, curved legs in which, when

the legs are level, water climbs up one side and flows down the other. See Reisel, "Concerning the Sipho"; see also the article "Syphon de Wirtemberg" in Diderot and d'Alembert, *Encyclopédie*, 15:766.

2. *Journal des sçavans* (1680), 34.

3. Leibniz, "Drôle de pensée, touchant une nouvelle sorte de représentations (septembre 1675)," in *SSB*, series 4, 1: 562–68, cited passages on 562 and 564. See also Smith, *Divine Machines*, 59.

4. Christian Philipp to Leibniz, March 1, 1679, and Leibniz to Christian Philipp, March 14, 1679, in *SSB*, series 1, 2:436, 445; on Reyselius's artificial man, see also 2:316, 325, and on a talking head machine, 2:296, 303.

5. *Journal des sçavans* (1680), 12. On Weigel, see Schielicke et al., *Erhard Weigel*.

6. One might call these machines "simulations" in its current sense (which originated around the middle of the twentieth century) to mean an experimental model from which one can discover properties of the natural subject. Simulation in its seventeenth- and eighteenth-century sense meant "artifice" and had a negative connotation, implying fakery. However, the practice of using machinery to approximate nature, then experimenting on the model and drawing conclusions about its prototype (simulation in the current sense) emerged as a general practice in science during the eighteenth century. For arguments that eighteenth-century machines were simulative in the modern sense, see Doyon and Liaigre, "Méthodologie"; Canguilhem, "Role of Analogies," 510–12; Price, "Automata"; and Fryer and Marshall, "Motives of Jacques Vaucanson."

7. On the legend of Albertus Magnus's artificial man, see Battisti, *L'antirinascimento*, 226; and Higley, "Legend."

8. El Tostado notably adapted a story associated with Pope Sylvester. See Jones, "Historical Materials," 91–92.

9. El Tostado quoted in Naudé, *Apologie*, 382–83. For the android's dictation of Albert's writings, see the article "Androïdes" in Chambers et al., *Supplement*.

10. Naudé, *Apologie*, 385–88.

11. Ibid.

12. Ibid., 389–90. Naudé is referring to the statue of Memnon on the Nile River that ancient authors said made sounds with the rising sun. See, for example, Callistratus, *Descriptions*, no. 9: "On the Statue of Memnon." On the speaking statue of Memnon, see also Stanhope, "Statue of Memnon."

13. Naudé, *Apologie*, 385–88.

14. Bayle, *Dictionnaire*, 1:131.

15. Ibid., 1:130.

16. See the article "Androïdes," in Scott, *Supplement*.

17. Vaucanson, *Mécanisme du fluteur automate*, 10–20. "One can substitute another flute entirely in the place of the one he plays": Luynes, *Mémoires*, 2:12–13. Similarly, the abbé Desfontaines emphasized that it was "the fingers positioned variously on the holes of the flute that vary the tones. . . . In a word art has done here all that nature does in those who play the flute well. That is what can be seen and heard, beyond a doubt," in Desfontaines, "Lettre," 339. On audiences' initial disbelief that the flute player was actually playing

his flute, see the reports cited in Chapuis and Droz, *Automata*, 274; Buchner, *Mechanical Musical Instruments*, 85–86; and David Lasocki's preface to Vaucanson, *Mécanisme du fluteur automate*, [ii].

18. Condorcet, "Eloge de Vaucanson" (1782), in *Œuvres*, 2:643–60; Doyon and Liaigre, *Vaucanson*, chaps. 1–2.

19. AS, Registre des procès-verbaux des séances for April 26 and 30, 1739.

20. "Nouvelles à la main au marquis de Langaunay," letter dated May 12, 1738, BNF, Ms Fr. 13700, n° 2; Doyon and Liaigre, *Vaucanson*, 30–34, 41.

21. *Mercure de France* (April 1738), 739; Doyon and Liaigre, *Vaucanson*, 51. On the vogue that Vaucanson launched, see also Metzner, *Crescendo*, chap. 5.

22. *Le pour et contre* (1738), 213; Doyon and Liaigre, *Vaucanson*, 53, 61.

23. Desfontaines, "Lettre," 340. The review of Vaucanson's treatise on the flute player in the *Journal des sçavans* also emphasized the role of anatomical and physical research in informing the android's design. See *Journal des sçavans* (1739), 441. See also Doyon and Liaigre, *Vaucanson*, 51.

24. See d'Alembert's 1751 article "Androïde," in Diderot and d'Alembert, *Encyclopédie*, 1:448.

25. Doyon and Liaigre, *Vaucanson*, 41.

26. See Vaucanson, *Mécanisme du fluteur automate*, 10–20. This process was the precursor of the procedure by which the first musical recordings were made, during the second and third decades of the twentieth century, when pianists such as Claude Debussy, Sergei Rachmaninoff, George Gershwin, Arthur Rubinstein, and Scott Joplin marked out rolls for player pianos. See Givens, *Re-enacting*.

27. Vaucanson, *Mécanisme du fluteur automate*, 10.

28. Ibid., 4, 16–17.

29. This was in conflict with the recommendations of some contemporary published flute tutors. Johann Quantz, in particular, denied that pitch was controlled by blowing pressure. See Quantz, *Versuch*.

30. "Doyon and Liaigre, *Vaucanson*, 53, 61.

31. Vaucanson, "Letter to the Abbé Desfontaines," in *Mécanisme du fluteur automate*, 23–24.

32. Hermann von Helmholtz explained the effects of partials in his *Lehre von den Tonempfindungen*. I am grateful to Myles Jackson for helping me to figure out the causes underlying Vaucanson's acoustical discovery.

33. Voltaire, "Discours en vers sur l'homme" (1738), in *OCV*, 17:455–535, at 521. For Frederick the Great's invitation, see Condorcet, "Eloge de Vaucanson," in *Œuvres*, 2:650–51.

34. Tax farmers (*fermiers généraux*) were the widely loathed members of the Tax Farm, the corporate body to whom the Crown contracted tax collection under the Old Regime.

35. See Doyon and Liaigre, *Vaucanson*, 43.

36. Marmontel, *Mémoires*, 1:237, 238–39. The story may of course be apocryphal.

37. Ibid., 1:247.

38. See Buffon, *Correspondance inédite*, 1:254 (46n12).

39. Doyon and Liaigre, *Vaucanson*, 308, 142–45.

40. On Kempelen's chess-playing Turk, see Carroll, *Chess Automaton*; Schaffer, "Babbage's Dancer" and "Enlightened Automata"; and Standage, *Turk*.

41. See George Walker, "Anatomy of a Chess Automaton," *Fraser's Magazine* (June 1839), 725; Aleck Abrahams, "Dr. Kempelen's Automaton Chess-player," *Notes and Queries* (January 28, 1922), 72; and responses in other issues of *Notes and Queries*: (February 11, 1922), 113; (February 15, 1922), 155. The response on February 15, by Barton R. A. Mills, contains the story of Napoleon's encounter with the Turk. See also Strauss, "Automata," 134; Schaffer, "Enlightened Automata," 162; and Evans, *Edgar Allan Poe*, 14.

42. A Knight's Tour entails moving a knight, starting on any square and using the rule governing knights' moves, to all the other squares in succession without touching any square twice. See Windisch, *Inanimate Reason*, 23–24; and Anonymous, *Observations on the Automaton Chess Player*, 24.

43. See Windisch, *Inanimate Reason*, 15, 18; and Carroll, *Chess Automaton*.

44. For early, prominent debunkings of the Turk, see Willis, *Attempt to Analyse*; and Poe, "Maelzel's Chess-Player."

45. Bradford, *History and Analysis*, 5. See also Windisch, *Inanimate Reason*, 10.

46. Kempelen—Sir R. M. Keith, August 14, 1774, BL Add MS 35507, fol. 275.

47. Horace Walpole, Earl of Oxford to William Wentworth, Earl of Strafford, December 11, 1783, in Walpole, *Letters*, 13:11.

48. Windisch, *Inanimate Reason*, 39, 13, 34, v.

49. Thicknesse was the acerbic and tyrannical governor of Landguard Fort on the Suffolk coast until the War Office found him too quarrelsome. His will gives a measure of his vindictiveness: he left instructions for his right arm to be severed and given to his son "in hopes that such a sight may remind him of his duty to God, after having for so long abandoned the duty he owed his father who once affectionately loved him." On "Dr. Viper," see Levitt, *Turk*, 202; and Gosse, *Dr. Viper*. On the severed arm, see also Chambers, *Book of Days*, entry for October 16.

50. Thicknesse, *Speaking Figure*, quoted in Levitt, *Turk*, 202. Thicknesse's refusal to believe in a responsive machine was part of a more general skepticism about mechanical innovations and scorn for their enthusiasts. By his own account, Thicknesse had earlier rejected what he now took for granted, that a machine could move autonomously in complex ways:

> Forty years since, I found three hundred people assembled, to see, at a shilling each, a coach which went without horses. . . . Many persons present, were angry with me, for saying it was trod round by a man within a hoop, or hinder wheels: but a small paper of snuff, put into the wheel, soon convinced every person present, that it could not only move, but sneeze too, *perfectly like a Christian*. That machine was not a wheel within a wheel, but a Man within a wheel: The *Speaking Figure* is a man in a closet above, and the Automaton Chess-Player is *a man within a man*: for whatever his outward form be composed of, he bears a living soul within. Thicknesse, *Speaking Figure*, 204.

51. Decremps, *Magie*, 67.

52. Racknitz, *Schachspieler*, quoted in translation in Levitt, *Turk*, appendix G.

53. See Aleck Abrahams, "Dr. Kempelen's Automaton Chess-player," *Notes and Queries*

(January 28, 1922), 72. On Babbage's encounter with the Turk, see Abrahams's follow-up note in *Notes and Queries* (February 25, 1922), 155–56.

54. Anonymous, *Observations on the Automaton Chess Player*, 30, 32.

55. Babbage, *Passages*, chap. 3; Schaffer, "Babbage's Dancer," 54–55; Jacob, *John Joseph Merlin*.

56. Babbage, *Passages*, 349–53. See also Schaffer, "Babbage's Dancer," 60–61.

57. See Schaffer, "OK Computer."

58. Standage, *Turk*, 103–5, 112, 245–46.

59. Willis, *Attempt to Analyse*. On Willis's study of the Turk, see also Carroll, *Chess Automaton*, 51–55; Schaffer, "Babbage's Dancer," 70–73; and Standage, *Turk*, 128–35.

60. See especially George Walker, "Anatomy of the Chess Automaton," *Fraser's Magazine* (June 1839), 717–31; Brewster, *Letters on Natural Magic*, 321–33; and Poe, "Maelzel's Chess-Player."

61. Bradford, *History and Analysis*, 9.

62. On Poe's encounter with the Turk, see Evans, *Edgar Allan Poe*; Carroll, *Chess Automaton*, 81–85; and Standage, *Turk*, chap. 10.

63. Poe, "Maelzel's Chess-Player," 178–79. Five years later, however, Poe opened his "Murders in the Rue Morgue" (1841) by using the game of chess in a seemingly opposite way: to exemplify something like mindless calculation in contrast with the genuine analysis required by whist (or detection).

64. Robert-Houdin, *Memoirs*, 117–26.

65. "The first player was a Polish patriot, Worousky, who had lost both legs in a campaign; as he was furnished with artificial limbs when in public, his appearance, together with the fact that no dwarf or child travelled in Kempelen's company, dispelled the suspicion that any person could be employed inside the machine." *Encyclopaedia Britannica*, 11th ed., s.v. "conjuring." See also Carroll, *Chess Automaton*, 62–63, 94; Chapuis and Droz, *Automata*, 365, 368 (fig. 462); Fechner, *Magie de Robert-Houdin* (hereafter *MRH*), 1:175–76; and Standage, *Turk*, 92–96.

66. On the Neuchâtel automata, see Voskuhl, *Androids*.

67. Doyon and Liaigre, *Vaucanson*, 56n13.

68. See Altick, *Shows of London*; Schaffer, "Enlightenment Automata," 138; and Chapuis and Droz, *Automata*, 280–82.

69. On the Writer's mechanism, see Perregaux and Perrot, *Jaquet-Droz*, 181–90.

70. Chapuis and Gélis, *Monde*, 2:270–78; Chapuis and Droz, *Automata*, 280–81; Jaquet-Droz, *Œuvres*. On the Jaquet-Droz family, see Perregaux and Perrot, *Jaquet-Droz*; Chapuis and Droz, *Jaquet-Droz*; Carrera et al., *Androïdes*; and Voskuhl, *Androids*.

71. See Perregaux and Perrot, *Jaquet-Droz*, 31–34.

72. Isaac Droz (solicitor of Locle) to Baron de Lentulus (gouverneur de Neuchâtel), July 9, 1774, quoted in Perregaux and Perrot, *Jaquet-Droz*, 102.

73. Bachaumont, *Mémoires*, 7:323; Chapuis and Gélis, *Monde*, 2:192; Metzner, *Crescendo*, 171.

74. Perregaux and Perrot, *Jaquet-Droz*, 110–11.

75. Bachaumont, *Mémoires*, 7: 310–11; Perregaux and Perrot, *Jaquet-Droz*, 108; Metzner, *Crescendo*, 171.

76. This brief sketch of Grimod de La Reynière's life relies on Denise Gigante's elegant account in *Gusto,* 1–4. See also Rival, *Grimod;* and MacDonogh, *Palate.*

77. See Perregaux and Perrot, *Jaquet-Droz,* 109, 166; and "DROZ, Pierre Jaquet," in Weiss, *Biographie universelle,* 2:414. In 1786, a cousin and perhaps collaborator of the Jaquet-Droz family, the director of the French Mint in Paris, a man named Jean-Pierre Droz, designed an artificial hand to improve the safety of workers at the mint, who, because they had to slide metal strips under the balance arm of the machine that stamped them, frequently had bad accidents. Droz's artificial hand was intended to take on this dangerous task. On the Jaquet-Droz artificial limbs, see Perregaux and Perrot, *Jaquet-Droz,* 140; Benhamou, "Cover Design"; and Metzner, *Crescendo,* chap. 5, section 1 and note 21.

78. Helvétius, *De l'homme,* 28–29; Diderot, *Réfutation de l'ouvrage d'Helvétius intitulé De l'homme* (1773–74), in *Œuvres complètes,* 2:283–84.

79. Lespinasse, *Lettres,* 214. On the Jaquet-Droz automata in Paris, see Bachaumont, *Mémoires,* 7: 310–11, 323; and Campardon, *Spectacles de la foire,* 1:276–77.

80. Vaucanson, *Mécanisme du fluteur automate,* 21–24.

81. Ibid., 22–23; see Chapuis and Gélis, *Monde,* 2:149–51; and Chapuis and Droz, *Automata,* 233–42.

82. Vaucanson, *Mécanisme du fluteur automate,* 21–24; Godefroy Bereis, letter dated November 2, 1785, quoted in Chapuis and Droz, *Automata,* 234, see also, 233–38 and note 14.

83. Vaucanson, *Mécanisme du fluteur automate,* 21–22. By claiming that his Duck digested by dissolution, Vaucanson entered a contemporary debate among physiologists over whether digestion was a chemical or a mechanical process.

84. See Doyon and Liaigre, *Vaucanson,* 479.

85. Condorcet, "Eloge de Vaucanson," in *Œuvres,* 2:648.

86. See Nicolai, *Chronique,* 1:284. Robert-Houdin claimed to have made the same discovery in 1845, while repairing the Duck's mechanism. See Robert-Houdin, *Memoirs,* 114–17. However, this debunking has itself been debunked. The Duck that Robert-Houdin punctured was probably not Vaucanson's, but a copy. On this question, see Chapuis and Gélis, *Monde,* 2:151–52; and Chapuis and Droz, *Automata,* 248, 404n17. On the Duck's fraudulence in general and its unveiling, see Doyon and Liaigre, *Vaucanson,* 125–29; and Stafford, *Artful Science,* 193–94.

87. "Diverses machines inventées par M. Maillard: Cygne artificial," in *Machines et inventions,* 1:133–35.

88. The machine is described as an "anatomie mouvante" in *Commission extraordinaire du Conseil,* Plumitif no. 10, AN V7 582, cited in in Doyon and Liaigre, *Vaucanson,* 110; see also 18, 34. The second description is from Acte de société Colvée-Vaucanson, 26-1-1734, ANF Minutier Central, Notaire CXVIII, also quoted in Doyon and Liaigre, *Vaucanson,* 18.

89. "Registre contenant le Journal des Conférences de l'Académie de Lyon," quoted in Doyon and Laigre, *Vaucanson,* 148; and in translation in Beaune, "Classical Age of Automata," at 457. See also Doyon and Liaigre, "Méthodologie," 298.

90. Bachaumont, *Mémoires,* 23:306–8.

91. On the plans for a model of the circulatory system, see Condorcet, "Eloge de Vaucanson," in *Œuvres,* 2:655; Doyon and Liaigre, *Vaucanson,* 55–56, 118–19, 133–35, 141, 152–61; and Strauss, "Automata," 71–72.

92. See Borgnis, *Traité complet*, 8:118.

93. Quesnay, *Essai physique*, 219–23.

94. Quesnay, *Observations*, iv–vi. See also Quesnay, *Art de guérir*, and *Traité des effets*. The *Traité* is a later version of the *Observations* and of *Art de guérir*. See Doyon and Liaigre, "Méthodologie," 297.

95. This description appeared in conjunction with Le Cat's *Traité de la saignée* (1739) as its "experimental part," in order "to confirm by experience" Le Cat's theory of bleeding; see Doyon and Liaigre, "Méthodologie," 298–99. See also Ballière de Laisement, *Eloge*, 53.

96. Le Cornier de Cideville to Fontenelle, December 15, 1744, in Tougard, *Documents*, 1:52–54, at 53. See Doyon and Liaigre, "Méthodologie," 300.

97. Registre-Journal des Assemblées et Déliberations de l'Académie des sciences . . . établie on 1744: 3 (manuscript non classé de la Bibliothèque publique de Rouen), cited in Doyon and Liaigre, "Méthodologie," 300.

98. Cervantes, *Don Quixote*, chap. 62. On this episode in the novel, see Jones, "Historical Materials," 101–2.

99. Del Río, *Disquisitionum*, 26.

100. Kircher, *Phonurgia Nova*, 161, quoted in Jones, "Historical Materials," 99.

101. Jones, "Historical Materials," 99.

102. Desfontaines, "Lettre," 341.

103. Dodart, "Sur les causes"; Dodart, "Supplément au Mémoire" (1706) and "Suite"; and Dodart, "Supplément au Mémoire" (1707). For Fontenelle's commentary on Dodart's memoirs, see Fontenelle's three articles "Sur la formation de la voix" (1700, 1706, and 1707). See also Séris, *Langages et machines*, 231–35.

104. Fontenelle, "Sur la formation de la voix" (1707), 20.

105. Court de Gébelin, *Monde primitif*, 2:83–84. See Séris, *Langages et machines*, 239.

106. La Mettrie, *L'homme machine*. 190.

107. Wilkins, *Mathematicall Magick*, 177–78. See also Hankins and Silverman, *Instruments and the Imagination*, 181.

108. Darwin, *Temple of Nature*, 119–20. See Hankins and Silverman, *Instruments and the Imagination*, 199.

109. *Têtes parlantes inventées et exécutées par M. l'abbé Mical. (Extrait d'un ouvrage qui a pour titre: Système de prononciation figurée, applicable à toutes les langues et exécuté sur les langues française et anglaise)*, VZ-1853, BNF; Rivarol, "Lettre à M. le president de," 20–24; Rivarol, *Discours*, 79–82; Bachaumont, *Mémoires*, 11: 237, 13: 270, 26: 214–216; Séris, *Langages et machines*, 245; Chapuis and Gélis, *Monde*, 2:204–206. According to Rivarol, Mical also built "an entire Concert in which the figures, as big as life, made Music from morning till evening," "Lettre à M. le president de," 29.

110. AS, Registre des procès-verbaux des séances for September 3, 1783; Bachaumont, *Mémoires*, 26: 214–16.

111. Hankins and Silverman, *Instruments and the Imagination*, 188–89, 198; Séris, *Langages et machines*, 247.

112. Kempelen, *Mécanisme de la parole*, 394–464. On Kempelen's and others' attempts to simulate human speech in the last third of the eighteenth century, see also Hankins

and Silverman, *Instruments and the Imagination*, chap. 8; and Séris, *Langages et machines*, 245–46.

113. Kempelen, *Mécanisme de la parole*, 395–400, 405, 415–59.

114. Ibid., 401.

115. Ibid., 463.

116. Windisch, *Inanimate Reason*, 47.

117. Strauss, "Automata," 123.

118. Windisch, *Inanimate Reason*, 49.

119. On Wheatstone's and Bell's reproductions, see Flanagan, "Voices of Men and Machines"; Flanagan, *Speech Analysis*, 166–71; Schroeder, "Brief History"; and Hankins and Silverman, *Instruments and the Imagination*, 218–19.

120. Hermann von Helmholtz, for example, built a machine using tuning forks and resonance chambers to produce the vowel sounds, described in Helmholtz, *On the Sensations of Tone*, 399. On Helmholtz's speech synthesizer, see Flanagan, *Speech Analysis*, 172–74; Lenoir, "Helmholtz"; Schroeder, "Brief History," 232–33; and Hankins and Silverman, *Instruments and the Imagination*, 203–5. On the transition from speech simulation to speech synthesis in the nineteenth century, see Hankins and Silverman, *Instruments and the Imagination*, 199, 209; and on the exceptions, see 210–13.

121. Willis, "On the Vowel Sounds." See Hankins and Silverman, *Instruments and the Imagination*, 203–5.

122. Bernard, *Cahier*, 171. This was probably in response to the assertion of Bernard's mentor, François Magendie, that "I see in the lung a bellows, in the trachea, a vent, in the glottis a reed. . . . We have for the eye an optical apparatus, for the voice a musical instrument, for the stomach a living retort." Magendie, *Phénomènes*, 2:20. On Bernard's probable reference to Magendie, see Canguilhem, *Études*, 332; and Séris, *Langages et machines*, 248.

123. See Lindsay, "Talking Head"; and Hankins and Silverman, *Instruments and the Imagination*, 214–16.

124. Cf. Hankins and Silverman, *Instruments and the Imagination*, 216, where the authors identify a partial and passing return to "more humanoid apparatus" in the "last years of the nineteenth century." On the early history of electrical speech synthesis, see Flanagan, *Speech Analysis*, 171–72; Flanagan, "Voices of Men and Machines," 1381–83; Klatt, "Review," 741–42; and Schroeder, "Brief History."

125. Levot, *Biographie bretonne*, 1:772–74; Doyon and Liaigre, *Vaucanson*, 128, 213–14; Dubuisson, Payen, and Pilisi, "Textile Industry," 216–17.

126. De Gennes was captured by the English when they conquered the island in 1702 and died in captivity. See Levot, *Biographie bretonne*, 1:774; and Doyon and Liaigre, *Vaucanson*, 128, 213–14.

127. Labat, *Nouveau voyage*, 2:298. For a version of the same story, see Catrou et al., ed., *Journal de Trévoux* (March 1722), 441–42.

128. Doyon and Liaigre, *Vaucanson*, 308, 142–45. For an argument that "before thinking of automating manual labor, one must conceive of mechanically representing the limbs of man," see Beaune, *L'automate et ses mobiles*, 257. Beaune takes Vaucanson's career as his central case. He returns to this trajectory from automata to industrial automation, simulation to replacement, in "Classical Age of Automata."

129. Doyon and Liaigre, *Vaucanson,* 206, 225–35; Bohnsack, *Jacquard-Webstuhl,* 27–28; CNAM, *Jacques Vaucanson,* 16.

130. *Mercure de France* (November 1745), 116–20.

131. Vaucanson, *Mémoire sur un nouveau métier,* 468–69.

132. Productive vs. sterile: Quesnay, "Analyse," 305–9; replacing human workers with machines: Quesnay, "Questions," 266, 303; "Maximes," 334; "Sur les travaux," 532, 542, 545. On the Physiocrats more generally, see Riskin, *Science,* ch. 4; and Riskin, "Spirit of System."

133. See the articles for "Artisan" and "Artiste," in Diderot and d'Alembert, *Encyclopédie,* 1:745. See also Sewell, *Work and Revolution in France,* 23.

134. Vaucanson, *Mémoire sur un nouveau métier,* 471.

135. Doyon and Liaigre, *Vaucanson,* 142–45.

136. Ibid., 456, 462. Charles Gillispie has pointed out that this was an early example of a combination that would be characteristic of the post-Revolutionary French economy: expert consulting, private money, and government guarantee and oversight. See Gillispie, *Science and Polity,* 416.

137. Doyon and Liaigre, *Vaucanson,* 191–203. See Alder, *Engineering.*

138. See Daston, "Enlightenment Calculations." Simon Schaffer has written similarly that "enlightened science imposed a division between subjects that could be automated and those reserved for reason. Such a contrast between instinctual mechanical labor and its rational analysis accompanied processes of subordination and rule." See Schaffer, "Enlightened Automata," 164.

139. Pascal, "Calculating Machine," 169.

140. Leibniz, "Calculating Machine," 181.

141. Babbage, *Passages,* 88.

142. Lovelace, *Ada,* 182.

143. Babbage, *Passages,* 30–31.

144. Buxton, *Memoir,* 46, quoted in Schaffer, "Babbage's Intelligence," 204.

145. Smith, *Wealth of Nations,* vol. 1, book 1, chap. 1, paragraph 3.

146. Babbage, *The Economy of Machinery and Manufactures* (1822), in *Works,* 8:136, 137. See Daston, "Enlightenment Calculations"; and Campbell-Kelly and Aspray, *Computer,* chap. 1. The tables were computed by the method of differences, the relevant theorem being that for a polynomial of degree *n,* the *n*th difference is a constant. On Babbage's notion of human and machine intelligence and mental labor, see also Schaffer, "Babbage's Intelligence," "Babbage's Dancer," and "OK Computer."

147. Diderot and d'Alembert, *Encyclopédie,* 9:794.

148. *Dictionnaire de l'Académie,* 1st ed., 2:1; *Dictionnaire de l'Académie,* 4th ed., 2:65; *Dictionnaire de l'Académie,* 5th ed., 2:48.

149. *Dictionnaire de l'Académie,* 6th ed., 2:140.

Chapter 5

1. I use the words "man"-machine and "his" because the Enlightenment model was indeed male. Its authors applied it primarily to men and only secondarily to women, though they were interested, as we shall see, in the specifics of the female man-machine and their implications.

2. Frederick II, "Eulogy," 6.

3. Ibid.

4. Carlyle, *History*, 4:386.

5. Frederick II, "Eulogy," 6.

6. La Mettrie, *L'homme machine*, 152–56.

7. Ibid., 183.

8. Frederick II, "Eulogy," 8.

9. Frederick to Wilhelmina, November 21, 1751, in his *Œuvres*, 27:230.

10. La Mettrie, *L'homme machine*, 180, 183.

11. On La Mettrie, Haller, and Boerhaave, see Vartanian, *L'Homme Machine*, 75–89. Despite his policy of ambiguity and overt criticisms of Spinoza, Boerhaave was reputed to be a Spinosist, a charge against which Samuel Johnson defended him. See Johnson, "Life," 73; and Israel, *Radical Enlightenment*, xviii, lx.

12. See Israel, *Radical Enlightenment*, 704.

13. On Haller's physiology, see Duchesneau, *Physiologie des lumières*, chaps. 5 and 6; Roe, "*Anatomia animata*"; and Steinke, *Irritating Experiments*. On the distinction between irritability and sensibility, see Boury, "Irritability and Sensibility."

14. Indeed, the line between vitalism and materialism was often very blurry in this period, as historians of the eighteenth and early nineteenth centuries have noted. I have taken up this blurriness in "Eighteenth-Century Wetware."

15. Haller, *Elementa physiologiae*, v, quoted in translation in Roe's introduction to Haller, *Natural Philosophy*, n.p.

16. Vartanian, "Trembley's Polyp," 271.

17. Carlyle, *History*, 4:387.

18. On La Mettrie's influence on his contemporaries, see Vartanian, "Trembley's Polyp"; and Israel, *Radical Enlightenment*, chap. 37.

19. Carlyle writes of Frederick: "It is certain he could, especially in his younger years, put up with a great deal of zanyism, ingenious foolery and rough tumbling, if it had any basis to tumble on. . . . By far his chief Artist in this kind, indeed properly the only one, was La Mettrie," in *History*, 4:385–86.

20. Thibault, *Mes souvenirs*, 5:405.

21. See Voltaire to Madame Denis, in November 14, 1751, in *OCV*, 96:314–15. Voltaire's letters from the Prussian court to his niece and lover, Madame Denis (Marie Louise Mignot) are in fact rewritten literary versions of the original genuine correspondence. For the full story of this rewriting and its belated discovery, see Mallinson, "What's in a Name?"

22. Voltaire's report of La Mettrie's account of Frederick's remark, and Voltaire's initial reaction, are in Voltaire to Madame Denis, September 2, 1751, in *OCV*, 96:277–78. Carlyle recounts the story in *History*, 396. The orange-skin affair has been retold by generation after generation of Voltaire biographers. For nineteenth-century examples, see Morley, *Voltaire*, 194–95; and Hamley, *Voltaire*, 154–55.

23. Voltaire to Madame Denis, October 29, 1751, in *OCV*, 96:305–6.

24. See Vartanian, *L'Homme Machine*, 8–9; Israel, *Radical Enlightenment*, 574, 671, 722; and Israel, *Enlightenment Contested*, 803.

25. Hollmann, Untitled letter, 411; see La Mettrie, *L'homme machine*, 101–2nn18–19.

26. La Mettrie, *Epître à Mlle. A.C.P. ou la 'Machine terrassée* (1749), in his *Œuvres philosophiques*, 2: 215–16. This *Epître* was one of three anonymous, self-satirizing pamphlets, all of which appeared in 1749. The others were *Epître à mon Esprit, ou l'Anonyme persiflé* and *Réponse à l'auteur de la Machine terrassée.* See La Mettrie, *L'homme machine*, 102n20; and Lemée, *La Mettrie*, 206–19. Lemée recounts how he found the only surviving copy of the last of these, "Réponse à l'auteur . . . ," in Paris during the Occupation and tried to have it reprinted, but his editor sent him the prints just at the moment of the Normandy landing and they never arrived.

27. Voltaire to Madame Denis, September 2, 1751, in *OCV*, 96:277–78.

28. Voltaire to Madame Denis, November 14, 1751, in *OCV*, 96:314–15. For versions of the story of La Mettrie's death, see also Frederick to Wilhelmina, November 21 1751, in Frederick II, *Œuvres*, vol. 27, part 1, letter #232, pp. 229–230, at 230; and Nicolai, *Anekdoten*, 6:197–201. For an analysis of the various versions of this story, see Dressler, "La Mettrie's Death."

29. Voltaire to Madame Denis, November 14, 1751, in *OCV*, 96:314–15.

30. Voltaire to Richelieu, November 13, 1751, in *OCV*, 96:313–14. The two meanings of *farce* may in fact be related, their common ancestry being the medieval comic performances that were stuffed in between religious presentations to entertain the hoi polloi.

31. Voltaire to Madame Denis, December 24, 1751, in *OCV*, 96:332–33.

32. Voltaire to Madame Denis, November 14, 1751, in *OCV*, 96:314–15.

33. See, for example, Thibault, *Mes souvenirs*, 5:407. Thibault also reports that La Mettrie had an uncontrolled imagination, and that as an "absolute and pronounced materialist" he was "afraid of almost everything"; see 5:406.

34. Voltaire to Madame Denis, December 24, 1751, in *OCV*, 96:332–33.

35. Nicolai, *Anekdoten*, 1:20, as recounted in Carlyle, *History*, 4:399–400.

36. La Mettrie, *Système*. To Epicurus, however, pleasure did not derive from bodily things but from living wisely and justly. La Mettrie's philosophy was also Epicurean in its materialism; Epicurus was a materialist and an atomist who rejected the idea of divine intervention in nature. Enlightenment authors including La Mettrie took a great interest in Epicurus, and in the rendition of his philosophy by the Roman poet Lucretius (1st century BCE), who presented an Epicurean view of the world in his *De rerum naturae* [On the nature of things]. On Enlightenment Epicureanism, see Jones, *Epicurean Tradition*, Ch. 7 and Leddy and Lifschitz, *Epicurus*.

37. La Mettrie, *L'homme machine*, 154, 182–83, 186, 190. Vaucanson appears at 190.

38. Ibid., 186, 189–190.

39. Regarding La Mettrie's influence on Diderot, see Crocker, "Diderot"; Vartanian, "Trembley's Polyp"; and Vartanian, "L'Homme-Machine since 1748," in La Mettrie's *L'homme machine*, chap. 6, at 115–21. Vartanian attributes Diderot's about-face between the deistic *Pensées philosophiques* (1746) and the materialistic *Pensées sur l'interprétation de la nature* (1754) importantly to La Mettrie's *L'homme machine*; see Vartanian, "Trembley's Polyp," 274.

40. Diderot, *Eléments*, 20–21.

41. La Mettrie, *L'homme machine*, 192.

42. La Mettrie, *L'Homme-Plante* (1748), in his *Œuvres philosophiques*, 1:281–306.

43. La Mettrie, *L'homme machine,* 173–75, 196. See also La Mettrie, *Système,* paragraph 76.

44. Diderot, *Eléments,* 32.

45. La Mettrie, *L'homme machine,* 145, 152, 164–65, 175.

46. Ibid., 176, 191–92, 194.

47. Ibid., 189, 192, 194, 196.

48. La Mettrie, *Système,* paragraph 46.

49. Diderot, *Eléments,* 308.

50. La Mettrie, *L'homme machine,* 150. "Soul of mud" was a paraphrase of the abbé Pluche's characterization of John Locke's view of the soul. See Pluche, *Spectacle de la nature,* 5:176–77.

51. Priestley and Price, *Free Discussion,* 75–76. See also Priestley, *Disquisitions,* 1:iv: "What we call mind . . . [is] the result of corporeal organization"; and 1:117: "A certain quantity of nervous system is necessary to such complex ideas and affections as belong to the human mind; and the idea of self, or the feeling that corresponds to the pronoun I (which is what some may mean by consciousness)."

52. See, for example, Israel, *Enlightenment Contested,* 803–5, 809, 810. Also at 804–5.

53. For instance: Voltaire, Rousseau, and Diderot.

54. Diderot, *Rêve,* 312.

55. See Hankins, *Jean d'Alembert,* chap. 4; and Riskin, *Science in the Age of Sensibility,* 56–58, 123.

56. Diderot, *Entretien,* 271–74.

57. Dulac, "Version déguisée," 147; Vartanian, "La Mettrie and Diderot Revisited," 177fn45; Cohen, "Enlightenment and the Dirty Philosopher," 408.

58. On d'Alembert's much-gossiped-about relations with Julie de Lespinasse, see Hankins, *Jean d'Alembert,* chap. 6.

59. Bordeu, *Recherches anatomiques,* 1:187, quoted in Moravia, "Homme Machine," 56.

60. Diderot, *Rêve,* 288–89, 291–95, 306, 339, 342.

61. Diderot, *Entretien,* 279.

62. Diderot, *Rêve,* 314–18, 330–33.

63. Diderot, *Eléments,* in *Œuvres philosophiques,* 306, 50, 59–60, 265.

64. Frederick was referring to Voltaire's *Dictionnaire philosophique* article on the soul. Frederick II to Voltaire, October/November 1752, in Voltaire, *Correspondence,* 13:225–26, letter #D5056.

65. Bonnet, *Contemplation,* 1:xl. (Note: a great confusion reigns among editions and printings of this work, which differ significantly and are not consistently labeled. This citation and all following are to the 1770 Amsterdam edition listed in the bibliography, and not, for example, to the 1769 Amsterdam edition, which has "Seconde Édition" on its title page. Other editions, such as the 1782 Hamburg edition, simply call themselves "new edition.") See also Bonnet, *Palingénésie philosophique,* 2:78–80, where he explains that the soul is necessary not for moral or theological but only philosophical reasons: without it, "good philosophy" cannot make sense of "the clear and simple sentiment [man] has of his *me*."

66. Buffon, "Discours sur la nature des animaux," 28; see also 23.

67. Buffon, *Histoire naturelle,* 3:364–70.

68. Buffon, "Discours sur la nature des animaux," 41, 86, 15–16, 23–24.

69. Montesquieu, *Lettres persanes* (1721), in his *Œuvres complètes de Montesquieu* [hereafter *OCM*], 1: entire, on 219–20 [letter #31]. Montesquieu elsewhere attributed suicide to a "fault in the filtration of the nervous sap" causing "the machine" to grow "weary of itself." See his *De l'esprit des lois*, 216.

70. Montesquieu, "Discours sur l'usage des glandes rénales" (1718), in *OCM*, 8:157–73, at 165–66.

71. See, for example, Rousseau, "Economie," 338, 343; Rousseau, *Émile*, 39, 317; and Rousseau, *Du contrat social*, 257.

72. Rousseau, *Confessions*, 1:229, 230, 245. In his correspondence too Rousseau dwelled extensively on "my machine," often "my poor machine." To the baron de Malesherbes, one of the few sweet-tempered enough to tolerate Rousseau, he complained bitterly of "the deplorable state" and "the decay" of "my poor machine." See Rousseau to Malesherbes, January 12, 1762, in his *Correspondance*, 10:24–29, letter #1633; and January 26, 1762, in *Correspondance*, 10:52–58, letter #1650. See also Rousseau to Laliaud, November 28, 1768, in *Correspondance* 36:195–97, letter #6497.

73. Rousseau, *Discours sur l'origine*, 47.

74. Rousseau, *Émile*, 330.

75. Ibid., 39.

76. Rousseau, *Confessions*, 2:80–82.

77. Rousseau, *Émile*, 39.

78. Buffon, Diderot, and Bonnet all presented thought experiments around midcentury featuring artificially composed people. I have written about these thought experiments in Riskin, *Science in the Age of Sensibility*, chap. 2. See Diderot, *Essai sur le mérite et la vertu* (1745), in *Œuvres complètes*, 1:3–121, at 24–25, 35, 50; Diderot, *Lettre*, 111–12; Buffon, *Histoire naturelle*, 3:364–70; and Bonnet, *Essai analytique*.

79. Condillac to Samuel Formey, February 22, 1756, BNF NAF 15551 F.19; Bongie, "Documents," 86–87. For Helvétius's admiration of Vaucanson, see Helvétius, *De l'esprit*, 28 where he invokes Vaucanson first in a list of illustrious men. On Condillac's priority dispute with Buffon, see also Condillac, "Réponse à un reproche" (1754), in his *Œuvres philosophiques*, 1:318–19; and Riskin, *Science in the Age of Sensibility*, 48–49.

80. Tourneux, *Correspondance littéraire*, 3:112; see Bongie, "Documents," 87n12. Buffon had not actually described his creature as a statue; that Grimm assumed he had indicates how tightly connected this sort of reflection had become with mechanical statues.

81. See Tourneux, *Correspondance littéraire*, 2:204. Elie-Catherine Fréron made the same charge (that he had plagiarized Diderot) in his *L'année littéraire*, 7:297–98. See Bongie, "Documents," 86.

82. On the plagiarism charges, see Condillac, "Réponse à un reproche," in *Œuvres philosophiques*, 1:318–19; *Traité des animaux*, préface and n1, 311–12; La Harpe, *Philosophie*, 1:193–94; Tourneux, *Correspondance littéraire*, 3:112–13; Lebeau, *Condillac*, 28–30; Riskin, *Science in the Age of Sensibility*, 48; and Bongie, "Documents." For Bonnet's excuse, see Bonnet, *Essai analytique*, chaps. 3 and 26.

83. Rousseau, *Confessions*, 2:69.

84. Ibid., 2:69–70.

85. Diderot, *Essai sur le mérite et la vertu*, in *Œuvres complètes*, 1:24–25, 35, 50.

86. Diderot, *Lettre*, 62. He committed himself more explicitly to rigorous materialism in his later writings including, as discussed above, the *Rêve de d'Alembert*.

87. Condillac, *Traité des sensations*, part 1, chap. 1, section 2; chap. 7, section 1; chap. 8, section 1; chap. 11, sections 1 and 2; chap. 12. This and the following is derived from the discussion in Riskin, *Science in the Age of Sensibility*, chap. 2. For other recent work on the animated statues thought experiments, see Douthwaite, *Wild Girl*, chap. 2. Douthwaite examines how the thought experiments served as rhetorical devices for marshaling experimental science to both egalitarian and inegalitarian social purposes.

88. Condillac, *Traité des sensations*, part 2, chap. 1.

89. Ibid., chaps. 4 and 5.

90. Ibid., part 4, chap. 8, section 6.

91. Ibid., part 2, chap. 4, section 3. See also Condillac, *Traité des animaux*, part 1, chaps. 1 and 2. On the *Traité des animaux*, see also Dagognet, *L'animal*.

92. La Mettrie, *L'homme machine*, 195.

93. Montesquieu, "Discours sur l'encouragement des sciences" (1725), in *Œuvres complètes*, Caillois ed., 1:53–57, at 53. Montesquieu's analysis of the Aztec and Inca defeats is akin to Montaigne's in "Des coches" (Of Coaches), in his *Essais*, book 3, chap. 6.

94. La Mettrie, *L'homme machine*, 157, 159, 163.

95. Ibid., 157–58.

96. Montesquieu, *De l'esprit des lois*, vol. 2, part 3, book 14, chap. 2.

97. Rousseau, *Émile*, 267.

98. La Mettrie, *L'homme machine*, 163, 166, 168, 184, 196.

99. Diderot, *Rêve*, 354.

100. Buffon, *Histoire naturelle*, 2:486.

101. Diderot, *Eléments*, 297.

102. Lavater, *Essays*, 188–89, 206.

103. Rousseau, *Émile*, 477.

104. Smith, *Theory of Moral Sentiments*, part 4, section 2.

105. Montesquieu, *Lettres persanes*, in *Œuvres*, 1: entire, on 419 [letter #103].

106. Buffon, *Histoire naturelle*, 14:30–32.

107. Ibid., 14:32.

108. La Mettrie, *L'homme machine*, 195, 170.

109. Bonnet, *Essai analytique*, 495; Bonnet, *Contemplation*, 1:lxii–lxiii.

110. See chapter 3; Riskin, *Science in the Age of Sensibility*, chaps. 2 and 7; and Spary, *Utopia's Garden*, chap. 5.

111. Bonnet, *Palingénésie philosophique*, 1:267–76, cited passage on 268. On Bonnet's Leibnizianism, see Duchesneau, "Charles Bonnet's Neo-Leibnizian Theory."

112. Leibniz, *Nouveaux essais*, in *PS*, 5:306; see also 293.

113. Bonnet, *Contemplation*, 1:85–87.

114. Leibniz, *Consilium Aegyptiacum* (1671) (the army of slave-warriors is proposed in the addendum, "Modus instituendi militam novam invictam"), in *SSB*, series 4, vol. 1:217–412. Here and in what follows, I am drawing upon Justin Smith's "A Series of Generations."

115. Smith also usefully summarizes the complicated history of writing on Leibniz's views of race in "A Series of Generations."

116. Leibniz, "Epistola . . . ad A. C. Cackenholz" (April 23, 1701), in *Opera Omnia*, vol. 2, part 2, 169–74, at 171. Here again I am indebted to Smith, see " Series of Generations," 331–32.

117. Sales, *Philosophie*, 326, 364, 151.

118. Maupertuis, *Vénus physique*, 140, 167–168, 158.

119. Ibid., 140, 144, 158. Maupertuis actually uses the phrase "confirm the race" in reference to "white-Negroes" (albinos) who, he argues, if they produced several generations of descendents together, "would confirm the race," 140.

120. Helvétius, *De l'esprit*, 207, 198.

121. Helvétius, *De l'homme*, 116–19.

122. Helvétius, *De l'esprit*, 441.

123. Helvétius, *De l'homme*, 570; Diderot, *Réfutation de l'ouvrage d'Helvétius intitulé De l'homme* (1773–74), in *Œuvres complètes*, 2:455–56.

124. Diderot to Louise Henriette Volland, October 4, 1767, in Diderot, *Œuvres complètes*, 19:256–57; Helvétius, *Correspondance générale*, 3: 288.

125. Lamarck, *Système analytique*, 306–7; see also Lamarck, "Idée," 86; and Lamarck, *Recherches*, 127–28.

126. Although "organization" became an alternative to "design" after the middle of the eighteenth century, authors of arguments from design had occasionally used the word "organization" as synonymous with "design" before then. See Ray, *Wisdom of God*, 48–49; Cudworth, *Intellectual System*, 1:149; Toland, *Letters to Serena*, 235; Grew, *Cosmologia sacra*, chap. 5; and La Touche Boesnier, *Préservatif*.

127. La Mettrie, *L'homme machine*, 149, 188.

128. Boerhaave, *Chemistry*, 142, 150.

129. It was also unlike the human-machinery announced by some polemical materialists, such as the baron d'Holbach, who described human beings as purely passive, puppetlike devices acted upon by unknown mechanical causes. See, for instance, Holbach, *Système*, 225.

130. La Mettrie, *L'homme machine*, 166–67, 180.

131. Priestley, *Disquisitions*, 1:150, 152; see also iv and (for a reference to La Mettrie) 202.

132. Ibid., 1:264–65, 269, 206.

133. La Mettrie, *L'homme machine*, 177–78. On Abraham Trembley and his study of the hydra *Chlorohydra viridissima*, see Vartanian, "Trembley's Polyp"; Dawson, *Nature's Enigma*; and Lenhoff and Lenhoff, *Hydra*.

134. La Mettrie, *Système*, paragraphs 18and 21; La Mettrie, *L'homme machine*, 359–60.

135. La Mettrie, *L'homme machine*, 177.

136. Buffon, "Discours sur la nature des animaux," 94–95.

137. Hume, *Dialogues*, part 7, paragraph 13.

138. Ibid.

139. It was possible, for example, in a treatise on reproduction in plants and animals, to refer to the mammalian fetus as a "little organization." See Parsons, *Philosophical observa-*

tions, 9. Diderot identified "organization" as the difference between living and dead matter in his *De l'interprétation de la nature* (1754), in *Œuvres philosophiques*, 242. The concept is ubiquitous in the naturalist Jean-Baptiste Robinet's monumental *De la nature*, esp. 395, 399, 402. Delisles de Sales, who sharply rejected La Mettrie's man-machine manifesto, nevertheless filled his clandestine bestseller *De la philosophie de la nature* (1769) with references to animals and humans as greater and lesser "organized machines." See Sales, *Philosophie*. Occurrences are rife, but see, for example, 297, 414, 256. The repudiation of La Mettrie is at 138. On the commercial status of Sales's book, see Darnton, *Forbidden Bestsellers*, 48, 49, 70, 397n.

140. For examples in poetry, moralist writing, and the arts, see Marmontel, *Poétique françoise*, 316; Anonymous, *Wisdom and Reason*; and Dubos, *Réflexions critiques*, part 1, section 49; part 2, section 7; part 3, section 10. Differences in organization between men and women acted in early theories of the novel. See Aubert de la Chesnaye-Debois, *Lettre sur les romans*, 227–29. By the early 1770s, the notion that the human body was an "organized machine" had become so stylish that Jean-Paul Marat, who had an insatiable fondness for popular science and considerable success at mobilizing its flashier components for political purposes, took up the idea as his own in Marat, *Philosophical Essay*, 33. On Marat, popular science and radical politics, see Darnton, *Mesmerism*, chap. 3. Evidence of the popularity of "organization" is that it caught on across the political and philosophical spectrum. At the opposite end from Marat, there were those such as the writer and philosophe Jacques-Henri Bernardin de Saint-Pierre who grafted the new notion of "organization" to an older tradition of arguments from design. See esp. Saint-Pierre, *Études de la nature*. But such adaptations were exceptional: the materialist, anti-design usage was typical.

141. Buffon, *Histoire naturelle*, 2:37.

142. Ibid., 2:19, 24; "organic molecules" and "interior mold" are presented in chap. 3.

143. Haller, preface to *Allgemeine Historie*, quoted in translation in Roe, *Matter, Life, and Generation*, 28–29.

144. Buffon, *Histoire naturelle*, 4:40.

145. Ibid., 2:486.

146. Bonnet, *Palingénésie philosophique*, 1:321–23; on "animal-machines" see Bonnet, *Contemplation*, vol. 1, part 3, chap. 14, 52–54.

147. Bonnet, *Palingénésie philosophique*, 1:322–23; on "organic machines" see also Bonnet, *Principes philosophiques*, 344, 363, 380–81; Bonnet, *Considérations*, 1:61, 157–58, 162; and Bonnet, *Contemplation*, 1: 91, 288.

148. Bonnet, *Considérations*, 1:65–66; Bonnet, *Palingénésie philosophique*, 1:278–79; cited passage: Bonnet, *Contemplation*, 1:23. See also Bonnet, *Palingénésie philosophique*, 1:320.

149. Buffon, *Histoire naturelle*, 2:44–46, 51–53, 486–87.

150. Ibid., 1:12.

151. On La Mettrie and Maupertuis, see Vartanian, *L'Homme Machine*, 5, 7–8. On Maupertuis's presidency of the Prussian Royal Academy of Sciences, see Terrall, *Man Who Flattened the Earth*, chap. 8. On Maupertuis's Leibnizianism, see Canguilhem, *Connaissance*, appendix 2; Roger, *Sciences de la vie*, 484; and Wolfe, "Endowed Molecules."

152. Maupertuis, "Accord." Leibniz had been working on the basis of an earlier, similar principle of the French lawyer and mathematician Pierre de Fermat. On Maupertuis's principle of least action, see also Terrall, *Man Who Flattened the Earth*, 176–78. On the controversy over whether light traveled faster or slower in a denser medium, see Sabra, *Theories*, 144–150.

153. Maupertuis, *Essai sur la formation*, 13, 29.

154. Maupertuis, *Vénus physique*, 111–14.

155. Maupertuis, *Essai sur la formation*, 57–58, 67.

156. Bonnet, *Contemplation*, 1:259–60.

157. Bonnet, *Principes philosophiques*, 304, 340. The Italian physiologist Lazzaro Spallanzani likewise characterized the "delicate organization of living beings" as a woven structure. See his *Nouvelles recherches*, ii, 6.

158. Diderot, *Rêve*, 325–26.

159. Bonnet, *Considérations*, 1:105–6. On the eternal, perpetual, and universal "circulation of organized matter," see also Smellie, *Philosophy*, 72.

160. Bonnet, *Contemplation*, 1:258.

161. Bonnet, *Palingénésie philosophique*, 2:98.

162. Diderot, *Entretien*, 258, 276; Diderot, *Rêve*, 313.

163. For discussion of La Mettrie's proto-evolutionism, see Boissier, *La Mettrie*; Vartanian, "Trembley's Polyp"; Richards, *Darwin and the Emergence*, 25, 30, 32; Richards, *Meaning of Evolution*, 64; and Richards, "Emergence of Evolutionary Biology." For a theory of transformation of species that was published the year after La Mettrie's *L'homme machine* (but had been written several decades earlier), see Maillet, *Telliamed*. See also Cohen, *Science, libertinage et clandestinité*.

164. Darwin, *Zoonomia*, 1:183.

165. On the Lunar Society, see Uglow, *Lunar Men*.

166. Darwin, *Zoonomia*, 1:499. On Erasmus Darwin's ideas about species-change, see Richards, *Meaning of Evolution*, esp. 64–65; Richards, "Influence of Sensationalist Tradition"; and Glass et al., *Forerunners of Darwin*, chaps. 3–5.

167. Darwin, *Botanic Garden*, part 2 ("The loves of plants").

168. La Mettrie, *L'homme machine*, 158–61; and La Mettrie, *Système*, paragraphs 32, 37. For an in-depth discussion of what La Mettrie had in mind when he spoke of "le grand Singe" and where he got his information on these animals, see Vartanian's commentary in *L'Homme Machine*, 213–26nn45–48.

169. Diderot, *Suite* (1784), 385.

170. La Mettrie, *Système*, paragraphs 27 and 28.

171. La Mettrie, *Histoire naturelle*, 151. La Mettrie attributed the image of man as a ship without a pilot to Spinoza, but in fact it came from a repudiation of Spinoza by Thémiseul de Saint-Hyacinthe (Hyacinthe Cordonnier). See Israel, *Radical Enlightenment*, 723.

172. La Mettrie, *L'homme machine*, 163; Mettrie, *Système*, paragraphs 16 and 13.

173. Maupertuis, *Essai de cosmologie*, 25–26.

174. La Mettrie, *L'homme machine*, 178. See Vartanian, *L'Homme Machine*, 25.

175. Broca, "Sur le transformisme."

176. Lamarck, *Philosophie zoologique*, 111, 445–46.

177. See Lamarck, *Système analytique*, 9–10, 12, 33. On the opposition between Lamarckism and the natural theology of Paley's tradition, see Mayr, "Lamarck Revisited"; Jordanova, "Nature's Powers"; Bowler, *Evolution*, 127; and Fyfe, *Science and Salvation*, chap. 2, esp. at 94.

178. For recent scholarship correcting the misapprehension of Lamarck as the first transformist, see Corsi, "Before Darwin."

179. One might believe it, Buffon added, if Revelation had not specified that all animals participated equally in the grace of Creation. See *Histoire naturelle*, 4:6, 11, 382–83. The only sort of species-change that Buffon explicitly avowed in writing was degeneration. See Buffon, *Histoire naturelle*, 14:311–74; 5:26–27. The secondary literature on Buffon's views about species-change is voluminous, but see, for example, Sloan, "Buffon"; Roger, *Buffon*, chap. 18; and Roger, *Life Sciences*, 460–74.

180. Maupertuis, *Essai sur la formation*, 41. On Maupertuis's transformism, see Roger, *Sciences de la vie*, 484; and Wolfe, "Endowed Molecules."

181. Hume, *Dialogues*, part 7, paragraphs 3, 7.

182. Diderot, *De l'interprétation de la nature* (1754), in *Œuvres philosophiques*, 241–42.

183. Diderot, *Entretien*, 267–68. On Diderot and transformism, see Gregory, *Evolutionism*, chap. 6.

184. Diderot, *Rêve*, 299, 302.

185. Barruel, *Helviennes*, 1:205.

186. Diderot, *Entretien*, 267–68.

187. Ibid., 268.

188. Diderot, *Rêve*, 303.

189. Ibid., 298.

Chapter 6

1. See, for example, Cunningham and Jardine, eds., *Romanticism and the Sciences*; Fulford, *Romanticism and Science*; Richards, *Romantic Conception*; Richardson, *British Romanticism*; and Tresch, *Romantic Machine*.

2. On the intimacy of science and poetry, see, for example, Lawrence, "Power and the Glory"; Levere, "Coleridge and the Sciences"; and King-Hele, "Romantic Followers." Robert J. Richards has mapped the union of poetry, philosophy, and life science in the German Romantic movement in his *Romantic Conception of Life*. Denise Gigante, in *Life*, and Alan Richardson, in *British Romanticism*, have done the same for the English Romantics. On the role of mechanist science in the culture of Romanticism, and specifically on the falsity of the opposition between mechanism and Romanticism, see Tresch, *Romantic Machine*, esp. chap. 1. Tresch argues that the idea of an opposition between mechanism and Romanticism was a conceit of the early twentieth century.

3. Kant, *Critique of Judgement*, 24–25, 28. See Kant, *Kritik der Urteilskraft*, in the standard Academy Edition (Akademie-Ausgabe) of Kant's works (hereafter cited as AE), 5:165–485, on pp. 189, 194.

4. Goethe, " Influence of Modern Philosophy," in *Collected Works*, 12:29.

5. Goethe, "History of the Printed Brochure" (1817), in *Botanical Writings*, 171–72.

6. Heine, "Concerning the History of Religion and Philosophy in German," in *Selected Works*, 274–420 at 368–69.

7. Timothy Lenoir has influentially characterized Kant's complicated stance with regard to the science of living beings as "teleomechanism." See Lenoir, *Strategy of Life*, chap. 1. Robert Richards has more recently disagreed with Lenoir's account, arguing that Kant excluded biology from the realm of science. See Richards, *Romantic Conception of Life*, 229–37. Here I am reading Kant's *Critique of Judgement* against the background of the animal-machine tradition with its escalating contradiction between teleological understandings of life and a mechanist model of science. In this context it seems to me that Kant's account of the study of life is unclassifiable either as science or nonscience, teleological or mechanist. Rather, it enacts the dialectic, swinging from one pole to the other.

8. Christian Friedrich Puttlich, diary for April 30, 1785, reproduced in Kant, *Rede und Gespräch*, 263.

9. Rousseau, *Discours sur les sciences et les arts*.

10. Kant, *Gedanken*; Coleridge, *Aids to Reflection*, 394–95n.

11. Kant, *Critique of Pure Reason*, 569–78; see also 366–83, 604, 680–81. Kant's relation to Leibniz's philosophy, and more proximately to Leibnizian principles in the philosophy of Christian Wolff, has commanded much philosophical, and some historical attention. For a recent overview and analysis suggesting that, contrary to the traditional view, Leibnizianism remained a crucial touchstone for Kant throughout his career, see Jauernig, "Kant's Critique." See also Wilson, "Reception," 457–60.

12. See, for example, Scruton, *Kant*, 97.

13. Here I am reading as a historian, not as a philosopher. My main purpose is not to demonstrate Kant's influence on the Romantic movement, or on nineteenth-century life sciences, nor is it to arrive at a rigorous Kantian program by interpreting his writings so as to eliminate any apparent tensions or contradictions. I am not interested in trying to free Kant's ideas from the messiness of historical context. Rather, I aim to situate his thinking, including its tensions and contradictions, in time and place, to see what it signifies regarding the intellectual world in which he was working, and the one we have inherited.

The *Critique of Judgement* has been the subject of a great deal of philosophical analysis and debate, but rather less historical consideration (important exceptions are the works of Timothy Lenoir and Robert Richards mentioned in note 7, above). I have studied the philosophical literature in this area and it has informed my own reading, but in a transdisciplinary and therefore indirect way. Philosophers read very differently from historians. Philosophers' primary purpose in reading Kant, or any philosopher, is to define a philosophical approach of their own in keeping with (or against) his principles. They therefore seek ways to "rescue" and "defend" key aspects of what he wrote, to resolve tensions and ambiguities, to solve apparent contradictions, to arrive at something they feel able to endorse. As a historian, in contrast, I am reading for the whole of what Kant wrote in a given text, including—indeed, especially—the tensions and ambiguities, because these reflect the preoccupations of the world in which he was working. For recent philosophical discussions of Kant's understanding of the role of teleology and mechanism in the sciences of life, see the works listed below. These offer ways to resolve the central tension of the *Critique of Judgement*:

the conflict between the requirement of contemporary mechanist science that scientific explanations attribute no intrinsic purposefulness to natural phenomena, and the apparent purposefulness of living forms. For my (historian's) purposes, these philosophical readings of Kant serve to confirm the presence and importance of this central tension, since what I am interested in is the tension itself, and in the fault line it represents in contemporary science, not in finding a philosophical resolution to it. See Steigerwald, *Kantian Teleology*; Breitenbach, "Teleology in Biology"; Ginsborg, "Kant's Biological Teleology"; Ginsborg, "Kant on Aesthetic and Biological Purposiveness"; Ginsborg, "Lawfulness"; Ginsborg, "Understanding Organisms"; Ginsborg, "Oughts without Intentions"; Guyer, "Organisms"; and McLaughlin, *Kant's Critique of Teleology*.

14. Kant, *Critique of Judgement*, 201, 198, 201. In AE at 5:373, 371, 374.

15. Ibid., 202. In AE at 5:374. Chapter 8 considers Kant's understanding of living beings in relation to mechanism, teleology, and agency.

16. Ibid., 222. In AE at 5:394.

17. Ibid., 240. In AE at 5:412.

18. Ibid., 265–70. In AE at 5:436–42. On Kant's rejection of the argument from design, see also *Beweisgrund*, esp. Beträchtungen 5–7 and section 68, where he argues that the argument from design is a vicious circle.

19. Kant, *Universal Natural History*, 115. See also Kant, *Critique of Judgement*, 237–43. In AE at 5:410–15.

20. Kant, *Critique of Judgement*, 250. In AE at 5:422.

21. Ibid., 203, 191. In AE at 5:375, 364.

22. Ibid., 204. In AE at 5:376.

23. Ibid., 211. In AE at 5:383.

24. Ibid., 197. In AE at 5:369.

25. Ibid., 247–48. In AE at 5:419–20.

26. Ibid., 247. In AE at 5:419.

27. Ibid., 256. In AE at 5:428.

28. Ibid., 256n2. In AE at 5:428.

29. See, for example, the article on "Histoire naturelle" in Diderot and d'Alembert, *Encyclopédie*, 8:225–30. For secondary work on the history of natural history, see Jardine et al., *Cultures*; Farber, *Finding Order*; and Spary, *Utopia's Garden*.

30. Kant, Critique of *Judgement*, 256n2. In AE at 5:428.

31. Voltaire, "Histoire," 223, 224.

32. Ibid., 221–22.

33. Babbage's phrase is reported in Stokes et al., "Report," quoted in Babbage, *Calculating Engines*, 268; see also Babbage, "Mechanical Arrangements," quoted in Babbage, *Calculating Engines*, 331.

34. Babbage, *Passages*, 93–94.

35. Kant, *Universal Natural History*, 94.

36. Kant, *Critique of Judgement*, 253. In AE at 5:424.

37. See Maienschein, "Epigenesis and Preformationism." Occasionalism as a general philosophical stance means that all events in nature are caused directly by God. Certain Cartesians were Occasionalists, notably Nicolas Malebranche. See Nadler, *Occasionalism*.

38. Kant, *Critique of Judgement*, 253. In AE at 5:424.

39. Blumenbach, *Essay on Generation*, 61. For an analysis of the nature and status of Blumenbach's *nisus* in relation to contemporary notions of vital forces, see Larson, "Vital Forces."

40. Kant, *Critique of Judgement*, 252. In AE at 5:424. Kant to J. F. Blumenbach, August 5, 1790, in *Correspondence*, 354.

41. Kant, *Critique of Judgement*, 253. In AE at 5:424.

42. Goethe, "Vorarbeiten zu einer Physiologie der Pflanzen" (1797), in *Werke*, section 2, 6:303–4. Published in partial English translation as "Excerpt."

43. Goethe, *Metamorphosis of Plants*, paragraph 113.

44. Goethe, "Commentary," 6.

45. Lamarck, *Hydrogéologie*, 8, 188. For *biologie*, see also Lamarck, *Histoire naturelle*, 49–50; Lamarck, *Recherches*, vi, 186, 202; Lamarck, *Philosophie zoologique*, 1:xviii; and Lamarck, "Biologie, ou considérations sur la nature, les facultés, les développements et d'origine des corps vivants" (ca. 1809–15), a manuscript plan for a never-written book found in the Muséum national d'histoire naturelle, Bibliothèque centrale. Ms 742, tome 1. Lamarck was not the only person to coin the term: several authors arrived at it independently around the same time. See Corsi, "Biologie."

46. Lamarck, *Philosophie zoologique*, 2:95, 127; Lamarck, *Histoire naturelle*, 50, 134; Lamarck, *Hydrogéologie*, 188. On Lamarck's "ceaseless tendency" of living beings to compose and complexify themselves, a "continually active cause," see also his *Philosophie zoologique*, 1:132; 2:69, 100, 101, 104. For *monade*, see 1:285; 2:67, 212. "Animated point" is in Lamarck, *Histoire naturelle*, 16. On Lamarck's relations to Leibniz and the possibly Leibnizian origins of his term *monade*, see Canguilhem, *Connaissance*, 188 (for Maupertuis's role in transmitting the Leibnizian monad into theories of life); Smith, "Leibniz's Hylomorphic Monad," 24; and Burkhardt, Jr., *Spirit of System*, 233n36.

47. Lamarck, *Recherches*, 50–62.

48. Lamarck, *Philosophie zoologique*, 1:401–3.

49. Ibid., 2:310–11.

50. For analyses of Cuvier's denigration of Lamarck, see Gould, "Foreword," viii; and Rudwick, *Georges Cuvier*, 83.

51. Georges Cuvier, "Eloge," xx.

52. Cuvier, *Règne animal*, 1:7.

53. Lamarck, *Histoire naturelle*, 184.

54. Lamarck, "Espèce," 441.

55. Coleridge, *Aids to Reflection*, 257.

56. Coleridge, *Biographia Literaria*, 264.

57. Ibid., 273.

58. See Micheli, *Early Reception of Kant's Thought*, chap. 3; and Edwards, *Statesman's Science*, 144.

59. Robinson, *Diary*, 1:305.

60. Coleridge, *Aids to Reflection*, 395, 393.

61. Ibid., 397, 396.

62. Ibid., 389–92.

63. First reported in Rossetti, *Mrs. Shelley,* 28. See also Seymour, *Mary Shelley,* 58.

64. Shelley, "Introduction," 170, 171.

65. Richardson, *British Romanticism,* 17.

66. Cabanis, *Rapports,* 1:128.

67. Ibid., 1:118.

68. Ibid., 1:243.

69. Darwin, *Temple of Nature,* 62–64. The talking head is reported at 98.

70. Shelley, "Introduction," 171; Darwin, *Temple of Nature,* note 12. Spontaneous generation was disproven several decades later by the French biologist Louis Pasteur, who performed experiments demonstrating that no organisms appeared in sterilized beef broth kept in an airtight container. See Geison, *Private Science,* chap. 5.

71. Davy, "Sons of Genius," in *Collected Works,* 1:26.

72. Davy, "Discourse Introductory to a Course of Lectures on Chemistry" (1802), in *Collected Works,* 2:311–26, paragraph 3.

73. Darwin, *Zoonomia,* vol. 2, section 2, paragraphs 1–2. See also Darwin, *Botanic Garden,* part 1, canto I, line 363ff; Darwin, *Temple of Nature,* 74–89; Abernethy, "Extracts"; and Humboldt, *Expériences.*

74. Galvani, *Commentary.*

75. Darwin, *Temple of Nature,* canto 3, lines 111–12.

76. Volta, *Electricity.*

77. Hogg, *Life of Percy Bysshe Shelley,* 1:56.

78. Shelley, "Introduction," 171–72.

79. Aldini, *Account.*

80. "Horrible Phenomena! – Galvanism," *The Times* (February 11, 1819), 3; see also *The Examiner,* (February 15, 1819), 103; and Keddie, *Anecdotes,* 3–4.

81. Ure, "Account," 290; see also Golinski, "Literature," and Sleigh, "Life, Death and Galvanism."

82. Shelley, "Introduction," 172.

83. Saumarez, *New System of Physiology,* 2:8. For Coleridge's admiration of this work, see Coleridge, *Biographia Literaria,* chap. 9 and note 31.

84. Davy, "Discourse," in *Collected Works.*

85. Lawrence, *Lectures,* 57.

86. Hoffmann, "Automata," 81, 95. Similarly Jacques-Henri Bernardin de Saint Pierre found an automaton talking head to be "horrifying" and likened it to "an articulate voice coming from a dead body." See Saint-Pierre, *Études,* 2:276–79.

87. Goethe, *Annals,* 112.

88. Ibid., 113; on Beireis's acquisition of Vaucanson's automata, see Chapuis and Droz, *Automata,* 233–34.

89. Goethe to Herzog Carl August, June 12, 1797, in his *Briefe,* 2:276–77; quoted in translation in Hankins and Silverman, *Instruments and the Imagination,* 196.

90. Kant, *Critique of Practical Reason,* 128.

91. Bichat, *Recherches physiologiques,* 2.

92. Darwin, *Botanic Garden,* part 1, canto 1, line 368.

93. Lamarck, "Biologie, ou considérations sur la nature, les facultés, les développements et d'origine des corps vivants" (ca. 1809–15), Muséum national d'histoire naturelle, Bibliothèque centrale. Ms 742, tome 1, 10–11.

94. Lamarck, *Mémoires,* 248–49; see also Lamarck, *Recherches sur les causes,* 289.

95. Lamarck, *Recherches sur les causes,* 289–90.

96. Saint-Hilaire, *Philosophie anatomique,* 1:208–9.

97. Darwin, *Temple of Nature,* canto 2, lines 41–42.

98. Herder, *Outlines,* 59, 109.

99. Darwin, *Temple of Nature,* canto 4, lines 379–80.

100. Coleridge, "On the Passions," 1442.

101. Abernethy, "Extracts," 51.

102. Because of the central role of gravitational force in the Newtonian system, some commentators from the seventeenth century onward have seen Newton's physics as intrinsically active in contrast with Cartesian cosmology. For a recent representation of this view of Newtonianism, see Dear, *Intelligibility of Nature,* chap. 1, section 4. As Dear points out, Newton did invoke "active principles" in his last major work, the *Opticks.* See Newton, *Opticks,* 398.

103. Schelling, *Ideen.* On Schelling and the origins of *Naturphilosophie,* see S. R. Morgan, "Schelling and the Origins of his Naturphilosophie," in Cunningham and Jardine, eds., *Romanticism and the Sciences,* chap. 2; and Richards, *Romantic Conception,* chap. 3.

104. See Schelling, *Ideen,* page 241, for the "inner absolute formative power" (*... innern absoluten Bildungsvermögen ...*) and its manifestations. For some overviews of the large subject of *Naturphilosophie* and Romantic science, see Cunningham and Jardine, "Age of Reflexion," in Cunningham and Jardine, eds., *Romanticism and the Sciences;* Richards, *Romantic Conception,* 6–17; and Tresch, *Romantic Machine,* part 1.

105. The principal primary texts for the emergence of energy conservation include Mayer, "Bemerkungen," an early formulation of the law of energy conservation with a physiological foundation; Helmholtz, *Erhaltung der Kraft;* and Du Bois-Reymond, "Über die Lebenskraft." For a primary account of the development of the principle of energy conservation, see Mach, *Conservation of Energy.* Important secondary treatments include Kuhn "Energy Conservation"; Cardwell, *Watt to Clausius;* and Smith, *Science of Energy.* On Helmholtz, see also Bevilacqua, "Helmholtz."

106. Johannes Müller first presented his doctrine of specific nerve energies in *Vergleichenden Physiologie.* For an overview of the emergence of ideas about energy in physiology, see Rothschuh, *History of Physiology,* chap. 6. For an analysis of the instrumental connections between Helmholtz's work in physics and in physiology, see Brain and Wise, "Muscles and Engines."

107. Müller, *Elements of Physiology,* 27, 31–35, 712, 714, 719.

108. Cabanis, *Rapports,* 2:423–24.

109. Müller, *Elements of Physiology,* 285.

110. Helmholtz, "Interaction of the Natural Forces," 19, 36.

111. On this disagreement over a life force or *Lebenskraft,* and, more generally, on the generational differences of approach that divided the community of Müller's students among themselves and from their teacher, see Lenoir, *Instituting Science,* 138–40.

112. See Rabinbach, *Human Motor*, 52–61. On Helmholtz's debt to Romantic science in general and to specific Romantic-era natural philosophers, see also Richards, *Romantic Conception*, 327–28, 407–8; and Tresch, *Romantic Machine*, 85–86.

113. Helmholtz, "Conservation of Force," 124.

114. Helmholtz, "Interaction of the Natural Forces," 37, 38.

115. Coleridge, "Eolian Harp" (1796), line 27.

116. Darwin, *Temple of Nature*, canto 2, lines 19–22.

117. Rabinbach, *Human Motor*, 92. Rabinbach uses the phrase "transcendental materialism" specifically in connection with the theory of energy conservation, which made energy the common basis of all things, and the corresponding social and technological concepts of work, labor power, and industry.

118. Herder, *Outlines*, 113.

119. Coleridge, *Aids to Reflection*, 393.

120. Wordsworth, "Lines Composed a Few Miles above Tintern Abbey, on Revisiting the Banks of the Wye during a Tour" (1798), in Wordsworth, *Complete Poetical Works*, lines 101, 102.

121. Herder, *Outlines*, 99.

122. Darwin, "Extracts," 44.

123. Wordsworth, " Recluse" (1800), in Wordsworth, *Complete Poetical Works*, Book Second, line 793.

124. Coleridge, "Eolian Harp" (1796), lines 44–47.

Chapter 7

1. Darwin, "Preliminary notice," 27, 92, 95.

2. Darwin, *Zoonomia*, 1:505.

3. This is the thesis of Robert J. Richards, who has traced the roots of Darwin's theory in Romantic philosophy and poetry. He has pointed in particular to the writings of Alexander von Humboldt. See Richards, "Darwin's Romantic Biology"; and Richards, *Romantic Conception of Life*, chap. 14. Darwin's Romanticism is evident for Richards in a Darwinian nature imbued with beauty, meaning, and moral import, resembling an organism rather than a machine, and also in a concept of natural selection as anything but mechanistic.

4. Lamarck, *Système des animaux sans vertèbres*, 16.

5. Darwin, *Notebook B*, 18–19, 78, 216–17. Other references to "monads" at 22–23, 29e, and in *Notebook C*, 206e.

6. Lyell, *Principles of Geology*, vol. 2, chap. 1.

7. Darwin, *Autobiography*, 77, 101; see *Darwin Correspondence Project* (hereafter cited as DCP), letter 171, note 3.

8. Lyell, *Principles of Geology*, 2:8.

9. Lyell to Darwin, October 3, 1859, in DCP, letter 2501; October 4, 1859, in DCP, letter 3132.

10. Lyell to Darwin, October 4, 1859, in DCP, letter 3132.

11. Darwin to Lyell, October 11, 1859, in DCP, letter 2503. Lyell had also supposed that the gulf between animal and human intellectual capabilities could only be accounted

for by a creative power that acted discontinuously. Darwin retorted, "I would give absolutely nothing for theory of nat. selection, if it require miraculous additions at any one stage of descent."

12. Lyell, *Geological Evidences,* 3 and chap. 20.

13. Darwin to Lyell, March 12–13, 1863, in DCP, letter 4038.

14. Lyell to Darwin, March 15, 1863, in DCP, letter 4041.

15. Lyell, *Geological Evidences,* chap. 20, at 385, 390, 391–92, 394.

16. On Haeckel's particular mode of Darwinism, see Richards, *Tragic Sense of Life.* On Haeckel's popularization of Darwinism, see ibid., 2.

17. Haeckel's best-selling *History of Creation* had as its subtitle: *A Popular Exposition of the Doctrine of Evolution in General, and of that of Darwin, Goethe, and Lamarck in Particular.*

18. Haeckel, *History of Creation,* 1:93.

19. Ibid., 1:114, 116, 118, 411.

20. Spencer, *Principles* 1:291; Darwin, *Origin of Species* (1869), 72.

21. Spencer, "Development Hypothesis" (1852), in *Essays,* 1:5; Spencer, "Progress: Its Law and Cause" (1857), in *Essays,* 1:46; Spencer, *Principles,* vol. 1, part 3; Darwin, *Origin of Species* (1876), 189, 201, 202, 215, 282, 424. "Evolution" does appear a single time in Darwin, *Variation* (1868), 2:60.

22. Darwin, *Origin of Species* (1869), 72.

23. Spencer, "Progress," in *Essays,* 1:4, 17–19.

24. Spencer, *Principles,* 1:492.

25. Ibid., 1:311, 602, 606.

26. Darwin to Lyell, March 25, 1865, in DCP, letter 4794.

27. Darwin to J. D. Hooker, November 3, 1864, in DCP, letter 4650.

28. Darwin to Lyell, February 25, 1860, in DCP, letter 2714.

29. Darwin to J. D. Hooker, June 23, 1863, in DCP, letter 4218.

30. Hawkins, *Social Darwinism,* 82.

31. Darwin to J. D. Hooker, June 30, 1866, in DCP, letter 5135; Darwin to J. D. Hooker, December 10, 1866 in DCP, letter 5300.

32. Darwin to J. D. Hooker, October 2, 1866, in DCP, letter 5217.

33. Darwin to A. R. Wallace, February 19, 1872, in DCP, letter 8211.

34. Darwin to F. M. Balfour, September 4, 1880, in DCP, letter 12706 (see also Darwin, *More Letters,* 2:424).

35. Darwin to R. L. Tait, February 22, 1876, in DCP, letter 10406; Darwin to G. J. Romanes, July 28, 1874, in DCP, letter 9569 (original in the Charles Darwin Papers, American Philosophical Society, letter 446).

36. Darwin to J. D. Hooker, June 23, 1863, in DCP, letter 4218.

37. Darwin, *Origin of Species* (1876), 189, 201, 202, 215, 282, 424.

38. Gray's letter itself has been lost, but this much can be inferred from Darwin's response to it: Darwin to Gray, November 29, 1857, in DCP, letter 2176.

39. Darwin to Gray, September 5, 1857, in DCP, letter 2136, enclosure.

40. Lyell to Darwin, June 15, 1860, in DCP, letter 2832a; Lyell to Darwin, June 19,

1860. in DCP, letter 2837a. For Darwin's response to the charge of "deification," see Darwin to Lyell, June 17, 1860, in DCP, letter 2833.

41. Darwin to Gray, November 29, 1857, in DCP, letter 2176.

42. Darwin, *Origin of Species* (1859), 83. Richards has emphasized the godlike agency of natural selection in Darwin's formulation and traced it, along with other core features of Darwin's theory, not least its poetic presentation, to roots in German Romanticism. See Richards, "Darwin's Romantic Biology."

43. Darwin, *Origin of Species* (1860), 83.

44. Darwin, *Origin of Species* (1861), 87.

45. Darwin, *Origin of Species* (1866), 94.

46. Darwin, *Origin of Species* (1869), 95.

47. See Richards, *Romantic Conception of Life*, 536.

48. Darwin, "Essay of 1844," 85–86.

49. Huxley made the remark to the American paleontologist and director of the American Museum of Natural History, Henry Fairfield Osborn. See Osborn, *Impressions*, 79. On Huxley and his defense of Darwinism, see Desmond, *Huxley*, chaps. 14–17 and part 2.

50. Huxley, "Reception," 551; Huxley to Darwin, November 23, 1859, in DCP, letter 2544.

51. Darwin to Huxley, November 24, 1859, in DCP, letter 2553.

52. Huxley, *Lectures and Essays*, 24.

53. Huxley, " Genealogy of Animals" (1869), in his *Darwiniana*, 115, 116.

54. Darwin, *Variation* (1875), 2:283–84.

55. Spencer, *Principles*, 1:537, 614, 674.

56. Darwin to J. D. Hooker, November 3, 1864, in DCP, letter 4650; *Variation*, 2nd ed. (1875), vol. 2:284.

57. Darwin, *Variation* (1875), vol. 2, chap. 27.

58. For some of Darwin's treatments of use and disuse as a source of variation, see *Origin of Species* (1876), 108–14; *Variation* (1875), vol. 2, chap. 14; and *Descent of Man* (1874), chap. 2.

59. Darwin, *Descent of Man* (1874), 209. Darwin did also consider the reciprocal possibility of females competing for a male, and of the male selecting among females— or indeed of a process of mutual sexual selection—but he thought the likelihood of such cases was remote, 225–26.

60. Darwin, *Origin of Species* (1876), 70, 161–62, 414.

61. Darwin, *Descent of Man* (1874), 617.

62. For Darwin's own account of how sexual selection became so important in his thinking that it overran much of his treatment of human evolution, see *Descent of Man* (1874), 3.

63. Darwin, *Origin of Species* (1861), note on xiv.

64. Ibid., 135.

65. Darwin to Wallace, August 28, 1872, in DCP, letter 8488 (also Darwin, *Life and Letters*, 3:168).

66. Wallace, "Tendency of Varieties."

67. Wallace, "Limits."

68. Ibid., 363, 366–68.

69. Wallace to Edward Bagnall Poulton, May 28, 1912, in Wallace, *Letters*, 344.

70. Darwin to Wallace, January 26, 1870, in Wallace, *Letters*, 250–51 (see also DCP, letter 8488). "Eheu" is Latin for "Alas." The "best paper" that Darwin refers to is Wallace, "Origin of Human Races," in which Wallace argued that human intellectual, moral, and social faculties would have been beneficial to the survival of the community of early humans and were therefore explicable by natural selection. For Darwin's response to Wallace's change of heart, see also Darwin to Wallace, April 14, 1869, in DCP, letter 6706.

71. On Butler's tangled relations to Darwin and Darwinism, see Dyson, *Darwin*, chap. 2.

72. Butler, *Unconscious Memory*, 12; Dyson, *Darwin*, 24.

73. Butler, *Erewhon*, 97, 199, 205, 209–11.

74. Butler to Darwin, May 11, 1872, in Jones, *Samuel Butler*, 1:156–57.

75. Butler, *Evolution*; Butler, *Unconscious Memory*, chap. 1.

76. Butler, "Lamarck and Mr. Darwin," 261.

77. Ibid., 272; Butler, "Mr. Mivart and Mr. Darwin," 282; Butler, *Life and Habit*, 295–97, 300.

78. Butler, *Evolution*, 31, 6, 30. See also Butler, *Luck*, 5.

79. Butler, *Evolution*, 31–32; Butler, *Luck*, 9.

80. Darwin to Huxley, February 4, 1880, in Jones, *Samuel Butler*, 2:454–55 (also in DCP, letter 12458).

81. Darwin to Krause, May 14, 1879, in DCP, letter 12052. Krause was the author of a biography of Erasmus Darwin to which Charles Darwin wrote the introduction for the English translation in 1879. This further angered Butler, who felt a passage in the introduction to be a snub directed at him. See Dyson, *Darwin*, 23.

82. Darwin, *Notebook E*, 95, 127.

83. Darwin, *Torn Apart Notebook*, 41.

84. Darwin, *Origin of Species* (1859), 152–53, 154.

85. Darwin, "Essay of 1842," 37.

86. See, e.g., Darwin, "Essay of 1842," 2, 32, 46; Darwin, "Essay of 1844," 60–61, 64, 69, 74, 79, 80, 84, 85, 108, 221, 227, 228, 234, 240. See also Darwin, *Natural Selection*, 242.

87. Darwin, "Essay of 1844," 60–61; Darwin, *Origin of Species* (1859), 80.

88. Darwin, *Origin of Species* (1876), 10, 11, 19, 52–53, 62, 64, 72, 80, 91, 107.

89. Darwin, *Origin of Species* (1859), 134.

90. Darwin, *Origin of Species* (1866), 160.

91. Darwin, *Origin of Species* (1869), 168.

92. Darwin, *Origin of Species* (1876), 107.

93. See, for example, Argyll, *Works*. Critics who were not partisans of the argument from design also raised the issue of the source of variability in the theory of natural selection, among them the botanist Alfred William Bennett. Whatever force or agency caused organisms to vary, Bennett argued, "what right have we to say that this principle then ceases to act, instead of being the main agent in all the other subsequent changes?" See Bennett, "Natural Selection."

94. Argyll, *Works*, 130, 70.

95. Ibid., 152, 65.

96. Boyle, *Disquisition*, 17.

97. See, for example, Darwin, *Variation* (1868), 1:8; Darwin, *Origin of Species* (1876), 170. Relevant passages first introduced in the fifth edition of 1869: see Darwin, *Origin of Species* (1869), 150–52, 155.

98. Darwin, *Variation* (1868), 2:252–53, 255.

99. Nägeli, *Entstehung*.

100. See Darwin, *More Letters*, 2:375n5 (letter 697). Darwin's son Francis and Francis's coeditor Steward were more dogmatic than Darwin himself, writing in their annotation that Nägeli's argument "seems to us to be of the nature of a truism, for it is clear that any structure whose evolution can be believed to have come about by Natural Selection must have a function."

101. Darwin to Nägeli, June 12, 1866, in DCP, letter 5119.

102. Darwin to Hooker, December 5, 1868, in Darwin, *More Letters*, 2:375–76 (letter 697). See also DCP, letter 6512.

103. See Darwin to Hooker, December 29, 1868; January 16, 1869; January 22, 1869; all in Darwin, *More Letters*, 2:377–79 (letters 698–700). See also in DCP, letters 6515, 6550, 5729, 6554, 6557, 6560, 6568, 6608.

104. Darwin to Farrer (Lord Farrer), August 10, 1869, in Darwin, *More Letters*, 2:380n1 (letter 701); see also DCP, letter 6859.

105. Darwin, *Origin of Species* (1869), 151–57.

106. Darwin, *Origin of Species* (1876), 171, 175.

107. Darwin, *Effects*, 451–52.

108. Butler, "Lamarck and Mr. Darwin," 261.

109. Bernard, *Leçons*, 1:112–24. The Harvard physiologist Walter Cannon later coined the term "homeostasis" to accompany Bernard's concept. See my chapter 9, "Outside In," and Cannon, *Wisdom of the Body*, 24. On Bernard's notion of the internal environment, see Grmek, *Claude Bernard*, chap. 4; and Canguilhem, *Études*, 323–27.

110. Bernard, *Introduction*, 75. See also ibid., 64, 78, 87, 90, 94, 180; Bernard, *Leçons*, 1:293, 316.

111. Bernard, *Leçons*, 1:46, 49; Bernard, *Cahier*, 124.

112. Bernard, *Leçons*, 1:67.

113. Bernard, *Introduction*, 78, 87, 94.

114. Ibid., 90, 94.

115. Bernard, *Cahier*, 223–24n72, 131. For "force vitale" and Bernard's protestations against it, see *Introduction*, 60–63, 67, 68, 70, 90, 94, 184–85; and Bernard, *Leçons*, 1:48–55.

116. Bernard, *Leçons*, 1:39, 42–43. On Bernard's complicated relation to vitalism, see Holmes, "Claude Bernard"; Canguilhem, *Études*, 149–60; Grmek, *Claude Bernard*, chap. 3; and Virtanen, *Claude Bernard*, chap. 4.

117. "*Principe de la spécialité des mécanismes vitaux.* Ce n'est pas l'action qui est vitale et d'essence particulière, c'est le mécanisme qui est spécifique, particulier, sans être d'un ordre distinct. La doctrine que je professe pourrait être appelée le vitalisme physique; je crois qu'elle est l'expression la plus complète de la vérité scientifique." See Bernard, *Leçons*, 2:219.

118. Darwin, *Variation* (1868), 2:368–69.

119. Marey, *Machine animale*, 62, 85–88.

120. Marey, *Méthode graphique*, iii. See Braun, *Picturing Time*.

121. Marey, *Machine animale*, 86 and chap. 9.

122. "Prof. Marey's discussion on the power of co-adaptation in all parts of the organisation is excellent." See Darwin, Variation, 2nd ed., 1875, 2:284n7.

123. Huxley, "Hypothesis," 200, 201–202, 237–38.

124. Ibid., 239–42, 244.

125. Darwin, *Origin of Species* (1859), 186.

126. Lyell to Darwin, October 3, 1859, in DCP, letter 2501.

127. Darwin to Gray, February 8–9, 1860, in DCP, letter 2701.

128. Darwin to Asa Gray, April 3, 1860, in DCP, letter 2743.

129. Natural theologians who promulgated arguments from design often rejected the doctrine of original sin and concomitant notion that humans were fundamentally frail and flawed. But the attribution of perfection to human structures such as the eye could also be rendered compatible with the doctrine of original sin in various ways. For one thing, Scholastic tradition had long held that man had retained his intellectual and physical gifts through the Fall. Then, although sixteenth- and seventeenth-century Protestants returned to a more severe, Augustinian view of the consequences of original sin, everything still depended upon where one located these consequences. Empiricists such as Francis Bacon located them in the human intellect, and treated the senses as the means of redemption and restoration to a prelapsarian state of knowledge and power. See Harrison, "Original Sin."

130. On Paley's reception, see Brooke, *Science and Religion*, 192; and Davidson, "'Identities Ascertained,'" 328. For a contrasting view, see Fyfe, "Reception."

131. Paley, *Natural Theology*, 1.

132. Darwin, *Autobiography*, 59.

133. Paley, *Natural Theology*, 451.

134. Darwin to Gray, May 22, 1860, in DCP, letter 2814. On Darwin and the question of moral agency, see Levine, *Darwin Loves You*.

135. Darwin to Gray, July 3, 1860, in DCP, letter 2855.

136. Paley, *Natural Theology*, 440–41.

137. Ibid., 439–40.

138. Ibid., 441–42. Some aspects of the mechanisms of visual accommodation in the avian eye remained unexplained two centuries later. See Glasser, "History of Studies."

139. Paley, *Natural Theology*, 442. Paley also, however, argued that perfection was not necessary to the case, at 436: "It is not necessary that a machine be perfect, in order to show with what design it was made: still less necessary, where the only question is, whether it were made with any design at all." Also at 447:

> When we are inquiring simply after the existence of an intelligent Creator, imperfection, inaccuracy, liability to disorder, occasional irregularities, may subsist in a considerable degree, without inducing any doubt into the question: just as a watch may frequently go wrong, seldom perhaps exactly right. . . . Irregularities and imperfections are of little or no weight in the consideration, when that consideration relates simply to the existence of a Creator. When the argument respects his attributes, they are of weight; but are then to be taken in conjunction (the attention is not to rest upon them, but they are to be taken in conjunction)

with the unexceptionable evidences which we possess, of skill, power, and benevolence, displayed in other instances. . . .

140. Gray to Darwin, March 6, 1862, in DCP, letter 3467.

141. Darwin to Lyell, June 17, 1860, in DCP, letter 2833.

142. Huxley, "Reception," 551; Huxley to Darwin, November 23, 1859, in DCP, letter 2544.

143. Huxley, "Genealogy of Animals," in his *Darwiniana*, 110.

144. Huxley, "Obituary of Darwin" (1888), in his *Darwiniana*, 288. Also at 280: "The process of natural selection is, in fact, dependent on adaptation."

145. Huxley, "Criticisms on the 'Origin of Species'" (1864), in his *Darwiniana*, 86.

146. Du Bois-Reymond first presented these at the July 8, 1880, meeting of the Berlin Academy of Sciences. They were then published as "Die sieben Welträthsel."

147. Darwin, *Autobiography*, 87.

148. Boyle, *Disquisition*, 55.

149. Paley, *Natural Theology*, 443.

150. Darwin, *Origin of Species* (1859), 188–89.

151. Helmholtz, "Recent Progress," 135, 137, 144–45. The original German version was *Fortschritte*.

152. Helmholtz, "Recent Progress," 140–43, 147.

153. Ibid., 141.

154. Ibid., 146, 197; on the role of unconscious inductive inference in sensory perception, see also Helmholtz, "Concerning the Perceptions in General" (1866) and "The Origin of the Correct Interpretation of our Sensory Impressions" (1894), in Helmholtz, *Helmholtz on Perception*, parts 5 and 7. See esp. 195, 255.

155. Darwin, *Origin of Species* (1876), 163. The passage in question is in Helmholtz, "Recent Progress," 173 (the translation differs somewhat from Darwin's version).

156. Darwin to Hooker, July 13, 1856, in DCP, letter 1924.

157. Darwin, *Origin of Species* (1876), 429.

158. Darwin, *Origin of Species* (1860), 490.

159. Darwin, *Origin of Species* (1876), 429.

Chapter 8

1. See Ward, *Neo-Darwinism*; and Romanes, "Darwinism."

2. On Weismann, see Churchill, "Break from Tradition"; Churchill, "Weismann-Spencer Controversy"; and Churchill, "Developmental Evolutionist." The most comprehensive work on Weismann is Weissman (no relation), *First Evolutionary Synthesis*.

3. "Vitalist" was coined around 1800 but did not come into common usage until the 1870s and '80s. On the history of the term "vitalism," and related words, see Rey, *Naissance*; Wolfe and Terada, "Animal Economy," 538–43n4; and Williams, *Medical Vitalism*.

4. Weismann, *Germ-Plasm*, 39.

5. See the frontispiece to Weismann, *Ausgewählte Briefe*. Thanks to Charlotte Weissman for pointing this out to me.

6. Wheeler, "On Instincts," 303. This was the case in America as well, where "neo-

Lamarckism" flourished in the last couple decades of the nineteenth century. See Pfeifer, "Genesis." Wheeler called himself an unrepentant sinner, i.e., a Lamarckian, in large part because he rejected the ahistoricity of Weismann's approach. See Sleigh, "Ninth Mortal Sin," especially 154–56.

7. The following discussion of the Prussian origins of the modern scientific research university is closely informed by Howard, *Protestant Theology*; and Clark, *Academic Charisma*. See also Zimmerman, *Anthropology*. In *Biology Takes Form*, Lynn Nyhart presents a different perspective on German biology in this period by focusing on museums, schools, zoos, and other public institutions rather than on elite university science.

8. See Howard, *Protestant Theology*, 4–5; Ringer, *German Mandarins*, 102–27; and Lenoir, *Instituting Science*. On the influence of the German model of research university in America, see also Turner, "Humboldt in North America," especially 292–93.

9. Paulsen, *German Universities*, 226. Frederick Gregory, in *Nature Lost*, addresses this development from the theological side: "Theology lost its capacity to speak of nature, but at the same time, it also denied natural science the right to usurp the old theological prerogative of making assertions about the ontological status of the cosmos," 22. On the endurance of theology faculties, see also Schwinges, *Humboldt International*, 6.

10. Howard, *Protestant Theology*, 6–7, 14, 29, 34, 292, 297–98. The phrase "Protestantische *Wissenschaft*" appears for example in Vogel, *Semisäcularfeier*, 94. On *Voraussetzungslosigkeit*, see Mommsen, "Universitätsunterricht und Konfession," 432, 433; Baumgarten, *Voraussetzungslosigkeit*; and esp. Weber, "Science as a Vocation."

11. Weber, "Science as a Vocation," 43.

12. Weber, *Protestant Ethic*, 24–25, 80, 136, 149, 224n30, 249n145.

13. On the turn away from the holistic approach of *Naturphilosophie*, see Howard, *Protestant Theology*, 281–83; and Clark, *Academic Charisma*, 449.

14. On the dominance of Prussian Protestant institutions in establishing the culture of the modern research university, see Howard, *Protestant Theology*, 4–7; Clark, *Academic Charisma*, 3, 13, 28, 147, 475; Charle, "Patterns," 47–53; Rüegg, "Theology and the Arts," 398, 405; Paulsen, *German Universities*, 226–27; and von Hehl, "Universität und Konfession," 286, 289–90. Hehl observes that this dominance imposed a "permanent, uniquely Protestant imprint on the German university," 290. On the dominance of Protestantism in the sciences in particular, see Schröder, *Naturwissenschaft und Protestantismus*, 487. On the "marginalization of the Catholic universities" see Höflechner, "Universität, Religion und Kirchen," 564.

15. The sources of history: Ranke, Introduction (1874) to *Geschichte der romanischen und germanischen Völker von 1494 bis 1514* (1824), in translation in *SWH*, 55–59, on 58. Particulars: Introduction (1874), 58–59; and fragment (1830s), in translation in SWD, 102–4, at 103. Facts: Introduction (1874), 58. Impartiality and objectivity: *Englische Geschichte, vornehmlich im sechzehnten und siebzehnten Jahrhundert* (1859–1869), excerpt in translation in *SWH*, 242–43, at 243; and Ranke—Otto von Ranke, May 25, 1873, in translation in *SWH*, 259–60, at 259.

16. Unity and progress: Ranke, Introduction (1874), 58. Form and beauty: *Französische Geschichte, vornehmlich im sechzehnten und siebzehnten Jahrhundert* (1852–61), translated excerpt in *SWH*, 258. Recent scholarship has emphasized this aspect of Ranke's

writing, his embrace of ideals such as unity and progress, form and beauty, in order to debunk an earlier vision of Ranke as a positivist-empiricist who sacrificed historical narrative to a dry rehearsal of facts. The debunking turned in part upon the meaning and translation of *eigentlich* in Ranke's much-cited injunction to write history "wie es eigentlich gewesen." Traditionally this injunction was translated into the English phrase "as it actually was," but Georg Iggers argued, and Peter Novick influentially developed the argument, that the word *eigentlich* in its historical context should more properly be translated as "essentially," and that Ranke's idea of the historical search for *essential* (rather than *actual*) truth was fundamentally intuitionist rather than empiricist. See Iggers, "Introduction," in Ranke, *Theory and Practice*, xi–xlv; Iggers, "Image of Ranke"; and Novick, *That Noble Dream*, 28–30. The degree of intuitionism in Ranke's methodology is not relevant to my purpose here, however; rather, the salient point is his announcement of a science of history whose method was founded in factual particulars, in the local and the proximate, and his suggestion that these, in their very particularity, made manifest a divine order. On the establishment of the discipline of history in the new research universities, see also Howard, *Protestant Theology*, 116–17, 275–76, 278–79; Iggers, *German Conception of History*, 63–89; and Briggs, "History and the Social Sciences," 464–69. On the history of the discipline of history in Germany, see also Reill, *German Enlightenment*; and Kelley, *Fortunes of History*.

17. Ranke, Introduction (1874), 58–59; Ranke—Otto von Ranke, May 25, 1873, in translation in *SWH*, 259–160, at 259.

18. Howard, *Protestant Theology*, 117. On the secularization of church history, see also ibid., 344.

19. See Gregory, *Nature Lost*, 63: "The German theologians frequently, though not always, examined the relationship between natural science and theology in the context of the Darwinism debate." On the radical political implications of Lamarckism, see Desmond, *Politics of Evolution*, especially chapter 7. Desmond writes that during the 1830s in Britain, "Lamarck's notion that an animal could, through its own exertions, transform itself into a higher being and pass on its gains—all without the aid of a deity— appealed to the insurrectionary working classes. His ideas were propagated in their illegal penny prints, where they mixed with demands for democracy and attacks on the clergy. Clearly Lamarckism had some disreputable associations" 4. See also p. 319: "Lamarckism with its democratic consequences had to be shown to be scientifically bankrupt."

20. Du Bois-Reymond, "Über die Grenzen," 464. See Richards, *Tragic Sense of Life*, 315; and Schröder, *Naturwissenschaft und Protestantismus*, 485–86. The unknowable things grew into the later list of seven mysteries presented in "Die sieben Welträthsel" in 1880. Gabriel Finkelstein argues against the view that Du Bois-Reymond intended these statements as "a sop to the Church" and characterizes Du Bois-Reymond's lectures as expressions of skepticism rather than conciliation. *Emil du Bois-Reymond*, 281–82. But either way, the slogan took on a life of its own as an expression of the boundaries of science's territory, inspiring countering storms of approval and disapproval, see 269–80; and Richards, *Tragic Sense of Life*, 315–18.

21. Spränger, *Wandlungen*, 23; quoted in translation in Howard, *Protestant Theology*, 30.

22. Hofmann, *Question*, 22; quoted in Howard, *Protestant Theology*, 277–78.

23. Weber, "Science as a Vocation," 31.

24. Ibid., 39, 47–48.

25. Bowler, *Eclipse of Darwinism*, 13: "For the first time, biologists appeared whose interests were so closely defined by a single discipline . . . that they had little understanding of, or sympathy with, the problems of other areas." See also Howard, *Protestant Theology*, chap. 5.

26. On the shifting of prestige between the humanities and the natural sciences in relation to the universities in Germany, see also Marchand, *Down from Olympus*, chaps. 2 and 3.

27. Cahan, "'Imperial Chancellor.'"

28. On His, see Gould, *Ontogeny and Phylogeny*, 189–93; Maienschein, "Origins," section 2.1; Hopwood, "'Giving Body'"; Hopwood, "Producing Development"; and Richards, *Tragic Sense of Life*, 280–91.

29. His, *Über die Bedeutung*, esp. 30–33; see Richards, *Tragic Sense of Life*, 282–83.

30. His, *Über die Bedeutung*, 30–32; His, *Über die Aufgaben*, 17–18; Richards, *Tragic Sense of Life*, 281–91.

31. The following discussion of His's experiences in Berlin and approach to his discipline is informed by Richards, *Tragic Sense of Life*, 281ff.

32. His, *Über die Aufgaben*, 17–18; Richards, *Tragic Sense of Life*, 283.

33. See Nyhart, *Biology Takes Form*, 201–2; Hopwood, "'Giving Body,'" 476.

34. His, *Unsere Körperform*, 93–104, esp. 97–99.

35. Ibid., 1–17; His, *Untersuchungen*, 138, 182; His also mentions these demonstrations in *Über die Aufgaben*, 17. See also Hopwood, "'Giving Body,'" 468–75.

36. See Hopwood, "'Giving Body,'" 469, 485.

37. Ziegler to Haeckel, November 30, 1875, quoted in translation in Hopwood, "'Giving Body,'" 485. On His's collaboration with Ziegler, see also His, *Untersuchungen*, 182.

38. Morgan, *Development*, 64.

39. His, *Unsere Körperform*, 19; Richards, *Tragic Sense of Life*, 284–85.

40. His, "Über mechanische Grundvorgänge," 2, quoted in translation in Hopwood, "'Giving Body,'" 492.

41. His, *Unsere Körperform*, esp. 165–76; see Richards, *Tragic Sense of Life*, 285–91; and Hopwood, "Pictures of Evolution."

42. See, for instance, Haeckel, *Evolution of Man*, vol. 1, chap. 1; Haeckel, *History of Creation*, 1:114, 116, 118.

43. Haeckel, *Ziele und Wege*, 27; Haeckel, *Evolution of Man*, 1:xix; see Hopwood, "'Giving Body,'" 474.

44. See, for instance, Haeckel, *Generelle Morphologie*, chap. 29; see also Richards's discussion of Haeckel's monism in *Tragic Sense of Life*, 125–28, 297–98.

45. See, for example, Haeckel's address to the Society of German Natural Scientists and Doctors (Gesellschaft deutscher Naturforscher und Ärzte) in September 1877, "Über die heutige Entwickelungslehre." See Richards, *Tragic Sense of Life*, 313–14.

46. Haeckel, *Generelle Morphologie*, 2:365. See Gould's discussion of this passage in *Ontogeny and Phylogeny*, 78–79.

47. Haeckel, *Evolution of Man*, 1:xix; See Gould's discussion of Haeckel's and His's different conceptions of causation in *Ontogeny and Phylogeny*, 193–94: "The greatest clash

between the two approaches took place on the battlefield of causality. Experimental embryologists relentlessly asserted that their kind of cause (proximate and efficient) exhausted the legitimate domain of causality. All that had come before them was merely descriptive; they had established the first causal science of embryology."

48. Haeckel, *Evolution of Man*, 1:83.

49. Ibid., 1:19, 20, 53; Gould, *Ontogeny and Phylogeny*, 78.

50. Haeckel, *Evolution of Man*, 2:881; Gould, *Ontogeny and Phylogeny*, 193.

51. Haeckel, *History of Creation*, vol. 2, Monistic religion 497–98; descent of humans from apes: 405 ff; process of historical development: 493. Haeckel believed more specifically that articulate language had constituted "the real and principal act of humanification" (410). Since linguists identified separate origins for the families of languages, "the different races of men . . . originated, independently of one another, from different branches of primaeval, speechless men directly sprung from apes" (411). Haeckel distinguished this view from the true "polygenist" opinion that the different human races arose from fully independent origins, since he believed the branches led back to a single trunk, and hence the monogenic or "monophyletic opinion is the right one" (410). Nevertheless, his assertion of distinct branches for each of the human races (or sometimes "species of the human race," e.g., 410–11), together with his characterizations of the differential physical, aesthetic, intellectual and moral qualities of these (408–46) has occasioned much analysis of Haeckel's position in the history of scientific racism. For an overview, see Richards, *Tragic Sense of Life*, 269–78. Richards points out that Haeckel's views about the hierarchy of the human races were not only unexceptional but essentially universal among nineteenth-century European naturalists. Thus the history of scientific racism in natural history and early evolutionary theory goes far beyond Haeckel. We have seen a host of eighteenth-century examples in chapter five of this book.

52. Goette, *Entwickelungsgeschichte*, quoted in translation in Richards, *Tragic Sense of Life*, 293. See also Richards's discussion of Goette in ibid., 291–93.

53. Haeckel, *Ziele und Wege*, 21. See Richards's discussion of this passage and Haeckel's notion of "historical natural science" in *Tragic Sense of Life*, 285, 299.

54. Haeckel, *Ziele und Wege*, 24; see Richards, *Tragic Sense of Life*, 299.

55. Haeckel, *Evolution of Man*, 2:507.

56. On Roux, see Nyhart, *Biology Takes Form*, chap. 9; Maienschein, "Origins," section 2.4; and Richards, *Tragic Sense of Life*, 189–92.

57. Morgan, *Development*, 124; Roux, *Beitrag*, 78; Roux, "Beiträge," 453.

58. This discussion is derived from Hopwood, "'Giving Body,'" 493–94. See Roux, "Meine entwicklungsmechanische Methodik," 601–7 (rubber models); Roux, *Entwickelungsmechanik*, 99–100 (bread dough balls).

59. Roux, "Über die Bedeutung 'geringer' Verschiedenheiten." See also Morgan, *Development*, 43–47.

60. Morgan, *Development*, 47.

61. Roux, "Einleitung," 1–2, 4, 5, 10; see also Hopwood, "'Giving Body,'" 493–94; and Churchill, "Chabry."

62. Roux, "Über den 'Cytotropismus'"; Przibram, *Embryogeny*, 41–43, at 41.

63. Roux, "Über den 'Cytotropismus,'" 51.

64. See Nyhart, *Biology Takes Form*, 288–92.

65. Haeckel, *Evolution of Man*, 1:57–58, 27; see also Nyhart, *Biology Takes Form*, 289–90.

66. Roux, "Autobiographie," 146, quoted in translation in Nyhart, *Biology Takes Form*, 295. See also Nyhart's discussion of Haeckel's reaction, which further polarized the situation, in *Biology Takes Form*, 295–96.

67. Roux, "Einleitung," 9, 28–30; see Nyhart, *Biology Takes Form*, 286–87.

68. Roux, "Beiträge"; Roux, "Entwickelungsmechanik der Organismen," 41–42; Maienschein, "Origins," section 2.4.

69. Roux, "Beiträge," 433–58, 519; Roux, "Entwickelungsmechanik der Organismen," 41–42; Maienschein, "Origins," section 2.4.

70. Roux, "Beiträge," 425, 519; Maienschein, "Origins," section 2.4.

71. The German biologist Walther Flemming proposed the term "chromatin" in his "Beiträge," published in English translation as "Contributions." The Berlin anatomist Heinrich Wilhelm Gottfried von Waldeyer-Hartz suggested "chromosome" in his "Über Karyokinese." The journal in which both neologisms were published later merged in 1923 with Roux's *Archiv für Entwicklungsmechanik* to become the *Archiv für mikroskopische Anatomie und Entwicklungsmechanik*: an institutional reflection of the fact that the notion of chromosomes emerged very much within the proximate-mechanist framework of Entwicklungsmechanik.

72. Roux, *Bedeutung der Kerntheilungsfiguren*, 4–5, 9–13, 16.

73. On Weismann, see Churchill, "Break from Tradition"; Churchill, "Weismann-Spencer Controversy"; Churchill, "Developmental Evolutionist."

74. On the University of Freiburg in Breisgau, see Schreiber, *Geschichte*. Von Hehl describes a similar trend toward Protestant ascendancy at the traditionally Catholic University of Munich, and cites this as an instance of "the shaping power of the Protestant University milieu," in "Universität und Konfession," 289.

75. For a summary of the theory, see Weismann, *Germ-Plasm*, 450–68; Weismann's association of qualitatively distinct units of heredity with Roux is at 26. See also Maienschein, "Origins," section 2.4.

76. An early use of the phrase, if not the first, is in Koestler, *Midwife Toad*, 126, 128, 131. The book is about the death of Paul Kammerer, an Austrian Lamarckian biologist who experimented upon midwife toads and claimed to have found evidence that they transmitted acquired changes. Kammerer was accused of falsification—most damningly by G. K. Noble, the curator of reptiles at the American Museum of Natural History in New York—and committed suicide in 1923.

77. Weismann, *Germ-Plasm*, xi.

78. Nägeli, *Mechanico-Physiological Theory*, 8–9, 23, 32, 36. For Nägeli's informing influence on Weismann as well as Weismann's departures from Nägeli's theory, see Weismann, *Essays upon Heredity*, 1:174, 180–83, 192–93, 204.

79. Weismann, *Essays upon Heredity*, 201, 257–59, 263, 269–71, 298–332; quotation at 258.

80. Spencer, *Principles*, 1:311: "Certain powers which mankind have gained in the course of civilization cannot, I think, be accounted for, without admitting the inheritance

of acquired modifications." Spencer argues that Darwin does not sufficiently acknowledge this phenomenon. See also Bain, "Review." Bain uses "inheritance of acquired modifications" in connection with Darwin only and does not mention Lamarck. Usage of the phrase and its variants rose steeply in the last decade of the nineteenth century, from a couple of appearances at most in any five-year period to many occurrences annually. See also Gayon, "Hérédité."

81. Spencer, *Principles*, appendix B, 1: 602–91. The appendix includes four pamphlets first published in the *Contemporary Review* between spring 1893 and autumn 1894 in which Spencer criticized Weismann's interpretation of Darwinism. See also Bowler, *Eclipse of Darwinism*, 41–42: "Weismann's more dogmatic selectionism became known as 'neo-Darwinism,' and it rapidly polarized the scientific community into two mutually hostile camps."

82. Morgan, *Critique*, 33. Morgan refers to the "brilliant essays" in which Weismann "laid bare the inadequacy of the supposed evidence" for inheritance of acquired characters, and to Weismann's mouse-tail experiments: "Weismann appealed to common sense. He made a few experiments to disprove Lamarck's hypothesis."

83. Weismann, "Supposed Transmission," 432–33.

84. Ibid., 434, 440–42.

85. In fact, Weismann himself acknowledged that the mouse experiments did not disprove Lamarckism. See Weismann, ibid., 445–46.

86. See chapter 6.

87. Darwin wrote that "habit, use, and disuse, have, in some cases, played a considerable part in the modification of the constitution and structure," but believed that these effects were secondary to natural selection. See Darwin, *Origin of Species* (1859), 134–43, at 142–43; this section remained from the first through the sixth and final edition in 1876. See also Darwin, *Variation* (1868), vol. 1, chaps. 4, 7, 8 and vol. 2, chap. 24; and Darwin, *Descent of Man*, vol. 1, chap. 4.

88. Lamarck, *Recherches sur l'organisation*, 50–62.

89. Darwin, *Variation* (1875), vol. 2, chap. 27. On "elective affinities," see 2:85n18, 2:164, and 2:374; on amputations, see 2:393.

90. Darwin, *Variation* (1868), 2:404.

91. Weismann, "On Heredity," 77.

92. Weismann, "Supposed Transmission," 447.

93. Weismann, "On Heredity," 95–96.

94. For examples, see Ghiselin, "Imaginary Lamarck."

95. Weismann, "Supposed Transmission," 447, 435.

96. Weismann, "On Heredity," 69. For similarly categorical statements, see Weismann, "Botanical Proofs," 387; Weismann, "Supposed Transmission," 423; and Weismann, *Germ-Plasm*, 395.

97. Weismann, "Mechanical Conception," 718.

98. Ibid., 638. Also at 642 and 637: a mechanist was "not justified in admitting directive forces," which were "directly opposed to the laws of natural science"; and Weismann, "On Heredity," 76: a *nisus formativus* or teleological force had "no value as a scientific explanation."

99. Guenther, *Darwinism*, 369. First published in 1904 as *Darwinismus*.

100. Russell, *Form and Function*, 307.

101. Thompson, *On Growth and Form*, 8, 10, 284, 618–19. On Thompson, see Gould, "D'Arcy Thompson"; and Keller, *Making Sense of Life*, chap. 2.

102. Thompson, *On Growth and Form*, 16, 1025, 1094–95.

103. Ibid., 1023.

104. Lysenko, *Science of Biology Today*, 13–14. On Lysenko and the reputation of Lamarckism, see Joravsky, *Lysenko Affair*, chaps. 7 and 8; and DeJong-Lambert, *Cold War Politics*, chaps. 3 and 4.

105. Biologists associating themselves with Lamarckism remained more strongly represented in some areas during the first two decades of the twentieth century, but "after 1926, it is hard to find anyone still defending the middle ground of old-school Darwinism," that is, Darwinism informed by Lamarckism. See Gliboff, "Golden Age," 53.

106. Weismann, "Mechanical Conception," 682; on the same page Weismann also refers to the "origination of transformations by the direct action of external conditions of the life." See also ibid., 687: "Transformation depends upon a double action of the environment, since the latter first induces small deviations in the organism by direct action, and then accumulates by selection the variations produced." Also Weismann, *Germ-Plasm*, 463: "Were it possible for growth to take place under absolutely constant external influences, variation would not occur."

107. Weismann, "On Heredity," 77, 103–104.

108. On nutrition, see also Weismann, *Essays upon Heredity*, 170, 179, 319. For a discussion of Weismann's embrace of the idea that certain environmental conditions had heritable effects, see Jablonka and Lamb, *Evolution*, 19–20.

109. Weismann, "On Heredity," 94.

110. Weismann, *Germinal Selection*, 11.

111. Roux, *Kampf*.

112. Weismann, *Germinal Selection*; Weismann, *Evolution Theory*, vol. 2, chaps. 25 and 26.

113. Weismann, *Germinal Selection*, 11, 15.

114. Weismann, *Evolution Theory*, 2:114ff; Weismann, *Germinal Selection*, 18–19. Weismann did say that once the determinants began passively to receive extra nutrients, they would begin actively to go after more nutrients and assimilate these more strongly. But this activeness on the part of the determinants nevertheless originated in a passive variation.

115. They include Peter Bowler and Ernst Mayr; see Weissman, *First Evolutionary Synthesis*, 342–43.

116. Weismann, *Evolution Theory*, 2:123.

117. Weismann, "Mechanical Conception," 716, 709; see also 697 on the necessity of teleology.

118. Ibid., 710, 711, 712; see also 716 on teleology and materialism.

119. Otto, *Naturalism and Religion*, 151, 14--48. On Otto's and other Protestant theologians' engagements with evolutionary theory, see also Schröder, *Naturwissenschaft und Protestantismus*, especially 317, 319, and 352; and on the theologians' particular antipathy for Haeckel, see 6, 12, 491. For a Weismann-like argument that a strict mechanical-causal

account of nature supported a supernaturalist teleology, see also Otto, "Darwinism and Religion."

120. On Weismann's ascendancy and his displacement of Haeckel, see Schröder, *Naturwissenschaft und Protestantismus*, 218n145, 254, 279, 317, and 317n571. On the marginalizaton of Haeckel more generally, see 6, 12, 491–94. Schröder discusses (on 449) the episode in which forty-six anatomists and zoologists signed an open letter in support of Haeckel against the "Keplerbund," an organization of Protestant scientists and lay people that attacked Haeckel for manipulating some of his embryo images to emphasize the similarities between human and other animal embryos. Although the forty-six signatories, who included Weismann, defended Haeckel by downplaying the importance of the manipulation, Schröder writes, the majority of them were not "Haeckelians" and had "long since declared . . . a number of Haeckel's hypotheses obsolete." On this episode see also Richards, *Tragic Sense of Life*, 373–83.

121. Stresemann, *Ornithology*, 330–31 (Stresemann ascribes the view he endorses here to Bernard Altum.)

122. Ibid., 339.

123. See Turner, "Humboldt in North America," 292–93.

124. The chief attacker was the conservative Lutheran theologian and antisemite Adolf Stöcker, later chaplain to Kaiser Wilhelm II. See Schröder, *Naturwissenschaft und Protestantismus*, 1–2. Schröder suggests that the venom of the attacks arose partly from the fact that Du Bois-Reymond had played the role of a comparative moderate since his *ignoramus-ignoramibus* speech, making his unequivocal embrace of Darwinism the more shocking. See *Naturwissenschaft und Protestantismus*, 6.

125. Huxley, *Evolution*, 22, 25, 412. See also Bowler, *Eclipse*.

126. Huxley, *Evolution*, 17–18.

127. Crick, "On Protein Synthesis," 152; see also Crick, "Central Dogma"; and Knudsen, "Nesting Lamarckism," 134. On the current fortunes of Crick's Central Dogma, see below, chapter 10.

128. Bowler, *Eclipse of Darwinism*, 41.

129. Stamhuis et al., "Hugo De Vries"; Van der Pas, "Vries, Hugo de"; Stamhuis, "Vries, Hugo De."

130. On the University of Amsterdam, see Knegtmans, *Illustrious School*, chap. 5.

131. De Vries, *Intracellular Pangenesis*; first published in 1889 as *Intracellulare Pangenesis*. (The word that De Vries proposed in the original German text was "pangene," although Gager changed it in the translation to "pangen.")

132. Ibid., 43.

133. Weismann, *Germ-Plasm*, 39.

134. De Vries, *Intracellular Pangenesis*, 48, 49, 62.

135. Ibid., 67–68, 70–71.

136. De Vries, *Mutation Theory*; first published as *Mutationstheorie*. For "latent characters" and "potentialities," see for example *Mutation Theory*, 1:491, 2:26–27, 73.

137. Ibid., 1:462–96, 418, 28–39.

138. Ibid., 1:468, 45, 53, 493.

139. Johannsen, "Genotype Conception," 130–33. See Keller, *Century of the Gene,* introduction, esp. 1–2.

140. Johannsen, "Genotype Conception," 132, 139.

141. On Driesch, see Churchill, "Machine-Theory to Entelechy"; Maienschein, "Origins," chap. 3.1, section 2.5; Sander, "Shaking a Concept"; Sander, "Hans Driesch"; Sander, "Hans Driesch's 'Philosophy,'"; and Sander, "Entelechy."

142. Driesch, "Entwicklungmechanische Studien," quoted in Willier and Oppenheimer, *Foundations,* 46; Driesch, *Science and Philosophy,* 1:61; Sander, "Shaking a Concept," 266; Maienschein, "Origins," chap. 3.1, section 2.5.

143. Driesch, *Problem of Individuality,* 10; Driesch, *Science and Philosophy,* 1:76–84; Maienschein, "Origins," chap. 3.1, section 2.5.

144. Driesch, *Science and Philosophy,* 1:62–64.

145. Driesch, *Problem of Individuality,* 11–12.

146. Quoted in translation in Churchill, "Machine-Theory to Entelechy," 173.

147. Driesch, *Analytische Theorie,* 139; Churchill, "Machine-Theory to Entelechy," 175–76; Driesch, "Maschinentheorie."

148. Driesch, "Lokalisation." See also Churchill, "Machine-Theory to Entelechy," 177: "That Driesch turned away from his earlier machine-theory of life and eventually from science altogether should not conceal the fact that he had been a highly ingenious experimenter." Maienschein, "Origins," chap. 3.1, section 2.5: "After 1900, Driesch turned increasingly from embryology toward philosophy and toward vitalistic views of life."

149. Driesch, *Lebenserinnerungen,* 108–11; Churchill, "Machine-Theory to Entelechy," 184.

150. Driesch, "Lokalisation," 39; quoted in Churchill, "Machine-Theory to Entelechy," 179.

151. Driesch, "Lokalisation," 74; quoted in Churchill, "Machine-Theory to Entelechy," 180.

152. Churchill, "Machine-Theory to Entelechy," 182.

153. Driesch, "Lokalisation," 109; Churchill, "Machine-Theory to Entelechy," 184.

154. Aristotle, *On the Soul,* 412a27–b29, in *CW,* vol. 1. "Entelechy" is here translated as "actuality."

155. Leibniz, *Monadologie,* paragraphs 18, 19, 62, 63, 70.

156. Driesch, "Lokalisation," 96, 103; quoted in Churchill, "Machine-Theory to Entelechy," 183.

157. Driesch, *Problem of Individuality,* 9, 12–13, 17–18; Driesch, *Science and Philosophy,* 1:66–68.

158. Driesch, *Science and Philosophy,* 1:55.

159. Driesch and Morgan, "Zur Analysis"; Driesch, *Science and Philosophy,* 1:66–67; Morgan, *Development,* 131.

160. Morgan, "Whole Embryos"; Driesch, *Science and Philosophy,* 1:66–68.

161. Driesch, *Science and Philosophy,* 1:68.

162. Ibid., 1:140. On Driesch's rejection of machinery as a model of living processes, see Harrington, *Reenchanted Science,* 48.

163. Driesch, *Problem of Individuality*, 18.

164. Driesch, *Science and Philosophy*, 66.

165. Ibid., 1:284–88, at 288.

166. Driesch, *Biologie*, 22; Driesch, *Science and Philosophy*, 1:300–305, 322, 324. On Driesch's ranking of "historical" sciences below "nonhistorical" ones, see also Driesch, *Betrachtung*, 57; Nyhart, *Biology Takes Form*, 291.

167. On Jacques Loeb, see Osterhout, *Biographical Memoir*; Pauly, *Controlling Life*; Rasmussen and Tilman, *Jacques Loeb*; and Fangerau, "Mephistopheles."

168. On Loeb and Mach, see Pauly, *Controlling Life*, 41–45; and Holton, *Science and Anti-Science*, 12.

169. Loeb, *Mechanistic Conception of Life*, 36.

170. Lewis, *Arrowsmith*, 29, 136. On *Arrowsmith*, Loeb, and Sinclair Lewis's rejection of the Pulitzer, see Lingeman, *Sinclair Lewis*, chaps. 14–15, 18.

171. Twain, "Dr. Loeb's Incredible Discovery."

172. Loeb to Edwin Ray Lankester, July 9, 1917, Jacques Loeb Papers [hereafter JLP], Box 8, cited in Fangerau, "Mephistopheles," 238.

173. Haeckel to Loeb, March 29, 1912, JLP Box 6, quoted in Fangerau, "Mephistopheles," 238.

174. Roux to Loeb, November 28, 1908, JLP Box 13; Nathan Zuntz to Loeb, December 4, 1908, JLP Box 16; Benjamin Wheeler to Loeb, January 28, 1910, JLP Box 15; all detailed in Fangerau, "Mephistopheles," 236–37.

175. Loeb, *Mechanistic Conception of Life*, 101–3.

176. Ibid., 72–73, 26–27.

177. Ibid., 26–27, 38, 41, 44.

178. Ibid., 48, 81.

179. Ibid., 30–31, 62.

180. Ibid., 14–15, 36.

181. Loeb to Morgan, May 3, 1914, JLP Box 9; Loeb to Svante Arrhenius, March 11, 1918, and August 1, 1922, JLP Box 1; and Loeb to Albert Einstein, September 4, 1922, JLP Box 4, on the "opposition of the old-fashioned type of colloid chemists" to "any application of physical chemistry to biological problems"; see Fangerau, "Mephistopheles," 242–43.

182. See Morgan, *Critique*, 30, 34–35, 193; Morgan, *Regeneration*, 105, 287–88; and Morgan, *Physical Basis*, 241.

183. See Morgan, *Development*, 129, 132–34; Morgan, *Regeneration*, 263.

184. See for instance Morgan, *Heredity and Sex*, 3, 13–14, 160, 184.

185. Morgan, *Regeneration*, 259, 281, 288.

186. Morgan, *Development*, 136.

187. Morgan, *Critique*, 2.

188. Morgan decided that these regulations were "added" by human designers and not "properties of the material of which the machine itself is composed." Thus, in this instance, he concluded against comparisons between "two entirely different things": "The adaptations of organisms are something peculiar to living things." See Morgan, *Evolution and Adaptation*, 26–29.

Chapter 9

1. Percy Hammond, "The Theaters," *New York Times* (October 10, 1922), 8; Burns Mantle, "Introducing a New Adam and Eve: By a Young Man with Imagination: Introducing the Synthetic Adam and Eve," *Chicago Daily Tribune* (October 15, 1922), E1.

2. Čapek, *R.U.R.*, 2, 9, 13.

3. Ibid., 6.

4. Botanists used the word "clone" from the turn of the twentieth century to refer to genetically identical plants propagated from a single parent, and biologists used "clone" from the late 1920s in reference to genetically identical bacterial and eukaryotic cells made by asexual reproduction from a single ancestor, but the term "clone" did not denote synthetic multicellular animals or humans until the 1970s (in science fiction) and 1980s (in biology). See *OED Online*, Oxford University Press, 2013.

5. James, " Automaton Theory," in *Principles of Psychology*, 1:134–35, 138.

6. Bergson, *Évolution*, chap. 1.

7. "For myself, I confess that as the author I was much more interested in men than in Robots." Čapek, "The Meaning of R.U.R.," *Saturday Review* 136 (July 21, 1923), 79.

8. Čapek, *R.U.R.*, 13.

9. Margaret O'Leary, "Plays from Bohemia," *New York Times* (September 10, 1922), Book Review and Magazine, 4; "Behind the Back Row: Theater Changes Rumored," *San Francisco Chronicle* (September 29, 1922), 11; John Corbin, "The Play," *New York Times* (October 10, 1922), 24; Hammond, "The Theaters," 8; Mantle, "Introducing," E1; Gilbert Seldes, "Great Drama Has Arrived," *Los Angeles Times* (October 15, 1922), section 3, 29; "R.U.R.: A Dramatic Indictment of Civilization," *Current Opinion* (January 1, 1923), 61.

10. D. H., "If There Really Were Robots!" *Life* (February 8, 1923), 20; cartoon in Waldemar Kaempffert, "Science Produces the 'Electrical Man,'" *New York Times* (October 23, 1927), 21.

11. Whether the Tin Woodman represented this argument straightforwardly, satirically, or by some combination of the two has been the subject of debate among historians of American culture and politics. See Ritter, "Silver Slippers," 172–73. At the same time, as founding editor of *The Show Window: A Monthly Journal of Practical Window Trimming*, the Tin Woodman's author, L. Frank Baum, also drew on his penchant for moving, mechanical window displays. Baum edited the journal between 1897 and 1902. See Culver, "What Manikins Want."

12. "Clothes in the Limelight," *Manchester Guardian* (May 11, 1923), 6.

13. See Peter Kussi's introduction to Čapek, *Toward the Radical Center*, 14; Harkins, *Karel Čapek*, chap. 10.

14. "The 'Robot' Office: Mechanical Aids to Business Efficiency," *Manchester Guardian* (February 11, 1926), 12.

15. See Maître, "Problème."

16. "Robot Dream Realized! Employed in Checking Reservoirs," *Times of India* (October 17, 1927), 17. See also "The Mechanical Man. Coming of the 'Robot.' America's Latest Wonder. The Slave Machine. Possibilities and Limits. The Voice as Motive Power," *Observer* (October 16, 1927), 17; Kaempffert, "Science Produces the 'Electrical Man,'"; H. I. Phillips,

"The Once Over: Have You a Little Robot in Your Home?" *Boston Daily Globe* (October 19, 1927); Herbert F. Powell, "Machines That Think: Electrical 'Men' Answer Phones, Do Household Chores, Operate Machinery and Solve Mathematical Problems," *Popular Science Monthly* 112 (January 1928), 12–13; "Televox Acquires a Voice," *Popular Science Monthly* 113 (October 1928), 70; Schaut, *Robots*.

17. "Wife Cooks by 'Televox,'" *Los Angeles Times* (October 14, 1927), 1.

18. David O. Woodbury, "Dramatizing the 'Robot,'" *New York Times* (November 6, 1927), E5.

19. R. J. Wensley, "Robot Seen as Boon to World," *Los Angeles Times* (February 26, 1928), 6.

20. "Steel Robot to Open London Exhibition," *New York Times* (September 11, 1928), 15. Image is from "A Robot to Open an Exhibition: The New Mechanical Man," *Illustrated London News* (London, England), Saturday, September 15, 1928; pg. 453; Issue 4665.

21. "Eric the Robot Opens Exposition. Mechanical Man Rises, Stretches Out Arm for Silence and Speaks. Small Boys are Awed. Slanting Eyes of Metal Clad Monster Glare Yellowly at Them as He Speaks," *New York Times* (September 16, 1928), 26.

22. On Philidog, see Latil, *Pensée*, 220–21; and "New Scientific 'Robots' to Give Most Everybody a Rest," *San Antonio Light Newspaper* (June 8, 1930), 59.

23. TSF stands for *transmission sans fil*. For examples of press accounts of Philidog, see "Robot Dog Brings Grins to Parisians," *Chicago Daily Tribune* (November 24, 1929), G6; and "Robot Dog Is Latest Plaything of Paris," *Washington Post* (November 24, 1929), A6.

24. Latil, *Pensée*, 220–21; "New Scientific 'Robots,'" 59.

25. "New Scientific 'Robots.'"

26. On Robert-Houdin's life and career, see Chavigny, *Le roman d'un Artiste*; During, *Modern Enchantments*, chap. 4; Evans, *Master of Modern Magic*; *MRH*; Manning, *Recollections of Robert-Houdin*; Sharpe, *Salutations to Robert-Houdin*; and Steinmeyer, *Hiding the Elephant*, chap. 7.

27. Robert-Houdin, *Secrets*, 8–9, 11.

28. On August 11, 1841, the Société d'encouragement pour l'industrie nationale granted Robert-Houdin a silver medal for his "nouveaux mécanismes d'horlogerie." He received a bronze medal for a clock with no visible motive force, the Pendule mystérieuse, and for an automaton, L'Escamoteur chinois, at the Exposition des produits de l'industrie française that opened on May 1, 1839. For two further automata, L'Ecrivain Dessinateur and La Leçon de chant, along with a set of tools and clockwork parts, he received a silver medal at the Exposition universelle (May 1—June 30, 1844). See *MRH*, 1:147. At the Exposition des produits de l'industrie of 1855, Robert-Houdin received a first-class medal. See Metzner, *Crescendo*, 208.

29. See *MRH*, 2:28–33, 229–32. See also *MRH*, 1:353; Metzner, *Crescendo*, 208.

30. See *MRH*, 2:231–32, 371, 375.

31. Robert-Houdin, *Memoirs*, 327; *MRH*, 2:32–37, 229, 232.

32. See *MRH*, 2:32, 232–33.

33. See ibid., 1:143.

34. Robert-Houdin, *Secrets*, 2. This is actually a three-way pun on *attraper* (to catch), *la trappe* (a trap or trap door), and *abbaye de la Trappe* (a Trappist monastery).

35. See *MRH*, 2:83–84.

36. Robert-Houdin, *Secrets*, 2–8; see also *MRH*, 2:83–84.

37. Robert-Houdin, *Secrets*, 8–9, 11.

38. Ibid., 13–16.

39. *MRH*, 2:396.

40. Robert-Houdin, *Secrets*, 12–13.

41. *MRH*, 2:396.

42. Robert-Houdin, *Secrets*, 16–18.

43. Robert-Houdin's primary English translator, "Professor Lewis Hoffman" (Angelo John Lewis), seems to have been the originator of the phrase "modern magic" as applied to Robert-Houdin's tradition. See Hoffmann's preface in Robert-Houdin, *Secrets*, v; and Hoffmann, *Modern Magic*, introduction and 522–30. But Robert-Houdin had already presented himself as such in all other ways.

44. Walter B. Cannon, "Physiological Regulation of Normal States: Some Tentative Postulates concerning Biological Homeostatics," in Pettit, *Charles Richet*, 91.

45. Cannon, "Organization," 399, 424, 400.

46. On the immediate prehistory of cybernetics, see Mindell, *Between Human and Machine*.

47. On Craik's wartime activities, see Walter, *Living Brain*, 125.

48. On Grey Walter, see Holland, "Grey Walter"; Hayward, "Tortoise and the Love-Machine"; and Pickering, *Cybernetic Brain*, chap. 3.

49. On Ross Ashby, see Asaro, "Mechanisms"; and Pickering, *Cybernetic Brain*, chap. 4.

50. On Norbert Wiener, see Heims, *John von Neumann and Norbert Wiener*; and Conway and Siegelman, *Dark Hero*. On Wiener's wartime activities, see Galison, "Ontology of the Enemy."

51. See von Neumann, "Role of High and Extremely High Complication" (1949), in his *Theory of Self-Reproducing Automata*, 64, 73; and Burks's introduction to ibid., 18, 19, 21. On von Neumann and cellular automata, see Aspray, *Origins of Modern Computing*, chap. 8.

52. Wiener, *Cybernetics*, 42.

53. Ibid., 11; and Wiener, *Human Beings*, 15, 151–52.

54. Wiener, *Cybernetics*, 40.

55. Ibid.

56. Ibid., 2, 12, 125, 41.

57. Leibniz, *Monadologie*, paragraphs 7, 18, 59.

58. Wiener, *Cybernetics*, 41.

59. Leibniz, "Considerations sur les principes de vie et sur les natures plastiques" (1705), in *PS*, 6:539–46, at 541.

60. Wiener, *Cybernetics*, 43.

61. See Mayr, *Feedback Control*.

62. Rosenblueth et al., "Behavior"; Wiener, *Human Beings*, 174.

63. Rosenblueth et al., "Behavior," 23.

64. Ibid., 19. In all feedback, "some of the output energy of an apparatus or machine is returned as input." In positive feedback, "the fraction of the output which reenters the

object has the same sign as the original input signal. Positive feed-back adds to the input signals, it does not correct them." In negative feedback, by contrast, "signals from the goal are used to restrict outputs which would otherwise go beyond the goal."

65. Ibid., 19–21.

66. Mayr, "Multiple Meanings"; first published as "Teleological and Teleonomic."

67. For example Taylor, *Action and Purpose*; see Cordeschi, "Cybernetics," 191.

68. Wiener, *Cybernetics*, 115–16.

69. Rosenblueth et al., "Behavior," 18, 22; Wiener, *Human Beings*, 174.

70. Canguilhem, "Machine et organisme," in his *Connaissance*, 129–64. First given as a lecture in 1946, published in 1952.

71. Skinner, *Science and Human Behavior*, 47.

72. Skinner, *Behavior of Organisms*, 55, 78; see also Skinner, *Science and Human Behavior*, esp. chap. 4. For other examples, see Watson, "Psychology"; and Pavlov, *Conditioned Reflexes*, 4, 8, 14.

73. Wiener, *Human Beings*, 48.

74. For secondary treatments of the Homeostat, see Dyson, *Darwin*, 176; and Pickering, *Cybernetic Brain*, chap. 4.

75. On Ashby, see Asaro, "Mechanisms"; and Pickering, *Cybernetic Brain*, chap. 4.

76. Ashby, *Design*, chap. 8.

77. Ibid., section 2.

78. Ibid., sections 2–3.

79. Ashby, "Design for a Brain"; Ashby, *Design*, chap. 8, section 3.

80. Ashby, *Design*, chap. 8, sections 4–5.

81. Ibid., chap. 7; chap. 8, section 6.

82. See Ashby, "Principles of the Self-Organizing Dynamic System"; Ashby, "Principles of the Self-Organizing System."

83. Ashby, "Principles of the Self-Organizing Dynamic System," 128.

84. Ashby, *Design*, chap. 8, section 9.

85. Walter, *Living Brain*, 123.

86. Ashby, *Design*, chap. 8, section 2–3.

87. See Walter, *Living Brain*, 124: "A very curious and impressive fact about [the Homeostat] . . . is that, although the machine is man-made, the experimenter cannot tell at any moment exactly what the machine's circuitry is without 'killing' it and 'dissecting' out the 'nervous system'—that is, switching off the current and tracing out the wires to the relays."

88. Ashby, *Introduction to Cybernetics*, 84.

89. Wiener, *Human Beings*, 37–38.

90. Ashby, "Mechanical Chess-Player," 44, 50, 52–53.

91. Ashby, "Biography," at the W. Ross Ashby Digital Archive (hereafter cited as ADA).

92. "Science: The Thinking Machine," *Time* (January 24, 1949).

93. Walter, *Living Brain*, 125.

94. Walter, "Imitation of Life"; Walter, *Living Brain*, chaps. 5 and 7. See also Holland, "Grey Walter."

95. Walter, *Living Brain*, 126.

96. Ibid., 127–28.

97. Ibid., 129.

98. Ibid., 130.

99. Walter to Wiener, cited in Wiener, *Human Beings*, 166.

100. Wiener, *Human Beings*, 174.

101. Walter, *Living Brain*, 130.

102. On Wiener, Wiesner, the cybernetics community at MIT and the supper club, see Heims, *John Von Neumann and Norbert Wiener*, 177; Conway and Siegelman, *Dark Hero*, 205–6, 193, 327; and Rosenblith, *Jerry Wiesner*.

103. Wiener, *Human Beings*, 164–65.

104. Ibid., 165–66.

105. Ibid.

106. Shannon, *Collected Papers*, xiv; Shannon, "Maze Solving Machine"; Shannon, "Potentialities"; Shannon, "Computers and Automata"; Shannon, "Game Playing Machines" ; "Better Mouse: A Robot Rodent Masters Maze," *Life* (July 28, 1952), 45–46.

107. Shannon, *Collected Papers*, xiv; Shannon, "Computers and Automata," 706–8; Shannon, "Game Playing Machines," 791–92; Shannon, "Maze Solving Machine," 684; "Better Mouse: A Robot Rodent Masters Maze," *Life* (July 28, 1952), 45–46.

108. Shannon, "Maze Solving Machine," 684.

109. Ibid., 681.

110. Ibid., 684.

111. Ducrocq, *Logique générale*, 281–83; Beaune, *L'automate et ses mobiles*, 334; and "Job, le renard électronique."

112. Ducrocq, *Logique générale*, 282.

113. On Turing's life and career, see Turing, *Alan M. Turing*; Hodges, *Turing*; Teuscher, *Alan Turing*; and Leavitt, *Man Who Knew Too Much*.

114. Michie, "Mind Machines," 61, 66.

115. Turing, "On Computable Numbers." On the ACE, see Copeland, *Automatic Computing Engine*.

116. For information regarding Ashby's letter to Darwin, see Turing to Ashby [November 19?] 1946, ADA (original at the British Library, Manuscripts Division, donated in 2003 and not yet catalogued); Husbands and Holland, "Ratio Club," 135.

117. Turing to Ashby, [November 19?] 1946, ADA.

118. An early and influential expression of this contrast was Hayles, *How We Became Posthuman*, xi, 62–63. For some more recent instances, see Johnston, *Allure of Machinic Life*, 61, 286–87; Asaro, "Computers as Models"; and Pickering, *Cybernetic Brain*, 328.

119. See Turing's report to Sir Charles Darwin after a year off at Cambridge, "Intelligent Machinery: A Report," 416.

120. On the importance of learning to machine intelligence, see Turing, "Lecture," 393–94; Turing, "Intelligent Machinery: A Report," 421–23; Turing, "Computing Machinery and Intelligence"; Turing, "Intelligent Machinery, A Heretical Theory," 473; and, from a BBC broadcast on May 15, 1951, Turing, "Can Digital Computers Think?" 485.

121. Darwin to Sir Edward V. Appleton, July 23, 1947, The National Archives, DSIR 10/385 (NA); see digital facsimile at the Turing Archive for the History of Computing,

http://www.alanturing.net/turing_archive/archive/p/p30/p30php, accessed March 2015. See also Turing, "Intelligent Machinery: A Report," 400.

122. Turing, "Computing Machinery and Intelligence," 456.

123. See Copeland and Proudfoot, "Anticipation."

124. Turing, "Intelligent Machinery: A Report", 416–20, 421–23.

125. Ibid., 425, 428. On the use of "punishments" and "rewards" to train the machine, see also Turing, "Computing Machinery and Intelligence," 457; and Turing, "Intelligent Machinery, A Heretical Theory," 474.

126. Wiener, *Cybernetics*, 43; Wiener, *Human Beings*, 22–23, 157. Wiener attributes the distinction to Henri Bergson.

127. Turing, "Lecture," 394.

128. For an elucidation of Gödel's theorem, its background and consequences, see Goldstein, *Incompleteness*.

129. Turing, "Intelligent Machinery: A Report," 410–11.

130. Ibid. On the importance of fallibility, see also Turing, "Lecture," 394; Turing, "Computing Machinery and Intelligence," 448-49; and Turing, "Intelligent Machinery, A Heretical Theory," 472.

131. Holland, *Adaptation*.

132. Turing, "Intelligent Machinery: A Report," 429–30, 431.

133. Turing, "Computing Machinery and Intelligence," 456, 459. On the importance of a random element in the learning machine, see also Turing, "Intelligent Machinery, A Heretical Theory," 475.

134. See the panel discussion recorded by the BBC on January 12, 1952, and broadcast on January 14 and 23, 1952: Turing et al., " Automatic Calculating Machines," 497–99, 502–503, 505.

135. Turing, "Can Digital Computers Think?" 482; Turing, "Computing Machinery and Intelligence," 438.

136. Turing, "Can Digital Computers Think?" 482; see also Turing, "Computing Machinery and Intelligence," 451, 454; and Lovelace, "Notes by the Translator," 722, note G.

137. Turing, "Computing Machinery and Intelligence," 451.

138. Turing, "Can Digital Computers Think?" 485.

139. Turing et al., "Automatic Calculating Machines," 494.

140. Turing, "Computing Machinery and Intelligence," 446.

141. Turing, "Intelligent Machinery: A Report," 431. See also Turing et al., "Automatic Calculating Machines," 500.

142. On the matter of Turing's behaviorism, or not, see Block, "Software of the Brain"; and Turing, "Computing Machinery and Intelligence," 434–35.

143. Marvin Minsky makes a similar claim in *The Society of Mind*, 71; see also Turing et al., "Automatic Calculating Machines," 491–92.

144. Turing et al., "Automatic Calculating Machines," 500.

145. Ashby, "What Is an Intelligent Machine?"; see also Asaro, "Mechanisms," 174.

146. Turing, "Computing Machinery and Intelligence," 434–35, 446.

147. Ibid., 456-60, cited passage on 458.

148. Turing et al., "Automatic Calculating Machines," 500.

149. Turing, "Computing Machinery and Intelligence," 452.

150. See Haugeland, *Artificial Intelligence*, esp. chap. 3; Dennett, "Logical Geography"; Franchi and Bianchini, "Introduction," xv.

151. See, for instance, Franchi and Bianchini, "Introduction," xxiii; Franchi, "Life, Death, and Resurrection," 5, 33, 34, 41; Asaro, "Information and Regulation"; Lakoff and Nuñez, *Where Mathematics*; Lakoff, *Women*; and Varela et al., *Embodied Mind*. Andrew Pickering writes that early cyberneticists construed the brain as an organ, not of representation, but of action: "an immediately embodied organ, intrinsically tied into bodily performances." See Pickering, *Cybernetic Brain*, 6. Pickering further suggests that while modern science describes the world as an assembly of things to be represented cognitively and thereby known, cybernetics, in contrast, was a kind of performance, a way of acting in the world as opposed to representing and knowing it. Pickering, *Cybernetic Brain*, 18–23.

152. Turing, "Intelligent Machinery: A Report," 425; Turing, "Computing Machinery and Intelligence," 457; Turing, "Intelligent Machinery, A Heretical Theory," 474.

Chapter 10

1. A recent book by the philosopher Thomas Nagel defends teleology, arguing that a "fully mechanistic" neo-Darwinist account of evolution cannot be adequate to the phenomena of life. See Nagel, *Mind and Cosmos*, quotation at 7. Nagel's book exemplifies the ongoing importance of this deep, historical rift at the heart of Darwin's theory. So does another recent book challenging Darwinism, by the philosopher Jerry Fodor and cognitive scientist Massimo Piattelli-Palmarini. The authors write that although they think Darwinism may be a fatally flawed theory, they believe neither in divine causes nor in "final causes, *élan* vital, entelechies, the intervention of extraterrestrial aliens and so forth" and propose that "there can be no general theory of evolution." See Fodor and Piattelli-Palmarini, *What Darwin Got Wrong*, quotations at xiii and xx. In short, neo-Darwinism and its critics remain in deep conflict over the problem of agency. See also Ghiselin, *Metaphysics*; Ruse, *Darwin and Design*; Ruse, *Darwinism and Its Discontents*; and Lennox, "Darwinism and Neo-Darwinism."

2. Brooks, "Elephants." Brooks has invoked a scourge of the AI community, the UC Berkeley philosopher Hubert Dreyfus, who has argued that classical AI is founded on a mistake. The brain is no electronic digital computer, operating according to formal rules. Thinking is a continual, nonformalizable, physical engagement with the world. See Dreyfus, "Alchemy," 48; *What Computers Still Can't Do*, ch. 7.

3. Brooks, *Intelligence without Reason*, 1.

4. Brooks, "Elephants," section 5.3.

5. Ibid., section 3.

6. Ibid., section 1; Brooks, *Intelligence without Reason*, 16.

7. Brooks, "Elephants," sections 1, 3.2.2, 6.2.

8. Brooks, "Intelligence without Representation," section 3.

9. Brooks, *Intelligence without Reason*, 16–17.

10. Brooks, "Elephants," section 4.3.

11. Brooks, *Intelligence without Reason*, 19.

12. Brooks, "Elephants," section 4.3.

13. Ibid., section 4.4; Brooks, "Intelligence without Representation," section 3.

14. Brooks, "Elephants," section 4.6. On Toto, see also Brooks, *Intelligence without Reason*, 19–20.

15. Brooks, *Cambrian Intelligence*, 37.

16. Brooks, "Elephants," section 5.3; Brooks, *Intelligence without Reason*, 16.

17. Brooks, *Cambrian Intelligence*, x–xi.

18. Brooks, "Intelligence without Representation," section 3.

19. Ibid.

20. Brooks, *Intelligence without Reason*, 14, 22. See also Brooks et al., "Alternate Essences of Intelligence."

21. On Cog, see Brooks et al., "Cog Project"; and Brooks, "Cog Project."

22. On the problem of coherence, see Brooks et al., "Cog Project," sections 6, 6.1. Following behaviorist robotics, there arose another, more moderate approach called "probabilistic robotics," which does not exclude internal models altogether. A group of this approach's practitioners (Sebastian Thrun, Wolfram Burgard, and Dieter Fox) have described it as a "*hybrid control*" architecture: "In probabilistic robotics, there are models, but they are assumed to be incomplete and insufficient for control. There are also sensor measurements, but they too are assumed to be incomplete and insufficient for control." Because "sensors are noisy, and there are usually many things that cannot be sensed directly," Thrun, Burgard, and Fox describe their robots as having "internal beliefs" about the world, which involve inference from sensory data. See Thrun, Burgard, and Fox, *Probabilistic Robotics*, 12, 19, 25, 26.

23. Brooks, "Elephants," section 4.1. Emphasis added.

24. Minsky, *Society of Mind*, 290. Minsky's landmark papers include "Steps toward Artificial Intelligence" and "Matter, Mind and Models," and he is coauthor (with Seymour Papert) of the book *Perceptrons*. The neural network simulator was called SNARC: Stochastic Neural-Analog Reinforcement Computer.

25. Minsky, "Thoughts about Artificial Intelligence," 217.

26. Minsky, *Society of Mind*, 20.

27. Humphrey and Dennett, "Speaking," 39.

28. See, for example, Dennett, *Content and Consciousness*, chap. 8.

29. Dennett, "Postscript [1985]," 21.

30. Dennett, "Logical Geography," 216.

31. Dennett, "Postscript [1985]," 21.

32. Dennett, "Real Consciousness," 136.

33. Dennett, *Brainchildren*, 58.

34. Pinker, *How the Mind Works*, 79; see also 111.

35. Dennett, "Artificial Intelligence," 123–24. See also Fodor, "Appeal," 629: "We refine a psychological theory by replacing global little men by less global little men, each of whom has fewer unanalyzed behaviors to perform than did his predecessors."

36. Kurzweil, *Age of Intelligent Machines*, 15.

37. Minsky, "Thoughts about Artificial Intelligence," 214. See also Minsky, "Steps toward Artificial Intelligence," 27: "To me, 'intelligence' seems to denote little more than the complex of performances which we happen to respect, but do not understand."

38. Minsky, *Society of Mind*, 23, 290, 323.

39. See Dennett, "Intentional Systems"; Dennett, *Content and Consciousness*, chap. 2; and Dennett, *Intentional Stance*.

40. Dennett, "Intentional Systems," 220, 221, 225. If this sounds familiar, it might be because of its striking similarities with Kant's argument in the *Critique of Judgement*. See chapter 8 and Ratcliffe, "Kantian Stance."

41. Dennett, *Content and Consciousness*, 40–41.

42. Dennett, *Darwin's Dangerous Idea*, 74–75.

43. Simon, *Sciences of the Artificial*, 22–25.

44. Simon, *Administrative Behavior*; Simon, "Behavioral Model"; Simon, *Models of Man*; Newell and Simon, *Human Problem Solving*; Simon, *Models of Discovery*; Simon, *Models of Thought*; Simon, *Models of Bounded Rationality*; Simon, *Reason in Human Affairs*.

45. Simon, *Sciences of the Artificial*, 25.

46. Pinker, *How the Mind Works*, 64–65, 77.

47. The section epigraph, in this form, is from Mayr, *Toward a New Philosophy*, 63n1, in which Mayr cites a letter from the Stanford biologist Colin Pittendrigh dated February 26, 1970. Pittendrigh attributes the remark to the English geneticist and evolutionary biologist J. B. S. Haldane. But it had earlier been attributed to the German physiologist Ernst Wilhelm Ritter von Brücke in the form: "Teleology is a lady without whom no biologist can live. Yet he is ashamed to show himself with her in public." See Krebs, "Excursion," 45.

48. Godfrey-Smith, *Darwinian Populations*, 5.

49. Maynard Smith and Price, "Logic of Animal Conflict"; Maynard Smith, *Evolution*.

50. See Dawkins, *Selfish Gene*, 39, 84–87 and 196 for the "language of purpose" as a "shorthand."

51. Haig, " Social Gene," 285.

52. Godfrey-Smith, *Darwinian Populations*, 10; Haig, " Social Gene," 285. Godfrey-Smith's characterization of Haig's psychological justification for applying game theory to genes comes from personal conversations with Haig, but Haig confirmed for me that the characterization accurately reflects his view.

53. Haig, "What is a Marmoset?" esp. 289, 292. For more recent developments of Haig's game theoretic approach to genes, see Haig, "Gene Meme" and "Strategic Gene."

54. Maynard Smith, *Evolution*, 10.

55. Dawkins, *Selfish Gene*, 229.

56. Godfrey-Smith, *Darwinian Populations*, 10.

57. Dawkins, *Extended Phenotype*, 164.

58. Dawkins, *Selfish Gene*, 19–20, 47.

59. On the traditional ascription of agency to genes throughout the history of molecular genetics, see Keller, *Century of the Gene*, 46–48.

60. In a footnote to the passage cited above, Dawkins rejects charges of "radical determinism" that critics leveled in response to the same passage in earlier editions of *The Selfish Gene*. He argues that modern robots are no more "deterministic" than people, since they are capable of learning and they behave in all sorts of creative ways. See Dawkins, *Selfish Gene*, 270. Regardless of how modern robots actually work, the passage clearly implies a contrast between active genes and the passive organisms that are their unwitting tools.

61. Dawkins, *Selfish Gene*, 4, 18–19, 35, 47, 50.

62. Dennett, *Darwin's Dangerous Idea*, 76.

63. Ibid., 77–79; Dennett, "Baldwin Effect."

64. Baldwin, " New Factor in Evolution," 443–44, 549–50.

65. See Depew, "Baldwin."

66. Dennett, *Darwin's Dangerous Idea*, 77–79; Dennett, "Baldwin Effect," 72: "Design explorations by phenotypic trial and error are just as mechanical and nonmiraculous as explorations by genetic natural selection; they just occur more swiftly and at less cost and once design improvements are thereby discovered, genetic assimilation can incorporate them gradually into the genome."

67. Dawkins, *Selfish Gene*, 19–20.

68. Eldredge, *Reinventing Darwin*, esp. 35ff, 199ff.

69. Gould, "Darwinian Fundamentalism."

70. Ibid. On punctuated equilibrium, see Gould and Eldredge, "Punctuated Equilibria"; and Gould and Eldredge, "Tempo and Mode."

71. "Evolution and Cognition," graduate seminar at Harvard University taught by Stephen Jay Gould and Massimo Piattelli-Palmarini, class notes from April 6, 1987.

72. Dennett, *Darwin's Dangerous Idea*, 289, 298; see also 286 ("Gould's leap") and 287: "It is this 'creative role' of something other than selection that caught the skeptical attention of Gould's colleagues."

73. Gould, "Darwinian Fundamentalism."

74. Ibid.

75. Gould and Lewontin, "Spandrels of San Marco."

76. Gould, "Pleasures of Pluralism."

77. Chomsky, *Cartesian Linguistics*.

78. Piattelli-Palmarini, "Evolution, Selection, and Cognition," 2–3.

79. Ibid., 3, 19.

80. Gould and Lewontin, "Spandrels of San Marco," 581.

81. Piattelli-Palmarini, "Evolution, Selection, and Cognition," 10, 26; Chomsky, *Language*.

82. Dennett, *Darwin's Dangerous Idea*, 391, 309; see in general ibid., chap. 13, section 2. For a retrospective view of this controversy advocating the side of Dawkins and Dennett, see Sterelny, *Dawkins vs. Gould*.

83. For a description accessible to laypeople, see Francis, *Epigenetics*, 6–7. For a more technical explanation, see the introduction to Hallgrimsson and Hall, *Epigenetics*, 1–5; McEachern and Lloyd, "Epigenetics," 44; and Shapiro, *Evolution*, 31–36. The term "epigenetics" was first coined by the British biologist Conrad Hal Waddington. See esp. Waddington, *Introduction*; Waddington, *Organisers and Genes*; and Waddington, *Strategy*. An early manifesto of the epigenetics movement is Lewontin, *Triple Helix*. Although Lewontin does not use the word "epigenetics" in the book, he presents an argument for the approach that advocates of epigenetics continue to cite.

84. Jablonka and Lamb, *Evolution*. See also Jablonka and Raz, "Transgenerational Epigenetic Inheritance."

85. Jablonka and Lamb, *Evolution*, 40; Gorelick et al., "Asexuality," 87. Another definition, from Francis, *Epigenetics*, xi: "Epigenetics is the study of how . . . long-lasting, gene-regulating attachments are emplaced and removed."

86. See, for example, Jablonka and Lamb, "Expanded Evolutionary Synthesis," 458; and Gissis and Jablonka, *Transformations*, xi.

87. Jablonka and Lamb, *Evolution*, 126–31, 137–46, 329–32; Francis, *Epigenetics*, 59–60; Grunau, "Methylation Mapping."

88. Jablonka and Lamb, *Evolution*, 126–31, 137–46; Francis, *Epigenetics*, 59–60. Jablonka specifies that she and Lamb are not certain that the monstrous toadflax is not a result of recurrent induction rather than hereditary transmission. That the peloric variant is inherited for at least two generations has been established by Professor John Parker, the director of the University of Cambridge Botanical Garden.

89. Jablonka and Lamb, *Evolution*, 131–32.

90. Shapiro, *Evolution*, 69; Shapiro, "Bacteria."

91. Shapiro, *Evolution*, 2, 87–88, 47–49; on Barbara McClintock, see also Keller, *Feeling*; and Markel and Trut, "Behavior," 177–78.

92. Shapiro, "Bacteria," 814.

93. Shapiro, *Evolution*, 66, 25, 43, 82. On "gene rewiring," see also Braun and David, "Role."

94. Jablonka and Lamb, *Evolution*, 88–89, 94, 249.

95. Weismann, *Germ-Plasm*, 202.

96. J. B. S. Haldane wrote in 1932: "In plants Weismann's *a priori* argument is worth nothing." See Haldane, *Causes of Evolution*, 20.

97. Jablonka and Lamb, *Evolution*, 249; Shapiro, *Evolution*, 35.

98. McClintock, "Significance," 800.

99. Shapiro, "Barbara McClintock."

100. Shapiro, *Evolution*, 143, 7, 55ff, 84, 137–38, 1, 4, 5; Shapiro, "Bacteria," 809. On the ways in which biologists attribute agency to the entities they study, and the intellectual politics surrounding these attributions, see also Myers, *Rendering Life Molecular*, chap. 7.

101. Dawkins, *Extended Phenotype*, 164.

102. Ridley, *Problems of Evolution*, 25.

103. Dawkins, *Extended Phenotype*, 166–67.

104. Steele, *Somatic Selection*, 34–42.

105. Gorczynski and Steele, "Inheritance." On Steele's hypothesis, see Jablonka and Lamb, *Evolution*, 152–53.

106. For an overview of the controversy surrounding Steele's "Lamarckian" arguments, see Paracandola, "Philosophy in the Laboratory." For examples of the seriousness with which Steele's proposals have continued to be taken, see Brosius, "RNAs from All Categories"; Zhivotovsky, "Model"; Jablonka and Raz, "Transgenerational Epigenetic Inheritance," 158; Day and Bonduriansky, "Unified Approach," E19; and Bonduriansky, "Rethinking." Steele's career has been tumultuous largely for reasons external to his research: in 2001 he accused the University of Wollongong, in Australia, where he was teaching, of systematic grade inflation for international students who paid the highest fees. On the affair, see Martin, "Dilemmas."

107. Pigliucci and Müller, *Evolution*, 9. On Francis Crick's restating of the Weismann barrier as the "Central Dogma," see chapter 8.

108. Dawkins, *Extended Phenotype*, 168–73.

109. Ibid., 169.

110. Dennett, *Darwin's Dangerous Idea*, 323, 141–42. With regard to the example of the mite and the fruit flies, Dennett cites Houck et al., "Possible Horizontal Transfer."

111. Moxon et al., "Adaptive Evolution"; Jablonka and Lamb, *Evolution*, 94–96.

112. Jablonka and Lamb, *Evolution*, 97–98.

113. Dawkins, *Extended Phenotype*, 167, 173.

114. See Bowler, *Evolution*, chaps. 3 and 4;and Larson, *Evolution*, 13, 23, 66.

115. On Lamarck's rigorous naturalism, see Keller, "Self-Organization," 359.

116. Smail, *Deep History*, deep history founded in neurobiology: 11–12; neo-Darwinist rendition of "Darwinian" as "blind variation and selective retention": 120, 123; Lamarckian "Mind-first" fallacy of considering mind and intentionality to be causal factors in human history: 102, 120; "discounts the goals," 142. Curiously, Smail writes that Lamarck himself did not ascribe intentionality to organisms in any important way, since giraffes don't intend to lengthen their necks but merely to reach tastier leaves, and that the privileging of intentions was the later innovation of neo-Lamarckian historians and sociologists: "Lamarckism, as I noted earlier, is innocent of intentionalism," 118, see also 112. Yet, as we saw in chapter 6, Lamarck wrote precisely that higher animals transformed themselves in part through the exercise of will. Since Lamarck is not included in Smail's bibliography, his interpretation probably expresses his own intuitions about animal behavior.

117. Smail, *Deep History*, "it matters not one whit how traits are acquired," 121; a Darwinist model of historical change that allows a place for Lamarck while still banishing intention: 121, 123; "genius and forethought," "Lamarckian-style historians," "abandon .. [the] Mind-first model":150.

118. Dawkins, *Extended Phenotype*, 173.

119. Ibid., 116, 172.

120. Jablonka and Lamb, *Evolution*, chap. 9, quotation at 319.

121. Müller and Newman, "Origination"; Newman and Müller, "Genes and Form."

122. Newman, "Fall and Rise," 12.

123. Newman and Müller, "Genes and Form," 39. See also Goodwin, *How the Leopard Changed Its Spots*, chap. 3; and Solé and Goodwin, *Signs of Life*, 24–25.

124. On excitable media, see Schimansky-Geier et al., *Analysis and Control.*

125. Newman and Müller, "Genes and Form," 39, 41–42.

126. Ibid., 39, 63.

127. In particular the philosophers John Stuart Mill and George Henry Lewes.

128. On the history and definition of "emergence," including distinctions such as that between epistemic and ontological emergence, see Goldstein, "Emergence"; Clayton and Davies, *Re-Emergence*; Kim, "Emergence"; Feltz et al., *Self-Organization*; Bedau and Humphreys, *Emergence*; and Corradini and O'Connor, *Emergence.*

129. Von Neumann, "Role of High and Extremely High Complication," (1949), in his *Theory of Self-Reproducing Automata*, 65–66. See also Burks's introduction to ibid., 20.

130. See Solé and Goodwin, *Signs of Life*, 24–25.

131. See, for example, Müller and Newman, "Origination," 7; Newman and Müller, "Genes and Form," 42; and Müller, "Epigenetic Innovation," 316, 323, 325.

132. Philosophers and scientists have devoted much discussion to the relations be-

tween self-organization and emergence. For the history and definition of self-organization and on its relations to emergence, see Kauffman, *Origins of Order*, chap. 5; Goldstein, "Emergence"; Solé and Goodwin, *Signs of Life*, 12–18; Keller, "Organisms . . . Part One" and "Organisms . . . Part Two." A landmark text in the post-cybernetic phase of ideas about self-organization and emergence in biology is Maturana and Varela, *Autopoiesis and Cognition*.

133. See, for example, Kauffman, *Origins of Order*, chap. 5; Goldstein, "Emergence"; and Solé and Goodwin, *Signs of Life*, 12–18.

134. See Kauffman, *Origins of Order*.

135. Kauffman, "On Emergence," 501, 505, 510, 517.

136. Ibid., 504, 516–17, 518–19.

137. The literature on the Newtonian justification of force, and the debates among Newtonians, Leibnizians, and Cartesians regarding force, matter, and mechanism, is vast. For classic and synthetic discussions, see Schofield, *Materialism and Mechanism*; Westfall, *Force*; Hatfield, "Force"; Hall, *Philosophers at War*; Gabbey, "Force and Inertia"; Meli, *Equivalence and Priority*; Vailati, *Leibniz and Clarke*; and Gaukroger, *Descartes' System*. I discuss the subject in Riskin, *Science in the Age of Sensibility*, chap. 3.

138. See Symonds, "What Is Life?" in which Symonds argues that people misconstrue the importance of Schrödinger's book, but begins by saying it has become part of the "folklore of biology" for having directed the early course of molecular biology. See also Watson, *Avoid Boring People*, 28, 43, 98, 218; and Dronamraju, "Erwin Schrödinger." Schrödinger drew on an earlier landmark paper by the geneticist Nikolai Timoféeff-Ressovsky, the radiation physicist Karl G. Zimmer and the physicist Max Delbrück entitled "On the Nature of Gene Mutation and Gene Structure" (1935). For a collection of historical evaluations of this paper, see Sloan and Fogel, *Creating a Physical Biology*.

139. Schrödinger, *What Is Life*, 69.

140. Ibid., 70–73, 77, quotations at 73, 77.

141. Ibid., 80–82.

142. Ibid., 82–83.

143. Ibid., 85.

144. Ibid, 48.

145. Ibid., 50, 56, 65. (Schrödinger attributed this picture of the gene and of mutation largely to the German-American physicist Max Delbrück.) Evelyn Fox Keller, a researcher with a degree in theoretical physics who has worked in molecular biology and in the history and philosophy of science, and who is also McClintock's biographer, has recently made a proposal that is a sort of mirror image of Schrödinger's, attributing the activity of living forms to the structure of large molecules such as proteins. She has suggested that life might originate in "a kind of primitive activity" inherent in these molecules, which bind together by a process that might "embody a link between rudimentary forms of perception and action," if one regards the binding site as a kind of sensory receptor and the resulting behavior of the new complex as a kind of action. "Smart matter, or rather, smart molecules," she thinks, might be something like Lamarck's upward-striving force of life. See Keller, "Self-Organization" [Rausing Lecture version], 22, 23, 26. See also Keller, "Self-Organization" in Gissis and Jablonka, eds., *Transformations*.

146. Schrödinger, *What Is Life*, 60, 67–68, 85.

147. See Keller, *Century of the Gene,* esp. 21–27, 31, 46.

148. Schrödinger, *What Is Life,* 33–34, 110.

149. Schrödinger, "Mind and Matter," 106.

150. Ibid., 107–108, 111–12.

151. Ibid., 113.

152. Ibid., 109–10.

153. Ibid., 122.

BIBLIOGRAPHY

Archival, Manuscript, and Museum Collections

American Philosophical Society (APS), Philadelphia, PA, USA	Charles Darwin Papers, 1831–82
Archives de l'Académie des Sciences (AS), Paris, France	Registre des procès-verbaux des séances
Archives départementales du Nord (AdN), Lille, France	Série B: Registres de comptes de la recette générale des finances (1405–1530)
Archives départementales du Pas-de- Calais (AdPC), Dainville, France	Série A: Actes du pouvoir souverain et domain public
Archives nationales de France (ANF), Paris, France	Série F: Versements des ministères Série G7: Contrôle générale des finances Série KK: Monuments historiques Série 0: Maison du Roi
W. Ross Ashby Digital Archive (ADA), http://www.rossashby.info	
Bibliothèque centrale du Muséum national d'Histoire naturelle (BMHN), Paris, France	MS 742–56, Manuscrits de J.-B. Pierre-Ant. de Monet de Lamarck

Bibliothèque nationale de France (BNF), Paris, France	Estampes et Photographie, Nouvelles Acquisitions Françaises (NAF), 15551 Autographes du XVIIIe siècle
British Library (BL), London, England	Add MS 15950; Add MS 78610 A-M; Add MS 78298–429: Papers of John Evelyn Add MS 35507: Hardwicke Papers, political correspondence Add MS 37182–205: Correspondence and papers of Charles Babbage
Centre national des arts et métiers, (CNAM), Paris, France	
Darwin Correspondence Project (DCP), Cambridge University Library, http://www.darwinproject.ac.uk	
Deutsches Museum (DM), Munich, Germany	
Library of Congress, Manuscript Division, Washington, DC, USA	Jacques Loeb Papers (JLP)
Musée d'art et d'histoire (MAH), Neuchâtel, Switzerland	
Musée d'horlogerie, Le Locle, Switzerland	
Musée international d'horlogerie, La Chaux-de-Fonds, Switzerland	
The National Archives (NA), Kew, England	
Turing Archive for the History of Computing (TA), http://www.alanturing.net	
Wellcome Centre, London, England	

Printed Primary Sources

Abernethy, John. "Extracts from *Introductory Lectures Exhibiting Some of Mr. Hunter's Opinions Respecting Life and Diseases*." In *Romanticism and Science, 1773–1833*, edited by Tim Fulford, 5:45–52. London: Routledge, 2002.

Aldini, Giovanni. *An Account of the Late Improvements in Galvanism*. London: Cuthell and Martin, 1803.

———. *Essai théorique et expérimental sur le galvanisme*. Paris: Fournier fils, 1804.

Aleotti, Giovanni Battista. *Gli artificiosi e curiosi moti spirituali di Herrone*. Ferrara, Italy: Vittorio Baldini, 1589.

Anonymous. *Observations on the Automaton Chess Player, Now Exhibited in London, at 4, Spring Gardens*. London, 1819.

———. *Wisdom and Reason; Or, Human Understanding Consider'd, with the Organization:*

Or, with the Form and Nature of the Solids and Fluids of the Body. London: John Hook, 1714.

Aquinas, Thomas. *Commentary on Aristotle's "Metaphysics."* Edited and translated by John P. Rowan. Revised ed. Notre Dame, IN: Dumb Ox Books, 1995.

———. *Opera omnia*. Edited by S. E. Fretté and P. Maré. 34 vols. Paris: Ludovicus Vivès, 1871–80.

———. *Summa contra gentiles*. Translated by A. C. Pegis, J. F. Anderson, V. J. Bourke, and C. J. O'Neil. 4 vols. Notre Dame, IN: University of Notre Dame Press, 1975.

———. *The "Summa Theologica" of St Thomas Aquinas*. Translated by the Fathers of the English Dominican province. 20 vols. New York: Benziger, 1911–25.

Argyll, George Douglas Campbell, Duke of. *The Works of the Duke of Argyll*. 3 vols. New York: John B. Alden, 1884.

Aristotle. *The Complete Works of Aristotle: The Revised Oxford Translation*. Edited by Jonathan Barnes. 2 vols. Princeton, NJ: Princeton University Press, 1984. (*CW*)

Ashby, William Ross. "Can a Mechanical Chess-Player Outplay Its Designer?" *British Journal for the Philosophy of Science* 3, no. 9 (1952): 44–57.

———. "Design for a Brain." *Electronic Engineering* 20 (December 1948): 379–83.

———. *Design for a Brain: The Origin of Adaptive Behaviour*. 1952. 2nd ed. London: Chapman and Hall, 1960.

———. *An Introduction to Cybernetics*. New York: Wiley, 1957.

———. "Principles of the Self-Organizing Dynamic System." *Journal of General Psychology* 37 (1947): 125–28.

———. "Principles of the Self-Organizing System." In *Principles of Self-Organization: Transactions of the University of Illinois Symposium*, edited by H. Von Foerster and G. W. Zopf, Jr., 255–78. Oxford, UK: Pergamon, 1962.

———. "What Is an Intelligent Machine?" Biological Computer Laboratory, Technical Report no. 7.1. Urbana: University of Illinois, 1961.

Asimov, Isaac. *I, Robot*. 1950. New York: Bantam, 2005.

Aubert de la Chesnaye-Debois, François-Alexandre. *Lettre sur les romans*. Paris: Gissey, 1743.

Augustine. "Of True Religion (*De Vera Religione*)." In *Augustine's Earlier Writings*, edited and translated by John H. S. Burleigh, 218–83. London: SCM-Canterbury Press, 1953.

Babbage, Charles. *Charles Babbage: Passages from the Life of a Philosopher*. 1864. Edited by Martin Campbell-Kelly. New Brunswick, NJ: Rutgers University Press, 1994.

———. *The Works of Charles Babbage*. Edited by Martin Campbell-Kelly. 11 vols. London: W. Pickering, 1989.

Babbage, Henry Prevost, ed. *Babbage's Calculating Engines: A Collection of Papers*. Los Angeles: Tomash, 1982.

———. "On the Mechanical Arrangements of the Analytical Engine of the Late Charles Babbage, F.R.S." In *Report of the Fifty-Eighth Meeting of the British Association for the Advancement of Science*, 616–17. London: John Murray, 1889.

Bachaumont, Louis Petit de. *Mémoires secrets pour server à l'histoire de la république des lettres en France depuis MDCCLXII jusqu'à nos jours*. 36 vols. London: John Adamson, 1777–89.

Baillet, Adrien. *La vie de M. Des-Cartes*. 2 vols. Paris: D. Horthemels, 1691.

Bain, Alexander. "Review of 'Darwin on Expression': Being a Postscript to *The Senses and the Intellect*." In *The Senses and the Intellect*, 697–714. 3rd ed. New York: Appleton, 1874.

Baldwin, James Mark. "A New Factor in Evolution." *American Naturalist* 30 (1986): 441–51, 536–53.

Ballière de Laisement, Denis. *Eloge de Monsieur Le Cat, ecuyer, docteur en médecine, chirurgien en chef de l'Hôtel-Dieu de Rouen*. Rouen, France: Laurent Dumesnil, 1769.

Barruel, Augustin. *Les Helviennes, ou lettres provinciales philosophiques*. 7th ed. 4 vols. Paris: Pailleux, 1830.

Baumgarten, Otto. *Die Voraussetzungslosigkeit der protestantischen Theologie*. Kiel, Germany: Lipsius and Tischer, 1903.

Bayle, Pierre. *Dictionnaire historique et critique*. 4 vols. 1695. Amsterdam: P. Brunel, 1740.

Bégon, Michel. *Lettres de Michel Bégon*. Edited by Louis Delavaud and Charles Dangibeaud. 3 vols. Paris: Saintes, 1925–35.

Bennett, A.W. "On the Theory of Natural Selection looked at from a Mathematical Point of View." In *Report of the Fortieth Meeting of the British Association for the Advancement of Science*, 130–31. London: John Murray, 1871.

Bentham, Jeremy. *Introduction to the Principles of Morals and Legislation*. 1789. Oxford, UK: Clarendon, 1907.

Bergson, Henri. *L'évolution créatrice*. 1907. Paris: Presses universitaires de France, 1959.

———. *Le rire: Essai sur la signification du comique*. 1900. Paris: Félix Alcan, 1912.

Berkowitz, Jacob. *The Stardust Revolution: The New Story of Our Origin in the Stars*. Amherst, NY: Prometheus Books, 2012.

Bernard, Claude. *Cahier de notes, 1850–1860*. Edited by Mirko Dražen Grmek. Paris: Gallimard, 1965.

———. *Introduction à l'étude de la médecine expérimentale*. 1865. Paris: Flammarion, 1966.

———. *Leçons sur les phénomènes de la vie, communs aux animaux et aux végétaux*. 2 vols. Paris: Baillière, 1878–79.

Bibiena, Ferdinando Galli. *L'architettura civile*. 1711. Reprinted with an introduction by Diane M. Kelder. New York: B. Blom, 1971.

Bichat, Xavier. *Recherches physiologiques sur la vie et la mort*. 4th ed. Edited by François Magendie. Paris: Bechet jeune, 1822.

Block, Ned. "The Mind as the Software of the Brain." In *Thinking: An Invitation to Cognitive Science*, edited by Edward E. Smith and Daniel N. Osherson, 3:377–425. Cambridge, MA: MIT Press, 1995.

Blumenbach, Johann Friedrich. *An Essay on Generation*. Translated by A. Crichton. London: T. Cadell, Faulder, Murray, and Creech, 1792.

Boerhaave, Hermann. *A New Method of Chemistry*. London: Printed for J. Osborn and T. Longman, 1727.

Bonduriansky, R. "Rethinking Heredity, Again." *Trends in Ecology and Evolution* 27, no. 6 (2012): 330–36.

Bonnet, Charles. *Considérations sur les corps organisés*. 1762. Amsterdam: M. M. Rey, 1768.

———. *La contemplation de la nature*. 2nd ed. 2 vols. Amsterdam: Marc-Michel Rey, 1770.

———. *Essai analytique sur les facultés de l'ame*. Copenhagen: Frères Philibert, 1760.

———. *Essai de psychologie, ou, Considérations sur les opérations de l'âme, sur l'habitude et sur l'éducation: auxquelles on a ajouté des principes philosophiques sur la cause première et sur son effet.* 2 vols. London, 1755.

———. *Œuvres d'histoire naturelle et de philosophie.* Neuchâtel, Switzerland: Fauche, 1779–83.

———. *La palingénésie philosophique.* 2 vols. Geneva: Philibert et Chirol, 1769.

Bordeu, Théophile de. *Recherches anatomiques sur la position des glandes, et sur leur action.* New ed. Paris: Chez Brosson, 1800.

Borelli, Giovanni Alfonso. *On the Movement of Animals.* 1680–81. Translated by Paul Maquet. Berlin: Springer-Verlag, 1989.

Borgnis, J.-A. *Traité complet de mécanique appliquée aux arts.* 8 vols. Paris: Bachelier, 1818–20.

Bossuet, Jacques Bénigne. *Œuvres complètes.* Paris: L. Vivès, 1864.

Boullier, David-Renaud. *Essai philosophiques sur l'âme des bêtes: où l'on trouve diverses réflexions sur la nature de la liberté, sur celle de nos sensations, sur l'union de l'ame et du corps, sur l'immortalité de l'ame.* Amsterdam: Chez Francois Changuion, 1727.

Boyer, Jean Baptiste de. *Lettres chinoises, ou correspondance philosophique, historique et critique, entre un chinois voyageur et ses correspondants à la Chine, en Muscovie, en Perse et au Japon.* The Hague: Pierre Gosse, 1751.

Boyle, Robert. *A Disquisition about the Final Causes of Natural Things.* London: J. Taylor, 1688.

———. *A Free Enquiry into the Vulgarly Received Notion of Nature.* Edited by Edward B. Davis and Michael Hunter. Cambridge: Cambridge University Press, 1996.

Bradford, Gamaliel. *The History and Analysis of the Supposed Automaton Chess Player of M. de Kempelen, Now Exhibiting in This Country by Mr. Maelzel.* Boston: Hilliard, Gray, 1826.

Braun, Erez, and Lior David. "The Role of Cellular Plasticity in the Evolution of Regulatory Novelty." In *Transformations of Lamarckism: From Subtle Fluids to Molecular Biology,* edited by Snait Gissis and Eva Jablonka, 181–92. Cambridge, MA: MIT Press, 2011.

Brewer, J. S., ed. *Letters and Papers, Foreign and Domestic, of the Reign of Henry VIII.* 22 vols. London: Longman, 1862–1932.

Brewster, David. *Letters on Natural Magic, Addressed to Sir Walter Scott, Bart.* New ed. London: William Tegg, 1868.

Britain, Thomas of. *Le roman de Tristan.* Edited by Joseph Bédier. 2 vols. Paris: Firmin-Didot, 1902–5.

Broca, Paul. "Sur le transformisme." *Bulletins de la Société d'anthropologie de Paris* 5 (1870): 168–242.

Brooks, Rodney. *Cambrian Intelligence: The Early History of the New AI.* Cambridge, MA: MIT Press, 1999.

———. "The Cog Project." *Journal of the Robotics Society of Japan* 15 (1997): 968–70.

———. "Elephants Don't Play Chess." *Robotics and Autonomous Systems* 6 (1990): 3–15.

———. *Intelligence without Reason.* MIT AI Lab Memo 1293. April 1991.

———. "Intelligence without Representation." *Artificial Intelligence Journal* (1991): 139–59.

Brooks, Rodney, Cynthia Breazeal, Matthew Marjanović, Brian Scassellati, and Matthew

Williamson. "The Cog Project: Building a Humanoid Robot." In *Computation for Metaphors, Analogy, and Agent*, edited by C. Nehaniv, 52–87. New York: Springer, 1999.

Brooks, Rodney, Cynthia Breazeal (Ferrell), Robert Irie, Charles C. Kemp, Matthew Marjanović, Brian Scassellati, and Matthew M. Williamson. "Alternate Essences of Intelligence." *Proceedings of the Fifteenth National Conference on Artificial Intelligence* (AAAI–98), Madison, WI. Palo Alto, CA: AAAI Press, 1998.

Brosius, J. "RNAs from All Categories Generate Retrosequences That May Be Exapted as Novel Genes or Regulatory Elements." *Gene* 238, no. 1 (1999): 115–34.

Buffon, Georges Louis Leclerc, comte de. *Correspondance inédite de Buffon, à laquelle ont été réunies les lettres publiées jusqu'à ce jour*. Edited by Henri Nadault de Buffon. 2 vols. Paris: L. Hachette, 1860.

———. "Discours sur la nature des animaux." In *Histoire naturelle, générale et particulière, avec la déscription du Cabinet du Roy*, 4:3–112. Paris: Imprimerie Royale, 1753.

———. *Histoire naturelle, générale et particulière, avec la déscription du Cabinet du Roy*. 44 vols. Paris: Imprimerie Royale, 1749–1804.

Buonarroti the Younger, Michelangelo. *Descrizione delle felicissime nozze della Cristianissima Maesta' di Madama Maria Medici Regina di Francia e di Navarra*. Florence, 1600.

Butler, Samuel. *Erewhon*. 1872. Edited by Peter Mudford. London: Penguin, 1985.

———. *Evolution, Old and New, or The Theories of Buffon, Dr. Erasmus Darwin, and Lamarck, as Compared with That of Mr. Charles Darwin*. 2nd ed. London: D. Bogue, 1882.

———. "Lamarck and Mr. Darwin." In *Life and Habit*, 252–72. London: Trübner, 1878.

———. *Life and Habit*. London: Trübner, 1878.

———. *Luck, Or Cunning as the Main Means of Organic Modification? An Attempt to Throw Additional Light upon the Late Mr. Charles Darwin's Theory of Natural Selection*. London: Trübner, 1887.

———. "Mr. Mivart and Mr. Darwin." In *Life and Habit*, 273–93. London: Trübner, 1878.

———. *Unconscious Memory* (1880). Vol. 6 of *The Shrewsbury Edition of the Works of Samuel Butler*, edited by Henry Festing Jones and A. T. Bartholomew. London: Jonathan Cape, 1924.

Cabanis, Pierre Jean Georges. *Rapports du physique et du moral de l'homme*. 1808. 3rd ed. 2 vols. Paris: Caille et Ravier, 1815.

Callistratus. *Descriptions*. In *Elder Philostratus, Younger Philostratus, Callistratus*, translated by Arthur Fairbanks. Loeb Classical Library. Cambridge, MA: Harvard University Press, 1931.

Campardon, Émile. *Les spectacles de la foire*. 2 vols. 1877. Geneva: Slatkine Reprints, 1970.

Campbell, Neil A., Brad Williamson, and Robin Heyden. *Biology: Exploring Life*. London: Pearson Prentice Hall, 2006. Online. Accessed October 2013. http://apps.cmsfq.edu.ec/biologyexploringlife/text/.

Camus, Charles-Etienne-Louis de. *Traité des forces mouvantes pour la pratique des arts et des métiers*. Paris: C. Jombert, 1722.

Cannon, Walter. "Organization for Physiological Homeostasis." *Physiological Reviews* 9 (1929): 399–431.

———. *The Wisdom of the Body*. Revised ed. New York: W. W. Norton, 1939.

Čapek, Karel. *R.U.R.* 1921. Translated by David Wyllie. Rockville, MD: Wildside Press, 2010.

———. *Toward the Radical Center: A Karel Čapek Reader*. Edited by Peter Kussi. North Haven, CT: Catbird Press, 1990.

Carlyle, Thomas. *History of Friedrich II of Prussia, Called Frederick the Great*. 6 vols. London: Chapman and Hall, 1859–65.

Castellan, Antoine Laurent. *Letters on Italy*. London: Printed for R. Phillips, 1820.

Catrou, François et al., eds. *Journal de Trévoux, ou Mémoires pour l'histoire des sciences et des beaux arts*. Trévoux: Chez Jean Boudot, 1701–67.

Cervantes Saavedra, Miguel de. *The Ingenious Gentleman Don Quixote of La Mancha*. 1605–15. Translated by Samuel Putnam. New York: Viking, 1958.

Chambers, Ephraim, et al., *A Supplement to Mr. Chambers's Cyclopaedia, or, Universal Dictionary of Arts and Sciences*. 2 vols. London: W. Innys and J. Richardson, 1753.

Chambers, Robert, ed. *The Book of Days: A Miscellany of Popular Antiquities in Connection with the Calendar*. 2 vols. London: W. & R. Chambers, 1869.

Châtelet, Gabrielle-Émilie le Tonnier de Breteuil, marquise du. *Institutions de physique*. Paris: Prault fils, 1740.

Cheyne, George. *Philosophical Principles of Natural Religion*. London: Printed for George Strahan, 1705.

Chomsky, Noam. *Cartesian Linguistics: A Chapter in the History of Rationalist Thought*. 3rd ed. Cambridge: Cambridge University Press, 2009.

———. *Language and the Problems of Knowledge: The Managua Lectures*. Cambridge, MA: MIT Press, 1988.

Coleridge, Samuel Taylor. *Aids to Reflection in the Formation of a Manly Character on the Several Grounds of Prudence, Morality and Religion*. 2nd ed. London: Hurst, Chance, 1831.

———. *Biographia Literaria, Or, Biographical Sketches of My Literary Life and Opinions*. 1817. New York: William Gowans, 1852.

———. "The Eolian Harp." In *Selected Poetry*, edited by H. J. Jackson, 27–29. Oxford: Oxford University Press, 2009.

———. "On the Passions." In *Shorter Works and Fragments*, edited by H. J. Jackson and J. R. de J. Jackson, 2:1419–53. London: Routledge, 1995.

Commandino, Frederigo. *Aristarchi de Magnitudinus, et Distantiis Sollis, et Lunae, Liber*. Pesaro: apud Camillum Francischinum, 1572.

Condillac, Etienne Bonnot de. *Œuvres philosophiques*. Edited by Georges Le Roy. 3 vols. Paris: Presses universitaires de France, 1947–51.

———. *Traité des animaux*. 1755. In *Traité des sensations, Traité des animaux*. Paris: Fayard, 1984.

———. *Traité des sensations*. 1754. In *Traité des sensations, Traité des animaux*. Paris: Fayard, 1984.

Condorcet, Jean-Antoine-Nicolas de Caritat, marquis de. *Œuvres de Condorcet*. Edited by Arthur O'Connor and François Arago. 12 vols. Paris: Firmin Didot frères, 1847–49.

Constans, Léopold, ed. *Le roman de Troie*. 6 vols. Paris: Fermin-Didot, 1906.

Cooper, Anna Julia, and Eduard Koschwitz, eds. *Le pèlerinage de Charlemagne*. Paris: A. Lahure, 1925.

Corley, Corin, ed. and trans. *Lancelot of the Lake*. Oxford World Classics. Oxford: Oxford University Press, 2000.

Court de Gébelin, Antoine. *Le monde primitif, analysé et comparé avec le monde moderne.* 9 vols. Paris: Chez l'auteur, 1773–82.

Coward, William. *Second Thoughts concerning Human Soul.* 2nd ed. London: A. Baldwin, 1704.

Crick, F. H. C. "The Central Dogma." *Nature* 227 (August 8, 1970): 562–63.

———. "On Protein Synthesis." *The Symposia of the Society for Experimental Biology* 12 (1958): 138–63.

Cudworth, Ralph. *The True Intellectual System of the Universe.* 2nd ed. 2 vols. London: J. Walhoe, 1743.

Cuvier, Georges. "Eloge de M. de Lamarck, lu a l'Académie des sciences, le 26 novembre 1832." *Mémoires de l'Académie Royale des Sciences de l'Institut de France* 8 (1935): i–xxxi.

———. *Le règne animal distribué d'après son organisation: Pour servir de base à l'histoire naturelle des animaux et d'introduction à l'anatomie comparée.* 3rd ed. 3 vols. Brussels: L. Hauman, 1836.

Daniel, Gabriel. *Voyage du monde de Descartes.* Paris: Vve de S. Bénard, 1690.

Darwin, Charles. *Autobiography.* Edited by Nora Barlow. London: Collins, 1958.

———. *Charles Darwin's Natural Selection: Being the Second Part of His Big Species Book Written From 1856 to 1858.* Edited by R. C. Stauffer. London: Cambridge University Press, 1975.

———. *The Descent of Man and Selection in Relation to Sex.* 2 vols. London: J. Murray, 1871.

———. *The Descent of Man, and Selection in Relation to Sex.* 2nd ed. London: J. Murray, 1874.

———. *The Effects of Cross and Self Fertilisation in the Vegetable Kingdom.* London: John Murray, 1876.

———. "Essay of 1842." In *The Foundations of the Origins of Species: Two Essays Written in 1842 and 1844,* edited by Francis Darwin, 1–56. Cambridge: Cambridge University Press, 1909.

———. "Essay of 1844." In *The Foundations of the Origins of Species: Two Essays Written in 1842 and 1844,* edited by Francis Darwin, 57–255. Cambridge: Cambridge University Press, 1909.

———. *The Life and Letters of Charles Darwin.* Edited by Francis Darwin. 2nd ed. 3 vols. New York: D. Appleton, 1887.

———. *More Letters of Charles Darwin.* Edited by Francis Darwin and A. C. Seward. 2 vols. London: J. Murray, 1903.

———. *Notebook B: Transmutation (1837–38).* CUL-DAR121. In *The Complete Work of Charles Darwin Online,* edited by John van Wyhe. Accessed February 2013. http://darwin-online.org.uk.

———. *Notebook C: Transmutation (1838).* CUL-DAR122. In *The Complete Work of Charles Darwin Online,* edited by John van Wyhe. Accessed February 2013. http://darwin-online.org.uk.

———. *Notebook E: Transmutation (1838–39).* CUL-DAR124. In *The Complete Work of Charles Darwin Online,* edited by John van Wyhe. Accessed February 2013. http://darwin-online.org.uk.

———. *On the Origin of Species by Means of Natural Selection, or the Preservation of Favoured Races in the Struggle for Life.* London: John Murray, 1859.

———. *On the Origin of Species by Means of Natural Selection, or the Preservation of Favoured Races in the Struggle for Life*. 2nd ed. London: John Murray, 1860.

———. *On the Origin of Species by Means of Natural Selection, or the Preservation of Favoured Races in the Struggle for Life*. 3rd ed. London: John Murray, 1861.

———. *On the Origin of Species by Means of Natural Selection, or the Preservation of Favoured Races in the Struggle for Life*. 4th ed. London: John Murray, 1866.

———. *On the Origin of Species by Means of Natural Selection, or the Preservation of Favoured Races in the Struggle for Life*. 5th ed. London: John Murray, 1869.

———. *On the Origin of Species by Means of Natural Selection, or the Preservation of Favoured Races in the Struggle for Life*. 6th ed. London: John Murray, 1876.

———. "Preliminary Notice." In Ernst Krause, *Erasmus Darwin*, translated by W. S. Dallas, 1–129. London: John Murray, 1879.

———. *Torn Apart Notebook (1839–41)*. CUL-DAR127. In *The Complete Work of Charles Darwin Online*, edited by John van Wyhe. Accessed February 2013. http://darwin-online.org.uk.

———. *The Variation of Animals and Plants under Domestication*. 2 vols. London: John Murray, 1868.

———. *The Variation of Animals and Plants under Domestication*. 2nd ed. 2 vols. London: John Murray, 1875.

Darwin, Erasmus. *The Botanic Garden: A Poem in Two Parts*. Lichfield, Staffordshire, England: J. Johnson, for J. Johnson, London, 1789.

———. "Extracts from *Zoonomia, or the Laws of Organic Life*." In *Romanticism and Science, 1773–1833*, edited by Tim Fulford, 5:24–44. London: Routledge, 2002.

———. *The Temple of Nature; or, The Origin of Society*. London: For J. Johnson by T. Bensley, 1803.

———. *Zoonomia, or the Laws of Organic Life*. 2 vols. London: J. Johnson, 1794–96.

Davy, Humphry. *The Collected Works of Sir Humphry Davy*. Edited by John Davy with an introduction by David Knight. New ed. 9 vols. Bristol: Thoemmes Press, 2001.

Dawkins, Richard. *The Extended Phenotype: The Long Reach of the Gene*. Oxford: Oxford University Press, 1989.

———. *The Selfish Gene*. 1976. 30th anniversary ed. Oxford: Oxford University Press, 2006.

Day, Troy, and Russell Bonduriansky. "A Unified Approach to the Evolutionary Consequences of Genetic and Nongenetic Inheritance." *American Naturalist* 178, no. 2 (2011): E18–36.

De Caus, Salomon. *Hortus Palatinus*. Frankfurt, 1620.

———. *Nouvelle invention de lever l'eau plus hault que sa source avec qualeques machines mouvantes par le moyen de l'eau, et uns discours de la conduit d'icelle*. London, 1644.

———. *Les raisons des forces mouvantes avec diverses machines tant utiles que plaisantes*. 3 vols. Frankfurt: J. Norton, 1615.

Decremps, Henri. *La magie blanche devoilée*. 1784. Paris: Desoer, 1789.

Dee, John. *Autobiographical Tracts of John Dee*. Edited by James Crossley. Whitefish, MT: Kessinger Publishing, 2005.

———. *The Mathematical Praeface to the Elements of Geometrie of Euclid of Megara*. 1570. Edited by Allen Debus. New York: Science History Publications, 1975.

Della Porta, Giambattista. *Pneumaticorum libri tres: Quibus accesserunt curvilineorum elementorum libri duo*. Naples: J. J. Carlinus, 1601.

De la Rivière, Pierre-Paul Mercier. *L'ordre natural et essentiel des sociétés politiques*. London: Jean Nourse, 1767.

Dellile, Jacques. *Épitre à M. Laurent . . . à l'occasion d'un bras artificiel qu'il a fait pour un soldat invalide*. 2nd ed. London, 1761.

Del Río, Martin. *Disquisitionum magicarum libri sex*. 1599–1600. Venice, 1640.

Dennett, Daniel C. "Artificial Intelligence as Philosophy and as Psychology." In *Brainstorms: Philosophical Essays on Mind and Psychology*, 109–26. Cambridge, MA: MIT Press, 1978.

———. "The Baldwin Effect: A Crane, Not a Skyhook." In *Evolution and Learning: The Baldwin Effect Reconsidered*, edited by Bruce H. Weber and David J. Depew, 69–106. Cambridge, MA: MIT Press, 2003.

———. *Brainchildren: Essays on Designing Minds*. Cambridge, MA: MIT Press, 1998.

———. *Content and Consciousness*. London: Routledge, 1969.

———. *Darwin's Dangerous Idea: Evolution and the Meanings of Life*. New York: Simon and Schuster, 1995.

———. *The Intentional Stance*. Cambridge, MA: MIT Press, 1987.

———. "Intentional Systems." In *Mind Design: Philosophy, Psychology, Artificial Intelligence*, edited by John Haugeland, 220–42. Cambridge, MA: MIT Press, 1981.

———. "The Logical Geography of Computational Approaches: A View from the East Pole." In *Brainchildren: Essays on Designing Minds*, 215–34. Cambridge, MA: MIT Press, 1998.

———. "Postscript [1985]." In *Brainchildren: Essays on Designing Minds*, 21–25. Cambridge, MA: MIT Press, 1998.

———. "Real Consciousness." In *Brainchildren: Essays on Designing Minds*, 131–40. Cambridge, MA: MIT Press, 1998.

Derham, William. *Physico-Theology; or, a Demonstration of the Being and Attributes of God*. London, 1713.

Descartes, René. *Le monde, l'homme*. Edited by Annie Bitbol-Hespériès and Jean-Pierre Verdet. Paris: Éditions de Seuil, 1996.

———. *Œuvres de Descartes*. Edited by Charles Adam and Paul Tannery. 11 vols. Paris: J. Vrin, 1974–89. (AT)

———. *The Philosophical Writings of Descartes*. Edited by John Cottingham, Robert Stoothoff, and Dugald Murdoch. 3 vols. Cambridge: Cambridge University Press, 1984–91.

Desfontaines, Pierre. "Lettre CLXXX sur le Flûteur automate et l'Aristipe moderne." *Observations sur les écrits modern* 12 (March 30, 1738): 337–42.

Deslisles de Sales, Jean-Baptiste-Claude. *De la philosophie de la nature*. 1769. Amsterdam: Arkstée et Merkus, 1770–74.

Desmarquets, Jean-Antoine-Samson. *Mémoires chronologiques pour server à l'histoire du Dieppe et à celle de la navigation françoise*. 2 vols. Paris: Desauges, 1785.

De Vries, Hugo. *Intracellulare Pangenesis*. Jena, Germany: G. Fischer, 1889.

———. *Intracellular Pangenesis: Including a Paper on Fertilization and Hybridization*. 1889. Translated by C. Stuart Gager. Chicago: Open Court, 1910.

———. *Die Mutationstheorie. Versuche und Beobachtungen über die Entstehung von Arten im Pflanzenreich.* 2 vols. Leipzig: Veit, 1901–3.

———. *The Mutation Theory: Experiments and Observations on the Origin of Species in the Vegetable Kingdom.* Translated by J. B. Farmer and A. D. Darbishire. 2 vols. Chicago: Open Court, 1909–10.

De Weck, Olivier L., and Daniel Roos, Christopher L. Magee. *Engineering Systems: Meeting Human Needs in a Complex Technological World.* Cambridge, MA: The MIT Press, 2011.

Dictionnaire de l'Académie française. 1st ed. Paris, 1694.

Dictionnaire de l'Académie française. 4th ed. Paris, 1762.

Dictionnaire de l'Académie française. 5th ed. Paris, 1798.

Dictionnaire de l'Académie française. 6th ed. Paris, 1832–35.

Diderot, Denis. *Eléments de physiologie.* 1784. Critical ed., with an introduction by Jean Mayer. Paris: Librairie M. Didier, 1964.

———. *Entretien entre Diderot et d'Alembert.* 1784. In *Œuvres philosophiques,* 257–84.

———. *Lettre sur les sourds et muets.* 1751. Edited by P. H. Meyer. Geneva: Droz, 1965.

———. *Œuvres complètes de Diderot.* Edited by Jean Assézat and Maurice Tourneux. 20 vols. Paris: Garnier frères, 1875–77.

———. *Œuvres philosophiques.* Edited by Paul Vernière. Paris: Garnier frères, 1961.

———. *Le rêve de d'Alembert.* 1784. In *Œuvres philosophiques,* 285–371.

———. *Suite de l'entretien.* In *Œuvres philosophiques,* 372–85.

Diderot, Denis, and Jean d'Alembert, eds. *L'Encyclopédie, ou, Dictionnaire raisonné des sciences, des arts, et des métiers, par une societé des gens de lettres.* Paris, 1751–72.

Digby, Kenelm. *Two Treatises. In the One of Which, the Nature of Bodies; in the Other, the Nature of Mans Soule; Is Looked Into: In Way of Discovery, of the Immortality of Reasonable Soules.* Paris: Gilles Blaizot, 1644.

Dodart, Denys. "Suite de la première partie du Supplément." In *Année 1706: Mémoires* of *Histoire de l'Académie royale des sciences,* 388–410. Amsterdam: Pierre de Coup, 1706.

———. "Supplément au Mémoire sur la voix et sur les tons." In *Année 1706: Mémoires* of *Histoire de l'Académie royale des sciences,* 136–48. Amsterdam: Pierre de Coup, 1706.

———. "Supplément au Mémoire sur la voix et sur les tons." In *Année 1707: Mémoires* of *Histoire de l'Académie royale des science,* 66–81. Amsterdam: Pierre de Coup, 1707.

———. "Sur les causes de la voix de l'homme et de ses différens tons." In *Année 1700: Mémoires* of *Histoire de l'Académie royale des sciences,* 244–93. Amsterdam: Gerard Kuyper, 1700.

Doppelmayr, Johann Gabriel. *Historische Nachricht von den nürnbergischen mathematicis und künstlern.* Nuremberg, Germany: P. C. Monath, 1730.

Dreyfus, Hubert L. "Alchemy and Artificial Intelligence." RAND paper P-3244. Santa Monica, CA: The RAND Corporation, 1965.

———. *What Computers (Still) Can't Do.* 1972. Cambridge, MA: MIT Press, 1992.

Driesch, Hans. *Analytische Theorie der organischen Entwicklung.* Leipzig: Wilhelm Engelmann, 1894.

———. *Die Biologie als selbständige Grundwissenschaft und das System der Biologie: Ein Beitrag zur Logik der Naturwissenschaften.* 1893. Leipzig: W. Engelmann, 1911.

———. "Entwicklungmechanische Studien. I. Der Werth der beiden ersten Furchungszellen in der Echinodermentwicklung. Experimentelle Erzeugen von Theil und Doppelbildung." *Zeitschrift für wissenschaftliche Zoologie* 53 (1891): 160–78.

———. *Lebenserinnerungen: Aufzeichnungen eines Forschers und Denkers in entscheidender Zeit*. Basel: Ernst Reinbardt, 1951.

———. "Die Lokalisation morphogenetischer Vorgänge: Ein Beweis vitalistischen Geschehens." *Archiv für Entwicklungsmechanik* 8 (1899): 35–111.

———. "Die Maschinentheorie des Lebens: Ein Wort zur Aufklärung." *Biologisches Centralblatt* 16, no. 9 (1986): 353–68.

———. *Die mathematisch-mechanische Betrachtung morphologischer Probleme der Biologie. Ein kritische Studie*. Jena, Germany: G. Fischer, 1891.

———. *The Problem of Individuality: A Course of Four Lectures Delivered Before the University of London*. London: Macmillan, 1914.

———. *The Science and Philosophy of the Organism: The Gifford Lectures Delivered before the University of Aberdeen in the Year 1907–1908*. 2 vols. London: Adam and Charles Black, 1908.

Driesch, Hans, and T.H. Morgan. "Zur Analysis der ersten Entwickelungsstadien des Ctenophoreneies." *Archiv für Entwicklungsmechanik* 2, no. 2 (1895): 204–24.

Du Bois-Reymond, Emil Heinrich. "Die sieben Welträthsel." In *Reden*, edited by Estelle du Bois-Reymond, 1:381–411. Leipzig: Veit, 1912.

———. "Über die Grenzen des Naturerkennens." (1872). In *Reden*, edited by Estelle du Bois-Reymond, 1: 441–73. Leipzig: Veit, 1912.

———. "Über die Lebenskraft." In *Reden*, edited by Estelle Du Bois-Reymond, 1:1–26. Leipzig: Veit, 1912.

Dubos, Jean-Baptiste. *Réflexions critiques sur la poésie et sur la peinture*. Paris: P.-J. Mariette, 1733.

Du Camp, Maxime. *Paris, ses organes, ses fonctions et sa vie dans la seconde moitié du XIXe siècle*. Paris: Hachette, 1873.

Du Chesne, André. *Les antiquités et recherches des villes, chasteaux, et places plus remarquables de toute la France*. Paris: Pierre Rocolet, 1637.

Ducrocq, Albert. *Logique générale des systèmes et des effets: Introduction à une physique des effets fondements de l'intellectique*. Paris: Dunod, 1960.

Eldredge, Niles. *Reinventing Darwin: The Great Debate at the High Table of Evolutionary Theory*. New York: John Wiley, 1995.

Evelyn, John. *Diary and Correspondence of John Evelyn, F.R.S.*. Edited by William Bray and John Forster. 4 vols. London: George Bell and Sons, 1884–84.

———. *Elysium Britannicum, or The Royal Gardens*. 1660. Edited by John E. Ingram. Philadelphia: University of Pennsylvania Press, 2001.

Fernel, Jean. *Jean Fernel's On the Hidden Causes of Things: Forms, Souls, and Occult Diseases in Renaissance Medicine*. Translated by John M. Forrester with an introduction and annotations by John Henry and John M. Forrester. Leiden: Brill, 2005.

———. *The Physiologia of Jean Fernel (1567)*. Translated and annotated by John M. Forrester with an introduction by John Henry and John M. Forrester. Philadelphia: American Philosophical Society, 2003.

Finocchiaro, Maurice, ed. *The Galileo Affair: A Documentary History*. Berkeley: University of California Press, 1989.

Figuier, Louis. *Les merveilles de la science, ou Description populaire des inventions modernes*. Paris: Jouvet, 1867.

Flemming, Walther. "Beiträge zur Kenntnis der Zelle und ihrer Lebenserscheinungen Theil II." *Archiv für Mikroskopische Anatomie* 18 (1880): 151–259.

———. "Contributions to the Knowledge of the Cell and Its Vital Processes: Part II." *Journal of Cell Biology* 25, no. 1 (1965): 1–69.

Fodor, Jerry. "The Appeal to Tacit Knowledge in Psychological Explanation." *Journal of Philosophy* 65 (1968): 625–40.

Fodor, Jerry, and Massimo Piattelli-Palmarini. *What Darwin Got Wrong*. New York: Farrar, Straus and Giroux, 2010.

Le fontane di Roma nelle piazze e lvoghi pvblici della città: Con li loro prospetti, come sono al presente. Part 2, *Le fontane delle ville di Frascati nel Tusculano*. Rome: G. G. de Rossi, 1691.

Fontenay, Louis-Abel. "Notice sur le P. Truchet." In *Dictionnaire des artistes*. Paris, 1772.

Fontenelle, Bernard de. *Œuvres diverses de M. de Fontenelle*. 3 vols. Amsterdam: Pierre Mortier, 1701.

———. *Suite des éloges des académiciens de l'Académie Royale des sciences*. Paris: Osmont, 1733.

———. "Sur la formation de la voix." In *Année 1700* of *Histoire de l'Académie royale des sciences*, 17–24. Amsterdam: Gerard Kuyper, 1700.

———. "Sur la formation de la voix." In *Année 1706* of *Histoire de l'Académie royale des sciences*, 136–48. Amsterdam: Pierre de Coup, 1706.

———. "Sur la formation de la voix." In *Année 1707* of *Histoire de l'Académie royale des sciences*, 18–20. Amsterdam: Pierre de Coup, 1707.

Franklin, Benjamin. "Dissertation on Liberty and Necessity." 1725. In *The Papers of Benjamin Franklin*, edited by Barbara B. Oberg et al., 1:57–71. New Haven, CT: Yale University Press, 1959–.

Frederick II. "Eulogy on Julien Offray de la Mettrie." 1751. In *Man a Machine*, edited and translated by Gertrude Carman Bussey, 1–10. Chicago: Open Court, 1912.

———. *Œuvres de Frédéric le Grand*. Edited by J. D. E. Preuss. 31 vols. Berlin: Imprimerie royale, 1846–57.

Fréron, Elie-Catherine. *L'année littéraire ou, Suite des lettres sur quelques écrits de ce temps*. Amsterdam: Michel Lambert, 1754.

Freud, Sigmund. *Jokes and Their Relation to the Unconscious*. 1905. Translated by James Strachey. New York: W. W. Norton, 1963.

Galen. *On the Natural Faculties*. Translated by Arthur John Brock. Cambridge, MA: Harvard University Press, 1916.

———. *On the Usefulness of the Parts of the Body*. Translated by Margaret Tallmadge May. Ithaca, NY: Cornell University Press, 1968.

Galileo Galilei. *Sidereus Nuncius, or, The Sidereal Messenger*. 1610. Translated by Albert Van Helden. Chicago: University of Chicago Press, 1989.

Galvani, Luigi. *Commentary on the Effects of Electricity on Muscular Motion*. 1791. Translated by Margaret Glover Foley with an introduction by I. Bernard Cohen. Norwalk, CT: Burndy Library, 1953.

Gassendi, Pierre. *Opera omnia*. 6 vols. Lyon: 1658–75.

Gennrich, Friedrich, ed. *Les romans de la Dame à la Lycorne et du beau Chevalier au Lyon*. Dresden: Gedruckt für die Gesellschaft für romanische Literatur, 1908.

Giorgi da Urbino, Alessandro. *Spirituali di Herone Alessandrino*. 1592. Urbino, 1595.

Gissis, Snait B. and Eva Jablonka, eds. *Transformations of Lamarckism: From Subtle Fluids to Molecular Biology*. Cambridge, MA: MIT Press, 2011.

Godfrey-Smith, Peter. *Darwinian Populations and Natural Selection*. Oxford: Oxford University Press, 2009.

Goethe, Johann Wolfgang von. *Annals, or Day and Year Papers*. Translated by Charles Nisbet with an introduction by Edward Dowden. Revised ed. New York: Colonial Press, 1901.

———. "A Commentary on the Aphoristic Essay 'Nature.'" In *Scientific Studies*, edited and translated by Douglas Miller, 6–7. Princeton, NJ: Princeton University Press, 1995.

———. "Excerpt from Studies for a Physiology of Plants." In *Scientific Studies*, edited and translated by Douglas Miller, 73–75. Princeton, NJ: Princeton University Press, 1995.

———. *Goethe's Botanical Writings*. Translated by Bertha Mueller. Woodbridge, CT: Ox Bow Press, 1989.

———. *Goethes Briefe*. Edited by Karl Robert Mandelkow and Bodo Morawe. 4 vols. Hamburg: Wegner, 1968–76.

———. *Goethe's Collected Works*. Edited and translated by Douglas Miller et al. 12 vols. Princeton, NJ: Princeton University Press, 1994–.

———. *The Metamorphosis of Plants*. Edited by Gordon L. Miller and translated by Douglas Miller. Cambridge, MA: MIT Press, 2009.

———. *Werke*. Edited by Grand Duchess Sophie of Saxony, Gustav von Loeper, and Paul Raabe. 152 vols. Weimar: H. Böhlau, 1887–.

Goette, Alexander. *Die Entwickelungsgeschichte der Unke (Bombinator igneus) als Grundlage einer vergleichenden Morphologie der Wirbelthiere*. Leipzig: Voss, 1875.

Goodwin, Brian. *How the Leopard Changed Its Spots: The Evolution of Complexity*. Princeton, NJ: Princeton University Press, 2001.

Gorelick, Root, Manfred Laublicher, and Rachel Massicotte. "Asexuality and Epigenetic Variation." In *Epigenetics: Linking Genotype and Phenotype in Development and Evolution*, edited by Benedikt Hallgrimsson and Brian K. Hall, 87–102. Berkeley: University of California Press, 2011.

Gorczynski, Reginald M., and Edward J. Steele. "Inheritance of Acquired Immunological Tolerance to Foreign Histocompatibility Antigens in Mice." *Proceedings of the National Academy of Sciences* 77 (1980): 2871–75.

Gould, Stephen Jay. "Darwinian Fundamentalism." *New York Review of Books* (June 12, 1997): 34–37.

———. "The Pleasures of Pluralism." *New York Review of Books* (June 26, 1997): 47–52.

———. *The Structure of Evolutionary Theory*. Cambridge, MA: Harvard University Press, 2002.

Gould, Stephen Jay, and Niles Eldredge. "Punctuated Equilibria: An Alternative to Phyletic Gradualism." In *Models in Paleobiology*, edited by T. J. M. Schopf, 82–115. San Francisco: Freeman Cooper, 1972.

———. "Punctuated Equilibria: The Tempo and Mode of Evolution Reconsidered." *Paleobiology* 3, no. 2 (1977): 115–51.

Gould, Stephen Jay, and Richard C. Lewontin. "The Spandrels of San Marco and the Panglossian Paradigm: A Critique of the Adaptationist Programme." *Proceedings of the Royal Society of London, Series B* 205, no. 1161 (1979): 581–98.

Grave, Jacques Salverda de, ed. *Eneas, roman du XIIe siècle*. Halle, Germany: Max Niemayer, 1891.

Gréban, Arnoul, and Simon Gréban. *Mystère des actes des apôtres, representé à Bourges en avril 1536, publié depuis le manuscript original*. Edited by Auguste-Théodore, Baron de Girardot. Paris: Didron, 1854.

Grew, Nehemiah. *Cosmologia sacra: Or a discourse of the universe as it is the creature and Kingdom of God*. London, 1701.

Grunau, Christoph. "Methylation Mapping in Humans." In *Epigenetics: Linking Genotype and Phenotype in Development and Evolution*, edited by Benedikt Hallgrimsson and Brian K. Hall, 70–86. Berkeley: University of California Press, 2011.

Guenther, Konrad. *Darwinism and the Problems of Life: A Study of Familiar Animal Life*. 1904. Translated by Joseph McCabe. London: Owen, 1906.

———. *Der Darwinismus und die Probleme des Lebens, zugleich eine Einführung in das einheimische Tierleben*. Freiburg im Breisgau, Germany: F. E. Fehsenfeld, 1904.

Guessard, François and Charles de Grandmaison, eds. *Huon de Bordeaux. Chanson de geste*. Paris: F. Vieweg, 1860.

Haeckel, Ernst. *The Evolution of Man*. 1874. Translated by Joseph McCabe. 2 vols. London: Watts, 1905.

———. *Generelle Morphologie der Organismen*. Berlin: Georg Reimer, 1866.

———. *The History of Creation, or The Development of the Earth and Its Inhabitants by the Actions of Natural Causes: A Popular Exposition of the Doctrine of Evolution in General, and of that of Darwin, Goethe, and Lamarck in Particular*. 1868. Translated by E. Ray Lankester. 6th ed. 2 vols. New York: Appleton, 1914.

———. "Über die heutige Entwickelungslehre im Verhältnisse zur Gesamtwissenschaft." In *Amtlicher Bericht der 50. Versammlung Deutscher Naturforscher und Ärzte*, 14–22. Munich: F. Straub, 1877.

———. *Ziele und Wege der heutigen Entwickelungsgeschichte*. Jena, Germany: Hermann Dufft, 1875.

Haig, David. "The Gene Meme." In *Richard Dawkins: How a Scientist Changed the Way We Think*, edited by A. Grafen and M. Ridley, 50–61. Oxford: Oxford University Press, 2006.

———. "The Social Gene." In *Behavioural Ecology: An Evolutionary Approach*, 4th ed., edited by John R. Krebs and Nicholas B. Davies, 284–304. Oxford, UK: Blackwell, 1997.

———. "The Strategic Gene." *Biology and Philosophy* 27, no. 4 (2012): 461–79.

———. "What Is a Marmoset?" *American Journal of Primatology* 49 (1999): 285–96.

Haldane, J. B. S. *The Causes of Evolution*. 1932. New ed. Princeton, NJ: Princeton University Press, 1990.

Haller, Albrecht von. *Elementa physiologiae corporis humani*. 8 vols. Lausanne: Apud Julium Henricum, 1757–78.

———. *The Natural Philosophy of Albrecht von Haller*. Edited by Shirley A. Roe. New York: Arno Press, 1981.

———. "Preface" in Georges Louis Leclerc, comte de Buffon, *Allgemeine Historie der Natur: Nach allen ihren besondern Theilen abgehandelt, nebst einer Beschreibung der Naturalienkammer Sr. Majestät des Königes von Frankreich*. 8 vols. Hamburg: Grund and Holle, 1750–74.

Hallgrimsson, Benedikt, and Brian K. Hall, eds. *Epigenetics: Linking Genotype and Phenotype in Development and Evolution*. Berkeley: University of California Press, 2011.

Harvey, William. *An Anatomical Disquisition on the Motion of the Heart*. In *The Works of William Harvey, M.D.*, translated by Robert Willis, 9–86. London: Sydenham Society, 1847.

———. *Anatomical Exercises on the Generation of Animals, to Which Are Added Essays on Parturition; on the Membranes, and Fluids of the Uterus; and on Conception*. In *The Works of William Harvey, M.D.*, translated by Robert Willis, 143–586. London: Sydenham Society, 1847.

———. *De motu locali animalium, 1627*. Edited and translated by Gweneth Whitteridge. Cambridge, UK: For the Royal College of Physicians at the University Press, 1959.

———. *Exercitationes de generatione animalium*. London, 1651.

———. *Lectures on the Whole of Anatomy: An Annotated Translation of Prelectiones anatomiae universalis*. Edited and translated by C. D. O'Malley, F. N. L. Poynter and K. F. Russell. Berkeley: University of California Press, 1961.

———. *A Second Disquisition to John Riolan, Jun., in Which Many Objections to the Circulation of the Blood are Refuted*. In *The Works of William Harvey, M.D.*, translated by Robert Willis, 107–41. London: Sydenham, Society, 1847.

———. *The Works of William Harvey, M.D.* Translated by Robert Willis. London: Sydenham Society, 1847.

Haugeland, John. *Artificial Intelligence: The Very Idea*. Cambridge, MA: MIT Press, 1985.

Hayles, N. Katherine. *How We Became Posthuman: Virtual Bodies in Cybernetics, Literature, and Informatics*. Chicago: University of Chicago Press, 1999.

Heine, Heinrich. *Selected Works*. Edited and translated by Helen M. Mustard and Max Knight. New York: Random House, 1973.

Helvétius, Claude-Adrien. *Correspondance générale d'Helvétius*. Edited by Alan Dainard et al. 5 vols. Toronto: University of Toronto Press, 1981–2004.

———. *De l'esprit*. Paris: Durand, 1758.

———. *De l'homme, de ses facultés intellectuelles et de son education*. London: Société typographique, 1773.

Helmholtz, Hermann von. *Helmholtz on Perception, Its Physiology, and Development*. Edited by Richard Warren and Roslyn P. Warren. New York: John Wiley, 1968.

———. *Die Lehre von den Tonempfindungen als physiologische Grundlage für die Theorie der Musik*. Braunschweig, Germany: Friedrich Vieweg und Sohn, 1863.

———. *Die neueren Fortschritte in der Theorie des Sehens*. Berlin, 1868.

———. "On the Conservation of Force (1862–63)." Translated by Edmund Atkinson. In

Science and Culture: Popular and Philosophical Essays, edited with an introduction by David Cahan, 96–126. Chicago: University of Chicago Press, 1995.

———. "On the Interaction of the Natural Forces (1854)." Translated by Edmund Atkinson. In *Science and Culture: Popular and Philosophical Essays*, edited with an introduction by David Cahan, 18–45. Chicago: University of Chicago Press, 1995.

———. *On the Sensations of Tone as a Physiological Basis for the Theory of Music*. Translated by Alexander J. Ellis. New York: Dover, 1954.

———. "The Recent Progress of the Theory of Vision (1868)." Translated by Edmund Atkinson. In *Science and Culture: Popular and Philosophical Essays*, edited with an introduction by David Cahan, 127–203. Chicago: University of Chicago Press, 1995.

———. *Über die Erhaltung der Kraft. Eine physikalische Abhandlung*. Berlin: Reimer, 1847.

Herbert of Cherbury, Lord Edward. *Life and Reign of King Henry the Eighth, Together with Which Is Briefly Represented a General History of the Times*. London: Mary Clark, 1683.

Herder, Johann Gottfried. *Outlines of a Philosophy of the History of Man*. 1784. Translated by T. Churchill. New York: Bergman, 1966.

Hérouard, Jean. *Journal de Jean Hérouard*. Edited by Madeleine Foisil. 2 vols. Paris: Fayard, 1989.

Hippocrates. *Hippocratic Writings*. Edited by G. E. R. Lloyd. London: Penguin, 1978.

His, Wilhelm. *Über die Aufgaben und Zielpunkte der wissenschaftlichen Anatomie*. Leipzig: F. C. W. Vogel, 1872.

———. *Über die Bedeutung der Entwickelungsgeschichte für die Auffassung der organischen Natur*. Leipzig: F. C. W. Vogel, 1870.

———.*Über mechanische Grundvorgänge thierischer Formenbildung*. Archiv für Anatomie und Physiologie. Anatomische Abtheilung. Leizig: Veit and Comp., 1894.

———. *Unsere Körperform und das physiologische Problem ihrer Entstehung. Briefe an einen befreundeten Naturforscher*. 1874. Leibniz: F. C. W. Vogel, 1875.

———. *Untersuchungen über die erste Anlage des Wirbelthierleibes: Die erste Entwickelung des Hühnchens im Ei*. Leipzig: F. C. W. Vogel, 1868.

Hobbes, Thomas. *Leviathan, Or, The Matter, Forme and Power of a Common-Wealth, Ecclesiaticall and Civill*. London: Andrew Crooke, 1651.

Hoffmann, E.T.A. "Automata." In *The Best Tales of Hoffmann*, edited by E. F. Bleiler, 71–103. New York: Dover, 1967.

Hoffmann, Professor [Angelo John Lewis]. *Modern Magic: A Practical Treatise on the Art of Conjuring*. 2nd ed. London: Routledge, 1877.

Hofmann, August Wilhelm von. *The Question of a Division of the Philosophical Faculty: Inaugural Address on Assuming the Rectorship of the University of Berlin*. 2nd ed. Boston: Ginn, Heath, 1883.

Holbach, Paul Henri Thiry, baron d'. *Système de la nature, ou des loix du monde physique et du monde moral*. London, 1771.

Holland, John. *Adaptation in Natural and Artificial Systems: An Introductory Analysis with Applications to Biology, Control, and Artificial Intelligence*. 1975. Cambridge, MA: MIT Press, 1992.

Hollmann, Samuel Christian. Untitled letter. *Göttingische Zeitungen von gelehrten Sachen* (May 1748): 409–12.

Hone, William. *The Every-Day Book and Table Book.* London: William Tegg, 1825–27.

Honnecourt, Villard de. *The Sketchbook of Villard de Honnecourt.* Edited by Theodore Bowie. Bloomington: Indiana University Press, 1959.

Hooke, Robert. *Micrographia, Or Some Physiological Descriptions of Minute Bodies Made by Magnifying Glasses with Observations and Inquiries Thereupon.* London: Royal Society, 1665.

Houck, Marilyn A., Jonathan B. Clark, Kenneth R. Peterson, and Margaret G. Kidwell. "Possible Horizontal Transfer of Drosophilia Genes by the Mite *Proctolaelaps Regalis.*" *Science* 253 (1991): 1125–29.

Houdard, Georges Louis. *Les Châteaux Royaux de Saint-Germain-en-Laye, 1124–1789.* 2 vols. Saint-Germain-en-Laye, France: M. Mirvault, 1909–11.

Humboldt, Alexander von. *Expériences sur le galvanisme, et en général sur l'irritation des fibres musculaires et nerveuses.* Paris: Didot jeune, 1799.

Hume, David. *Dialogues concerning Natural Religion: And Other Writings.* Cambridge: Cambridge University Press, 2007.

———. *An Enquiry concerning Human Understanding.* 1748. *The Clarendon Edition of the Works of David Hume,* edited by Tom L. Beauchamp. Oxford: Oxford University Press, 2000.

———. *Principal Writings on Religion.* With an introduction by J. C. A. Gaskin. Oxford: Oxford University Press, 1993.

———. *A Treatise of Human Nature.* 1762. Edited by L. A. Selby-Bigge. Oxford, UK: Clarendon, 1978.

Humphrey, Nicholas, and Daniel C. Dennett. "Speaking for Our Selves." In *Brainchildren: Essays on Designing Minds,* by Daniel C. Dennett, 31–56. Cambridge, MA: MIT Press, 1998.

Huxley, Julian. *Evolution: The Modern Synthesis.* 1942. New ed. Cambridge, MA: MIT Press, 2010.

Huxley, Thomas Henry. *Darwiniana: Essays.* New York: Appleton, 1896.

———. *Lectures and Essays.* New York: Macmillan, 1902.

———. *Method and Results.* Vol. 1 of *Collected Essays.* London: Macmillan, 1894.

———. "On the Hypothesis That Animals Are Automata." 1874. In *Collected Essays,* 1:199–250. London: Macmillan, 1894.

———. "On the Physical Basis of Life." In *The Fortnightly Review,* n.s., 5, no. 26 (February 1, 1869): 129–45.

———. "On the Reception of The Origin of Species." 1887. In *The Life and Letters of Charles Darwin,* edited by Francis Darwin, 1:533–58. New York: D. Appleton, 1896.

Huygens, Christiaan. *The Celestial Worlds Discover'd: Or, Conjectures concerning the Inhabitants, Plants and Productions of the Worlds in the Planets.* London, 1698.

———. *Œuvres complètes.* 22 vols. The Hague: M. Nikhoff, 1888.

Jablonka, Eva, and Marion J. Lamb. *Evolution in Four Dimensions: Genetic, Epigenetic, Behavioral, and Symbolic Variation in the History of Life.* Cambridge, MA: MIT Press, 2005.

———. "The Expanded Evolutionary Synthesis—A Response to Godfrey-Smith, Haig, and West-Eberhard." *Biology and Philosophy* 22 (2007): 453–72.

Jablonka, Eva and Gal Raz. "Transgenerational Epigenetic Inheritance: Prevalence, Mechanisms, and Implications for the Study of Heredity and Evolution." *Quarterly Review of Biology* 84, no. 2 (2009): 131–76.

James, William. *The Principles of Psychology*. 2 vols. New York: Henry Holt. 1890.

Jaquet-Droz, Pierre. *Les œuvres des Jaquet-Droz. Montres, pendules et automates*. La Chaux-de-Fonds, Switzerland: Courvoisier, 1971.

Jazari, Ismail ibn al-Razzaz. *The Book of Knowledge of Ingenious Mechanical Devices*. Translated by Donald R. Hill. Dordrecht, the Netherlands: Reidel, 1974.

Johannsen, Wilhelm. "The Genotype Conception of Heredity." *American Naturalist* 45, no. 531 (March 1911): 129–59.

Johnson, Samuel. "The Life of Dr. Hermann Boerhaave, con't." *Gentleman's Magazine* 9, no. 2 (February 1739): 72–73.

Jones, Henry Festing. *Samuel Butler, Author of Erewhon (1835–1902): A Memoir*. 2 vols. London: Macmillan, 1919.

Kant, Immanuel. *Correspondence*. Edited and translated by Arnulf Zweig. *The Cambridge Edition of the Works of Immanuel Kant*. Cambridge: Cambridge University Press, 1999.

———. *Critique of Judgement*. 1790. Translated by James Creed Meredith and edited by Nicholas Walker. Oxford: Oxford University Press, 2007.

———. *Critique of Practical Reason*. 1788. Translated by Werner S. Pluhar with an introduction by Stephen Engstrom. Indianapolis, IN: Hackett, 2002.

———. *Critique of Pure Reason*. 1781. Edited and translated by Paul Guyer. Cambridge: Cambridge University Press, 1999.

———. *Der einzig mögliche Beweisgrund zu einer demonstration des Daseyns Gottes*. Königsberg: Johann Jakob Kanter, 1763.

———. *Gedanken von der wahren Schätzung der lebendigen Kräfte*. Königsberg, 1746.

———. *Gesammelte Schriften* (Akademie-Ausgabe). Edited by the Königlish Preußischen (later Deutschen, and currently Berlin-Brandenburgischen) Akademie der Wissenschaften. 29 volumes. Berlin: Georg Reimer, later Walter de Gruyter, 1902–. (AE)

———. *Immanuel Kant in Rede und Gespräch*. Edited by Rudolf Malter. Hamburg: Felix Meiner Verlag, 1990.

———. *Kritik der Urteilskraft*. 1790. In *Gesammelte Schriften (Akademie-Ausgabe)*, edited by Wilhelm Windelband, 5:167–485. Berlin: De Gruyter, 1922.

———. *Universal Natural History and Theory of the Heavens*. 1755. Translated by Ian Johnston. Arlington, VA: Richer Resources, 2009.

Keddie, William, ed. *Anecdotes, Literary and Scientific: Illustrative of the Characters, Habits, and Conversation of Men of Letters and Science*. New York: Routledge, 1873.

Keller, Evelyn Fox, and L. Segel. "Initiation of Slime Mold Aggregation Viewed as an Instability." *Journal of Theoretical Biology* 26 (1970): 399–415.

Kempelen, Wolfgang von. *Le mécanisme de la parole, suivi de la description d'une machine parlante*. Vienna: J. V. Degen, 1791.

Kennedy, Elspeth, trans. *Lancelot du Lac: The Non-Cyclic Old French Prose Romance*. Oxford, UK: Clarendon Press, 1980.

Kepler, Johannes. *Ad vitellionem paralipomena, quibus astronomiae pars optica traditur*. Frankfurt, Germany, 1604.

Kircher, Athanasius. *Musurgia Universalis*. 2 vols. Rome: Francisi Corbelletti, 1650.

———. *Oedipus Aegyptiacus, hoc est, Vniuersalis hieroglyphicae veterum doctrinae temporum iniuria abolitae instauratio*. 3 vols. Rome: Ex typographia Vitalis Mascardi, 1652–54.

Knudsen, Thorbjørn. "Nesting Lamarckism within Darwinian Explanations: Necessity in Economics and Possibility in Biology?" In *Darwinism and Evolutionary Economics*, edited by John Laurent and John Nightingale, 121–59. Cheltenham, UK: Elgar, 2001.

Krebs, H.A. "Excursion into the Borderland of Biochemistry and Philosophy." *Bulletin of the Johns Hopkins Hospital* 95 (1954): 45–51.

Kurzweil, Ray. *The Age of Intelligent Machines*. Cambridge, MA: MIT Press, 1990.

Kussi, Peter, ed. *Toward the Radical Center: A Karel Čapek Reader*. Highland Park, NJ: Catbird Press, 1990.

Labat, Jean-Baptiste. *Nouveau voyage aux isles de l'Amérique*. 6 vols. The Hague: P. Husson, 1724.

Laborde, Léon. *Les ducs de Bourgogne, études sur les lettres, les arts, et l'industrie pendant le XVe siècle et plus particulièrement dans les Pays-Bas et le duché de Bourgogne*. Second part, 3 vols. Paris: Plon frères, 1849–52.

La Harpe, Jean-François de. *Philosophie du dix-huitième siècle*. 2 vols. Paris: Déterville, 1818.

Lakoff, George. *Women, Fire and Dangerous Things: What Categories Reveal About the Mind*. Chicago: University of Chicago Press, 1990.

Lakoff, George, and Rafael E. Nuñez. *Where Mathematics Comes From: How the Embodied Mind Brings Mathematics into Being*. New York: Basic Books, 2000.

Lamarck, Jean-Baptiste. "Espèce." In *Nouveaux dictionnaire d'histoire naturelle*, 10:441–51. Paris: Déterville, 1817.

———. *Histoire naturelle des animaux sans vertèbres*. Paris: Déterville, 1801.

———. *Hydrogéologie, ou Recherches sur l'influence qu'ont les eaux sur la surface du globe terrestre*. Paris: Chez l'auteur, 1802.

———. "Idée." In *Nouveaux dictionnaire d'histoire naturelle*, 16:78–94. Paris: Déterville, 1817.

———. *Mémoires de physique et de l'histoire naturelle*. Paris: Chez l'auteur, 1797.

———. *Philosophie zoologique, ou, Exposition des considérations relative à l'histoire naturelle des animaux*. 2 vols. Paris: Dentu, 1809.

———. *Recherches sur les causes des principaux faits physiques*. 2 vols. Paris: Maradan, 1794.

———. *Recherches sur l'organisation des corps vivans*. Paris: Chez l'auteur et Maillard, 1802.

———. *Système analytique des connaissances positives de l'homme, restreintes à celles qui proviennent directement ou indirectement de l'observation*. Paris: Chez l'auteur et Belin, 1820.

———. *Système des animaux sans vertèbres*. Paris: Déterville, 1801.

Lambarde, William. *Dictionarium Angliæ topographicum & historicum: An Alphabetical Description of the Chief Places in England and Wales; with an Account of the Most Memorable Events Which Have Distinguish'd Them*. London: F. Gyles, 1730.

———. *A Perambulation of Kent*. Edited by Richard Church. Bath, UK: Adams and Dart, 1970.

La Mettrie, Julien Offray de. *Histoire naturelle de l'âme*. The Hague: J. Neulme, 1745.

———. *L'homme machine*. 1747. In *L'Homme Machine: A Study in the Origins of an Idea*, edited by Aram Vartanian, 139–97. Princeton, NJ: Princeton University Press, 1960.

———. *Œuvres philosophiques*. 2 vols. Paris: Fayard, 1984–87.

———. *Système d'Epicure*. 1750. In *Œuvres philosophiques*, 1:351–86.

Lamy, Guillaume. *Discours anatomiques: Explication méchanique et physique des fonctions de l'âme sensitive.* 1677. Edited by Anna Minerbi Belgrado. Paris: Voltaire Foundation, 1996.

La Touche Boesnier, Pierre de. *Préservatif contre l'irreligion, ou démonstrations des véritex fondamentales de la religion chrétienne.* The Hague, 1707.

Lavater, Johann Caspar. *Lavater's Essays on Physiognomy.* 1775–78. London, 1797.

Lawrence, William. *Lectures on Physiology, Zoology, and the Natural History of Man.* 3rd ed. London: J. Smith, 1823.

Leeuwenhoek, Antonie van. Alle de brieven van Antoni van Leeuwenhoek. Amsterdam: N.v. Swets and Zeitlinger, 1939.

Leibniz, Gottfried Wilhelm. *Discours de métaphysique.* Edited by H. Lestienne. Paris: Félix Alcan, 1907.

———. "Leibniz on His Calculating Machine." Translated by Mark Kormes. In *A Source Book in Mathematics*, edited by David Eugene Smith, 173–81. New York: Dover, 1959.

———. *Leibniz's "New System" and Associated Contemporary Texts.* Edited and translated by R. S. Woolhouse and Richard Francks. Oxford: Oxford University Press, 1997.

———. *Mathematische Schriften.* Edited by Carl Immanuel Gerhardt. 7 vols. Berlin: A. Asher, 1849–63.

———. *Monadologie.* 1714. In *Principes de la nature et de la grâce, Monadologie et autres textes, 1703–1716*, edited by Christiane Frémont, 241–68. Paris: Flammarion, 1996.

———. *Opera Omnia, nunc primum collecta, in classes distributa, praefationibus & indicibus exornata.* Edited by Louis Dutens. 6 vols. Geneva: De Tournes, 1768.

———. *Philosophical Essays.* Edited by Roger Ariew and Daniel Garber. Indianapolis, IN: Hackett, 1989. (*PE*)

———. *Philosophical Papers and Letters.* Edited by Leroy E. Loemker. 2nd ed. Reidel: Dordrecht, the Netherlands, 1969. (*PPL*)

———. *Die philosophischen Schriften.* Edited by Carl Immanuel Gerhardt. 7 vols. Berlin: Weidmannsche Buchhandlung, 1875–90. (*PS*)

———. *Principes de la nature et de la grâce, Monadologie et autres textes, 1703–1716.* Edited by Christiane Frémont. Paris: Flammarion, 1996.

———. "Reflections on the Souls of Beasts." 1710? Translated by Donald Rutherford. In *G. W. Leibniz: Texts and Translations.* Last modified November 1, 2001. http://philosophyfaculty.ucsd.edu/faculty/rutherford/Leibniz/beasts.htm.

———. *Sämtliche Schriften und Briefe.* Berlin: Akademie-Verlag, 1950–. (*SSB*)

———. *Système nouveau de la nature et de la communication des substances et autres textes, 1690–1703.* Edited by Christiane Frémont. Paris: Flammarion, 1994.

———. *Writings on China.* Edited and translated by Daniel J. Cook and Henry Rosemont, Jr. Chicago: Open Court, 1994.

Lespinasse, Julie de. *Lettres.* Edited by Jacques Dupont. Paris: La Table Ronde, 1997.

Levot, Prosper Jean. *Biographie bretonne.* 2 vols. Vannes, France: Cauderan, 1852–57.

Lewis, Sinclair. *Arrowsmith.* 1925. New York: Penguin, 1998.

Lewontin, Richard. *The Triple Helix: Gene, Organism and Environment.* Cambridge, MA: Harvard University Press, 2002.

Locke, John. *An Essay concerning Human Understanding.* 1690. Edited by Peter H. Nidditch. Oxford, UK: Clarendon, 1975.

———. *Locke's Travels in France 1675–1679, as Related in his Journals, Correspondence and Other Papers.* Edited with an introduction by John Lough. Cambridge: Cambridge University Press, 1953.

———. *Two Treatises of Government.* Edited with an introduction by Peter Laslett. Cambridge: Cambridge University Press, 1988.

———. *The Works of John Locke.* 12th ed. 10 vols. London: Thomas Tegg, 1823.

Loeb, Jacques. *The Mechanistic Conception of Life: Biological Essays.* Chicago: University of Chicago Press, 1912.

———. "Mechanistic Science and Metaphysical Romance." *Yale Review* 4, no. 4 (July 1915): 766–85.

Lomazzo, Giovanni Paolo. *Idea del tempio della pittura.* 1590. Hildesheim, Germany: Olms, 1965.

———. *Scritti sulle arti.* Edited by Roberto Ciardi. 2 vols. Florence: Marchi and Bertolli, 1973.

Löseth, Eilert, ed. *Le roman en prose de Tristan.* Paris: Émile Bouillon, 1891.

Lovelace, Ada. *Ada, the Enchantress of Numbers.* Edited by Betty A. Toole. Mill Valley, CA: Strawberry Press, 1992.

———. "Notes by the Translator." Accompanying "Sketch of the Analytical Engine Invented by Charles Babbage by L. F. Menabrea of Turin, Officer of the Military Engineers." In *Scientific Memoirs, Selected from the Transactions of Foreign Academies of Science and Learned Societies and from Foreign Journals,* edited by Richard Taylor, 3:666–731. London: R. & J. E. Taylor, 1842.

Luynes, Charles Philippe d'Albert, duc de. *Mémoires du duc de Luynes sur la cour de Louis XV (1735–1758).* 17 vols. Paris: Firmin Didot frères, 1860–.

Lyell, Charles. *The Geological Evidences of the Antiquity of Man, with Remarks on Theories of the Origin of Species by Variation.* 3rd ed. London: John Murray, 1863.

———. *Principles of Geology, Being an Attempt to Explain the Former Changes of the Earth's Surface, by References to Causes Now in Operation.* 2nd ed. 3 vols. London: John Murray, 1832–33.

Lysenko, Trofim Denisovich. *The Science of Biology Today.* New York: International Publishers, 1948.

Mach, Ernst. *History and Root of the Principle of the Conservation of Energy.* 1871. Translated by Philip E. B. Jourdain. Chicago: Open Court, 1911.

Machines et inventions approuvées par l'Académie Royale des Sciences depuis son établissement jusqu'à present; avec leur Description. 7 vols. Paris: G. Martin, 1735–77.

Magendie, François. *Phénomènes physiques de la vie: Leçons professées au collège de France.* 4 vols. Paris: J.-B. Baillière, 1842.

Maillet, Benoît de. *Telliamed, ou entretiens d'un philosophe indien avec un missionnaire françois, sur la diminution de la Mer, la formation de la Terre, l'origine de l'Homme etc. mis en ordre sur les Mémoires de feu M. de Maillet par J. A. G* [J. Antoine Guers]. 2 vols. Amsterdam: L'Honoré et fils, 1748.

Maître, Henri-Bernard. "Le problème du robot scientifique." *Revue d'histoire des sciences et de leurs applications* 3, no. 4 (1950): 370–75.

Malebranche, Nicolas. *Œuvres de Malebranche.* New ed. Edited by Jules Simon. 4 vols. Paris: Charpentier, 1871–.

Malpighi, Marcello. *Opera omnia*. 2 vols. London: R. Littlebury, R. Scott, Tho. Sawbridge, and G. Wells, 1686.

———. *Opera posthuma*. Amsterdam: Georgium Gallet, 1700.

Marat, Jean-Paul. *A Philosophical Essay on Man*. London, 1773.

Marey, Etienne-Jules. *La machine animale: Locomotion terrestre et aérienne*. Paris: G. Baillière, 1873.

———. *La méthode graphique dans les sciences expérimentales et particulièrement en physiologie et en médecine*. Paris: G. Masson, 1878.

Mariotte, Edme. "A New Discovery Touching Vision." *Philosophical Transactions of the Royal Society* 3 (1668): 668–71.

———. *Nouvelle découverte touchant le veüe*. Paris: Frédéric Léonard, 1668.

Marmontel, Jean François. *Mémoires de Marmontel*. 1804. Edited by Maurice Tourneux. 3 vols. Paris: Librairie des bibliophiles, 1891.

———. *Poétique françoise*. 2 vols. Paris: Lesclapart, 1763.

Maturana, Humberto, and Francisco Varela. *Autopoiesis and Cognition: The Realization of the Living*. Dordrecht, the Netherlands: Reidel, 1980.

Maupertuis, Pierre-Louis Moreau de. "Accord de différentes lois de la nature qui avoient jusqu'ici paru incompatibles." In *Histoire de l'Académie royale des sciences de Paris* 1744. 417–26. Paris: Imprimerie royale, 1748.

———. *Essai de cosmologie*. Leiden: sn, 1751.

———. *Essai sur la formation des corps organisés*. Berlin, 1754.

———. *Vénus physique*. n.p. (France): n.p., 1745.

Mayer, Julius Robert. "Bemerkungen über die Kräfte der unbelebten Natur." *Justus Liebigs Annalen der Chemie* 42, no. 2 (1842): 233–40.

Maynard Smith, John. *Evolution and the Theory of Games*. Cambridge: Cambridge University Press, 1982.

Maynard Smith, John, and G. R. Price. "The Logic of Animal Conflict." *Nature* 246 (November 2, 1973): 15–18.

Mayr, Ernst. "Lamarck Revisited." In *Evolution and the Diversity of Life*, 222–50. Cambridge, MA: Harvard University Press, 1976.

———. "The Multiple Meanings of Teleology." In *Toward a New Philosophy of Biology: Observations of an Evolutionist*, 38–66. Cambridge, MA: Harvard University Press, 1988.

———. "Teleological and Teleonomic." *Boston Studies in the Philosophy of Science* 14 (1974): 91–117.

———. *Toward a New Philosophy of Biology: Observations of an Evolutionist*. Cambridge, MA: Harvard University Press, 1988.

Mayr, Otto. *The Origins of Feedback Control*. Cambridge, MA: MIT Press, 1970.

McClintock, Barbara. "The Significance of Responses of the Genome to Challenge." *Science* 226 (November 16, 1984): 792–801.

McEachern, Lori A., and Vett Lloyd. "The Epigenetics of Genomic Imprinting: Core Epigenetic Processes are Conserved in Mammals, Insects, and Plants." In *Epigenetics: Linking Genotype and Phenotype in Development and Evolution*, edited by Benedikt Hallgrimsson and Brian K. Hall, 43–69. Berkeley: University of California Press, 2011.

Michelant, Heinrich, ed. "Li Romans d'Alixandre." Vol. 13 of *Bibliothek des literarischen Vereins in Stuttgart*. Stuttgart: Anton, Hiersemann, 1846.

Minsky, Marvin. "Matter, Mind, and Models." *Proceedings of the IFIP Congress* 1 (1965): 45–49.

———. *The Society of Mind*. New York: Simon and Schuster, 1985.

———. "Steps toward Artificial Intelligence." *Proceedings of the IRE* 49, no. 1 (1961): 8–30.

———. "Thoughts about Artificial Intelligence." In *Age of Intelligent Machines*, edited by Ray Kurzweil, 214–19. Cambridge, MA: MIT Press, 1990.

Minsky, Marvin and Seymour Papert. *Perceptrons: An Introduction to Computational Geometry*. Cambridge: MIT Press, 1969.

Mommsen, Theodor. "Universitätsunterricht und Konfession." In *Reden und Aufsätze: Mit zwei Bildnissen*, 432–36. Berlin: Weidmann, 1905.

Montaigne, Michel de. *Les Essais: Édition Villey-Saulnier*. 1580. Paris: Presses universitaires de France, 2004. Online at http://www.lib.uchicago.edu/efts/ARTFL/projects/montaigne/.

———. *Journal de voyage*. Edited by Louis Lautrey. Paris: Hachette, 1909.

Montesquieu, Charles de Secondat, baron de. *De l'esprit des lois*. 1758. Edited by Laurent Versini. 2 vols. Paris: Gallimard, 1995.

———. *Œuvres complètes*. Edited by Roger Caillois. 2 vols. Paris: Gallimard, 1949–51.

———. *Œuvres complètes de Montesquieu*. Edited by J. Ehrard, C. Volpilhac-Auger et al. 22 vols. Oxford, UK: Voltaire Foundation, 1998–. (*OCM*)

Montpensier, Anne-Marie-Louise d'Orléans, duchesse de. *Mémoires de Mlle. de Montpensier*. Edited by Bernard Quilliet. Paris: Mercure de France, 2005.

More, Henry. *An Antidote against Atheism*. 2nd ed. London: J. Flesher, 1655.

Morgan, Thomas Hunt. *A Critique of the Theory of Evolution*. Princeton, NJ: Princeton University Press, 1916.

———. *The Development of the Frog's Egg: And Introduction to Experimental Embryology*. New York: Macmillan, 1897.

———. *Evolution and Adaptation*. New York: Macmillan, 1903.

———. *Heredity and Sex*. 2nd ed. New York: Columbia University Press, 1914.

———. *The Physical Basis of Heredity*. Philadelphia: Lippincott, 1919.

———. *Regeneration*. New York: Macmillan, 1901.

———. "Whole Embryos and Half Embryos from One of the First Two Blastomeres of the Frog's Egg." *Anatomischer Anzeiger* 10 (1894–95): 623–38.

Morley, John, Viscount. *Recollections*. 2 vols. London: Macmillan, 1917.

Moxon, E. R., P. B. Rainey, M. A. Nowak, and R. A. Lenski. "Adaptive Evolution of Highly Mutable Loci in Pathogenic Bacteria." *Current Biology* 4 (1994): 24–33.

Müller, Gerd B. "Epigenetic Innovation." In *Evolution: The Extended Synthesis*, edited by Massimo Pigliucci and Gerd B. Müller, 307–32. Cambridge, MA: MIT Press, 2010.

Müller, Gerd B., and Stuart A. Newman, eds. *The Origination of Organismal Form: Beyond the Gene in Developmental and Evolutionary Biology*. Cambridge, MA: MIT Press, 2003.

———, "Origination of Organismal Form: The Forgotten Cause in Evolutionary Theory." In *The Origination of Organismal Form: Beyond the Gene in Developmental and Evolution-*

ary Biology, edited by Gerd B. Müller and Stuart A. Newman, 3–10. Cambridge, MA: MIT Press, 2003.

Müller, Johannes. *Elements of Physiology*. Translated by William Baly and edited by John Bell. Philadelphia: Lea and Blanchard, 1843.

———. *Zur vergleichenden Physiologie des Gesichtssinnes des Menschen und der Thiere, nebst einem Versuch über die Bewegung der Augen und über den menschlichen Blick*. Leipzig: C. Cnobloch, 1826.

Nägeli, Carl. *Entstehung und Begriff der Naturhistorischen Art*. Munich: Königlich-Bayerische Akademie, 1865.

———. *A Mechanico-Physiological Theory of Organic Evolution*. 1884. Chicago: Open Court, 1914.

Naudé, Gabriel. *Apologie pour tous les grands hommes, qui ont esté accusez de magie*. 1625. Paris: Eschart, 1669.

Newell, Allen and Herbert A. Simon. *Human Problem Solving*. Englewood Cliffs, NJ: Prentice-Hall, 1972.

Newman, Stuart A. "The Fall and Rise of Systems Biology: Recovering from a Half-Century Gene Binge." *GeneWatch* 16, no. 4 (2003): 8–12.

Newman, Stuart A., and Gerd B. Müller. "Genes and Form: Inherency in the Evolution of Developmental Mechanics." In *Genes in Development: Re-reading the Molecular Paradigm*, edited by E. M. Newmann-Held and C. Rehmann-Sutter, 38–73. Durham, NC: Duke University Press, 2006.

Newton, Isaac. *Opticks, or A Treatise of the Reflections, Refractions, Inflections & Colours of Light*. 1706. New ed., based on the 4th ed. of 1730. New York: Dover, 1952.

Nicolai, Friedrich. *Anekdoten von König Friedrich II. von Preussen, und von einigen Personen, die um ihn waren*. 6 vols. Berlin, 1788–92.

———. *Chronique à travers l'Allemagne et la Suisse*. 2 vols. Berlin, 1783.

Oresme, Nicole. *Le livre du ciel du monde*. Edited by Albert D. Menut and Alexander J. Denomy and translated by Albert D. Menut. Madison: University of Wisconsin Press, 1968.

Osborn, Henry Fairfield. *Impressions of Great Naturalists: Reminiscences of Darwin, Huxley, Balfour, Cope, and Others*. New York: C. Scribner's Sons, 1924.

Otto, Rudolf. "Darwinism and Religion." In *Religious Essays: A Supplement to the Idea of the Holy*, translated by Brian Lunn, 121–39. Oxford: Oxford University Press, 1931.

———. *Naturalism and Religion*. Translated by J. Arthur Thomson and Margaret R. Thomson. Edited with an introduction by the Rev. W. D. Morrison, LLD. London: Williams and Norgate, 1907.

Paley, William. *Natural Theology: Or, Evidences of the Existence and Attributes of the Deity, Collected from the Appearances of Nature*. 12th ed. London: J. Faulder, 1809.

Paré, Ambroise. *The Collected Works of Ambroise Paré*. Translated by Thomas Johnson. 1634. Pound Ridge, NY: Milford House, 1968.

Parsons, James. *Philosophical Observations on the Analogy between the Propagation of Animals and That of Vegetables*. London: C. Davis, 1752.

Pascal, Blaise. "Pascal on His Calculating Machine." Translated by L. Leland Locke. In

A Source Book in Mathematics, edited by David Eugene Smith, 165–72. New York: Dover, 1959.

Paulsen, Friedrich. *The German Universities: Their Character and Historical Development*. Translated by Edward Delavan Perry. New York: Macmillan, 1895.

Pavlov, Ivan Petrovich. *Conditioned Reflexes: An Investigation of the Physiological Activity of the Cerebral Cortex*. Edited and translated by G. V. Anrep. London: Oxford University Press, 1927.

Perrault, Claude. *Essais de physiques: ou Recueil de plusieurs traités touchant les choses naturelles*. 3 vols. Paris: Jean Baptiste Coignard, 1680.

Perrault, Claude and Pierre Perrault. *Œuvres diverses de physique et de mechanique*. 2 vols. Leiden: Chez P. van der Aa, 1721.

Pettit, Auguste, ed. *À Charles Richet de ses amis, ses collègues, ses élèves. 22 Mai 1926*. Paris: Éditions médicales, 1926.

Piattelli-Palmarini, Massimo. "Evolution, Selection, and Cognition: From 'Learning' to Parameter Setting in Biology and in the Study of Language." *Cognition* 31 (1989): 1–44.

Pigliucci, Massimo, and Gerd B. Müller, eds. *Evolution: The Extended Synthesis*. Cambridge, MA: MIT Press, 2010.

Pinker, Steven. *How the Mind Works*. New York: Norton, 1997.

Plato. *Dialogues*. Translated by Benjamin Jowett. 4th ed. Oxford, UK: Clarendon Press, 1953.

Pluche, Noël Antoine, abbé. *Le spectacle de la nature, ou, Entretiens sur les particularités de l'histoire naturelle*. 8 vols. Paris: Frères Estienne, 1732–50.

Poe, Edgar Allan. "Maelzel's Chess-Player." In *The Works of Edgar Allan Poe*, edited by Edmund Clarence Stedman and George Edward Woodberry, 9:173–212. New York: 1914.

Priestley, Joseph. *Disquisitions Relating to Matter and Spirit. To Which Is Added the History of the Philosophical Doctrine concerning the Origin of the Soul* . . . 2nd ed. 2 vols. Birmingham, UK: Pearson and Rollason for J. Johnson, 1782.

Priestley, Joseph and Richard Price. *A Free Discussion of the Doctrines of Materialism, and Philosophical Necessity, in a Correspondence between Dr. Price and Dr. Priestley*. London: J. Johnson and T. Cadell, 1778.

Przibram, Hans. *Embryogeny: An Account of the Laws Governing the Development of the Animal Egg as Ascertained Through Experiment*. Cambridge: Cambridge University Press, 1908.

Quantz, Johann Joachim. *Versuch einer Anweisung die Flöte traversiere zu spielen*. Berlin: J. F. Voss, 1752.

Quesnay, François. "Analyse du tableau économique" (1758). In *Œuvres*, 305–29.

———. *L'art de guérir par la saignée*. Paris: G. Cavelier, 1736.

———. *Essai physique sur l'œconomie animale*. Paris, 1736.

———. "Maximes générales du gouvernement économique d'un royaume agricole" (1758). In *Œuvres*, 330–36.

———. *Observations sur les effets de la saignée*. Paris, 1730.

———. *Œuvres économiques et philosophiques*. Frankfurt: Joseph Baer and Paris, France: Jules Peelman, 1888.

———. "Questions intéressantes sur la population, l'agriculture et le commerce" (1758). In *Œuvres*, 250–304.

———. "Sur les travaux des artisans, second dialogue" (1766). In *Œuvres*, 526–54.

———. *Traité des effets et de l'usage de la saignée*. Paris: D'Houry père, 1750.

Ramelli, Agostino. *The Various and Ingenious Machines of Agostino Ramelli (1588)*. Edited by Eugene S. Ferguson and translated by Martha Teach Gnudi. Baltimore: Johns Hopkins University Press, 1976.

Ramus, Peter. *Scholarum mathematicarum libri unus et triginta*. 1569. Frankfurt, 1627.

Ranke, Leopold von. *The Secret of World History: Selected Writings on the Art and Science of History*. Edited and translated by Roger Wines. New York: Fordham University Press, 1981. (*SWH*)

———. *The Theory and Practice of History*. Edited with an introduction by Georg G. Iggers. Oxford: Routledge, 2011.

Ray, John. *The Wisdom of God Manifested in the Works of the Creation*. 1692. 7th ed. London, 1717.

Réaumur, Antoine René Ferchault de. *Mémoires pour server à l'histoire des insectes*. Vols. 1–6. Paris: Imprimerie royale, 1734–42. Vol. 7. Paris: Paul Lechevalier, 1928.

Régis, Pierre Sylvain. *Cours entier de philosophie, ou système général selon les principes de M. Descartes*. 3 vols. Amsterdam: Huguetan, 1691.

Reisel, Salomon. "Concerning the Sipho Wirtembergicus Stutgardiae." *Philosophical Transactions of the Royal Society of London* 15 (1685): 1272–73.

Ricci, Matteo. *China in the Sixteenth Century: The Journals of Matthew Ricci, 1583–1610*. 1615. Edited by Nicolas Trigault and translated by Louis J. Gallagher. New York: Random House, 1953.

Ridley, Mark. *The Problems of Evolution*. Oxford: Oxford University Press, 1985.

Rivarol, Antoine. *Discours sur l'universalité de la langue française*. In *Œuvres choisies de Rivarol*, 1:1–82. Paris: Librairie des Bibliophiles, 1880.

———. *Lettre à M. le Président de *** sur le globe airostatique, sur les têtes parlantes et sur l'état présent de l'opinion publique à Paris: Pour servir de suite à la Lettre sur le poème des Jardins*. London; Paris: Cailleau, 1783.

Robert-Houdin, Jean-Eugène. *Memoirs of Robert-Houdin, King of the Conjurers*. 1858. Translated by Lascelles Wraxall. New York: Dover, 1964.

———. *The Secrets of Conjuring and Magic, Or, How to Become a Wizard*. 1868. Edited and translated by Prof. Hoffmann. 5th ed. London: Routledge, 1877.

Robinet, Jean-Baptiste. *De la nature*. Amsterdam: E. van Harrevelt, 1761.

Robinson, Henry Crabb. *Diary, Reminiscences, and Correspondence of Henry Crabb Robinson*. Edited by Thomas Sadler. 2nd ed. 3 vols. London: Macmillan, 1869.

Romanes, George John. "The Darwinism of Darwin, and of the Post-Darwinian Schools." *Monist* 6, no. 1 (1895): 1–27.

Rosenblueth, Arturo, Norbert Wiener, and Julian Bigelow. "Behavior, Purpose, and Teleology." *Philosophy of Science* 10, no. 1 (1943): 18–34.

Rothschild, James de, ed. *Le mistère du viel testament*. 6 vols. Paris: Firmin Didot, 1878–91.

Rousseau, Jean-Jacques. *Les confessions*. 1770. 12 vols. Paris: Launette, 1889.

———. *Correspondance complète de Jean-Jacques Rousseau*. Edited by R. A. Leigh et al. 52 vols. Geneva: Institut et Musée Voltaire, 1965–98.

———. *Discours qui a remporté le prix a l'Academie de Dijon en l'année 1750: Sur cette question proposée par la meme académie, si le rétablissement des sciences & des arts a contribute à épurer les mœurs. (Discours sur les sciences et les arts)*. Geneva: Barillot et fils, 1750.

———. *Discours sur l'origine et les fondements de l'inégalité parmi les hommes*. 1755. Paris: Librairie de la Bibliothèque Nationale, 1899.

———. *Du contrat social*. 1762. Introduced by Bertrand de Jouvenel. Geneva: C. Bourquin, 1947.

———. "Economie ou Œconomie." In *L'Encyclopédie, ou, Dictionnaire raisonné des sciences, des arts, et des métiers, par une societé des gens de lettres*, edited by Denis Diderot and Jean d'Alembert, 5:337–49. Paris, 1755.

———. *Émile, ou De l'education*. 1762. Paris: Firmin Didot, 1854.

Roux, Wilhelm. "Autobiographie." In *Die Medizin in der Gegenwart in Selbstdarstellungen*, edited by L. R. R. Grote, 1:141–206. Leipzig: Felix Meiner, 1923.

———. *Beitrag zur Entwickelungsmechanik des Embryo*. Vienna: Hof- und Staatsdruckerei, 1892.

———. "Beiträge zur Entwickelungsmechanik des Embryo: Über die künstliche Hervorbringung halber Embryonen durch Zerstörung einer der beiden ersten Furchungskugeln, sowie über die Nachentwickelung (Postgeneration) der fehlenden Körperhälfte." In *Gesammelte Abhandlungen über Entwickelungsmechanik der Organismen*, 2:419–520. Leipzig: William Engelmann, 1895.

———. "Einleitung zum Archiv für Entwickelungsmechanik." *Archiv für Entwicklungsmechanik* 1 (1894): 1–42.

———. "Die Entwickelungsmechanik der Organismen: Eine anatomische Wissenschaft der Zukunft." In *Gesammelte Abhandlungen über Entwickelungsmechanik der Organismen*, 2:24–54. Leipzig: William Engelmann, 1895.

———. *Die Entwickelungsmechanik: Ein neuer Zweig der biologischen Wissenschaft*. Leipzig: William Engelmann, 1905.

———. *Der Kampf der Teile im Organismus: Ein Beitrag zur vervollständigung der mechanischen Zweckmässigkeitslehre*. Leipzig: Engelmann, 1881.

———. "Meine entwicklungsmechanische Methodik." In *Handbuch der biologischen Arbeitsmethoden*, edited by Emil Aberhalden, vol. 5, part 3A:539–616. Berlin: Urban and Schwarzenberg, 1923.

———. "Über den 'Cytotropismus' der Furchungszellen des Grasfrosches (Rana fusca)." *Archiv für Entwicklungsmechanik* 1 (1894): 43–68.

———. *Über die Bedeutung der Kerntheilungsfiguren: Eine hypothetische Erörterung*. Leipzig: Engelmann, 1883.

———. "Über die Bedeutung 'geringer' Verschiedenheiten der relative Größe der Furchungszellen für den Charakter des Furchungsschemas: Nebst Erörterung über die nächsten Ursachen der Anordnung und Gestalt der ersten Furchungszellen." *Archiv für Entwicklungsmechanik der Organismen* 4, no. 1 (1896): 5–40.

Russell, E. S. *Form and Function: A Contribution to the History of Animal Morphology*. London: John Murray, 1916.

Sabbattini, Nicola. *Manual for Constructing Theatrical Scenes and Machines*. Edited by Barnard Hewitt. Coral Gables, FL: University of Miami Press, 1958.

Saint-Hilaire, Étienne Géoffroy. *Philosophie anatomique*. 2 vols. Paris: J.-B. Baillière, 1818.

Saint-Pierre, Jacques-Henri-Bernardin de. *Études de la nature*. 4 vols. Paris: Didot, 1784–88.

Saumarez, Richard. *A New System of Physiology*. 2nd ed. 2 vols. London: J. Davis, 1799.

Scheiner, Christoph. *Oculus*. Innsbruck, 1619.

Schelling, Friedrich Wilhelm Joseph von. *Ideen zu einer Philosophie der Natur: Als Einleitung in das Studium dieser Wissenschaft*. Landshut, Germany: Philipp Krüll, 1805.

Schimansky-Geier, Lutz, Bernold Fiedler, J. Kurths, and Eckehard Schöll, eds. *Analysis and Control of Complex Non-Linear Processes in Physics, Chemistry and Biology*. London: World Scientific Publishing Company, 2007.

Schreiber, Heinrich. *Geschichte der Stadt und Universität Freiburg im Breisgau: Geschichte der Albert-Ludwigs-Universität zu Freiburg im Breisgau*. 1859. Charleston, SC: Nabu, 2011.

Schrödinger, Erwin. "Mind and Matter." 1956. In *What Is Life?* 91–164. Cambridge: Cambridge University Press, 1967.

———. *What Is Life?* 1944. Cambridge: Cambridge University Press, 1967.

Scudéry, Madeleine de. *Madeleine de Scudéry, sa vie et sa correspondance*. Edited by Edme Jacques Benoît Rathéry and Boutron. Paris: L. Techener, 1873.

Scudéry, Marie-Françoise de Martinvast de. *Lettres de mesdames de Scudéry, de Salvan de Saliez et de mademoiselle Descartes*. Paris: Léopold Collin, 1806.

Shannon, Claude E. *Claude Elwood Shannon: Collected Papers*. Edited by N. J. A. Sloane and Aaron D. Wyner. New York: IEEE Press, 1993.

———. "Computers and Automata." 1953. In *Claude Elwood Shannon: Collected Papers*, edited by N. J. A. Sloane and Aaron D. Wyner, 703–10. New York: IEEE Press, 1993.

———. "Game Playing Machines." 1955. In *Claude Elwood Shannon: Collected Papers*, edited by N. J. A. Sloane and Aaron D. Wyner, 786–92. New York: IEEE Press, 1993.

———. "The Potentialities of Computers." 1953. In *Claude Elwood Shannon: Collected Papers*, edited by N. J. A. Sloane and Aaron D. Wyner, 691–94. New York: IEEE Press, 1993.

———. "Presentation of a Maze Solving Machine." 1952. In *Claude Elwood Shannon: Collected Papers*, edited by N. J. A. Sloane and Aaron D. Wyner, 681–87. New York: IEEE Press, 1993.

Shapiro, James A. "Bacteria Are Small but Not Stupid: Cognition, Natural Genetic Engineering and Socio-Bacteriology." *Studies in the History and Philosophy of Biological and Biomedical Sciences* 38 (2007): 807–19.

———. "Barbara McClintock, X-Rays and Self-Aware, Self-Healing Cells." *Huffington Post*, March 8, 2012. Accessed October 2013. http://www.huffingtonpost.com/james-a-shapiro/barbara-mcclintock-x-rays_b_1322879.html.

———. *Evolution: A View from the 21st Century*. Upper Saddle River, NJ: FT Press Science, 2011.

Shelley, Mary. *Frankenstein, or the Modern Prometheus* (1818). In *Frankenstein: The 1818 Text, Contexts, Nineteenth-Century Responses, Modern Criticism*, edited by J. Paul Hunter, 169–73. New York: W. W. Norton, 1996.

———. "Introduction to *Frankenstein*, Third Edition (1831)." In *Frankenstein: The 1818 Text, Contexts, Nineteenth-Century Responses, Modern Criticism*, by Mary Shelley, edited by J. Paul Hunter, 169–73. New York: W. W. Norton, 1996.

Simon, Herbert A. *Administrative Behavior: A Study of Decision-Making Processes in Administrative Organization*. New York: Free Press, 1947.

———. "A Behavioral Model of Rational Choice." *Quarterly Journal of Economics* 69 (1955): 99–188.

———. *Models of Bounded Rationality*. 3 vols. Cambridge, MA: MIT Press, 1982–97.

———. *Models of Discovery: And Other Topics in the Methods of Science*. Dordrecht: Reidel, 1977.

———. *Models of Man: Social and Rational; Mathematical Essays on Rational Human Behavior in a Social Setting*. New York: Wiley, 1957.

———. *Models of Thought*. 2 vols. New Haven, CT: Yale University Press, 1979–89.

———. *Reason in Human Affairs*. Stanford, CA: Stanford University Press, 1983.

———. *The Sciences of the Artificial*. Cambridge, MA: MIT Press, 1969.

Skinner, B. F. *The Behavior of Organisms: An Experimental Analysis*. New York: Appleton-Century-Crofts, 1938.

———. *Science and Human Behavior*. 1953. Cambridge, MA: B. F. Skinner Foundation, 2005.

Smail, Daniel Lord. *On Deep History and the Brain*. Berkeley: University of California Press, 2008.

Smellie, William. *The Philosophy of Natural History*. 2 vols. Dublin, 1790.

Smith, Adam. *An Inquiry into the Nature and Causes of the Wealth of Nations*. 1776. Edited by Edwin Cannan. 2 vols. London: Methuen, 1904.

———. *The Theory of Moral Sentiments*. London: A. Millar, 1759.

Solé, Ricard V., and Brian Goodwin. *Signs of Life: How Complexity Pervades Biology*. New York: Basic Books, 2000.

Sommer, H. Oskar, ed. *Lestoire del Saint Graal*. Vol. 1 of *The Vulgate Version of the Arthurian Romances*. Washington, DC: Carnegie Institution, 1909.

Spallanzani, Lazzaro. *Nouvelles recherches sur les découvertes microscopiques, et la génération des corps organisés*. 2 vols. London, 1769.

Spencer, Herbert. *Essays: Scientific, Political, and Speculative*. 3 vols. New York: D. Appleton, 1891.

———. *The Principles of Biology*. 1864. Revised and expanded ed. 2 vols. New York: D. Appleton, 1898–1900.

Spinoza, Benedict. *Benedict de Spinoza: His Life, Correspondence, and Ethics*. Edited and translated by Robert Willis. London: Trübner, 1870.

———. *The Correspondence of Spinoza*. Edited and translated by Abraham Wolf. Reprint. Whitefish, MT: Kessinger, 2003.

Spränger, Eduard. *Wandlungen im Wesen der Universität seit 100 Jahren*. Leipzig: E. Wiegandt, 1913.

Stahl, Georg Ernst. *Œuvres médico-philosophiques et pratiques*. Edited and translated by T. Blondin. 5 vols. Paris: Baillière: 1859–64.

Steele, Edward J. *Somatic Selection and Adaptive Evolution: On the Inheritance of Acquired Characters*. 2nd ed. Chicago: University of Chicago Press, 1981.

Stokes, G. G., W. H. Miller, C. Wheatstone, and R. Willis. "Report of a Committee Appointed by the Council to Examine the Calculating Engine of M. Scheutz." *Proceedings of the Royal Society* 7 (1855): 499–509.

Stow, John, and Edmund Howes. *Annales, or, a Generall Chronicle of England: Begun by John Stow; Continued and Augmented with Matters Forraigne and Domestique, Ancient and Moderne, unto the End of This Present Yeere, 1631*. London: Printed by John Beale, Bernard Alsop, Thomas Fawcett, and Augustine Matthewes, impensis Richardi Meighen, 1632.

Strada, Famiano. *De bello belgico: The History of the Low-Countrey Warres*. Translated by Sr. Rob Stapylton. London: H. Moseley, 1650.

Stresemann, Erwin. *Ornithology from Aristotle to the Present*. 1951. Edited by G. William Cottrell, with a foreword and epilogue by Ernst Mayr. Translated by Hans J. Epstein and Cathleen Epstein. Cambridge, MA: Harvard University Press, 1975.

Swammerdam, Jan. *The Book of Nature: Or, the History of Insects*. 1676–79. Translated by Thomas Flloyd. London: C. G. Seyffert, 1758.

Swift, Jonathan. *The Writings of Jonathan Swift: Authoritative Texts, Backgrounds, Criticism*. Edited by Robert A Greenberg and William Bowman Piper. New York: W. W. Norton, 1973.

Taylor, Richard. *Action and Purpose*. Englewood Cliffs, NJ: Prentice-Hall, 1966.

Thibault, Dieudonné. *Mes souvenirs de vingt ans to séjour à Berlin: ou Frédéric le Grand, sa famille, sa cour, son gouvernement, son académie, ses écoles, et ses amis littérateurs et philosophes*. 2nd ed. Paris: F. Buisson, 1805.

Thicknesse, Philip. *The Speaking Figure and the Automaton Chess-Player*. London: J. Stockdale, 1784.

Thompson, D'Arcy Wentworth. *On Growth and Form*. 1917. New ed. Cambridge: Cambridge University Press, 1945.

Thrun, Sebastian, Wolfram Burgard, and Dieter Fox. *Probabilistic Robotics*. Cambridge, MA: MIT Press, 2005.

Toland, John. *Letters to Serena*. London: Bernard Lintot, 1704.

Tors, Lambert li, and Alexandre de Bernay. *Li romans d'Alixandre*. Edited by Heinrich Michelant. Vol. 13 of *Bibliothek des literarischen Vereins in Stuttgart*. Stuttgart: Anton Hiersemann, 1846.

Tougard, Albert, ed. *Documents concernant l'histoire littéraire du XVIIIe siècle*. 2 vols. Rouen, France: A. Lestringant, 1912.

Tourneux, Maurice, ed. *Correspondance littéraire, philosophique et critique par Grimm, Diderot, Raynal, Meister, etc.* 16 vols. Paris: Garnier frères, 1877–82.

Troyes, Chrétien de. *Perceval le Gallois ou le conte du Graal*. Edited by Charles Potvin. 6 vols. Mons, Belgium: Société des bibliophiles belges, 1866–71.

Trublet, abbé. *Mémoires pour servir à l'histoire de la vie et des ouvrages de M. de Fontenelle*. Amsterdam: Marc Michel Rey, 1759.

Turing, Alan. "Can Digital Computers Think?" 1951. In *The Essential Turing: Seminal Writings in Computing, Logic, Philosophy, Artificial Intelligence, and Artificial Life, plus the Secrets of Enigma*, edited by B. Jack Copeland, 476–86. Oxford: Oxford University Press, 2004.

———. "The Chemical Basis of Morphogenesis." *Philosophical Transactions of the Royal Society of London. Series B, Biological Sciences* 237, no. 641 (August 14, 1952): 37–72.

———. "Computing Machinery and Intelligence." *Mind* 59, no. 236 (1950): 433–60.

———. *The Essential Turing: Seminal Writings in Computing, Logic, Philosophy, Artificial Intelligence, and Artificial Life, plus the Secrets of Enigma*. Edited by B. Jack Copeland. Oxford: Oxford University Press, 2004.

———. "Intelligent Machinery, A Heretical Theory." 1951. In *The Essential Turing: Seminal Writings in Computing, Logic, Philosophy, Artificial Intelligence, and Artificial Life, plus the Secrets of Enigma*, edited by B. Jack Copeland, 465–75. Oxford: Oxford University Press, 2004.

———. "Intelligent Machinery: A Report by A.M. Turing." 1948. In *The Essential Turing: Seminal Writings in Computing, Logic, Philosophy, Artificial Intelligence, and Artificial Life, plus the Secrets of Enigma*, edited by B. Jack Copeland, 395–432. Oxford: Oxford University Press, 2004.

———. "Lecture on the Automatic Computing Engine." 1947. In *The Essential Turing: Seminal Writings in Computing, Logic, Philosophy, Artificial Intelligence, and Artificial Life, plus the Secrets of Enigma*, edited by B. Jack Copeland, 362–94. Oxford: Oxford University Press, 2004.

———. "On Computable Numbers, with an Application to the Entscheidungsproblem." *Proceedings of the London Mathematical Society*, series 2, vol. 42, no. 1 (1937): 230–65.

Turing, Alan, Richard Braithwaite, Geoffrey Jefferson, and Max Newman. "Can Automatic Calculating Machines Be Said to Think?" 1952. In *The Essential Turing: Seminal Writings in Computing, Logic, Philosophy, Artificial Intelligence, and Artificial Life, plus the Secrets of Enigma*, edited by B. Jack Copeland, 487–506. Oxford: Oxford University Press, 2004.

Turriano, Juanelo. *The Twenty-One Books of Devices and of Machines*. Edited by José A. García-Diego. Madrid: Colegio de ingenieros de Caminos, Canales y Puertos, 1983–84.

Twain, Mark. "Dr. Loeb's Incredible Discovery." 1910. In *Europe and Elsewhere*, 304–9. New York: Harper, 1923.

Ure, Andrew. "An Account of Some Experiments Made on the Body of a Criminal Immediately after Execution, with Physiological and Practical Observations." *Journal of Science and the Arts* 6 (1819): 283–94.

Valla, Giorgio. *Georgii Vallae Placentini viri clari de expetendis, et fugiendis rebus opus*. Venice: In aedibvs Aldi Romani, 1501.

Van Emden, Wolfgang, ed. *Jeu d'Adam*. Edinburgh: Société Rencesvals British Branch, 1996.

Varela, Francisco J., Evan Thomspon, and Eleanor Rosch. *The Embodied Mind: Cognitive Science and Human Experience*. Cambridge, MA: MIT Press, 1991.

Vasari, Giorgio. *Lives of the Most Eminent Painters, Sculptors and Architects*. Translated by Gaston de Vere. 10 vols. London: Macmillan and the Medici Society, 1912–15.

Vaucanson, Jacques. *Le mécanisme du fluteur automate*. Translated by J. T. Desaguliers. 1738. Buren, the Netherlands: F. Knuf, 1979.

———. *Mémoire sur un nouveau métier à tisser la soye*. 1749. AN F/12/642, reproduced in

André Doyon and Lucien Liaigre, *Jacques Vaucanson, mécanicien de génie*, 463–71. Paris: Presses universitaires de France, 1967.

Vitruvius: Ten Books on Architecture, The Corsini Incunabulum with the annotations and drawings of Giovanni Battista da Sangallo. Edited with an introduction by Ingrid D. Rowland. Rome: Edizioni dell'Elefante, 2003.

Vogel, Albrecht. *Die Semisäcularfeier der K.K. Evangelisch-theologischen Facultät in Wien am 25. April 1871. Im Auftrage des Professorencollegiums*. Vienna: W. Braumüller, 1872.

Volta, Alessandro. *On the Electricity Excited by the Mere Contact of Conducting Substances of Different Kinds*. 1800. Milan: U Hoepli, 1999.

Voltaire, François Marie Arouet de. *Les Cabales, un œuvre pacifique*. London: s.l., 1772.

———. *Candide, ou L'optimisme*. 1759. Paris: Hachette, 1913.

———. *Correspondence and Related Documents*. 51 vols. Edited by Theodore Besterman. Vols. 85–135 of *Œuvres complètes de Voltaire*, edited by Theodore Besterman et al. Geneva: Voltaire Foundation, 1968–77.

———. *Dictionnaire philosophique*. 1764. Edited by Julien Benda and Raymond Naves. Paris: Garnier, 1954.

———. "Histoire." In *L'Encyclopédie, ou Dictionnaire raisonné des sciences, des arts, et des métiers, par une societé des gens de lettres*, edited by Denis Diderot and Jean d'Alembert, 8:220–25. Paris, 1765.

———. *Lettres philosophiques*. 1734. Paris: Hachette, 1917.

———. *Œuvres complètes de Voltaire*. Edited by Theodore Besterman et al. 143 vols. Geneva, Banbury, Oxford, UK: Voltaire Foundation, 1968–. (*OCV*)

———. *Œuvres de Voltaire*. Edited by Adrien-Jean-Quentin Beuchot. Paris: Firmin Didot, 1831.

———. *Traité de métaphysique*. 1734. In *OCV*, 14:357–503.

Von Neumann, John. *Theory of Self-Reproducing Automata*. Edited by Arthur W. Burks. Urbana: University of Illinois Press, 1966.

Waddington, C.H. *An Introduction to Modern Genetics*. London: Allen and Unwin, 1939.

———. *Organisers and Genes*. Cambridge: Cambridge University Press, 1940.

———. *The Strategy of the Genes: A Discussion of Some Aspects of Theoretical Biology*. London: Allen and Unwin, 1957.

Waldeyer-Hartz, Heinrich Wilhelm Gottfried von. "Über Karyokinese und ihre Beziehungen zu den Befruchtungsvorgängen." *Archiv für Mikroskopische Anatomie* 32 (1888): 1–122.

Wallace, Alfred Russel. *Alfred Russel Wallace, Letters and Reminiscences*. Edited by James Marchant. New York: Harper, 1916.

———. "The Limits to Natural Selection as Applied to Man." In *Contributions to the Theory of Natural Selection*, 332–71. London: Macmillan, 1870.

———. "On the Origin of Human Races and the Antiquity of Man Deduced from the Theory of 'Natural Selection.'" *Anthropological Review* 2 (1864): clviii–clxxxv.

———. "On the Tendency of Varieties to Depart Indefinitely from the Original Type." *Journal of the Proceedings of the Linnean Society: Zoology* 3, no. 9 (August 20, 1858): 53–62.

Walpole, Horace. *The Letters of Horace Walpole, Fourth Earl of Oxford*. Edited by Paget Toynbee. 16 vols. Oxford, UK: Clarendon Press, 1903–05.

Walter, William Grey. "An Imitation of Life." *Scientific American* 182, no. 5 (1950): 42–45.

———. *The Living Brain*. New York: Norton, 1953.

Ward, Lester Frank. *Neo-Darwinism and Neo-Lamarckism: Annual Address of the President of the Biological Society of Washington*. Washington, DC: Gidney and Roberts, 1891.

Waterworth, James, ed. *The Canons and Decrees of the Sacred and Oecumenical Council of Trent*. London: Dolman, 1848.

Watson, James D. *Avoid Boring People: Lessons from a Life in Science*. New York: Knopf, 2007.

Watson, John B. "Psychology as the Behaviorist Views It." *Psychological Review* 20 (1913): 158–77.

Wayne, Randy O. *Plant Cell Biology: From Astronomy to Zoology*. San Diego, CA: Elsevier, 2009.

Weber, Bruce H., and David J. Depew, eds. *Evolution and Learning: The Baldwin Effect Reconsidered*. Cambridge, MA: MIT Press, 2003.

Weber, Max. *The Protestant Ethic and the Spirit of Capitalism*. Translated by Talcott Parsons. New York: Scribner, 1930.

———. "The Psychology of World Religions." In *From Max Weber: Essays in Sociology*, edited and translated by H. H. Gerth and C. Wright Mills, chap 12. New York: Oxford University Press, 1946.

———. "Science as a Vocation." 1917. In *Max Weber's Complete Writings on Academic and Political Vocations*, edited and with an introduction by John Dreijmanis, translated by Gordon C. Wells, 25–52. New York: Algora, 2008.

———. *The Sociology of Religion*. Translated by Ephraim Fischoff. Boston: Beacon Press, 1963.

Weismann, August. *August Weismann: Ausgewählte Briefe und Dokumente, Selected Letters and Documents*. Edited by Frederick B. Churchill and Helmut Risler. 2 vols. Freiburg im Breisgau, Germany: Universitätsbibliothek, 1999.

———. *Essays upon Heredity and Kindred Biological Problems*. Edited and translated by Edward B. Poulton, Selmar Schönland, and Arthur E. Shipley. 2 vols. Oxford, UK: Clarendon Press, 1889–92.

———. *The Evolution Theory*. Translated by J. Arthur Thomson and Margaret R. Thomson. 2 vols. London: Edward Arnold, 1904.

———. *The Germ-Plasm: A Theory of Heredity*. Translated by W. Newton Parker and Harriet Rönnfeldt. New York: Scribner's, 1893.

———. *On Germinal Selection as a Source of Definite Variation*. Translated by Thomas J. McCormack. Chicago: Open Court, 1896.

———. "On Heredity." 1886. In *Essays upon Heredity and Kindred Biological Problems*, edited and translated by Edward B. Poulton, Selmar Schönland, and Arthur E. Shipley, 1:69–106. Oxford, UK: Clarendon Press, 1889–92.

———. "On the Mechanical Conception of Nature." In *Studies in the Theory of Descent*, edited and translated by Raphael Meldola, 2:634–718. London: Sampson, Low, Marston, Searle, and Rivington, 1882.

———. "On the Supposed Botanical Proofs of the Transmission of Acquired Characters." 1888. In *Essays upon Heredity and Kindred Biological Problems*, edited and translated by

Edward B. Poulton, Selmar Schönland, and Arthur E. Shipley, 1:385–417. Oxford, UK: Clarendon Press, 1889.

———. *Studies in the Theory of Descent*. Edited and translated by Raphael Meldola, with a prefatory notice by Charles Darwin. 2 vols. London: Sampson, Low, Marston, Searle, and Rivington, 1882.

———. "The Supposed Transmission of Mutations." 1888. In *Essays upon Heredity and Kindred Biological Problems,* edited and translated by Edward B. Poulton, Selmar Schönland, and Arthur E. Shipley, 1:419–48. Oxford, UK: Clarendon Press, 1889–92.

Wharton, Edith. *Italian Villas and Their Gardens*. 1904. Reprint. New York: Da Capo Press, 1976.

Wheeler, William Morton. "On Instincts." *Journal of Abnormal Psychology* 15 (1920–21): 295–318.

Wiener, Norbert. *Cybernetics: or, Control and Communication in the Animal and the Machine*. 1948. 2nd ed. Cambridge, MA: MIT Press, 1961.

———. *The Human Use of Human Beings: Cybernetics and Society*. Boston: Houghton Mifflin, 1950.

Wilkins, John. *Mathematicall Magick, or, The Wonders That May Be Performed by Mathematicall Geometry*. London: Sa. Gellibrand, 1648.

Willier, Benjamin H., and Jane M. Oppenheimer, eds. *Foundations of Experimental Embryology*. 2nd ed. New York: Hafner Press, 1974.

Willis, Robert. *An Attempt to Analyse the Automaton Chess Player of Mr. de Kempelen with an Easy Method of Imitating the Movements of That Celebrated Engine*. London: J. Booth, 1821.

———. "On the Vowel Sounds, and on Reed Organ-Pipes." *Transactions of the Cambridge Philosophical Society* 3 (1830): 231–68.

Willis, Thomas. *Cerebri Anatome: Cui accessit nervorum desciprtio et usus*. Londini [London]: Typis Ja. Flesher: Impensis Jo. Martyn and Ja. Allestry, 1664.

———. *Two Discourses concerning the Soul of Brutes: Which Is That of the Vital and Sensitive of Man*. London: Thomas Dring, 1683.

Wimpheling, Jacob, ed. *Petri Schotti Argentinensis Patricii: Juris utriusque doctoris consultissimi; Oratoris et Poetae elegantissimi: graecaeque linguae probe aeruditi: Lucubraciunculae ornatissimae*. Strasbourg: Martin Schott, 1498.

Windisch, Karl Gottlieb. *Inanimate Reason; or, A Circumstantial Account of that Astonishing Piece of Mechanism, De Kempelen's Chess-Player*. London: S. Bladon, 1784.

———. *Lettres sur le joueur d'échecs de M. de Kempelen*. Basel: Chez l'éditeur, 1783.

Woodcroft, Bennet, trans. and ed. *The Pneumatics of Hero of Alexandria, from the Original Greek*. London: Taylor, Walton and Maberly, 1851.

Wordsworth, William. *The Complete Poetical Works of William Wordsworth*. Edited by John Morley. London: Macmillan, 1888.

Wriothesley, Charles. *A Chronicle of England during the Reign of the Tudors, 1485–1559*. Edited by William Hamilton. 2 vols. London: Camden Society, 1875.

Zahn, Johann. *Oculus artificialis teledioptricus sive telescopium*. Herbipoli [Würzburg]: Sumptibus, Querini, Heyl . . . , 1685–86.

Zhivotovsky, L. A. "A Model of the Early Evolution of Soma-to-Germline Feedback." *Journal of Theoretical Biology* 216, no. 1 (2002): 51–57.

Secondary Sources

Adelmann, Howard B. *Marcello Malpighi and the Evolution of Embryology*. 5 vols. Ithaca, NY: Cornell University Press, 1966.

Alder, Ken. *Engineering the Revolution: Arms and Enlightenment in France, 1763-1815*. Princeton: Princeton University Press, 1999.

Al-Hassan, A.Y., Maqbul Ahmed, and A.Z. Iskandar, eds. *Science and Technology in Islam: The Different Aspects of Islamic Culture*. Vol. 4. Paris: UNESCO, 2001.

Alquié, Ferdinand. *Le cartésianisme de Malebranche*. Paris: J. Vrin, 1974.

Alsace, Conseil Régional de. *Orgues Silbermann d'Alsace*. Strasbourg, France: A.R.D.A.M. 1992.

Altick, Richard Daniel. *The Shows of London*. Cambridge, MA: Belknap Press, 1978.

Antognazza, Maria Rosa. *Leibniz: An Intellectual Biography*. Cambridge: Cambridge University Press, 2009.

Ariès, Philippe. *L'enfant et la vie familiale sous l'Ancien Régime*. Paris: Plon, 1960.

Ariew, Roger. *Descartes and the Last Scholastics*. Ithaca, NY: Cornell University Press, 1999.

———. "G. W. Leibniz, Life and Works." In *The Cambridge Companion to Leibniz*, edited by Nicholas Jolley, 18–42. Cambridge: Cambridge University Press, 1995.

Ariew, Roger, et al. *Historical Dictionary of Descartes and Cartesian Philosophy*. Lanham, MD: Scarecrow Press, 2003.

Ariew, Roger, John Cottingham, and Tom Sorell, eds. and trans. *Descartes' Meditations: Background Source Materials*. Cambridge: Cambridge University Press, 1998.

Ariew, Roger, and Marjorie Grene, eds. *Descartes and his Contemporaries: Meditations, Objections, and Replies*. Chicago: University of Chicago Press, 1995.

Armogathe, Jean-Robert. "Cartesian Physics and the Eucharist in the Documents of the Holy Office and the Roman Index (1671–1676)." In *Receptions of Descartes: Cartesianism and Anti-Cartesianism in Early Modern Europe*, edited by Tad M. Schmaltz, 149–70. London: Routledge, 2005.

———. "Caterus' Objections to God." In *Descartes and his Contemporaries: Meditations, Objections, and Replies*, edited by Roger Ariew and Marjorie Grene, 34–43. Chicago: University of Chicago Press, 1995.

Asaro, Peter M. "Computers as Models of the Mind: On Simulations, Brains, and the Design of Computers." In *The Search for a Theory of Cognition: Early Mechanisms and New Ideas*, edited by Stefano Franchi and Francesco Bianchini, 89–113. Amsterdam: Rodopi, 2011.

———. "From Mechanisms of Adaptation to Intelligence Amplifiers: The Philosophy of W. Ross Ashby." In *The Mechanical Mind in History*, edited by Phil Husbands, Owen Holland, and Michael Wheeler, 149–84. Cambridge, MA: MIT Press, 2008.

———. "Information and Regulation in Robots, Perception, and Consciousness: Ashby's Embodied Minds." *International Journal of General Systems* 38, no. 2 (2009): 111–28.

Ashby, Jill. "Biography: W. Ross Ashby (1903–1972)." *The W. Ross Ashby Digital Archive*. Accessed October 2013. http://www.rossashby.info/biography.html.

Aspray, William. *John von Neumann and the Origins of Modern Computing*. Cambridge, MA: MIT Press, 1990.

Auguste, Alphonse. "Gabriel de Ciron et Madame de Mondonville." *Revue historique de Toulouse* 2 (1915–19): 20–69.

Baker, Gordon, and Katherine J. Morris. *Descartes' Dualism*. London: Routledge, 1996.

Baker, Keith. *Condorcet: From Natural Philosophy to Social Mathematics*. Chicago: University of Chicago Press, 1975.

Bates, David. "Cartesian Robotics." In *Representations* 124, no 1 (Fall 2013): 43–68.

Battisti, Eugenio. *L'antirinascimento*. Milan: Feltrinelli, 1962.

Battisti, Eugenio and Guiseppa Saccaro Battisti. *Le macchine cifrate di Giovanni Fontana*. Milan: Arcadia, 1984.

Beaune, Jean-Claude. *L'automate et ses mobiles*. Paris: Flammarion, 1980.

———. "The Classical Age of Automata: An Impressionistic Survey from the Sixteenth to the Nineteenth Century." In *Fragments for a History of the Human Body*, edited by Michael Feher, Romana Nadaff, and Nadia Tazi, 3:430–80. New York: Zone Books, 1989.

Bedau, Mark A., and Paul Humphreys, eds. *Emergence: Contemporary Readings in Philosophy and Science*. Cambridge, MA: MIT Press, 2008.

Bedini, Silvio A. "The Role of Automata in the History of Technology." *Technology and Culture* 5, no. 1 (1964): 24–42.

Bedini, Silvio, and Francis R. Maddison, *Mechanical Universe: The Astrarium of Giovanni De' Dondi*. Philadelphia: American Philosophical Society, 1966.

Benhamou, Reed. "The Cover Design: The Artificial Limb in Preindustrial France." *Technology and Culture* 35, no. 4 (1994): 835–45.

———. "From *Curiosité* to *Utilité*: The Automaton in Eighteenth-Century France." *Studies in Eighteenth Century Culture* 17 (1987): 91–105.

Bennett, Jane. *Vibrant Matter: A Political Ecology of Things*. Durham, NC: Duke University Press, 2010.

Benocci, Carla. *Villa Aldobrandini a Roma*. Rome: Àrgos, 1992.

Bernardi, Walter. *Le metafisiche dell'embrione: Scienze della vita e filosofia da Malpighi a Spallanzani (1672–1793)*. Florence: L. S. Olschki, 1986.

Berryman, Sylvia. "The Imitation of Life in Ancient Greek Philosophy." In *Genesis Redux: Essays in the History and Philosophy of Artificial Life*, edited by Jessica Riskin, 35–45. Chicago: University of Chicago Press, 2007.

Besançon, Alain. *The Forbidden Image: An Intellectual History of Iconoclasm*. Chicago: University of Chicago Press, 2001.

Bevilacqua, Fabio. "Helmholtz's Über *die Erhaltung der Kraft*: The Emergence of a Theoretical Physicist." In *Hermann von Helmholtz and the Foundations of Nineteenth-Century Science*, edited by David Cahan, 291–333. Berkeley: University of California Press, 1993.

Bitbol-Hespériès, Annie. "Cartesian Physiology." In *Descartes' Natural Philosophy*, edited by Stephen Gaukroger, John Schuster, and John Sutton, 349–82. London: Routledge, 2000.

———. *Le principe de vie chez Descartes*. Paris: J. Vrin, 1990.

Blackwell, Richard J. *Galileo, Bellarmine, and the Bible: Including a Translation of Foscarini's Letter on the Motion of the Earth*. South Bend, IN: University of Notre Dame Press, 1991.

Bohnsack, Almut. *Der Jacquard-Webstuhl*. Munich: Deutsches Museum, 1993.

Boissier, Raymond. *La Mettrie, médicin, pamphlétaire et philosophe (1709–1751)*. Paris: Société d'édition "Les Belles Lettres," 1931.

Bongie, Laurence L. "Documents: A New Condillac Letter and the Genesis of the Traité des Sensations." *Journal of the History of Philosophy* 16 (1978): 83–94.

Bourdette, Jean Julien. *Le monastère de Saint-Sabi de Labéda (ou Saint-Savin de Lavedan) et la vie de Saint-Sabi, ermite*. Saint-Savin, au Presbytère et Toulouse: l'Auteur, 1911.

Boury, Dominique. "Irritability and Sensibility: Key Concepts in Assessing the Medical Doctrines of Haller and Bordeu." *Science in Context* 21, no. 4 (2008): 521–35.

Bouveresse, Renée. *Spinoza et Leibniz: L'idée d'animisme universel*. Paris: J. Vrin, 1992.

Bowler, Peter J. "The Changing Meaning of 'Evolution.'" *Journal of the History of Ideas* 36 (1975): 95–114.

———. "Darwinism and the Argument from Design: Suggestions for a Reevaluation." *Journal of the History of Biology* 10, no. 1 (1977): 29–43.

———. *The Eclipse of Darwinism: Anti-Darwinian Evolution Theories in the Decades Around 1900*. Baltimore: Johns Hopkins University Press, 1983.

———. *Evolution: The History of an Idea*. 3rd ed. Berkeley: University of California Press, 2003.

———. "Preformation and Pre-existence in the Seventeenth Century: A Brief Analysis." *Journal of the History of Biology* 4, no. 2 (1971): 221–44.

Brain, Robert M., and M. Norton Wise. "Muscles and Engines: Indicator Diagrams and Helmholt's Graphical Methods." In *The Science Studies Reader*, edited by Mario Biagioli, 51–66. New York: Routledge, 1999.

Braun, Marta. *Picturing Time: The Work of Etienne-Jules Marey (1830–1904)*. Chicago: University of Chicago Press, 1992.

Breitenbach, Angela. "Teleology in Biology: A Kantian Approach." *Kant Yearbook* 1 (2009): 31–56.

Bremmer, Jan M., and Herman Roodenburg, eds. *A Cultural History of Humour: From Antiquity to the Present Day*. Cambridge, UK: Polity Press, 1997.

Briggs, Asa. "History and the Social Sciences." In *A History of the University in Europe*, edited by Walter Rüegg, 3:459–89. Cambridge: Cambridge University Press, 2004.

Bröer, Ralf. *Salomon Reisel (1625–1701): Barocke Naturforschung eines Leibarztes im Banne der mechanistischen Philosophie*. Leipzig: Barth, 1996.

Brooke, John Hedley. *Science and Religion: Some Historical Perspectives* Cambridge: Cambridge University Press, 1991.

———. "Science and the Fortunes of Natural Theology: Some Historical Perspectives." *Zygon: Journal of Religion and Science* 24, no.1 (1989): 3–22.

———. "Scientific Thought and Its Meaning for Religion: The Impact of French Science on British Natural Theology, 1827–1859." *Revue de synthèse* 4, no. 1 (1989): 33–59.

———. "'Wise Men Nowadays Think Otherwise': John Ray, Natural Theology and the Meanings of Anthropocentrism." *Notes and Records of the Royal Society of London* 54, no. 2 (2000): 199–213.

Brooke, John Hedley, and Ian Maclean, eds. *Heterodoxy in Early Modern Science and Religion*. Oxford: Oxford University Press, 2006.

Brooke, John Hedley, Margaret J. Osler, and Jitse M. van der Meer, eds. "Science in Theistic Contexts: Cognitive Dimensions." Published as *Osiris* 16 (2001).

Brown, Stuart. "The Seventeenth-Century Intellectual Background." In *The Cambridge Companion to Leibniz,* edited by Nicholas Jolley, 43–66. Cambridge: Cambridge University Press, 1995.

Brown, Theodore M. "From Mechanism to Vitalism in Eighteenth-Century English Physiology." *Journal of the History of Biology* 7, no. 2 (1974): 179–216.

———. "Physiology and the Mechanical Philosophy in Mid-Seventeenth-Century-England." *Bulletin of the History of Medicine* 51 (1977): 25–54.

Bruce, J. Douglas. "Human Automata in Classical Tradition and Medieval Romance." *Modern Philology* 10 (April 1913): 511–26.

Buccheri, Alessandra. *The Spectacle of the Clouds, 1439–1650: Italian Art and Theatre.* London: Ashgate, 2014.

Buchner, Alexander. *Mechanical Musical Instruments.* Translated by Iris Urwin. London: Batchworth Press, 1959.

Buckland, William Warwick. *The Main Institutions of Roman Private Law.* 1931. Reprint edition. Cambridge: Cambridge University Press, 2011.

———. *A Text-Book of Roman Law: From Augustus to Justinian.* 3rd ed. Cambridge: Cambridge University Press, 2007.

Burckhardt, Jacob. *The Civilization of the Renaissance in Italy.* 1860. London: Penguin, 1990.

Burke, Jill. "Meaning and Crisis in the Early Sixteenth Century: Interpreting Leonardo's Lion." *Oxford Art Journal* 29, no. 1 (2006): 77–91.

Burke, Peter. "Frontiers of the Comic in Early Modern Italy." In *Varieties of Cultural History,* 77–93. Ithaca, NY: Cornell University Press, 1997.

Burkhardt, Jr., Richard W. *The Spirit of System: Lamarck and Evolutionary Biology.* New ed. Cambridge, MA: Harvard University Press, 1995.

Burwick, Frederick, ed. *Approaches to Organic Form: Permutations in Science and Culture.* Dordrecht, the Netherlands: Reidel, 1987.

Bynum, Caroline Walker. *Christian Materiality: An Essay on Religion in Late Medieval Europe.* New York: Zone Books, 2011.

———. *The Resurrection of the Body in Western Christianity, 200–1336.* New York: Columbia University Press, 1995.

Cahan, David. "The 'Imperial Chancellor of the Sciences': Helmholtz between Science and Politics." *Social Research* 73, no. 4 (2006): 1093–1128.

Campbell-Kelly, Martin and William Aspray. *Computer: A History of the Information Machine.* New York: Basic Books, 1996.

Canguilhem, Georges. *La connaissance de la vie.* 2nd ed. Paris: J. Vrin, 1965.

———. *Études d'histoire et de philosophie des sciences.* Paris: J. Vrin, 1968.

———. "The Role of Analogies and Models in Biological Discovery." In *Scientific Change: Historical Studies in the Intellectual, Social, and Technical Conditions for Scientific Discovery and Technical Invention, from Antiquity to the Present,* edited by A. Crombie, translated by J. A. Z. Gardin and G. Kitchin, 510–12. New York: Heineman, 1963.

Cardwell, D.S.L. *From Watt to Clausius: The Rise of Thermodynamics in the Early Industrial Age.* London: Heinemann Educational, 1971.

———. "Some Factors in the Early Development of the Concepts of Power, Work and Energy." *British Journal for the History of Science* 3, no. 3 (1967): 209–24.

Carlin, Lawrence. "Leibniz on Conatus, Causation, and Freedom." *Pacific Philosophical Quarterly* 85, no. 4 (2004): 365–79.

Carrera, Roland, Dominique Loiseau, and Olivier Roux. *Androïdes. Les automates des Jaquet-Droz*. Lausanne: Scriptar, 1979.

Carroll, Charles Michael. *The Great Chess Automaton*. New York: Dover Publications, 1975.

Carvallo, Sarah, ed. *La controverse entre Stahl et Leibniz sur la vie, l'organisme et le mixte*. Paris: J. Vrin, 2004.

Ceccarelli, Leah. *Shaping Science with Rhetoric: The Cases of Dobzhansky, Schrödinger and Wilson*. Chicago: University of Chicago Press, 2001.

Chapuis, Alfred. "The Amazing Automata at Hellbrunn." *Horological Journal* 96, no. 6 (June 1954): 388–89.

Chapuis, Alfred, and Edmond Droz. *Automata: A Historical and Technological Study*. Translated by Alec Reid. Geneva: Editions du Griffon, 1958.

———. *The Jaquet-Droz Mechanical Puppets*. Neuchâtel, Switzerland: Historical Museum, 1956.

Chapuis, Alfred, and Edouard Gélis. *Le monde des automates*. 2 vols. Paris: Blondel La Rougery, 1928.

Charle, Christophe. "Patterns." In *A History of the University in Europe*, edited by Walter Rüegg, 3:33–75. Cambridge: Cambridge University Press, 2004.

Chavigny, Jean. *Le Roman d'un artiste: Robert-Houdin, Rénovateur de la magie blanche*. Orléans, France: Imprimerie industrielle, 1969.

Churchill, Frederick B. "August Weismann: A Developmental Evolutionist." In *August Weismann: Ausgewählte Briefe und Dokumente, Selected Letters and Documents*, edited by Frederick B. Churchill and Helmut Risler, 2:749–98. Freiburg im Breisgau, Germany: Universitätsbibliothek, 1999.

———. "August Weismann and a Break from Tradition." *Journal of the History of Biology* 1, no. 1 (1968): 91–112.

———. "Chabry, Roux, and the Experimental Method in Nineteenth-Century Embryology." In *Foundations of Scientific Method: The Nineteenth Century*, edited by Ronald N. Giere and Richard S. Westfall, 161–205. Bloomington: Indiana University Press, 1973.

———. "From Machine-Theory to Entelechy: Two Studies in Developmental Teleology." *Journal of the History of Biology* 2, no. 1 (1969): 165–85.

———. "The Weismann-Spencer Controversy over the Inheritance of Acquired Characters." In *Human Implications of Scientific Advancements: Proceedings of the XV International Congress for the History of Science*, edited by Eric G. Forbes, 451–68. Edinburgh: Edinburgh University Press, 1977.

Cipolla, Carlo M. *Clocks and Culture, 1300–1700*. 1967. Reprinted with an introduction by Anthony Grafton. New York: W. W. Norton, 2003.

Clark, William. *Academic Charisma and the Origins of the Research University*. Chicago: University of Chicago Press, 2006.

Clarke, Desmond. *Descartes: A Biography*. Cambridge: Cambridge University Press, 2006.

Clayton, Philip and Paul Davies, eds. *The Re-Emergence of Emergence: The Emergentist Hypothesis from Science to Religion*. Oxford: Oxford University Press, 2006.

Close, A. J. "Commonplace Theories of Art and Nature in Classical Antiquity and the Renaissance." *Journal of the History of Ideas* 10 (1969): 467–86.

Cohen, Claudine. *Science, libertinage et clandestiné à l'aube des lumières: Le transformisme de Telliamed*. Paris: Presses universitaires de France, 2011.

Cohen, Emily-Jane. "Enlightenment and the Dirty Philosopher." *Configurations* 5, no. 3 (1997): 369–424.

Cohen, Gustave. *Histoire de la mise en scène dans les théatre religieux francais du moyen âge*. Paris: Champion, 1926.

Cohen, Sarah R. "Chardin's Fur: Painting, Materialism, and the Question of the Animal Soul." *Eighteenth-Century Studies* 38, no.1 (2004): 39–61.

Conway, Flo, and Jim Siegelman. *Dark Hero of the Information Age: In Search of Norbert Wiener, the Father of Cybernetics*. New York: Basic Books, 2005.

Cook, Margaret G. "Divine Artifice and Natural Mechanism: Robert Boyle's Mechanical Philosophy of Nature." In "Science in Theistic Contexts: Cognitive Dimensions," edited by John Hedley Brooke, Margaret J. Osler, and Jitse M. van der Meer, published as *Osiris* 16 (2001): 133–50.

Copeland, B. Jack, ed. *Alan Turing's Automatic Computing Engine: The Master Codebreaker's Struggle to Build the Modern Computer*. Oxford: Oxford University Press, 2005.

Copeland, B. Jack, and Diane Proudfoot, "On Alan Turing's Anticipation of Connectionism." *Synthèse: International Journal for Epistemology, Methodology and Philosophy of Science* 108 (1996): 361–77.

Cordeschi, Roberto. "Cybernetics." In *The Blackwell Guide to the Philosophy of Computing and Information*, edited by Luciano Floridi, 186–96. Oxford, UK: Blackwell, 2004.

Corradini, Antonella and Timothy O'Connor, eds. *Emergence in Science and Philosophy*. New York: Routledge, 2010.

Corsi, Pietro. "Before Darwin: Transformist Concepts in European Natural History." *Journal of the History of Biology* 38, no. 1 (2005): 67–83.

———. "Biologie." In *Lamarck, philosophe de la nature*, edited by Pietro Corsi, Jean Gayon, Gabriel Gohau, and Stéphane Tirard, 37–64. Paris: Presses universitaires de France, 2006.

Costabel, Pierre. *La signification d'un débat sur trente ans (1728–1758): La question des forces vives*. Paris: Centre national de la recherche scientifique, 1983.

Cowie, Murray A., and Marian L. Cowie. "Geiler von Kayserberg and Abuses in Fifteenth Century Strassburg." *Studies in Philology* 58 (1961): 483–95.

Coudert, Allison P. "Henry More, the Kabbalah, and the Quakers." In *Philosophy, Science, and Religion in England, 1640–1700*, edited by Richard Kroll, Richard Ashcraft, and Perez Zagorin, 31–67. Cambridge: Cambridge University Press, 1992.

Crocker, Lester. "Diderot and Eighteenth-Century French Transformism." In *Forerunners of Darwin, 1745–1859*, edited by Bentley Glass, Owsei Temkin, and William L. Strauss, Jr., 114–43. Baltimore: Johns Hopkins University Press, 1959.

Crombie, A.C., ed. *Scientific Change: Historical Studies in the Intellectual, Social and Technical*

Conditions for Scientific Discovery and Technical Invention from Antiquity to the Present. New York: Basic Books, 1963.

Culver, Stuart. "What Manikins Want: The Wonderful Wizard of Oz and the Art of Decorating Dry Goods Windows." *Representations*, no. 21 (1988): 97–116.

Cunningham, Andrew and Nicholas Jardine, eds. *Romanticism and the Sciences.* Cambridge: Cambridge University Press, 1990.

Dagognet, François. *L'animal selon Condillac: Une introduction au "Traité des animaux" de Condillac.* Paris: Vrin, 2004.

Damasio, Antonio. *Descartes's Error: Emotion, Reason and the Human Brain.* New York: HarperCollins, 1994.

d'Ancona, Alessandro. *Origini del teatro in Italia.* 2 vols. Florence: Successori le Monnier, 1877.

Darnton, Robert. *The Forbidden Bestsellers of Prerevolutionary France.* New York: W. W. Norton, 1996.

———. *The Great Cat Massacre and Other Episodes in French Cultural History.* New York: Vintage, 1985.

———. *Mesmerism and the End of the Enlightenment in France.* Cambridge, MA: Harvard University Press, 1968.

Darrigol, Olivier. "God, Waterwheels, and Molecules: Saint-Venant's Anticipation of Energy Conservation." *Historical Studies in the Physical and Biological Sciences* 31, no. 2 (2001): 285–353.

Daston, Lorraine. *Classical Probability in the Enlightenment.* Princeton: Princeton University Press, 1988.

———. "Enlightenment Calculations." *Critical Inquiry* 21, no. 1 (1994): 182–202.

Daston, Lorraine and Katharine Park. *Wonders and the Order of Nature, 1150-1750.* Cambridge: Zone Books, 2001.

Davidson, Luke. "'Identities Ascertained': British Ophthalmology in the First Half of the Nineteenth Century." *Social History of Medicine* 9, no. 3 (1996): 313–33.

Dawson, Virginia P. *Nature's Enigma: The Problem of the Polyp in the Letters of Bonnet, Trembley, and Réaumur.* Philadelphia: American Philosophical Society, 1987.

Dear, Peter. *The Intelligibility of Nature: How Science Makes Sense of the World.* Chicago: University of Chicago Press, 2006.

———. "A Mechanical Microcosm: Bodily Passions, Good Manners, and Cartesian Mechanism." In *Science Incarnate: Historical Embodiments of Natural Knowledge*, edited by Christopher Lawrence and Steven Shapin, 51–82. Chicago: University of Chicago Press, 1998.

DeJong-Lambert, William. *The Cold War Politics of Genetic Research: An Introduction to the Lysenko Affair.* Dordrecht, the Netherlands: Springer, 2012.

Depew, David J. "Baldwin and His Many Effects." In *Evolution and Learning: The Baldwin Effect Reconsidered*, edited by Bruce H. Weber and David J. Depew, 3–31. Cambridge, MA: MIT Press, 2003.

De Robeck, Nesta. *The Christmas Crib.* Milwaukee: Bruce Publishing Co., 1996.

Des Chene, Dennis. "Abstracting from the Soul: The Mechanics of Locomotion." In *Genesis Redux: Essays in the History and Philosophy of Artificial Life*, edited by Jessica Riskin, 85–95. Chicago: University of Chicago Press, 2007.

———. "Animal as Concept: Bayle's 'Rorarius.'" In *The Problem of Animal Generation in Early Modern Philosophy*, edited by Justin E. H. Smith, 216–31. Cambridge: Cambridge University Press, 2006.

———. "Descartes and the Natural Philosophy of the Coimbra Commentaries." In *Descartes' Natural Philosophy*, edited by Stephen Gaukroger, John Schuster, and John Sutton, 29–45. London: Routledge, 2000.

———. *Life's Form: Late Aristotelian Conceptions of the Soul*. Ithaca, NY: Cornell University Press, 2000.

———. "Mechanisms of Life in the Seventeenth Century: Borelli, Perrault, Régis." *Studies in the History and Philosophy of Science Part C: Studies in History and Philosophy of Biological and Biomedical Sciences* 36, no. 2 (2005): 245–60.

———. *Physiologia: Natural Philosophy in Late Aristotelian and Cartesian Thought*. Ithaca, NY: Cornell University Press, 1996.

———. *Spirits and Clocks: Machine and Organism in Descartes*. Ithaca, NY: Cornell University Press, 2001.

Desmond, Adrian. *Huxley: From Devil's Disciple to Evolution's High Priest*. Reading, UK: Addison-Wesley, 1997.

———. *The Politics of Evolution: Morphology, Medicine and Reform in Radical London*. Chicago: University of Chicago Press, 1989.

Dijksterhuis, Fokko Jan. *Lenses and Waves: Christiaan Huygens and the Mathematical Science of Optics in the Seventeenth Century*. Dordrecht, the Netherlands: Kluwer, 2004.

Dobbs, Betty Jo. "Studies in the Natural Philosophy of Sir Kenelm Digby." Pts. 1–3. *Ambix* 18 (1971): 1–25; *Ambix* 20 (1973): 143–63; *Ambix* 21 (1974): 1–28.

Dolezal, Mary-Lyon, and Maria Mavroudi. "Theodore Hyrtakenos' *Description of the Garden of St Anna* and the Ekphrasis of Gardens." In *Byzantine Garden Culture*, edited by Anthony Littlewood, Henry Maguire, and Joachim Wolschke-Bulmahn, 105–58. Washington, D.C.: Dumbarton Oaks, 2002.

Douthwaite, Julia. *The Wild Girl, Natural Man, and the Monster: Dangerous Experiments in the Age of Enlightenment*. Chicago: University of Chicago Press, 2002.

Doyon, André, and Lucien Liaigre. *Jacques Vaucanson, mécanicien de génie*. Paris: Presses Universitaires de France, 1967.

Doyon, André and Lucien Liaigre. "Méthodologie comparée du biomécanisme et de la mécanique comparée." *Dialectica* 10 (1956): 292–335.

Dressler, Stephan. "La Mettrie's Death, or: The Nonsense of an Anecdote." *Neophilologus* 75, no. 2 (1991): 194–99.

Dronamraju, Krishna R. "Erwin Schrödinger and the Origins of Molecular Biology." *Genetics* 153, no. 3 (1999): 1071–76.

Drover, C. B. "Thomas Dallam's Organ Clock." *Antiquarian Horology* 1, no. 10 (1956): 150–52.

Dubuisson, Marguerite, Jacques Payen and Jean Pilisi, "The Textile Industry." In *A History of Technology and Invention*, edited by Maurice Daumas and translated by Eileen B. Hennessy, vol. 2, chap. 11. New York: Crown Publishers, 1964.

Duchesneau, François. "Charles Bonnet's Neo-Leibnizian Theory of Organic Bodies." In *The Problem of Animal Generation in Early Modern Philosophy*, edited by Justin E. H. Smith, chap. 13. Cambridge: Cambridge University Press, 2006.

———. "Leibniz's Model for Organizing Organic Phenomena." *Perspectives on Science* 11, no. 4 (2003): 378–409.

———. *Les modèles du vivant de Descartes à Leibniz.* Paris: J. Vrin, 1998.

———. *La physiologie des lumières: Empirisme, modèles et théories.* The Hague: M. Nijhoff, 1982.

Dulac, Georges. "Une version déguisée du Rêve du d'Alembert: le manuscript de Moscou (1774)." In *Recherches nouvelles sur quelques écrivains des Lumières II,* edited by Jacques Proust, 131–17. Montpellier, France: Université Paul Valéry, Centre d'étude du XVIIIe siècle, 1979.

Duret, Edmond. "L'horloge historique de Nyort en Poitou, fabriquée en 1750 par Jean Bouhin." *Revue Poitevine et Saintongeaise* 6 (1889): 432–34.

During, Simon. *Modern Enchantments: The Cultural Power of Secular Magic.* Cambridge, MA: Harvard University Press, 2004.

Dyrness, William A. *Reformed Theology and Visual Culture: The Protestant Imagination from Calvin to Edwards.* Cambridge: Cambridge University Press, 2004.

Dyson, George B. *Darwin among the Machines: The Evolution of Global Intelligence.* Reading, UK: Addison-Wesley, 1997.

Edwards, Pamela. *The Statesman's Science: History, Nature, and Law in the Political Thought of Samuel Taylor Coleridge.* New York: Columbia University Press, 2004.

Eire, Carlos M. N. *War against the Idols: The Reformation of Worship from Erasmus to Calvin.* Cambridge: Cambridge University Press, 1989.

Elias, Norbert. *The Civilizing Process: Sociogenetic and Psychogenetic Investigations.* 1939. Translated by Edmund Jephcott. London: Blackwell, 1994.

Erizzo, Nicolò. *Relazione storico-critica della Torre dell' Orologio di S. Marco in Venezia.* Venice: Tip. del commercio, 1860.

Evans, Henry Ridgely. *Edgar Allan Poe and Baron von Kempelen's Chess-Playing Automaton.* Kenton, OH: International Brotherhood of Magicians, 1939.

———. *A Master of Modern Magic: The Life and Adventures of Robert-Houdin.* New York: Macoy Publishing Company, 1932.

Fangerau, Heiner. "From Mephistopheles to Isaiah: Jacques Loeb, Technical Biology, and War." *Social Studies of Science* 39 (2009): 229–56.

Farber, Paul Lawrence. *Finding Order in Nature: The Naturalist Tradition from Linnaeus to E. O. Wilson.* Baltimore: Johns Hopkins University Press, 2000.

Faulkner, Quentin. *Wiser than Despair: The Evolution of Ideas in the Relation of Music and the Christian Church.* Westport, CT: Greenwood Press, 1996.

Fearing, Franklin. *Reflex Action: A Study in the History of Physiological Psychology.* Baltimore: The Williams and Wilkins Co., 1930.

Fechner, Christian. *La magie de Robert-Houdin: Une vie d'artiste.* 4 vols. Boulogne, France: Éditions FCF, 2002. (*MRH*)

Feltz, Bernard, Marc Crommelinck, and Philippe Goujon, eds. *Self-Organization and Emergence in Life Sciences.* Dordrecht, the Netherlands: Springer, 2006.

Fiedler, Wilfried. *Analogiemodelle bei Aristotles: Untersuchungen zu den Vergleichen zwischen den einzelen Wissenschaften un Künstern.* Amsterdam: B. R. Grüner, 1978.

Findlen, Paula. "Introduction: The Last Man Who Knew Everything . . . or Did He? Athanasius Kircher, S. J. (1602–1680) and His World." In *Athanasius Kircher: The Last Man Who Knew Everything*, 1–48. New York: Routledge, 2004.

———. "Scientific Spectacle in Baroque Rome: Athanasius Kircher and the Roman College Museum." *Roma moderna e contemporanea* 3, no. 3 (1995): 625–65.

Finkelstein, Gabriel. *Emil du Bois-Reymond: Neuroscience, Self and Society in Nineteenth-Century Germany*. Cambridge, MA: MIT Press, 2013.

Fisher, Saul. "Gassendi's Atomist Account of Generation and Heredity in Plants and Animals." *Perspectives on Sciences* 11, no. 4 (2003): 484–512.

Flanagan, James L. *Speech Analysis, Synthesis, and Perception*. Berlin: Springer, 1965.

———. "Voices of Men and Machines." *Journal of the Acoustical Society of America* 51 (1972): 1375–87.

Floridi, Luciano, ed. *The Blackwell Guide to the Philosophy of Computing and Information*. Oxford, UK: Blackwell, 2004.

Fowler, C. F. *Descartes on the Human Soul: Philosophy and the Demands of Christian Doctrine*. Dordrecht, the Netherlands: Kluwer Academic Publishers, 1999.

Franchi, Stefano. "Life, Death, and Resurrection of the Homeostat." In *The Search for a Theory of Cognition: Early Mechanisms and New Ideas*, edited by Stefano Franchi and Francesco Bianchini, 3–52. Amsterdam: Rodopi, 2011.

Franchi, Stefano, and Francesco Bianchini. "Introduction: On the Historical Dynamics of Cognitive Science: A View from the Periphery." In *The Search for a Theory of Cognition: Early Mechanisms and New Ideas*, edited by Stefano Franchi and Francesco Bianchini, xiii–xxviii. Amsterdam: Rodopi, 2011.

Francis, Richard C. *Epigenetics: The Ultimate Mystery of Inheritance*. New York: W. W. Norton, 2011.

Freedberg, David. *The Eye of the Lynx: Galileo, His Friends, and the Beginnings of Modern Natural History*. Chicago: University of Chicago Press, 2002.

———. *The Power of Images: Studies in the History and Theory of Response*. Chicago: University of Chicago Press, 1989.

Fryer, David M., and John C. Marshall. "The Motives of Jacques Vaucanson." *Technology and Culture* 20 (1979): 257–69.

Fuchs, Thomas. *The Mechanization of the Heart: Harvey and Descartes*. Translated by Marjorie Grene. Rochester, NY: University of Rochester Press, 2001.

Fulford, Tim, ed. *Romanticism and Science, 1773–1833*. London: Routledge, 2002.

Fyfe, Aileen. "The Reception of William Paley's Natural Theology in the University of Cambridge." *British Journal for the History of Science* 30 (1997): 321–35.

———. *Science and Salvation: Evangelical Popular Science Publishing in Victorian Britain*. Chicago: University of Chicago Press, 2004.

Gabbey, Alan. "Force and Inertia in the Seventeenth Century: Descartes and Newton." In *Descartes: Philosophy, Mathematics, and Physics*, edited by Stephen Gaukroger, 230–320. Sussex, UK: Harvester Press, 1980.

Galison, Peter. "The Ontology of the Enemy: Norbert Wiener and the Cybernetic Vision." *Critical Inquiry* 21 (1994): 228–66.

Galluzzi, Paolo. *Gli ingegneri del Rinascimento da Brunelleschi a Leonardo da Vinci*. Bologna: Instituto e Museo di storia della scienza, 1996.

———. "Leonardo da Vinci: From the 'Elementi Macchinali' to the Man-Machine." *History and Technology* 4 (1987): 235–65.

Gamba, Enrico, and Vico Montebelli, eds. *Macchine da teatro e teatri di macchine: Branca, Sabbatini, Torelli scenotecnici e meccanici del Seicento: Catalogo della mostra*. Urbino, Italy: Quattroventi, 1995.

Garber, Daniel. "Descartes and the Scientific Revolution: Some Kuhnian Reflections." *Perspectives on Science* 9, no. 4 (2001): 405–22.

———. *Descartes Embodied: Reading Cartesian Philosophy through Cartesian Science*. Cambridge: Cambridge University Press, 2000.

———. *Descartes' Metaphysical Physics*. Chicago: University of Chicago Press, 1992.

———. "J.-B. Morin and the Second Objections." In *Descartes and his Contemporaries: Meditations, Objections, and Replies*, edited by Roger Ariew and Marjorie Grene, 63–82. Chicago: University of Chicago Press, 1995.

———. "Leibniz and the Foundations of Physics: The Middle Years." In *The Natural Philosophy of Leibniz*, edited by K. Okruhlik and J. R. Brown, 27–130. Dordrecht, the Netherlands: Reidel, 1985.

———. *Leibniz: Body, Substance and Monad*. Oxford: Oxford University Press, 2009.

———. "Leibniz: Physics and Philosophy." In *The Cambridge Companion to Leibniz*, edited by Nicholas Jolley, 270–352. Cambridge: Cambridge University Press, 1995.

———. "Soul and Mind: Life and Thought in the Seventeenth Century." In *The Cambridge History of Seventeenth-Century Philosophy*, edited by Daniel Garber and Michael Ayers, 1:757–95. Cambridge: Cambridge University Press, 1998.

Garber, Daniel, and Béatrice Longuenesse, eds. *Kant and the Early Moderns*. Princeton, NJ: Princeton University Press, 2008.

García-Diego, José A. *Juanelo Turriano, Charles V's Clockmaker: The Man and His Legend*. Madrid: Editorial Castalia, 1986.

———. *Los relojes y autómatas de Juanelo Turriano*. Madrid: Tempvs Fvgit, Monografias Españolas de Relojería, 1982.

Gargano, Pietro. *Il presepio: Otto secoli di storia, arte, tradizione*. Milan: Fenice 2000, 1995.

Garside, Charles. *Zwingli and the Arts*. New Haven, CT: Yale University Press, 1966.

Gaskin, J. C. A. "Hume on Religion." In *The Cambridge Companion to Hume*, edited by David Fate Norton, 480–514. 2nd ed. Cambridge: Cambridge University Press, 2009.

Gass, Joseph. *Les Orgues de la Cathédrale de Strasbourg à travers les siècles: Etude historique, ornée de gravures et de planches hors texte, à l'occasion de la bénédiction des Grandes Orgues Silbermann-Roethinger, le 7 juillet 1935*. Reprint. [n.p.]: Librissimo-Phénix Éditions, 2002.

Gasté, Armand. *Les drames liturgiques de la Cathédrale de Rouen*. Evreux, France: Imprimerie de l'Eure, 1893.

Gaukroger, Stephen. *Descartes: An Intellectual Biography*. Oxford: Oxford University Press, 1995.

———. *Descartes' System of Natural Philosophy*. Cambridge: Cambridge University Press, 2002.

———. *Descartes: The World and other Writings*. Cambridge: Cambridge University Press, 1998.

———. "The Resources of a Mechanist Physiology and the Problem of Goal-Directed Processes." In *Descartes' Natural Philosophy*, edited by Stephen Gaukroger, John Schuster, and John Sutton, 383–400. London: Routledge, 2000.

Gaukroger, Stephen, John Schuster, and John Sutton, eds. *Descartes' Natural Philosophy*. London: Routledge, 2000.

Gayon, Jean. "Hérédité des caractères acquis." In *Lamarck, philosophe de la nature*, edited by Pietro Corsi, Jean Gayon, Gabriel Gohau, and Stéphane Tirard, 105–64. Paris: Presses universitaires de France, 2006.

Geison, Gerald. *The Private Science of Louis Pasteur*. Princeton, NJ: Princeton University Press, 1995.

Ghiselin, Michael T. "The Imaginary Lamarck: A Look at 'Bogus History' in Schoolbooks." *The Textbook Letter* (September–October 1994). Accessed October 2013. http://www.textbookleague.org/54marck.htm.

———. *Metaphysics and the Origin of Species*. Albany: SUNY Press, 1997.

———. *The Triumph of the Darwinian Method*. Berkeley: University of California Press, 1969.

Gigante, Denise, ed. *Gusto: Essential Writings in Nineteenth-Century Gastronomy*. New York: Routledge, 2005.

———. *Life: Organic Form and Romanticism*. New Haven, CT: Yale University Press, 2009.

Giglioni, Guido. "Automata Compared: Boyle, Leibniz and the Debate on the Notion of Life and Mind." *British Journal for the History of Philosophy* 3 (1995): 249–78.

Gillespie, Neal C. "Divine Design and the Industrial Revolution: William Paley's Abortive Reform of Natural Theology." *Isis* 81, no. 2 (1990): 214–29.

Gillispie, Charles Coulston. *Science and Polity in France at the End of the Old Regime*. Princeton, NJ: Princeton University Press, 1980.

Gilson, Etienne. *Études sur le rôle de la pensée médiévale dans la formation du système cartésien*. Paris: J. Vrin, 1930.

———. *Index scolastico-cartésien*. 1912. 2nd ed. Paris: J. Vrin, 1979.

Ginsborg, Hannah. "Kant on Aesthetic and Biological Purposiveness." In *Reclaiming the History of Ethics: Essays for John Rawls*, edited by Andrews Reath, Barbara Herman, and Christine M. Korsgaard, 329–60. Cambridge: Cambridge University Press, 1997.

———. "Kant on Understanding Organisms as Natural Purposes." In *Kant and the Sciences*, edited by Eric Watkins, 231–58. Oxford: Oxford University Press, 2001.

———. "Kant's Biological Teleology and its Philosophical Significance." In *A Companion to Kant*, edited by Graham Bird, 455–69. Malden, MA: Blackwell, 2006.

———. "Lawfulness without a Law: Kant on the Free Play of Imagination and Understanding." *Philosophical Topics* 25, no. 1 (1997): 37–81.

———. "Oughts without Intentions: A Kantian Perspective on Biological Teleology." In *Kant's Theory of Biology*, edited by Ina Goy and Eric Watkins. Berlin: De Gruyter, forthcoming.

Givens, Larry. *Re-enacting the Artist: A Story of the Ampico Reproducing Piano*. Vestal, NY: Vestal Press, 1970.

Glass, Bentley, Owsei Temkin, and William L. Straus, Jr., eds. *Forerunners of Darwin, 1745–1859*. Baltimore: Johns Hopkins University Press, 1968.

Glasser, Adrian. "A History of Studies of Visual Accommodation in Birds." *Quarterly Review of Biology* 71, no. 4 (1996): 475–509.

Gliboff, Sander. "The Golden Age of Lamarckism, 1866–1926." In *Transformations of Lamarckism: From Subtle Fluids to Molecular Biology*, edited by Snait Gissis and Eva Jablonka, 45–56. Cambridge, MA: MIT Press, 2011.

Goldstein, Jan. *The Post-Revolutionary Self: Politics and Psyche in France, 1750–1850*. Cambridge, MA: Harvard University Press, 2005.

Goldstein, Jeffrey. "Emergence as a Construct: History and Issues." *Emergence: Complexity and Organization* 1, no. 1 (1999): 49–72.

Goldstein, Rebecca. *Incompleteness: The Proof and Paradox of Kurt Gödel*. New York: W. W. Norton, 2005.

Golinski, Jan. "The Literature of the New Sciences." In *The Cambridge History of English Romantic Literature*, edited by James Chandler, 527–52. Cambridge: Cambridge University Press, 2008.

Gorman, Michael John. "Between the Demonic and the Miraculous: Athanasius Kircher and the Baroque Culture of Machines." In *The Great Art of Knowing: The Baroque Encyclopedia of Athanasius Kircher*, edited by Daniel Stolzenberg, 59–70. Stanford, CA: Stanford University Libraries, 2001.

Gosse, Philip. *Dr. Viper, The Querulous Life of Philip Thicknesse*. London: Cassell, 1952.

Gouhier, Henri. *Cartésianisme et Augustinisme au XVIIe siècle*. Paris: J. Vrin, 1978.

Gouk, Penelope. "Making Music, Making Knowledge: The Harmonious Universe of Athanasius Kircher." In *The Great Art of Knowing: The Baroque Encyclopedia of Athanasius Kircher*, edited by Daniel Stolzenberg, 71–84. Stanford, CA: Stanford University Libraries, 2001.

Gould, Stephen Jay. "D'Arcy Thompson and the Science of Form." *New Literary History* 2, no. 2 (1971): 229–58.

———. "Darwin and Paley Meet the Invisible Hand." *Natural History* 99, no. 11 (1990): 8–16.

———. "Foreword." In *Georges Cuvier: An Annotated Bibliography of His Published Works*, edited by Jean Chandler Smith, vii–xi. Washington, DC: Smithsonian, 1993.

———. "On Heroes and Fools in Science." *Natural History* 83 (1974): 30–32.

———. *Ontogeny and Phylogeny*. Cambridge, MA: Harvard University Press, 1977.

Grafton, Anthony. "The Devil as Automaton." In *Genesis Redux: Essays in the History and Philosophy of Artificial Life*, edited by Jessica Riskin, 46–62. Chicago: University of Chicago Press, 2007.

———. "From New Technologies to Fine Arts: Alberti among the Engineers." In *Leon Battista Alberti: Master Builder of the Italian Renaissance*, 71–109. Cambridge, MA: Harvard University Press, 2002.

———. *What Was History? The Art of History in Early Modern Europe*. Cambridge: Cambridge University Press, 2007.

Grattan-Guinness, Ivor. "Work for the Workers: Advances in Engineering, Mechanics, and Instruction in France, 1800–1830." *Annals of Science* 41 (1984): 1–33.

Gregory, Frederick. *Nature Lost? Natural Science and the German Theological Traditions of the Nineteenth Century*. Cambridge, MA: Harvard University Press, 1992.

Gregory, Mary Efrosini. *Evolutionism in Eighteenth-Century French Thought*. New York: Peter Lang, 2008.

Greene, Robert. "Henry More and Robert Boyle on the Spirit of Nature." *Journal of the History of Ideas* 23, no. 4 (1962): 451–74.

Grene, Marjorie. *Descartes among the Scholastics*. Milwaukee: Marquette University Press, 1991.

———. "The Heart and the Blood: Descartes, Plemp, and Harvey." In *Essays on the Philosophy and Science of René Descartes*, edited by Stephen Voss, 324–36. Oxford: Oxford University Press, 1993.

Grmek, Mirko D. *Les legs de Claude Bernard*. Paris: Fayard, 1997.

Guyer, Hannah. "Organisms and the Unity of Science." In *Kant and the Sciences*, edited by Eric Watkins, 259–81. Oxford: Oxford University Press, 2001.

Häberlein, Mark. *The Fuggers of Augsburg: Pursuing Wealth and Honor in Renaissance Germany*. Charlottesville: University of Virginia Press, 2012.

Hahn, Thomas, and Alan Lupack, eds. *Retelling Tales: Essays in Honor of Russell Pick*. Cambridge, UK: Boydell and Brewer, 1997.

Hall, A. Rupert. *Philosophers at War: The Quarrel between Newton and Leibniz*. Cambridge: Cambridge University Press, 1980.

Hamel, Marie Pierre. "Notice historique abregée pour l'histoire de l'orgue." In *Nouveau manuel complet du facteur d'orgues*, xxiv–cxxv. Paris: L. Mulo, 1903.

Hamley, Edward Bruce. *Voltaire*. Edinburgh: William Blackwood and Sons, 1877.

Hanafi, Zakiya. *The Monster in the Machine: Magic, Medicine, and the Marvelous in the Time of the Scientific Revolution*. Durham, NC: Duke University Press, 2000.

Hankins, Thomas L. *Jean d'Alembert: Science and the Enlightenment*. New York: Taylor and Francis, 1990.

Hankins, Thomas L., and Robert J. Silverman. *Instruments and the Imagination*. Princeton, NJ: Princeton University Press, 1995.

Harkins, William Edward. *Karel Čapek*. New York: Columbia University Press, 1962.

Harkness, Deborah E. *John Dee's Conversations with Angels: Cabala, Alchemy, and the End of Nature*. Cambridge: Cambridge University Press, 1999.

Harrington, Anne. *Reenchanted Science: Holism in German Culture from Wilhelm II to Hitler*. Princeton, NJ: Princeton University Press, 1996.

Harrison, Peter. "Original Sin and the Problem of Knowledge in Early Modern Europe." *Journal of the History of Ideas* 62, no. 2 (2002): 239–59.

———. "The Virtues of Animals in Seventeenth-Century Thought." *Journal of the History of Ideas* 59, no. 3 (1998): 463–84.

Haspels, Jan Jaap. *Automatic Musical Instruments: Their Mechanics and their Music, 1580–1820*. Koedijk, Netherlands: Nirota, Muiziekdruk C.V. 1987.

Hatfield, Gary. "Force (God) in Descartes' Physics." *Studies in History and Philosophy of Science* 10 (1979): 113–40.

Hawkins, Miles. *Social Darwinism in European and American Thought, 1860–1945: Nature as Model and Nature as Threat*. Cambridge: Cambridge University Press, 1997.

Hayward, Rhodri. "The Tortoise and the Love-Machine: Grey Walter and the Politics of Electro-Encephalography." *Science in Context* 14, no. 4 (2001): 615–42.

Hehl, Ulrich von. "Universität und Konfession im 19./20. Jahrhundert." In *Universität, Religion und Kirchen*, edited by Rainer Christoph Schwinges, 2:277–301. Basel: Schwabe AG, Verlag, 2011.

Heilbron, J.L. *Galileo*. Oxford: Oxford University Press, 2010.

———. Electricity in the Seventeenth and Eighteenth Centuries. Berkeley: University of California Press, 1979.

———. *The Sun in the Church: Cathedrals as Solar Observatories*. Cambridge, MA: Harvard University Press, 1999.

Heims, Steve J. *John von Neumann and Norbert Wiener: From Mathematics to the Technologies of Life and Death*. Cambridge, MA: MIT Press, 1980.

Hiebert, Erwin. *Historical Roots of the Principle of Conservation of Energy*. Madison: State Historical Society of Wisconsin for the Department of History at the University of Wisconsin, 1962.

Higley, Sarah. "The Legend of the Learned Man's Android." In *Retelling Tales: Essays in Honor of Russell Pick*, edited by Thomas Hahn and Alan Lupack, 127–60. Cambridge, UK: Boydell and Brewer, 1997.

Hillerbrand, Hans J., ed. *The Protestant Reformation*. 1968. Revised ed. New York: Harper Perennial, 2009.

Hodges, Andrew. *Alan Turing: The Enigma*. 1983. Princeton, NJ: Princeton University Press, 2012.

Höflechner, Walter. "Universität, Religion und Kirchen: Zusammenfassung." In *Universität, Religion und Kirche*, edited by Rainer Christoph Schwinges, 557–64. Basel: Schwabe Verlag, 2011.

Hogg, Thomas Jefferson. *The Life of Percy Bysshe Shelley*. 1832. 2 vols. London: J. M. Dent, 1933.

Holland, Owen. "Grey Walter: The Pioneer of Real Artificial Life." In *Proceedings of the 5th International Workshop on Artificial Life*, edited by Christopher Langton, 34–44. Cambridge, MA: MIT Press, 1997.

Holmes, Frederic L. "Claude Bernard and the Vitalism of his Time." In *Vitalisms from Haller to Cell Theory*, edited by Guido Cimino and François Duchesneau, 281–95. Florence: Olschki, 1997.

Holton, Gerald James. *Science and Anti-Science*. Cambridge, MA: Harvard University Press, 1993.

Hopwood, Nick. "'Giving Body' to Embryos: Modeling, Mechanism, and the Microtome in Late Nineteenth-Century Anatomy." *Isis* 90, no. 3 (1999): 462–96.

———. "Pictures of Evolution and Charges of Fraud: Ernst Haeckel's Embryological Illustrations." *Isis* 97 (2006): 260–301.

———. "Producing Development: The Anatomy of Human Embryos and the Norms of Wilhelm His." *Bulletin of the History of Medicine* 74, no. 1 (2000): 29–79.

Howard, Thomas A. *Protestant Theology and the Making of the Modern German University*. Oxford: Oxford University Press, 2006.

Hülsen, Christian with Henri Schickhardt. "Ein deutscher Architekt in Florenz (1600)." *Mitteilungen des Kunsthistorischen Institutes in Florenz* 2, no. 5/6 (1917): 152–93.

Hunter, Graeme K. *Vital Forces: The Discovery of the Molecular Basis of Life*. London: Academic Press, 2000.

Hunter, Michael, and David Wootton. *Atheism from the Reformation to the Enlightenment*. Oxford, UK: Clarendon Press, 1992.

Hunter, Richard A., and Ida Macalpine. "William Harvey and Robert Boyle." *Notes and Records of the Royal Society of London* 13, no. 2 (1958): 115–27.

Husbands, Philip, and Owen Holland. "The Ratio Club: A Hub of British Cybernetics." In *The Mechanical Mind in History*, edited by Philip Husbands, Owen Holland, and Michael Wheeler, 91–148. Cambridge, MA: MIT Press, 2008.

Hutton, Sarah. "Edward Stillingfleet, Henry More, and the Decline of *Moses Atticus*: A Note on Seventeenth-Century Anglican Apologetics." In *Philosophy, Science, and Religion in England, 1640–1700*, edited by Richard Kroll, Richard Ashcraft, and Perez Zagorin, 68–84. Cambridge: Cambridge University Press, 1992.

Iggers, Georg G. *The German Conception of History: The National Tradition of Historical Thought from Herder to the Present*. Rev. ed. Middletown, CT: Wesleyan University Press, 1983.

———. "The Image of Ranke in American and German Historical Thought." *History and Theory* 2, no. 1 (1962): 17–40.

Iltis, Carolyn. "Leibniz and the Vis Viva Controversy." *Isis* 62, no. 1 (1971): 21–35.

Israel, Jonathan I. *Enlightenment Contested: Philosophy, Modernity, and the Emancipation of Man, 1670–1752*. New York: Oxford University Press, 2006.

———. *Radical Enlightenment: Philosophy and the Making of Modernity*. Oxford: Oxford University Press, 2001.

Jacob, John, ed. *John Joseph Merlin: The Ingenious Mechanick*. London: Greater London Council, 1985.

Jacob, Margaret. *The Newtonians and the English Revolution, 1689–1720*. Ithaca, NY: Cornell University Press, 1976.

Jacques Vaucanson. Exhibition catalog, Musée national des techniques. Paris: Conservatoire nationale des arts et métiers, 1983.

Janacek, Bruce. "Catholic Natural Philosophy: Alchemy and the Revivication of Sir Kenelm Digby." In *Rethinking the Scientific Revolution*, edited by Margaret J. Osler, 89–118. Cambridge: Cambridge University Press, 2000.

Jardine, Nicholas, J. A. Secord, and E. C. Spary, eds. *Cultures of Natural History*. Cambridge: Cambridge University Press, 1996.

Jauernig, Anja. "Kant's Critique of the Leibnizian Philosophy: Contra the Leibnizians, but Pro Leibniz." In *Kant and the Early Moderns*, edited by Daniel Garber and Béatrice Longuenesse, 41–63. Princeton, NJ: Princeton University Press, 2008.

Jaynes, Julian. "The Problem of Animate Motion in the Seventeenth Century." *Journal of the History of Ideas* 31 (1970): 219–34.

Jenkins, Jane E. "Arguing about Nothing: Henry More and Robert Boyle on the Theological Implications of the Void." In *Rethinking the Scientific Revolution*, edited by Margaret J. Osler, 153–80. Cambridge: Cambridge University Press, 2000.

"Job, le renard électronique." *Musée des arts et métiers*. Published April 6, 2006. Accessed

October 2013. http://www.arts-et-metiers.net/musee.php?P=49&id=23&lang=fra&flash=f&arc=1.

Johnson, Mark. *The Body in the Mind: The Bodily Basis of Meaning, Imagination and Reason*. Chicago: University of Chicago Press, 1987.

Johnson, Steven. "Emergence." *New York Times*, September 9, 2001. Accessed February 2014. http://www.nytimes.com/2001/09/09/books/chapters/09–1stjohns.html.

Johnston, John. *The Allure of Machinic Life: Cybernetics, Artificial Life, and the New AI*. Cambridge, MA: MIT Press, 2008.

Jolley, Nicholas. *The Cambridge Companion to Leibniz*. Cambridge: Cambridge University Press, 1995.

———. *Locke: His Philosophical Thought*. Oxford: Oxford University Press, 1999.

Jones, Howard. *The Epicurean Tradition*. New York: Routledge, 1992.

Jones, Joseph R. "Historical Materials for the Study of the Cabeza Encantada Episode in Don Quijote II.62." *Hispanic Review* 41, no. 1 (1979): 87–103.

Jones, Michael. "Theatrical History in the Croxton Play of the Sacrament." *ELH* 66, no. 2 (1999): 223–60.

Joravsky, David. *The Lysenko Affair*. Cambridge, MA: Harvard University Press, 1970.

Jordanova, Ludmilla. "Nature's Powers: A Reading of Lamarck's Distinction between Creation and Production." In *History, Humanity, and Evolution: Essays for John C. Greene*, edited by James R. Moore, 71–98. Cambridge: Cambridge University Press, 1989.

Kauffman, Stuart A. "On Emergence, Agency and Organization." *Biology and Philosophy* 21 (2006): 501–21.

———. *The Origins of Order: Self-Organization and Selection in Evolution*. New York: Oxford University Press, 1993.

Keller, Eve. "Embryonic Individuals: The Rhetoric of Seventeenth-Century Embryology and the Construction of Early-Modern Identity." *Eighteenth-Century Studies* 33, no. 3 (2000): 321–48.

Keller, Evelyn Fox. *The Century of the Gene*. Cambridge, MA: Harvard University Press, 2002.

———. *A Feeling for the Organism: The Life and Work of Barbara McClintock*. San Francisco: W. H. Freeman, 1983.

———. *Making Sense of Life: Explaining Biological Development with Models, Metaphors, and Machines*. Cambridge, MA: Harvard University Press, 2002.

———. "Organisms, Machines, and Thunderstorms: A History of Self-Organization, Part One." *Historical Studies in the Natural Sciences* 38, no. 1 (2008): 45–75.

———. "Organisms, Machines, and Thunderstorms: A History of Self-Organization, Part Two." *Historical Studies in the Natural Sciences* 39, no. 1 (2009): 1–31.

———. "Self-Organization, Self-Assembly, and the Inherent Activity of Matter." In *Transformations of Lamarckism: From Subtle Fluids to Molecular Biology*, edited by Snait Gissis and Eva Jablonka, 357–64. Cambridge, MA: MIT Press, 2011.

———. "Self-Organization, Self-Assembly, and the Inherent Activity of Matter." The Hans Rausing Lecture 2009, Uppsala University. Salvia Småskrifter, no. 12. Stockholm: Författaren, 2009.

Kelley, Donald R. *Fortunes of History: Historical Inquiry from Herder to Huizinga*. New Haven, CT: Yale University Press, 2003.

Kim, Jaegwon. "Emergence: Core Ideas and Issues." *Synthèse* 151, no. 3 (2006): 347–54.

King, Elizabeth. "Clockwork Prayer: A Sixteenth-Century Mechanical Monk." *Blackbird: An Online Journal of Literature and the Arts* 1, no. 1 (2002).

———. "Perpetual Devotion: A Sixteenth-Century Machine that Prays." In *Genesis Redux: Essays in the History and Philosophy of Artificial Life*, edited by Jessica Riskin, 263–92. Chicago: University of Chicago Press, 2007.

King-Hele, Desmond. "Romantic Followers: Wordsworth, Coleridge, Keats and Shelley." In *The Essential Writings of Erasmus Darwin*, edited by Desmond King-Hele, 163–75. London: MacGibbon and Kee, 1968.

Klatt, Dennis H. "Review of Text-to-Speech Conversion for English." *Journal of the Acoustical Society of America* 82, no. 3 (1987): 737–93.

Knegtmans, Peter Jan. *From Illustrious School to University of Amsterdam*. Translated by Paul Andrews. Amsterdam: Amsterdam University Press, 2007.

Koerner, Joseph Leo. *The Reformation of the Image*. Chicago: University of Chicago Press, 2004.

Koestler, Arthur. *The Case of the Midwife Toad*. New York: Random House, 1972.

Koyré, Alexandre. *From the Closed World to the Infinite Universe*. Baltimore: Johns Hopkins University Press, 1957.

Kroll, Peter, Richard Ashcraft, and Peter Zagorin, eds. *Philosophy, Science, and Religion in England, 1640–1700*. Cambridge: Cambridge University Press, 1992.

Kuhn, Thomas S. "Energy Conservation as an Example of Simultaneous Discovery." In *Critical Problems in the History of Science*, edited by Marshall Clagett, 321–56. Madison: University of Wisconsin Press, 1959.

———. *The Structure of Scientific Revolutions*. Chicago: University of Chicago Press, 1962.

Laborde, Léon Emmanuel Simon. *Notice des emaux, bijoux et autres objets divers, exposés dans les galeries du Musée du Louvre*. 2 vols. Paris: Vinchon, 1853.

Lamalle, Edmond. "La propagande de P. Nicolas Trigault en faveur des missions de Chine (1616)." *Archivum Historicum Societatis Jesu* 9, no. 1 (1940): 49–120.

Lamb, Marion. "Attitudes to Soft Inheritance in Britain, 1930s–1970s." In *Transformations of Lamarckism: From Subtle Fluids to Molecular Biology*, edited by Snait Gissis and Eva Jablonka, 109–20. Cambridge, MA: MIT Press, 2011.

Landes, David. *A Revolution in Time: Clocks and the Making of the Modern World*. Rev. ed. Cambridge, MA: Harvard University Press, 2000.

Langford, Jerome J. *Galileo, Science, and the Church*. South Bend, IN: St Augustine's Press, 1998.

Larson, Edward. *Evolution: The Remarkable History of a Scientific Theory*. New York: Modern Library, 2004.

Larson, James L. "Vital Forces: Regulative Principles or Constitutive Agents? A Strategy in German Physiology, 1786–1802." *Isis* 70, no. 2 (1979): 235–49.

Latil, Pierre de. *La pensée artificielle. Introduction à la cybernétique*. Paris: Gallimard, 1953.

Latour, Bruno. *We Have Never Been Modern*. Translated by Catherine Porter. Cambridge, MA: Harvard University Press, 1993.

Latour, Bruno, and Peter Weibel, eds. *Iconoclash: Beyond the Image Wars in Science, Religion, and Art*. Cambridge, MA: MIT Press, 2002.

La Tourrasse, Léonel de. *Le Château-neuf de Saint-Germain-en-Laye, ses terrasses et ses grottes*. Paris: Édition de la Gazette des beaux-arts, 1924.

Lawrence, Christopher. "The Power and the Glory: Humphry Davy and Romanticism." In *Romanticism and the Sciences*, edited by Andrew Cunningham and Nicholas Jardine, 213–27. Cambridge: Cambridge University Press, 1990.

Lawrence, Christopher, and Stephen Shapin, eds. *Science Incarnate: Historical Embodiments of Natural Knowledge*. Chicago: University of Chicago Press, 1998.

Leavitt, David. *The Man Who Knew Too Much: Alan Turing and the Invention of the Computer*. New York: W. W. Norton, 2006.

Lebeau, Auguste. *Condillac, économiste*. Paris: Guillaumin, 1903.

LeBuffe, Michael. "Spinoza's Psychological Theory." *Stanford Encyclopedia of Philosophy*. Stanford University, 1997–. Article published August 9, 2010. Accessed July 2012. http://plato.stanford.edu/entries/spinoza-psychological/.

Leddy, Neven, and Avi S. Lifschitz. *Epicurus in the Enlightenment*. Oxford, UK: Voltaire Foundation, 2009.

Le Goff, Jacques. *The Medieval Imagination*. Translated by Arthur Goldhammer. Chicago: University of Chicago Press, 1988.

Lemée, Pierre. *Julien Offray de la Mettrie, St Malo 1709–Berlin, 1751. Médicin, philosophe, polémiste. Sa vie, son œuvre*. Mortain, France: Mortainais, 1954.

Lenhoff, Sylvia G., and Howard M. Lenhoff. *Hydra and the Birth of Experimental Biology, 1744: Abraham Trembley's Mémories Concerning the Polyps*. Pacific Grove, CA: Boxwood Press, 1986.

Lennox, James G. *Aristotle's Philosophy of Biology: Studies in the Origins of Life Science*. Cambridge: Cambridge University Press, 2001.

———. "The Comparative Study of Animal Development: William Harvey's Aristotelianism." In *The Problem of Animal Generation in Early Modern Philosophy*, edited by Justin E. H. Smith, 21–46. Cambridge: Cambridge University Press, 2006.

———. "Darwinism and Neo-Darwinism." In *A Companion to the Philosophy of Biology*, edited by Sahotra Sarkar and Anya Plutynski, 77–98. Malden, MA: Blackwell, 2008.

Lennox, James G., and Mary Louise Gill, eds. *Self-Motion from Aristotle to Newton*. Princeton, NJ: Princeton University Press, 1994.

Lenoir, Timothy. "Helmholtz and the Materialities of Communication." *Osiris* 9 (1994): 185–207.

———. *Instituting Science: The Cultural Production of Scientific Disciplines*. Stanford, CA: Stanford University Press, 1997.

———. *The Strategy of Life: Teleology and Mechanics in Nineteenth Century German Biology*. Studies in the History of Modern Science 13. Dordrecht, the Netherlands: Reidel, 1982.

Levere, Trevor H. "Coleridge and the Sciences." In *Romanticism and the Sciences*, edited by Andrew Cunningham and Nicholas Jardine, 295–306. Cambridge: Cambridge University Press, 1990.

Levine, George. *Darwin Loves You: Natural Selection and the Re-enchantment of the World.* Princeton, NJ: Princeton University Press, 2008.

Levitt, Gerald M. *The Turk, Chess Automaton.* Jefferson, NC: McFarland, 2000.

Lindberg, David C., and Ronald L. Numbers, eds. *God and Nature: Historical Essays on the Encounter between Christianity and Science.* Berkeley: University of California Press, 1986.

Lindsay, David. "Talking Head." *Invention and Technology* (Summer 1997): 57–63.

Lingeman, Richard R. *Sinclair Lewis: Rebel from Main Street.* New York: Random House, 2002.

Lovejoy, Arthur. *The Great Chain of Being: A Study of the History of an Idea.* Cambridge, MA: Harvard University Press, 1936.

MacCulloch, Diarmaid. *Reformation: Europe's House Divided, 1490–1700.* London: Allen Lane, 2003.

MacDonogh, Giles. *A Palate in Revolution: Grimod de La Reynière and the Almanach des gourmands.* London: Robin Clark, 1987.

Mahoney, Michael S. "Mariotte, Edme." In *Dictionary of Scientific Biography,* edited by Charles Coulston Gillispie, 9:114–22. New York: Scribner, 1970–.

Maienschein, Jane. "Competing Epistemologies and Developmental Biology." In *Biology and Epistemology,* edited by Richard Creath and Jane Maienschein, 122–37. Cambridge: Cambridge University Press, 2000.

———. "Epigenesis and Preformationism." In *Stanford Encyclopedia of Philosophy.* Stanford University, 1997–. Article published October 11, 2005, and accessed October 2013. http://plato.stanford.edu/archives/spr2012/entries/epigenesis/.

———. "The Origins of Entwicklungsmechanik." In *A Conceptual History of Modern Embryology,* edited by Scott F. Gilbert, 43–61. Baltimore: Johns Hopkins University Press, 1994.

Maindron, Ernest. *Marionnettes et guignols: Les poupées agissantes et parlantes à travers les ages.* Paris: Félix Juven, 1897.

Maisano, Scott. "Infinite Gesture: Automata and the Emotions in Descartes and Shakespeare." In *Genesis Redux: Essays in the History and Philosophy of Artificial Life,* edited by Jessica Riskin, 63–84. Chicago: University of Chicago Press, 2007.

Mallinson, Jonathan. "What's in a Name? Reflections on Voltaire's Pamela." *Eighteenth-Century Fiction* 18, no. 2 (2005): 157–68.

Manning, William. *Recollections of Robert-Houdin, Clockmaker, Electrician, Conjuror.* London: Chiswick Press, 1891.

Marchand, Suzanne L. *Down from Olympus: Archaeology and Philhellenism in Germany, 1750–1970.* Princeton, NJ: Princeton University Press, 1996.

Marie, Alfred. *Jardins français crées à la Renaissance.* Paris: V. Fréal, 1955.

Marion, Jean-Luc. *Descartes's Grey Ontology: Cartesian Science and Aristotelian Thought in the Regulae.* Translated by Sarah E. Donohue. South Bend, IN: Saint Augustine's Press, 2004.

———. *Sur la théologie blanche de Descartes: Analogie, création des vérités éternelles, fondement.* 1981. Rev. ed. Paris: Presses universitaires de France, 1991.

Markel, Arkady L., and Lyudmila N. Trut. "Behavior, Stress, and Evolution in Light of the Novosibirsk Selection Experiments." In *Transformations of Lamarckism: From Subtle Fluids to Molecular Biology*, edited by Snait Gissis and Eva Jablonka, 171–80. Cambridge, MA: MIT Press, 2011.

Martin, Brian. "Dilemmas of Defending Dissent: The Dismissal of Ted Steele from the University of Wollongong." *Australian Universities' Review* 45, no. 2 (2002): 7–17.

Mates, Benson. *The Philosophy of Leibniz: Metaphysics and Language*. New York: Oxford University Press, 1986.

McGrath, Alister E. *The Intellectual Origins of the European Reformation*. 2nd ed. Malden, MA: Blackwell, 2004.

McLaughlin, Peter. *Kant's Critique of Teleology in Biological Explanation: Antinomy and Teleology*. Lewiston, NY: E. Mellen Press, 1990.

Meli, Domenico Bertoloni. *Equivalence and Priority: Newton versus Leibniz*. Oxford: Oxford University Press, 1993.

Menn, Stephen. *Descartes and Augustine*. Cambridge: Cambridge University Press, 1998.

———. "The Greatest Stumbling Block: Descartes' Denial of Real Qualities." In *Descartes and his Contemporaries: Meditations, Objections, and Replies*, edited by Roger Ariew and Marjorie Grene, 182–207. Chicago: University of Chicago Press, 1995.

Mercer, Christia. *Leibniz's Metaphysics: Its Origins and Development*. Cambridge: Cambridge University Press, 2001.

Méril, Edelestand du. *Origines latines du théâtre moderne*. Paris: Franck, 1849.

Merton, Robert K. *Science, Technology & Society in Seventeenth-Century England*. 1938. New York: H. Fertig, 1970.

Metzner, Paul. *Crescendo of the Virtuoso: Spectacle, Skill, and Self-Promotion in Paris during the Age of Revolution*. Berkeley: University of California Press, 1998.

Michalski, Sergiusz. *The Reformation and the Visual Arts: The Protestant Image Question in Western and Eastern Europe*. New York: Routledge, 1993.

Micheli, Giuseppe. *The Early Reception of Kant's Thought in England: 1785–1805*. 1931. London: Routledge, 1999.

Michie, Donald. "Alan Turing's Mind Machines." In *The Mechanical Mind in History*, edited by Philip Husbands, Owen Holland, and Michael Wheeler, 61–74. Cambridge, MA: MIT Press, 2008.

Mindell, David A. *Between Human and Machine: Feedback, Control and Computing Before Cybernetics*. Baltimore: Johns Hopkins University Press, 2004.

Monnier, Philippe. *Le Quattrocento: Essai sur l'histoire littéraire du XVe siècle italien*. 2 vols. Paris: Perrin, 1908.

Montañes Fonteñla, Luis. "Los relojes del Emperador: Los relojes de la exposición 'Carlos V y su ambiente.'" *Cuadernos de Relojería* 18 (1959): 3–22.

Moravia, Sergio. "From Homme Machine to Homme Sensible: Changing Eighteenth Century Models of Man's Image." *Journal of the History of Ideas* 39 (1978): 45–60.

Moreau, Denis. *Deux cartésiens: La polémique entre Antoine Arnauld et Nicolas Malebranche*. Paris: J. Vrin, 1999.

Morel, Philippe. *Les grottes maniéristes en Italie au XVIe siècle: Théâtre et alchimie de la nature*. Paris: Macula, 1998.

Morley, John. *Voltaire*. London: Macmillan, 1886.

Morris, Katherine. "Bêtes-machines." In *Descartes' Natural Philosophy*, edited by Stephen Gaukroger, John Schuster, and John Sutton, 401–19. London: Routledge, 2000.

Mousset, Albert. *Les Francine: Créateurs des eaux de Versailles, intendants des eaux et fontaines de France de 1623–1784*. Paris: E. Champion, 1930.

Muir, Edward. *Ritual in Early Modern Europe*. Cambridge: Cambridge University Press, 1997.

Mukerji, Chandra. *Territorial Ambitions and the Gardens of Versailles*. Cambridge: Cambridge University Press, 1997.

Murray, Glenn. "Génesis del Real Ingenio de la Moneda de Segovia." Pts 1–4. *Revista Nvmisma*, no. 228 (1991): 59–80; *Revista Nvmisma*, no. 232 (1993): 177–222; *Revista Nvmisma*, no. 234 (1994): 111–53; *Revista Nvmisma*, no. 235 (1994): 85–119.

Myers, Natasha. *Rendering Life Molecular: Models, Modelers and Excitable Matter*. Durham, NC: Duke University Press, 2015.

Nadler, Steven M. *Arnauld and the Cartesian Philosophy of Ideas*. Princeton, NJ: Princeton University Press, 1989.

———. *Malebranche and Ideas*. New York: Oxford University Press, 1992.

———. *Occasionalism: Causation among the Cartesians*. Oxford: Oxford University Press, 2011.

Nagel, Thomas. *Mind and Cosmos: Why the Materialist Neo-Darwinian Conception of Nature is Almost Certainly False*. Oxford: Oxford University Press, 2012.

Newman, Stuart A., and Ramray Bhat. "Lamarck's Dangerous Idea." In *Transformations of Lamarckism: From Subtle Fluids to Molecular Biology*, edited by Snait Gissis and Eva Jablonka, 157–69. Cambridge, MA: MIT Press, 2011.

Newman, William R. *Promethean Ambitions: Alchemy and the Quest to Perfect Nature*. Chicago: University of Chicago Press, 2004.

Novick, Peter. *That Noble Dream: The "Objectivity Question" and the American Historical Profession*. Cambridge: Cambridge University Press, 1988.

Nyhart, Lynn. *Biology Takes Form: Animal Morphology and the German Universities, 1800–1900*. Chicago: University of Chicago Press, 1995.

———. *Modern Nature: The Rise of the Biological Perspective in Germany*. Chicago: University of Chicago Press, 2009.

Olson, Richard. "On the Nature of God's Existence, Wisdom and Power: The Interplay Between Organic and Mechanistic Imagery in Anglican Natural Theology, 1640–1740." In *Approaches to Organic Form: Permutations in Science and Culture*, edited by Frederick Burwick, 1–48. Dordrecht, the Netherlands: Reidel, 1987.

Osler, Margaret J. *Divine Will and the Mechanical Philosophy: Gassendi and Descartes on Contingency and Necessity in the Created World*. Cambridge: Cambridge University Press, 1994.

———. "Mixing Metaphors: Science and Religion or Natural Philosophy and Theology in Early Modern Europe." *History of Science* 36 (1998): 91–113.

———, ed. *Rethinking the Scientific Revolution*. Cambridge: Cambridge University Press, 2000.

———. "Whose Ends? Teleology in Early Modern Natural Philosophy." In "Science in Theistic Contexts: Cognitive Dimensions," edited by John Hedley Brooke, Margaret J. Osler, and Jitse M. van der Meer, published as *Osiris* 16 (2001): 151–68.

Ospovat, Dov. *The Development of Darwin's Theory: Natural History, Natural Theology, and Natural Selection, 1838–1859*. Cambridge: Cambridge University Press, 1981.

Osterhout, W. J. V. *Biographical Memoir of Jacques Loeb, 1859–1924*. Washington, DC: National Academy of Sciences, 1930.

Page, William, ed. *The Victoria History of the County of Kent*. 3 vols. London: Constable, 1908.

Pagel, Walter. "Medieval and Renaissance Contributions to Knowledge of the Brain and Its Functions." In *The History and Philosophy of Knowledge of the Brain and Its Functions: An Anglo-American Symposium, London, July 15–17, 1957*, edited by F. N. L. Poynter, 95–114. Oxford, UK: Blackwell Scientific, 1958.

Pagels, Elaine H. *The Gnostic Gospels*. New York: Random House, 1979.

Paracandola, M. "Philosophy in the Laboratory: The Debate over Evidence for E. J. Steele's Lamarckian Hypothesis." *Studies in the History and Philosophy of Science* 26 (1995): 469–92.

Parkinson, G. H. R. "Philosophy and Logic." In *The Cambridge Companion to Leibniz*, edited by Nicholas Jolley, 199–223. Cambridge: Cambridge University Press, 1995.

Pauly, Philip J. *Controlling Life: Jacques Loeb and the Engineering Ideal in Biology*. New York: Oxford University Press, 1987.

Pedretti, Carlo. *Leonardo architetto*. Milan: Electa, 1978.

———. *Leonardo: A Study in Chronology and Style*. Berkeley: University of California Press, 1973.

———. "Leonardo at Lyon." *Raccolta Vinciana* XIX (1962): 267–72.

Penzer, N. M. "A Fourteenth-Century Table Fountain." *Antique Collector* (June 1957): 112–17.

Perkins, Franklin. *Leibniz and China: A Commerce of Light*. Cambridge: Cambridge University Press, 2004.

Perregaux, Charles, and François-Louis Perrot. *Les Jaquet-Droz et Leschot*. Neuchâtel, Switzerland: Attinger Frères, 1916.

Pfeifer, Edward J. "The Genesis of American Neo-Lamarckism." *Isis* 56, no. 2 (1965): 156–67.

Pfister, Louis. *Notices biographiques et bibliographiques sur les jésuites de l'ancienne mission de Chine, 1552–1773*. 2 vols. Shanghai: Imprimerie de la Mission catholique, 1932–34.

Phipps, William E. "Darwin and Cambridge Natural Theology." *Bios* 54, no. 4 (1983): 218–27.

Pickering, Andrew. *The Cybernetic Brain: Sketches of Another Future*. Chicago: University of Chicago Press, 2010.

Prager, Frank D., and Gustina Scaglia. *Brunelleschi: Studies of His Technology and Inventions*. Cambridge, MA: MIT Press, 1970.

Price, Derek J. de Solla. "Automata and the Origins of Mechanism and Mechanistic Philosophy." *Technology and Culture* 5, no. 1 (1964): 9–23.

Poynter, F. N. L. *The History and Philosophy of Knowledge of the Brain and Its Functions: An Anglo-American Symposium, London, July 15–17, 1957*. Oxford, UK: Blackwell Scientific, 1958.

Pyle, Andrew. *Hume's Dialogues Concerning Natural Religion: Reader's Guide*. London: Continuum, 2006.

Rabinbach, Anson. *The Human Motor: Energy, Fatigue, and the Origins of Modernity*. Berkeley: University of California Press, 1992.

Raimondo, Guarino. "Torelli a Venezia: l'ingegnere teatrale tra scena e apparato." *Teatro e storia* 7, no. 12 (1992): 35–72.

Rasmussen, Charles, and Rick Tilman. *Jacques Loeb: His Science and Social Activism and their Philosophical Foundations*. Philadelphia: American Philosophical Society, 1998.

Ratcliffe, Matthew. "A Kantian Stance on the Intentional Stance." *Biology and Philosophy* 16, no. 1 (2001): 29–52.

Reill, Peter. *The German Enlightenment and the Rise of Historicism*. Berkeley: University of California Press, 1975.

———. *Vitalizing Nature in the Enlightenment*. Berkeley: University of California Press, 2005.

Rey, Roseleyne. *Naissance et développement du vitalisme en France de la deuxième moitié du 18e siècle à la fin du Premier Empire*. Oxford, UK: Voltaire Foundation, 2000.

Richard, Jules-Marie. *Une petite-nièce de saint Louis. Mahaut, comtesse d'Artois et de Bourgogne (1302–1329): Etude sur la vie privée, les Arts et l'Industrie, en Artois et à Paris au commencement du XIVe siècle*. Paris: Champion, 1887.

Richards, Robert J. *Darwin and the Emergence of Evolutionary Theories of Mind and Behavior*. Chicago: University of Chicago Press, 1987.

———. "Darwin's Romantic Biology: The Foundation of his Evolutionary Ethics." In *Biology and the Foundation of Ethics*, edited by Jane Maienschein and Michael Ruse, 113–53. Cambridge: Cambridge University Press, 1999.

———. "The Emergence of Evolutionary Biology of Behaviour in the Early Nineteenth Century." *British Journal for the History of Science* 15, no. 3 (1982): 241–80.

———. "Influence of Sensationalist Tradition on Early Theories of the Evolution of Behavior." *Journal of the History of Ideas* 40 (1979): 85–105.

———. "Instinct and Intelligence in British Natural Philosophy: Some Contributions to Darwin's Theory of the Evolution of Behavior." *Journal of the History of Biology* 14, no. 2 (1981): 193–230.

———. *The Meaning of Evolution: The Morphological Construction and Ideological Reconstruction of Darwin's Theory*. Chicago: University of Chicago Press, 1992.

———. *The Romantic Conception of Life: Science and Philosophy in the Age of Goethe*. Chicago: University of Chicago Press, 2002.

———. *The Tragic Sense of Life: Ernst Haeckel and the Struggle over Evolutionary Thought*. Chicago: University of Chicago Press, 2008.

Richardson, Alan. *British Romanticism and the Science of the Mind*. Cambridge: Cambridge University Press, 2001.

Ringer, Fritz. *The Decline of the German Mandarins: The German Academic Community, 1890–1933*. Cambridge, MA: Harvard University Press, 1969.

Riskin, Jessica. "The Defecating Duck; or, The Ambiguous Origins of Artificial Life." *Critical Inquiry* 29, no. 4 (Summer 2003): 599–633.

———. "Eighteenth-Century Wetware." *Representations* 83 (2003): 97–125.

———, ed. *Genesis Redux: Essays in the History and Philosophy of Artificial Life*. Chicago: University of Chicago Press, 2007.

———. *Science in the Age of Sensibility: The Sentimental Empiricists of the French Enlightenment*. Chicago: University of Chicago Press, 2002.

———. "The 'Spirit of System' and the Fortunes of Physiocracy." In *Oeconomies in the Age of Newton*, edited by Neil De Marchi and Margaret Schabas. Durham, NC: Duke University Press, 2003: 42–73.

Ritter, Gretchen. "Silver Slippers and a Golden Cap: L. Frank Baum's *The Wonderful Wizard of Oz* and Historical Memory in American Politics." *Journal of American Studies* 31, no. 2 (1997): 171–203.

Rival, Ned. *Grimod de La Reynière. Le gourmand gentilhomme*. Paris: Le Pré aux clercs, 1983.

Rochemonteix, Camille de. *Un collège de jésuites aux XVIIe & XVIIIe siècles: Le Collège Henri IV de la Flèche*. 4 vols. Le Mans, France: Leguicheux, 1889.

Roe, Shirley A. "*Anatomia animata*: The Newtonian Physiology of Albrecht von Haller." In *Transformation and Tradition in the Sciences: Essays in Honor of I. Bernard Cohen*, edited by Everett Mendelsohn, 273–300. Cambridge: Cambridge University Press, 1981.

———. *Matter, Life, and Generation: Eighteenth-Century Embryology and the Haller-Wolff Debate*. Cambridge: Cambridge University Press, 1981.

Roger, Jacques. *Buffon: A Life in Natural History*. Edited by L. Pearce Williams. Translated by Sarah Lucille Bonnefoi. Ithaca, NY: Cornell University Press, 1997.

———. "Leibniz et les sciences de la vie." *Studia Leibnitiana* supplementa 2, no. 2 (Wiesbaden, Germany: Steiner, 1969).

———. *The Life Sciences in Eighteenth-Century French Thought*. Edited by Keith R. Benson. Translated by Robert Ellrich. Stanford, CA: Stanford University Press, 1997.

———. "The Mechanistic Conception of Life." In *God and Nature: Historical Essays on the Encounter Between Christianity and Science*, edited by David C. Lindberg and Ronald L. Numbers, 277–95. Berkeley: University of California Press, 1986.

———. *Les sciences de la vie dans la pensée française au dix-huitième siècle*. Paris: Albin Michel, 1993.

Roman, F. "The Discovery of Accommodation." *British Journal of Ophthalmology* 79, no. 4 (1995): 375.

Rosenblith, Walter A., ed. *Jerry Wiesner: Scientist, Statesman, Humanist: Memories and Memoirs*. Cambridge, MA: MIT Press, 2003.

Rosenfield, Leonora. *From Beast-Machine to Man-Machine: Animal Soul in French Letters from Descartes to La Mettrie*. New York: Octagon Books, 1940.

Rosheim, Mark Elling. *Leonardo's Lost Robots*. Berlin: Springer, 2006.

Rossetti, Lucy Madox Brown. *Mrs. Shelley*. London: W. H. Allen, 1890.

Rossi, Paolo. *Philosophy, Technology, and the Arts in the Early Modern Era*. Edited by Benjamin Nelson and translated by Salvator Attanasio. New York: Harper and Row, 1970.

Rothschuh, Karl E.. *History of Physiology*. 1953. Edited and translated by Guenter B. Risse. Huntington, NY: Krieger, 1973.

Rudwick, M.J.S. *Georges Cuvier, Fossil Bones, and Geological Catastrophes: New Translations and Interpretations of the Primary Texts*. Chicago: University of Chicago Press, 1997.

Rüegg, Walter. "Theology and the Arts." In *A History of the University in Europe*, edited by Walter Rüegg, 3:393–457. Cambridge: Cambridge University Press, 2004.

Runciman, David. "Debate: What Kind of Person is Hobbes's State? A Reply to Skinner." *Journal of Political Philosophy* 8, no. 2 (2000): 268–78.

———. *Pluralism and the Personality of the State*. Cambridge: Cambridge University Press, 1997.

Ruse, Michael. *Darwin and Design: Does Evolution Have a Purpose?* Cambridge, MA: Harvard University Press, 2003.

———. *The Darwinian Paradigm: Essays on Its History, Philosophy and Religious Implications*. London: Routledge, 1989.

———. *Darwinism and Its Discontents*. Cambridge: Cambridge University Press, 2006.

Sabra, A. I. *Theories of Light from Descartes to Newton*. Cambridge: Cambridge University Press, 1981.

Salies, Alexandre de. "Lettre sur une tête automatique autrefois attachée à l'orgue des Augustins de Montoire." *Bulletin de la Société archéologique, scientifique et littéraire du Vendômois* 6, no. 2 (1867): 97–118.

Sander, Klaus. "Entelechy and the Ontogenetic Machine: Work and Views of Hans Driesch from 1895–1910." *Roux's Archives of Developmental Biology* 202, no. 2 (1993): 67–69.

———. "Hans Driesch's 'philosophy really *ab ovo*,' or why to Be a Vitalist." *Roux's Archives of Developmental Biology* 202, no. 1 (December 1992): 1–3.

———. "Hans Driesch the Critical Mechanist: *Analytische Theorie der organischen Entwicklung*." *Roux's Archives of Developmental Biology* 201, no. 6 (1992): 331–33.

———. "Shaking a Concept: Hans Driesch and the Varied Fates of Sea Urchin Blastomeres." *Roux's Archives of Developmental Biology* 201, no. 5 (1992): 265–67.

Sanhueza, Gabriel. *La pensée biologique de Descartes dans ses rapports avec la philosophie scolastique: Le cas Gomez-Péreira*. Paris: L'Harmattan, 1997.

Santillana, Giorgio de. *The Crime of Galileo*. Chicago: University of Chicago Press, 1955.

Schaber, Wilfried. *Hellbrunn. Schloss, Park, und Wasserspiele*. Salzburg: Schlossverwaltung Hellbrunn, 2004.

Schaffer, Simon. "Babbage's Dancer and the Impresarios of Mechanism." In *Cultural Babbage: Technology, Time, and Invention*, edited by Francis Spufford and Jennifer S. Uglow, 53–80. London: Faber, 1996.

———. "Babbage's Intelligence: Calculating Engines and the Factory System." *Critical Inquiry* 21, no. 1 (1994): 203–27.

———. "Enlightened Automata." In *The Sciences in Enlightened Europe*, edited by William Clark, Jan Golinski, and Simon Schaffer, 126–64. Chicago: University of Chicago Press, 1999.

———. "OK Computer." 1999. Accessed February 2013. http://www.hrc.wmin.ac.uk/theory-okcomputer.html.

———. "The Political Theology of Seventeenth-Century Natural Science." *Ideas and Production: A Journal in the History of Ideas* 1 (1983): 2–14.

Schaut, Scott. *Robots of Westinghouse, 1924–Today*. Mansfield, OH: Mansfield Memorial Museum, 2007.

Schielicke, Reinhard E., Klaus-Dieter Herbst, and Stefan Kratchowil, eds. *Erhard Weigel, 1625 bis 1699: Barocker Erzvater der deutschen Frühaufklärung; Beiträge des Kolloqui-*

ums anlässlich seines 300. Todestages am 20. März 1999 in Jena. Thun, Switzerland: H. Deutsch, 1999.

Schmaltz, Tad M., ed. *Receptions of Descartes: Cartesianism and Anti-Cartesianism in Early Modern Europe*. London: Routledge, 2005.

———. "What Has Cartesianism to Do with Jansenism?" *Journal of the History of Ideas* 60, no. 1 (1999): 37–56.

Schofield, Malcolm. *Plato: Political Philosophy*. Oxford: Oxford University Press, 2006.

Schofield, Robert E. *Materialism and Mechanism: British Natural Philosophy in the Age of Reason*. Princeton, NJ: Princeton University Press, 1970.

Schroeder, M.R. "A Brief History of Synthetic Speech." *Speech Communication* 13 (1993): 231–37.

Schröder, Tilman Matthias. *Naturwissenschaft und Protestantismus im Deutschen Kaiserreich. die Versammlungen der Gesellschaft Deutscher Naturforscher und Ärzte und ihre Bedeutung für die evangelische Theologie*. Stuttgart: Steiner, 2008.

Schwinges, Rainer Christoph, ed. *Humboldt International: Der Export des deutschen Universitätsmodells im 19. und 20. Jahrhundert*. Basel: Schwabe, 2001.

———, ed. *Universität, Religion und Kirche*. Basel: Schwabe, 2011.

Scruton, Roger. *Kant: A Very Short Introduction*. Rev. ed. Oxford: Oxford University Press, 2001.

Séris, Jean-Pierre. *Langages et machines à l'âge classique*. Paris: Hachette, 1995.

———. *Machine et communication: Du théâtre des machines à la mécanique industrielle*. Paris: J. Vrin, 1987.

Sewell, Jr., William H. *Work and Revolution in France: The Language of Labor from the Old Regime to 1848*. Cambridge: Cambridge University Press, 1980.

Seymour, Miranda. *Mary Shelley*. London: John Murray, 2000.

Shapin, Steven. *A Social History of Truth: Civility and Science in Seventeenth-Century England*. Chicago: University of Chicago Press, 1994.

Shapin, Steven, and Simon Schaffer. *Leviathan and the Air-Pump: Hobbes, Boyle, and the Experimental Life*. Princeton, NJ: Princeton University Press, 1985.

Shapiro, Alan. "Newton and Huygens' Explanation of the 22° Halo." *Centaurus* 24, no. 1 (1980): 273–87.

Sharpe, Sam H. *Salutations to Robert-Houdin: His Life, Magic and Automata*. Calgary: Micky Hades International, 1983.

Shea, William R., and Mariano Artigas. *Galileo in Rome: The Rise and Fall of a Troublesome Genius*. Oxford: Oxford University Press, 2004.

Sherwood, Merriam. "Magic and Mechanics in Medieval Fiction." *Studies in Philology* 44, no. 4 (1947): 567–92.

Simmons, A. John. *The Lockean Theory of Rights*. Princeton, NJ: Princeton University Press, 1992.

Skinner, Quentin. "Hobbes and the Purely Artificial Person of the State." *Journal of Political Philosophy* 7, no. 1 (1999): 1–29.

———. *Visions of Politics*. 3 vols. New York: Cambridge University Press, 2002.

Sleigh, Charlotte. "Life, Death and Galvanism." *Studies in History and Philosophy of Science Part C* 29, no. 2 (1998): 219–48.

———. " 'The Ninth Mortal Sin': The Lamarckism of W. M. Wheeler." In *Darwinian Heresies*, edited by Abigail Lustig, Robert Richards and Michael Ruse, 151–72. Cambridge: Cambridge University Press, 2004.

Sloan, Phillip R. "Buffon, German Biology, and the Historical Interpretation of Biological Species." *British Journal for the History of Science* 12, no. 2 (1979): 109–53.

———. "Descartes, the Sceptics and the Rejection of Vitalism in Seventeenth-Century Physiology." *Studies in the History and Philosophy of Science* 8 (1977): 1–28.

Sloan, Phillip R., and Brandon Fogel, eds. *Creating a Physical Biology: The Three-Man Paper and Early Molecular Biology*. Chicago: University of Chicago Press, 2011.

Smets, Alexis. "The Controversy between Leibniz and Stahl on the Theory of Chemistry." In *Neighbours and Territories: The Evolving Identity of Chemistry, Proceedings of the 6th International Conference on the History of Chemistry*, edited by José Ramón Bertomeu-Sánchez, Duncan Thorburn Burns, and Brigitte Van Tiggelen, 291–306. Louvain-la-Neuve, Belgium: Mémosciences, 2008.

Smith, Crosbie. *The Science of Energy: A Cultural History of Energy Physics in Victorian Britain*. Chicago: University of Chicago Press, 1998.

Smith, Edward E., and Daniel N. Osherson, eds. *Thinking*. Vol. 3 of *An Invitation to Cognitive Science*. Cambridge, MA: MIT Press, 1995.

Smith, Justin E. H. *Divine Machines: Leibniz and the Sciences of Life*. Princeton, NJ: Princeton University Press, 2011.

———. "Leibniz's Hylomorphic Monad." *History of Philosophy Quarterly* 19, no. 1 (2002): 21–42.

———, ed. *The Problem of Animal Generation in Early Modern Philosophy*. Cambridge: Cambridge University Press, 2006.

———. " 'A Series of Generations': Leibniz on Race." *Annals of Science* 70, no. 3 (2013): 319–35.

Soergel, Philip M. *Miracles and the Protestant Imagination*. Oxford: Oxford University Press, 2012.

Sortais, Gaston. "Le cartésianisme chez les Jésuites français au XVIIe et XVIIIe siècles." *Archives de philosophie* 6, no. 3 (1929): 37–40.

Spary, Emma. *Utopia's Garden: French Natural History from Old Regime to Revolution*. Chicago: University of Chicago Press, 2000.

Spence, Jonathan D. *The Memory Palace of Matteo Ricci*. New York: Penguin, 1985.

Stafford, Barbara Maria. *Artful Science: Enlightenment, Entertainment, and the Eclipse of Visual Education*. Cambridge, MA: MIT Press, 1994.

Stamhuis, Ida H. "Vries, Hugo de." In *Complete Dictionary of Scientific Biography*, edited by Frederic L. Holmes, 25:189–92. New York: Scribner, 1970–.

Stamhuis, Ida H., Onno G. Meijer, and Erik J. A. Zevenhuizen. "Hugo De Vries on Heredity, 1889–1903: Statistics, Mendelian Laws, Pangenes, Mutations." *Isis* 90, no. 2 (1999): 238–67.

Standage, Tom. *The Turk: The Life and Times of the Famous Eighteenth-Century Chess-Playing Machine*. New York: Walker, 2002.

Stanhope, Paul Henry. "The Statue of Memnon." *London Quarterly Review* (April 1875): 278–84.

Statham, F. Reginald. "The Real Robert Elsmere." *National Review* 28, no. 164 (October 1896): 252–61.

Steigerwald, Joan, ed. *Kantian Teleology and the Biological Sciences*. Special issue of *Studies in the History and Philosophy of Science Part C: Studies in History and Philosophy of Biological and Biomedical Sciences* 37, no. 4 (2006).

Steinke, Hubert. *Irritating Experiments: Haller's Concept and the European Controversy on Irritability and Sensibility, 1750–90*. Amsterdam: Rodopi, 2005.

Steinmeyer, Jim. *Hiding the Elephant: How Magicians Invented the Impossible and Learned to Disappear*. New York: Carroll and Graf, 2003

Sterelny, Kim. *Dawkins vs. Gould: Survival of the Fittest*. New ed. Thriplow, UK: Icon, 2007.

Sterling-Maxwell, William. *The Cloister Life of the Emperor Charles V*. 4th ed. London: J. C. Nimmo, 1891.

Strauss, Linda. "Automata: A Study in the Interface of Science, Technology, and Popular Culture." PhD diss., University of California, San Diego, 1987.

Symonds, Neville. "What is Life? Schrödinger's Influence on Biology." *Quarterly Review of Biology* 61, no. 2 (1986): 221–26.

Tabbaa, Yasser. "The Medieval Islamic Garden: Typology and Hydraulics." In *Garden History: Issues, Approaches, Methods*. Dumbarton Oaks Colloquium on the History of Landscape Architecture, edited by John Dixon Hunt, 13:303–29. Washington, D.C.: Dumbarton Oaks Research Library and Collection, 1992.

Taylor, Charles. *Sources of the Self: The Making of the Modern Identity*. Cambridge, MA: Harvard University Press, 1989.

Terrall, Mary. *The Man Who Flattened the Earth: Maupertuis and the Sciences in the Enlightenment*. Chicago: University of Chicago Press, 2002.

Teuscher, Christof, ed. *Alan Turing: Life and Legacy of a Great Thinker*. Berlin: Springer, 2004.

Thomas, Keith. *Religion and the Decline of Magic*. 1971. New ed. New York: Penguin, 2012.

Thomson, Ann. *Bodies of Thought: Science, Religion, and the Soul in the Early Enlightenment*. Oxford: Oxford University Press, 2008.

Toulmin, Stephen. *Cosmopolis: The Hidden Agenda of Modernity*. Chicago: University of Chicago Press, 1992.

Tresch, John. *The Romantic Machine: Utopian Science and Technology after Napoleon*. Chicago: University of Chicago Press, 2012.

Tripps, Johannes. *Handelnde Bildwerk in der Gotik: Forschungen zu den Bedeutungsschichten und der Funktion des Kirchengebaudes und seiner Ausstattung in der Hoch- und Spätgotik*. Berlin: Gebr. Mann, 1998.

Truitt, Elly. "'Trei poëte, sages dotors, qui mout sorent di nigromance': Knowledge and Automata in Twelfth-Century French Literature." *Configurations* 12, no. 2 (2004): 167–93.

Tronzo, William. *Petrarch's Two Gardens: Landscape and the Image of Movement*. New York: Italica Press, 2013.

Turing, Sara. *Alan M. Turing*. 1959. Centenary ed. Cambridge: Cambridge University Press, 2012.

Turner, Roy Steven. "Humboldt in North America." In *Humbold Interntional: Der Export des deutschen Universitätsmodells im 19. und 20. Jahrhundert*, edited by Christoph Schwinges, 289–312. Basel: Schwabe, 2001.

Uglow, Jennifer S. *The Lunar Men: The Friends Who Made the Future, 1730–1810*. London: Faber, 2002.

Vailati, Ezio. *Leibniz and Clarke: A Study of Their Correspondence*. New York: Oxford University Press, 1997.

Vanderjagt, Arno Johan, and Klaas van Berkel, eds. *The Book of Nature in Antiquity and the Middle Ages*. Louvain, Belgium: Peeters, 2005.

Van der Pas, Peter W. "Vries, Hugo de." In *Complete Dictionary of Scientific Biography*, edited by Charles Coulston Gillispie, 14:95–105. New York: Scribner, 1970–.

Van Helden, Albert. "The Development of Compound Eye Pieces, 1640–1670." *Journal for the History of Astronomy* 8, no. 1 (February 1977): 26–37.

———. *The Invention of the Telescope*. *Transactions of the American Philosophical Society* 67, part 4 (1977).

Van Helden, Albert, Sven Dupré, and Rob van Gent, eds. *The Origins of the Telescope*. Amsterdam: Royal Netherlands Academy of Arts and Sciences, 2010.

Van Nouhuys, Tabitta. "Copernicanism, Jansenism, and Remonstrantism in the Seventeenth-Century Netherlands." In *Heterodoxy in Early Modern Science and Religion*, edited by John Brooke and Ian Maclean, 145–68. Oxford: Oxford University Press, 2005.

Van Roosbroeck, Gustave L. "The 'Unpublished' Poems of Mlle. de Scudéry and Mlle. Descartes." *Modern Language Notes* 40, no. 3 (1925): 155–58.

Vartanian, Aram, ed. *L'Homme Machine: A Study in the Origins of an Idea*. Princeton, NJ: Princeton University Press, 1960.

———. "La Mettrie and Diderot Revisited: An Intertextual Encounter." *Diderot Studies* 21 (1983): 155–97.

———. "Trembley's Polyp, La Mettrie and Eighteenth-Century French Materialism." *Journal of the History of Ideas* 11, no. 3 (1950): 259–86.

Verbeek, Theo. "The First Objections." In *Descartes and his Contemporaries: Meditations, Objections, and Replies*, edited by Roger Ariew and Marjorie Grene, 21–33. Chicago: University of Chicago Press, 1995.

Vermij, Rienk. *The Calvinist Copernicans: The Reception of the New Astronomy in the Dutch Republic, 1575–1750*. Amsterdam: Edita, 2003.

Vidal, Auguste-Michel. *Notre-Dame du Montement à Rabastens: Projet pour la construction d'un appareil destiné à figurer l'Assomption*. Paris: Imprimerie nationale, 1910.

Vidal, Fernando. "Brains, Bodies, Selves and Science: Anthropologies of Identity and the Resurrection of the Body." *Critical Inquiry* 28 (2002): 930–74.

Virtanen, Reino. *Claude Bernard and His Place in the History of Ideas*. Lincoln: University of Nebraska Press, 1960.

Vitet, Ludovic, *Histoire de Dieppe*. Paris: Ch. Gosselin, 1844.

Voskuhl, Adelheid. *Androids in the Enlightenment: Mechanics, Artisans, and Cultures of the Self*. Chicago: University of Chicago Press, 2013.

Voss, Stephen, ed. *Essays on the Philosophy and Science of René Descartes*. New York: Oxford University Press, 1993.

Wade, Nicholas J., and Stanley Finger. "The Eye as an Optical Instrument: From Camera Obscura to Helmholtz's Perspective." *Perception* 30, no. 10 (2001): 1157–77.

Wahrman, Dror. *The Making of the Modern Self: Identity and Culture in Eighteenth-Century England*. New Haven, CT: Yale University Press, 2004.

Walsham, Alexandra. "The Reformation and the 'Disenchantment of the World' Reassessed." *Historical Journal* 51, no. 2 (2008): 497–528.

Weiss, Charles, ed. *Biographie universelle, ou dictionnaire historique*. 6 vols. Paris: Furne, 1853.

Wellman, Kathleen. *La Mettrie: Medicine, Philosophy and Enlightenment*. Durham, NC: Duke University Press, 1992.

Westfall, Richard S. *Essays on the Trial of Galileo*. South Bend, IN: University of Notre Dame Press, 1990.

———. *Force in Newton's Physics: The Science of Dynamics in the Seventeenth Century*. New York: Elsevier, 1971.

———. *Science and Religion in Seventeenth-Century England*. Ann Arbor: University of Michigan Press, 1973.

Williams, Elizabeth A. *A Cultural History of Medical Vitalism in Enlightenment Montpellier*. Burlington, VT: Ashgate, 2003.

Wilson, Catherine. "Descartes and the Corporeal Mind: Some Implications of the Regius Affair." In *Descartes' Natural Philosophy*, edited by Stephen Gaukroger, John Schuster, and John Sutton, 659–79. London: Routledge, 2000.

———. *The Invisible World: Early Modern Philosophy and the Invention of the Microscope*. Princeton, NJ: Princeton University Press, 1995.

———. *Leibniz's Metaphysics: A Historical and Comparative Study*. Princeton, NJ: Princeton University Press, 1989.

———. "The Reception of Leibniz in the Eighteenth Century." In *The Cambridge Companion to Leibniz*, edited by Nicholas Jolley, chap. 13. Cambridge: Cambridge University Press, 1995.

Wilson, Raymond N. *Reflecting Telescope Optics I: Basic Design Theory and its Historical Development*. 2nd ed. Berlin: Springer, 2007.

Wins, Alphonse. *L'horloge à travers les âges*. Mons, Belgium: Léon Dequesne, 1924.

Wolfe, Charles T. "Endowed Molecules and Emergent Organization: The Maupertuis-Diderot Debate." *Early Science and Medicine* 15 (2010): 38–65.

Wolfe, Charles T., and Motoichi Terada. "The Animal Economy as Object and Program in Montpellier Vitalism." *Science in Context* 21 (2008): 537–79.

Wolfson, H.A. *Philo: Foundations of Religious Philosophy in Judaism, Christianity, and Islam*. Cambridge, MA: Harvard University Press, 1947.

Woloch, Nathaniel. "Christiaan Huygens's Attitude toward Animals." *Journal of the History of Ideas* 61, no. 3 (2000): 415–32.

Woolhouse, R. S. *Descartes, Spinoza, Leibniz: The Concept of Substance in Seventeenth-Century Metaphysics*. London: Routledge, 1993.

Woolley, Benjamin. *The Queen's Conjurer: The Science and Magic of Dr. John Dee, Advisor to Queen Elizabeth I*. New York: Henry Holt, 2001.

Weissman, Charlotte. "The First Evolutionary Synthesis: August Weismann and the Origins of Neo-Darwinism." PhD diss., Tel Aviv University, 2011.

———. "Germinal Selection: A Weismannian Solution to Lamarckian Problematics." In *Transformations of Lamarckism: From Subtle Fluids to Molecular Biology*, edited by Snait Gissis and Eva Jablonka, 57–66. Cambridge, MA: MIT Press, 2011.

Wright, Thomas. *Circulation: William Harvey's Revolutionary Idea*. London: Chatto and Windus, 2012.

———. *William Harvey: A Life in Circulation*. Oxford: Oxford University Press, 2013.

Yates, Frances A. *The Rosicrucian Enlightenment*. London: Routledge and Kegan Paul, 1972.

Yolton, John W. *Thinking Matter: Materialism in Eighteenth-Century Britain*. Minneapolis: University of Minnesota Press, 1983.

Zamberlan, Renato, and Franco Zamberlan. "The St Mark's Clock, Venice." *Horological Journal* 143, no. 1 (January 2001): 11–14.

Zimmerman, Andrew. *Anthropology and Antihumanism in Imperial Germany*. Chicago: University of Chicago Press, 2001.

~ INDEX ~

Page numbers in italic refer to figures.